Advanced Chemical Methods for Soil and Clay Minerals Research

NATO ADVANCED STUDY INSTITUTES SERIES

Proceedings of the Advanced Study Institute Programme, which aims at the dissemination of advanced knowledge and the formation of contacts among scientists from different countries

The series is published by an international board of publishers in conjunction with NATO Scientific Affairs Division

A	Life Sciences	Plenum Publishing Corporation
B	Physics	London and New York
C	Mathematical and Physical Sciences	D. Reidel Publishing Company Dordrecht, Boston and London
D	Behavioural and Social Sciences	Sijthoff & Noordhoff International Publishers
E	Applied Sciences	Alphen aan den Rijn and Germantown U.S.A.

Series C – Mathematical and Physical Sciences

Volume 63 – Advanced Chemical Methods for Soil and Clay Minerals Research

Advanced Chemical Methods for Soil and Clay Minerals Research

Proceedings of the NATO Advanced Study Institute held at the University of Illinois, July 23 – August 4, 1979

edited by

J. W. STUCKI and W. L. BANWART

University of Illinois, Urbana, Illinois, U.S.A.

D. Reidel Publishing Company

Dordrecht : Holland / Boston : U.S.A. / London : England

Published in cooperation with NATO Scientific Affairs Division

Library of Congress Cataloging in Publication Data
Main entry under title:

CIP

Advanced chemical methods for soil and clay minerals research.

(NATO advanced study institutes series : Series C, Mathematical and physical sciences ; v. 63)
"Published in cooperation with NATO Scientific Affairs Division."
Includes index.
1. Soil mineralogy–Methodology–Congresses. 2. Clay minerals–Research–Congresses. 3. Soils–Analysis–Congresses. 4. Clay–Analysis–Congresses. I. Stucki, J. W. II. Banwart, Wayne L. III. Illinois. University at Urbana-Champaign. IV. North Atlantic Treaty Organization. Division of Scientific Affairs. V. Series.
S592.55.A38 631.4'16 80-23081
ISBN 90-277-1158-5

Published by D. Reidel Publishing Company
P.O. Box 17, 3300 AA Dordrecht, Holland

Sold and distributed in the U.S.A. and Canada
by Kluwer Boston Inc.,
190 Old Derby Street, Hingham, MA 02043, U.S.A.

In all countries, sold and distributed
by Kluwer Academic Publishers Group,
P.O. Box 322, 3300 AH Dordrecht, Holland

D. Reidel Publishing Company is a member of the Kluwer Group

Printed in The Netherlands

TABLE OF CONTENTS

Chapter 1

MÖSSBAUER SPECTROSCOPY

Bernard A. Goodman
Department of Spectrochemistry
The Macaulay Institute for Soil Research
Aberdeen AB9 2QJ, United Kingdom

1-1. INTRODUCTION TO THE MÖSSBAUER EFFECT

The γ-radiation emitted by nuclei in excited states, formed as a result of radioactive decay of unstable parent nuclei, may subsequently be reabsorbed by other nuclei of the same type. If the emitting nucleus is assumed to be moving with a velocity, V, so that the linear momentum of the system is mV, where m is the mass of the nucleus, then, after emission of the γ-ray, the linear momentum of the system, which comprises the γ-ray plus de-excited nucleus, must still equal mV (conservation of momentum). Thus the momentum of the γ-ray, E_γ/c, must be balanced by a change in the velocity of the nucleus so that,

$$mV = E_\gamma/c + m(V+v) \qquad [1\text{-}1]$$

and v is thus equal to $-E_\gamma/mc$ and is independent of the initial velocity of the atom.

Also considering the conservation of energy, the kinetic energy of the nucleus before emission of the γ-ray is $\frac{1}{2}mV^2$ and after emission is $\frac{1}{2}m(V+v)^2$. Thus the difference in energy, δE, between the nuclear transition energy and that of the emitted γ-ray is given by

$$\delta E = \tfrac{1}{2}mv^2 + mVv \qquad [1\text{-}2]$$

$$\tfrac{1}{2}mv^2 = E_\gamma^2/2mc^2 = E_r \qquad [1\text{-}3]$$

where E_r is the free atom recoil energy and is independent of the velocity of the nucleus. Recoil of the nucleus also occurs on absorption of radiation and resonant absorption can only occur if overlap exists between the energy profiles of the emitted and absorbed γ-rays.

With free atoms the recoil energy, E_r, is much greater than the widths of these absorption profiles (Fig. 1-1). If the nuclei are held in a lattice in which the characteristic energy of the lattice vibrations (the phonon energy) is greater than

J. W. Stucki and W. L. Banwart (eds.), Advanced Chemical Methods for Soil and Clay Minerals Research, 1–92.

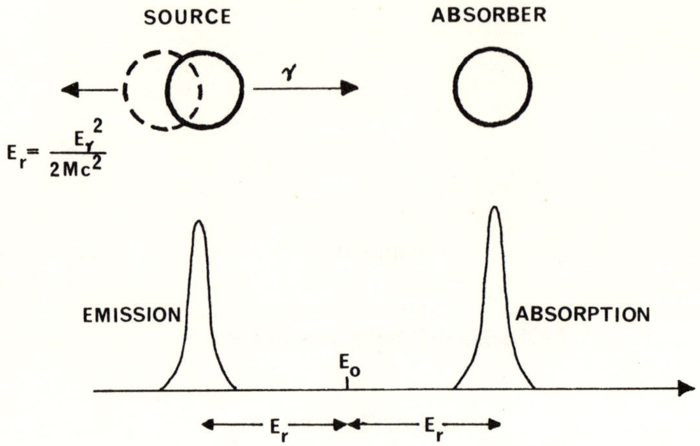

Figure 1-1. Energy profiles for the emission and absorption of γ-radiation.

the recoil energy, there is a finite probability that emission and absorption will occur without recoil. This is because the lattice is a quantized system and energy can only be transferred to the lattice in multiples of the phonon energy (Fig. 1-2). The fraction, f, of the decays that produce no change in the quantum state of the lattice is known as the "recoil-free fraction" or the "f-factor" and it is these γ-rays that account for the resonance. The full width at half height, Γ, of the energy profiles of the γ-rays is determined by the mean life time of the nuclear excited state (τ) such that

$$\Gamma\tau = \hbar \qquad [1\text{-}4]$$

where $\hbar$ = Planck's constant divided by 2π, i.e. $\hbar = h/2\pi$, and $\tau = t_{1/2}/0.693$, where $t_{1/2}$ is the half life of the nuclear excited state. Equation [1-4] can be evaluated for the case of the 14.4 KeV γ-ray for ^{57}Fe (see Fig. 1-3 which illustrates a simplified scheme for ^{57}Co decay to ^{57}Fe). Using a value of 0.6626×10^{-33} joule sec for h, and the relationship 1 joule = 0.624×10^{19} eV, Γ is found to be 0.467×10^{-8} eV. This is very small compared to the value of E_γ and it is because these nuclear energy levels are so sharply defined that the γ-rays can only be reabsorbed by the same type of nucleus. Mössbauer spectroscopy is, therefore, completely specific to a particular isotope.

The second term on the right hand side of equation [1-2] depends upon the initial velocity of the source nucleus and is known as the Doppler effect. This provides the principle by which the energy of the γ-ray is modulated in order that a region of the spectrum near to the unperturbed energy of the γ-ray can be examined; the source is moved and the percentage transmission as a function of source velocity provides the Mössbauer spectrum. Because of this method used for obtaining a spectrum it is usual to express the absorption energies relative to the velocity of the source, usually in mm sec^{-1} as a unit of convenience. Conversion factors are given in Table 1-11 for commonly-used energy units.

Many nuclei have spins, which arise from the resultant angular momentum of their protons and neutrons. The spin states are quantized so that for a nucleus with

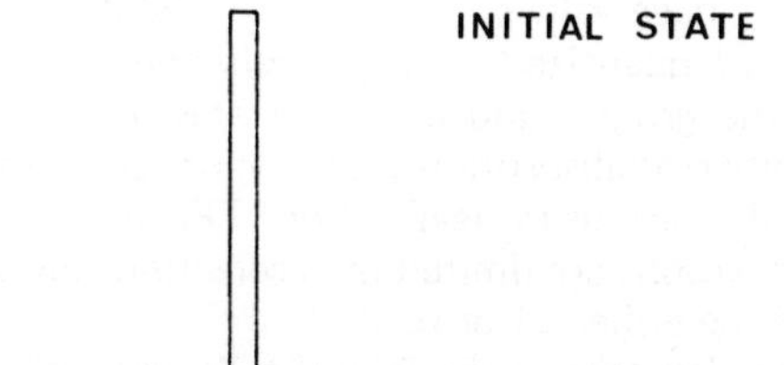

Figure 1-2. Emission or absorption of γ-radiation for a nucleus held in a crystal lattice and the origin of the recoil-free fraction.

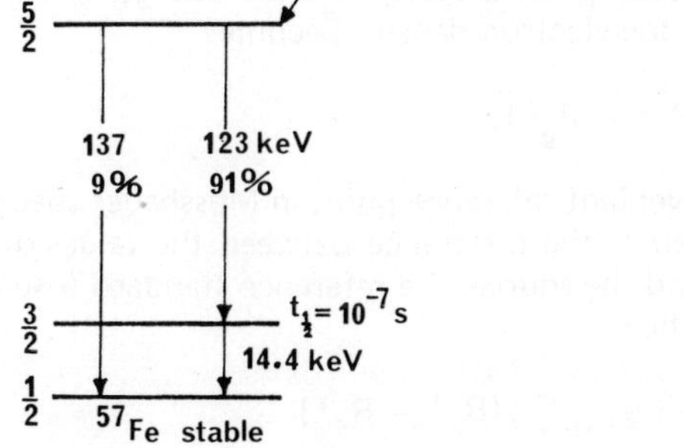

Figure 1-3. A simplified radioactive decay scheme for ^{57}Co to ^{57}Fe.

spin, I, the component levels, m_I, have values I, I-1, ..., −I. States with non-zero nuclear spins can have a series of quantized energy levels and a multiplicity of transitions can occur between the ground and excited states. A Mössbauer spectrum may thus consist of a number of absorption peaks whose separation depends upon the energy separations of the various m_I levels. For ^{57}Fe and for many other nuclei the number of allowed transitions are limited by a selection rule which states that the change in m_I, Δm_I, must be either ±1 or 0.

Approximately half of the elements in the Periodic Table have isotopes that have been shown to exhibit the Mössbauer effect (Table 1-1) but these tend to be concentrated among the heavier elements. The characteristics that determine a useful Mössbauer isotope are: (i) The parent source should have a half-life sufficiently long to allow convenient use, otherwise access to a nuclear reactor is required. (ii) The energy of the emitted γ-ray should be small enough for there to be a significant recoil-free fraction at temperatures conveniently obtainable in the laboratory. The separation of adjacent phonon energy levels decreases with increasing n (Fig. 1-2). Thus by lowering the temperature, which increases the population of the levels with lowest n, the energy required for excitation to an unpopulated level is increased. Hence the Mössbauer f-factor is similarly increased by decreasing the temperature. (iii) The lifetime of the nuclear excited state in the daughter nucleus should be sufficiently long for the line width of the transition to be small enough to allow resolution in the spectrum. Also, for the environmental scientist, there should be added a fourth requirement that the Mössbauer isotope should occur naturally at appreciable concentrations. Thus almost all isotopes are eliminated as being unsuitable for most investigation on soils and clay minerals and ^{57}Fe is left as by far the most important nucleus. Consequently, the remainder of this Chapter will be concerned almost entirely with this isotope.

The principal parameters that can be obtained from a Mössbauer spectrum are the isomer shift, δ, the quadrupole splitting, Δ, and the magnetic hyperfine field, H. The isomer shift originates from changes in the electron density at the nucleus as the chemical environment of that nucleus is varied. Thus for a uniformly charged spherical nucleus of radius R, the energy, E, due to electron density at the nucleus is given by

$$E = \left(\frac{2\pi}{5}\right) Ze^2 |\psi_{(o)}|^2 R^2 \qquad [1\text{-}5]$$

where Z is the atomic number, e is the electronic charge and $|\psi_{(o)}|^2$ is the electron density at the nucleus. Since the radius of the nuclear excited state, R_e, is usually different from the radius of the ground state, R_g, the energy shift, δE, in the Mössbauer effect as a result of the electron density becomes

$$\delta E = \left(\frac{2\pi}{5}\right) Ze^2 |\psi_{(o)}|^2 (R_e^{\,2} - R_g^{\,2}). \qquad [1\text{-}6]$$

The bare nucleus is not a convenient reference point in Mössbauer spectroscopy so the isomer shift, δ, is measured as the difference between the values of δE for the absorber under investigation and the source or a reference standard (usually sodium nitroprusside or iron metal). Thus

$$\delta = \left(\frac{2\pi}{5}\right) Ze^2 \{|\psi_{(o)}|_a^2 - |\psi_{(o)}|_r^2\} (R_e^{\,2} - R_g^{\,2}) \qquad [1\text{-}7]$$

Table 1-1. A section of the periodic table showing the atoms with isotopes that exhibit the Mössbauer effect.

K	Ca	Sc	Ti	V	Cr	Mn	**Fe**	Co	*Ni*	Cu	*Zn*	Ga	*Ge*	As	Se	Br	*Kr*
Rb	Sr	Y	Zr	Nb	Mo	Tc	*Ru*	Rh	Pd	Ag	Cd	In	**Sn**	**Sb**	*Te*	**I**	*Xe*
Cs	*Ba*	*La*	*Hf*	*Ta*	*W*	*Re*	*Os*	*Ir*	*Pt*	*Au*	*Hg*	Tl	Pb	Bi	Po	At	Rn
Fr	Ra	Ac															

Ce	*Pr*	*Nd*	*Pm*	*Sm*	**Eu**	*Gd*	*Tb*	**Dy**	*Ho*	*Er*	*Tm*	*Yb*	*Lu*
Th	*Pa*	*U*	*Np*	Pu	*Am*	Cm	Bk	Cf	Es	Fm	Md	No	Lw

Nuclei easily studied by Mössbauer spectroscopy

Other nuclei exhibiting the Mössbauer effect

where $|\psi_{(o)}|^2_a$ and $|\psi_{(o)}|^2_r$ are the electron densities at the nuclei of the absorber and reference materials, respectively.

Changes in isomer shift are thus due to variations in s-electron density and it might be expected that in iron-containing samples there would be little change in δ with oxidation state, since this usually just involves a change in d-electron density. However, there is a substantial indirect contribution to δ from changes in the number of d electrons because of altered shielding of the outermost s electrons. This will be dealt with in more detail in section 1-2.

The quadrupole splitting, Δ, arises if a nucleus has a non-uniform charge density, which is the case when $I > ½$. The energy levels may then be split by interaction of the nuclear charge density with a non-cubic electric field gradient. With ^{57}Fe, this results in a splitting of the excited state into two levels with $m_I = \pm 3/2$ and $\pm 1/2$, respectively, while the ground state is unaffected (Fig. 1-4). The separation of the two transitions arising from the splitting illustrated in Fig. 1-4 is known as the quadrupole splitting, Δ, and is proportional to the principal component of the electric field gradient tensor. Thus,

$$\Delta = ½e^2qQ\left(1 + \frac{\eta^2}{3}\right)^{½} \qquad [1\text{-}8]$$

where Q is the nuclear quadrupole moment, e is the charge on the proton, $-eq$ is the z component of the electric field gradient tensor (V_{zz}) and η is an asymmetry parameter that equals $(V_{xx} - V_{yy})/V_{zz}$. Axes are chosen so that $0 \leqslant \eta \leqslant 1$. The electric field gradient tensor is composed of electronic and lattice terms (q_{val} and q_{latt}), and a more detailed discussion of these will be carried out in section 1-2.

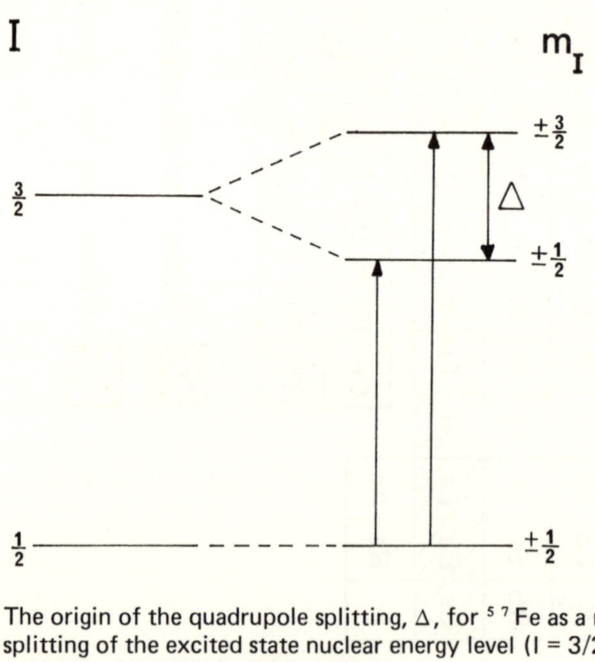

Figure 1-4. The origin of the quadrupole splitting, Δ, for ^{57}Fe as a result of the splitting of the excited state nuclear energy level (I = 3/2) by the presence of an electric field gradient.

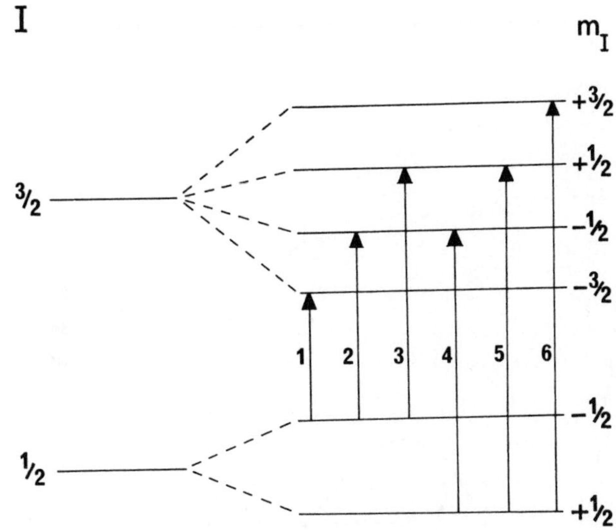

Figure 1-5. The splitting of the nuclear energy levels of ^{57}Fe (ground state I = 1/2, excited state I = 3/2) in the presence of a magnetic field.

The magnetic hyperfine interaction or nuclear Zeeman effect arises from the coupling of the nuclear magnetic moment with local or applied magnetic fields at the nucleus. The degeneracy of the nuclear energy levels is completely removed so that in ^{57}Fe the excited state is split into 4 levels and the ground state into 2 levels (Fig. 1-5). This gives eight possible transitions, six of which are allowed because of the selection rule $\Delta m_I = \pm 1,0$. The splitting of the energy levels is directly related to the combined magnetic and quadrupole interactions, so that the magnitude of any magnetic field at the iron nucleus can be determined.

1-2. BASIC PRINCIPLES OF MÖSSBAUER SPECTROSCOPY

1-2.1. The Isomer Shift, δ

As stated in the previous section, the isomer shift originates from changes in the electron density at the nucleus as the chemical environment of that nucleus is varied. In iron there are two mechanisms by which this electron density can be varied: (i) Direct changes in 4s electron density through the involvement of 4s orbitals in molecular orbitals. Although this mechanism may be important in highly covalent low spin compounds, it is usually small when the iron is in the high spin state, which is the state in almost all silicate minerals. (ii) Indirect changes as a result of changes in 3d electron density. This mechanism is effective because there is a fraction of the time when the 3s electrons are further from the nucleus than the 3d electrons (Fig. 1-6). As an example one can compare high spin Fe^{2+} and Fe^{3+} ions, where the outermost electronic configurations are $3s^2 3p^6 3d^6$ and $3s^2 3p^6 3d^5$, respectively. When the 3s electrons are further from the nucleus than the 3d electrons, the attractive coulomb potential between the s electrons and the

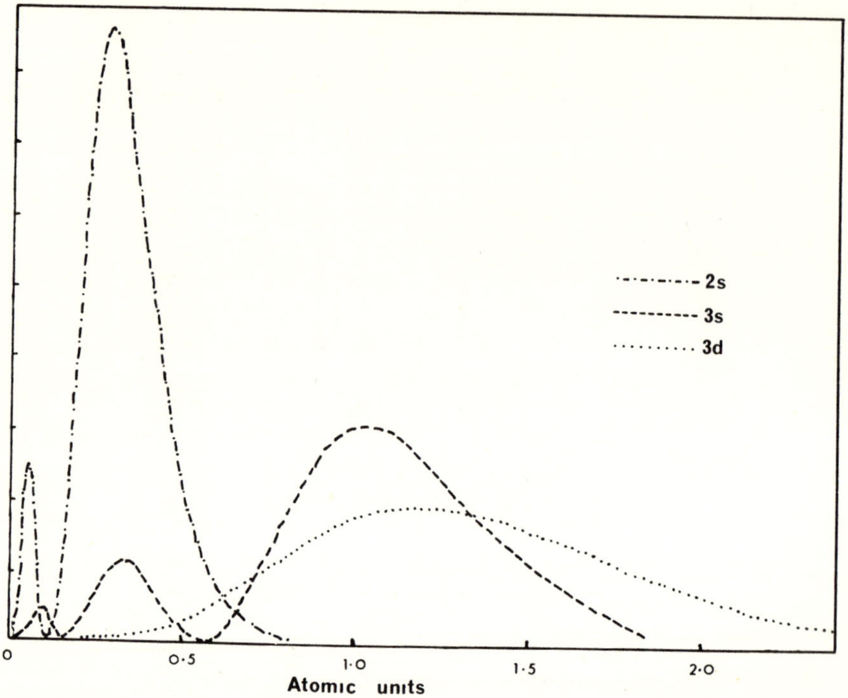

Figure 1-6. The radial distribution of 2s, 3s and 3d orbitals for a first row transition metal.

nucleus will be inversely proportional to the number of 3d electrons. The presence of d electrons, therefore, causes the 3s wavefunction to expand and reduces its charge density at the nucleus. Consequently, the removal of a d electron on going from Fe^{2+} to Fe^{3+} actually increases the charge density at the nucleus and produces a sizeable isomer shift.

Relative values of the isomer shift in ^{57}Fe, as a function of the 3d- and 4s-electron densities, have been calculated (51) and are shown in (Fig. 1-7). δ decreases with increasing 4s electron density and with decreasing 3d electron density because the term $R_e^2 - R_g^2$ in equation [1-7] is negative, the nucleus being smaller in its excited state than in its ground state.

In addition to being able to distinguish the oxidation states of high spin ions, the isomer shift is also able to provide information on the coordination number. With high spin ions there is a progressive decrease in δ with decrease in coordination number (Fig. 1-8), as a result of an increasing degree of covalency which effectively removes electron density from the metal. There is a certain amount of variation with type of ligand but, with a knowledge of coordination group, there is usually little difficulty in distinguishing between 4- and 6-coordination.

With low spin ions there is no systematic variation of isomer shift with oxidation state or coordination number. For example, the ferricyanide and ferrocyanide ions have similar values of δ. In these cases, therefore, Mössbauer spectroscopy is not a suitable structural probe, but fortunately for the present topic, low spin ions are extremely rare in soils and clay minerals.

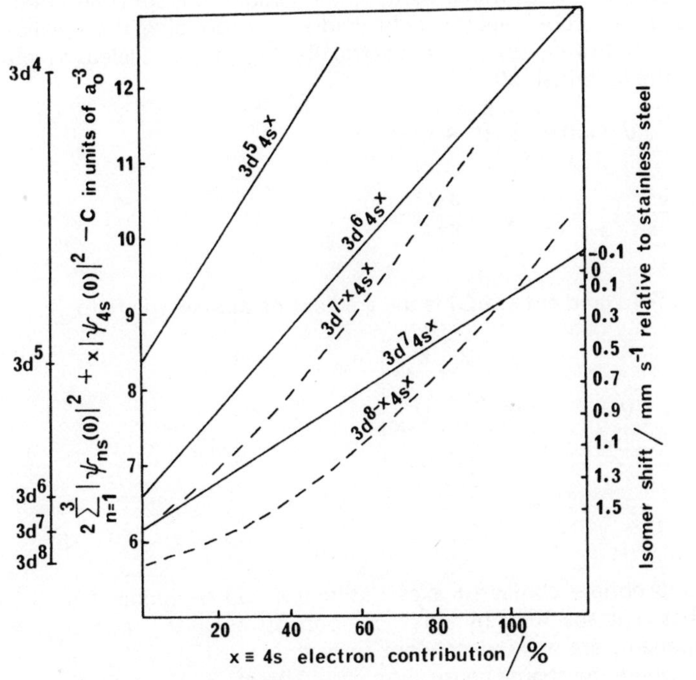

Figure 1-7. Variation of the isomer shift for ^{57}Fe with electron density (adapted from Walker *et al.*, 1961).

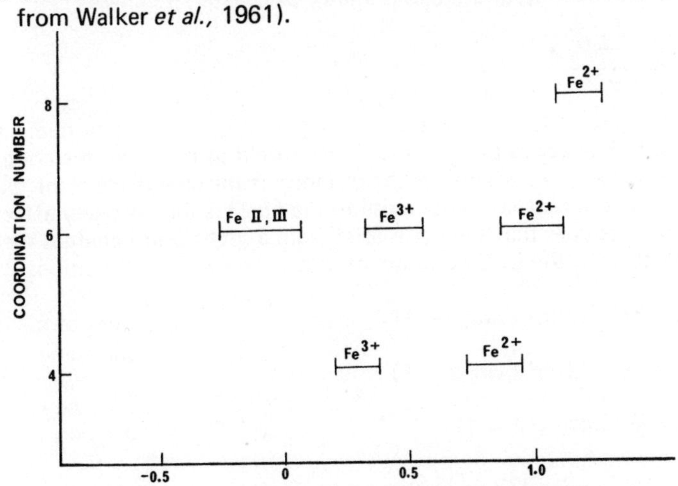

Figure 1-8. The relationship of isomer shift for ^{57}Fe with oxidation state and coordination number (adapted from Bancroft, 1973). Fe II, III refer to low spin and Fe^{2+}, Fe^{3+} to high spin ions.

1-2.2. The quadupole splitting, Δ

As noted in the previous section, the quadrupole splitting arises from the presence of a non-cubic electric field gradient surrounding the nucleus and this arises in the following way. The electric field, E, at the nucleus is the negative gradient of the potential, V

$$E = -\nabla V = -(\mathrm{i}V_x + \mathrm{j}V_y + \mathrm{k}V_z) \qquad [1\text{-}9]$$

where
$$V_x = \frac{\partial V}{\partial x},\; V_y = \frac{\partial V}{\partial y},\; V_z = \frac{\partial V}{\partial z}.$$

The electric field gradient (EFG) is the gradient of this electric field

$$\mathrm{EFG} = \nabla V = -\begin{bmatrix} V_{xx} & V_{xy} & V_{xz} \\ V_{yx} & V_{yy} & V_{yz} \\ V_{zx} & V_{zy} & V_{zz} \end{bmatrix} \qquad [1\text{-}10]$$

where
$$V_{ij} = \frac{\partial^2 V}{\partial i \partial j}.$$

By appropriate choice of axes this tensor can be reduced to the diagonal form so that it is specified by three components V_{xx}, V_{yy} and V_{zz}. Also these three components are not independent since they must obey the Laplace equation in a region where the charge density vanishes. Therefore, $V_{xx} + V_{yy} + V_{zz} = 0$ and EFG is completely determined by two independent parameters, usually chosen as V_{zz} (also known as $-eq$) and the asymmetry parameter, η, where

$$\eta = \frac{V_{xx} - V_{yy}}{V_{zz}}. \qquad [1\text{-}11]$$

Since the potential varies as r^{-1}, the electric field as r^{-2}, and the components of the EFG tensor as r^{-3}, where r is the distance of the charge from the nucleus, it is only those changes that are quite close to the nucleus that strongly affect the EFG tensor. If one assumes that the fields arise from a set of point charges then for each charge at distance r, the EFG components are

$$V_{xx} = qr^{-3}(3\sin^2\theta\,\cos^2\phi - 1)$$

$$V_{yy} = qr^{-3}(3\sin^2\theta\,\sin^2\phi - 1)$$

$$V_{zz} = qr^{-3}(3\cos^2\theta - 1) \qquad [1\text{-}12]$$

$$V_{xy} = V_{yx} = qr^{-3}(3\sin^2\theta\,\sin\phi\,\cos\phi)$$

$$V_{xz} = V_{zx} = qr^{-3}(3\sin\theta\,\cos\theta\cos\phi)$$

$$V_{yz} = V_{zy} = qr^{-3}(3\sin\theta\,\cos\theta\sin\phi)$$

where the polar coordinates r, θ and ϕ have the usual meaning (see Fig. 1-9).

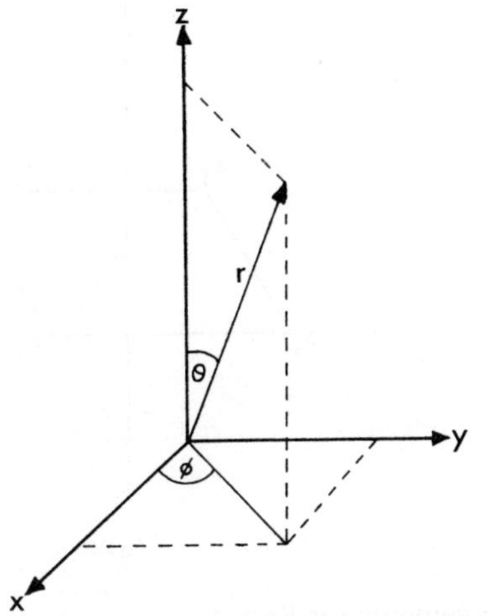

Figure 1-9. The relationship between Cartesian and polar coordinates.

The charge distribution responsible for the EFG tensor is made up of contributions from the electrons on the iron (q_{val}) and the charges on the surrounding atoms (q_{latt}), so that

$$q = (1 - \gamma_\infty)\, q_{latt} + (1 - R)\, q_{val}, \qquad [1\text{-}13]$$

where γ_∞ and R are the Sternhemier factors (49). The contributions of the various 3d wave functions to V_{zz} are given in Table 1-2. By using these values it can be readily seen that the q_{val} terms for high spin Fe^{3+} (1 electron in each d orbital) and low spin Fe^{2+} (2 electrons in each of the orbitals d_{xy}, d_{xz}, d_{yz}) are zero, and that q_{val} for high spin Fe^{2+} (i.e. the high spin Fe^{3+} case plus 1 electron) and low spin Fe^{3+} (i.e. the low spin Fe^{2+} case minus 1 electron) is non-zero.

Table 1-2. The contributions to V_{zz} of an electron in each of the 3d orbitals (in their usual forms each of these orbitals has η = 0)

Wavefunction	V_{zz}*
d_{z^2}	4/7
$d_{x^2-y^2}$	–4/7
d_{xy}	–4/7
d_{xz}	2/7
d_{yz}	2/7

*in units of $e\langle r^{-3}\rangle$, where e is the charge of an electron and $\langle r^{-3}\rangle$ is the mean value of r^{-3} for the 3d orbitals.

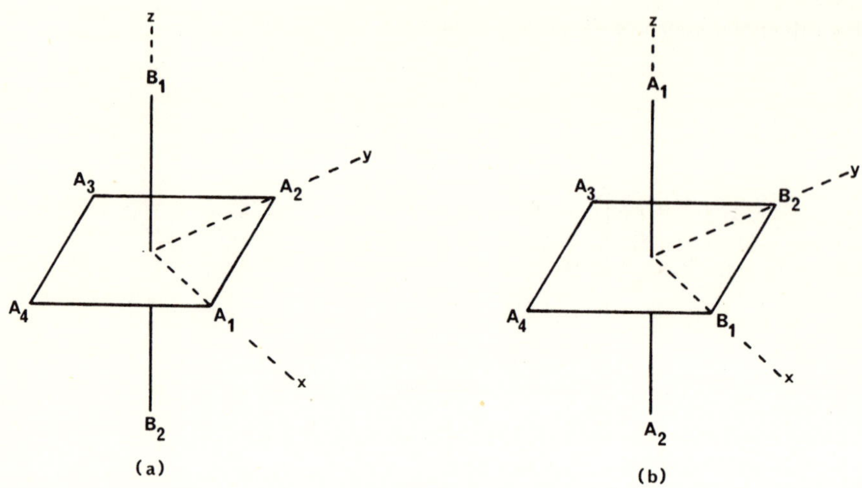

Figure 1-10. (a) *trans* and (b) *cis* arrangements for FeA_4B_2.

Qualitative evaluations of the lattice contributions to the EFG can also be readily made and, in certain circumstances, can give information on the arrangement of groups around the iron. This will be illustrated by considering the *trans* and *cis* isomers FeA_4B_2 (Fig. 1-10), where the charges on A and B are denoted by q_A and q_B, respectively. For *trans* FeA_4B_2, the components of the EFG tensor are:

$$V_{xx} = V_{yy} = 2\left(\frac{q_A}{r_A{}^3} - \frac{q_B}{r_B{}^3}\right),$$

$$V_{zz} = 4\left(\frac{q_B}{r_B{}^3} - \frac{q_A}{r_A{}^3}\right), \qquad [1\text{-}14]$$

$$V_{xy} = V_{xz} = V_{yz} = 0,$$

and for *cis* FeA_4B_2:

$$V_{xx} = V_{yy} = \frac{q_B}{r_B{}^3} - \frac{q_A}{r_A{}^3},$$

$$V_{zz} = 2\left(\frac{q_A}{r_A{}^3} - \frac{q_B}{r_B{}^3}\right), \qquad [1\text{-}15]$$

$$V_{xy} = V_{xz} = V_{yz} = 0,$$

where r_A and r_B are the Fe-A and Fe-B bond lengths, respectively.

Thus the magnitude of the EFG (and hence the quadrupole splitting) is twice as great for *trans* FeA_4B_2 as for *cis* FeA_4B_2. The signs are also opposite. This example, however, assumes that all bond angles are 90° and all bond lengths are equal for each type of group, so caution should be exercised when attempting to apply these arguments to real systems.

In most systems containing high spin Fe^{2+}, where there is a combination of lattice and valence terms, the two contributions are of opposite sign, so that increasing values of q_{latt} cause decreases in Δ from the value obtained for q_{val} alone (~3.7 mm sec^{-1}). Also, except when very large distortions are involved, this value of Δ is temperature dependent, with lower values of Δ being obtained at higher temperatures because of thermal population of excited electronic states.

So far no mention has been made of the relative intensities of the two transitions producing the quadrupole splitting. The angular dependence of the intensities of the two lines are given in Table 1-3. For $\theta=0$ the intensity ratio of the peaks is 3:1, whereas for $\theta = 90°$ the ratio is 3:5. In a randomly oriented, polycrystalline sample a summation over all angles is required and an intensity ratio of 1:1 is obtained.

Table 1-3. Angular dependence of the intensities of the peaks in a quadrupole-split spectrum.

Transition	Relative intensity *
$\pm\frac{1}{2} \rightarrow \pm\frac{3}{2}$	$1 + \cos^2\theta$
$\pm\frac{1}{2} \rightarrow \pm\frac{1}{2}$	$\frac{5}{3} - \cos^2\theta$

*θ is the angle between the EFG z-axis and the direction of the γ-ray.

1-2.3. Magnetic Hyperfine Interaction

It was shown in section 1-1 that in the presence of a magnetic field, the various nuclear energy levels are completely separated. If the z-axis is taken as the direction of the magnetic field, then in the absence of any EFG, the solution of the Hamiltonian

$$\mathcal{H}_m = -g\beta_n \, I \cdot H \qquad [1\text{-}16]$$

produces 2I + 1 eigenvalues given by

$$E_{m_I} = -\, g\beta_n \, H m_I; \; m_I = I, I\text{-}1, \ldots, -I \qquad [1\text{-}17]$$

where β is the nuclear magneton and g is the gyromagnetic ratio. The ground state energy levels in Fig. 1-5 are separated by an amount $g_o\beta_n H$ and the excited state sublevels by an amount $g_e\beta_n H$, where g_o and g_e are the ground and excited state g factors, respectively. The relative energies and intensities of the transitions in Fig. 1-5 are given in Table 1-4. In a single crystal, or for an externally applied field making an angle θ with the γ-beam, the peak ratios are 3:0:1:1:0:3 for $\theta = 0°$ and 3:4:1:1:4:3 for $\theta = 90°$. In a randomly oriented magnetically ordered material the intensity ratio becomes 3:2:1:1:2:3.

Table 1-4. Relative energies and intensities of the transitions illustrated in Fig. 1-5 for ^{57}Fe in the presence of a magnetic field.

	Transition	Relative energy	Relative intensity
1	$(-\frac{1}{2} \to -\frac{3}{2})$	$-\frac{1}{2}\beta_n H\,(3g_e + g_o)$	$\frac{9}{4}(1 + \cos^2\theta)$
2	$(-\frac{1}{2} \to -\frac{1}{2})$	$-\frac{1}{2}\beta_n H\,(g_e + g_o)$	$3\sin^2\theta$
3	$(-\frac{1}{2} \to +\frac{1}{2})$	$\frac{1}{2}\beta_n H\,(g_e - g_o)$	$\frac{3}{4}(1 + \cos^2\theta)$
4	$(+\frac{1}{2} \to -\frac{1}{2})$	$-\frac{1}{2}\beta_n H\,(g_e - g_o)$	$\frac{3}{4}(1 + \cos^2\theta)$
5	$(+\frac{1}{2} \to +\frac{1}{2})$	$\frac{1}{2}\beta_n H\,(g_e + g_o)$	$3\sin^2\theta$
6	$(+\frac{1}{2} \to +\frac{3}{2})$	$\frac{1}{2}\beta_n H\,(3g_e + g_o)$	$\frac{9}{4}(1 + \cos^2\theta)$

1-2.4. Combined Quadrupole and Magnetic Interactions

A description of the general form of the combined magnetic dipolar and electric quadrupolar interactions is rather complex. Consequently this section will consider three special cases of the combined interactions. In the first two the quadrupole interaction will be assumed to be axially symmetric and small compared to the magnetic dipolar term. Cases with the principal axis of the EFG parallel to and at an angle θ with the magnetic field will then be considered. Finally, the case where the magnetic field makes a fixed angle with the γ-ray direction will be discussed.

For an axially symmetric EFG tensor with symmetry axis parallel to the magnetic field, H, the eigenvalues for the I = 3/2 state are

$$E_{m_I} = -g\beta_n H m_I + (-I)^{[|m_I|+\frac{1}{2}]}\, e^2qQ/4 \qquad [1\text{-}18]$$

where all terms have the same meaning as before. All four magnetic sublevels are displaced by the same amount, with the ±3/2 levels being increased by $\Delta/2$ and the ± 1/2 levels decreased by $\Delta/2$. (Fig. 1-11).

For an axially symmetric EFG tensor with symmetry axis at an angle θ with respect to the magnetic axis and with $e^2qQ << g\beta_n H$, the I = 3/2 energy levels are

$$E_{m_I} = -g\beta_n H m_I + (-I)^{[|m_I|+\frac{1}{2}]}\frac{e^2qQ}{4}\frac{(3\cos^2\theta - 1)}{2} \quad . \qquad [1\text{-}19]$$

Thus it is not possible without additional knowledge to determine either the magnitude or the sign of e^2qQ from a magnetically-ordered spectrum.

In the third case of interest, a magnetic field is applied to a sample at a fixed angle θ to the γ-ray with the magnitude of $g\beta_n H$ comparable to e^2qQ. For a

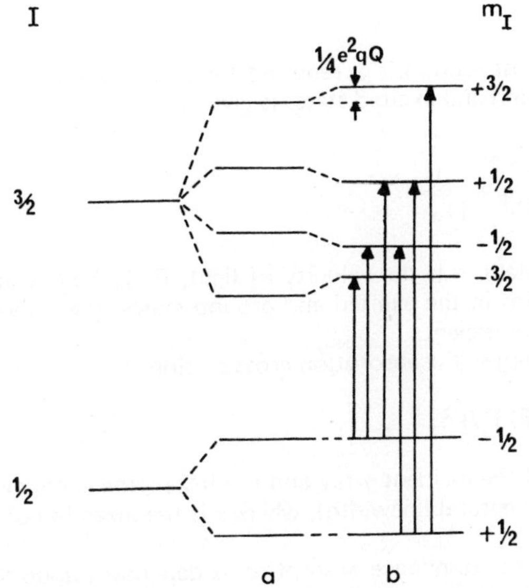

Figure 1-11. The splitting of the nuclear energy levels of ^{57}Fe by (a) interaction with a magnetic field (see Fig. 1-5) and (b) combined magnetic and electric quadrupole interactions.

polycrystalline sample the EFG axes take up all possible orientations with respect to the magnetic field and a large number of superimposed spectra are observed. For zero or small η, the zero field spectrum, initially with only 2 lines, is split into a doublet and a triplet, the former arising from the $+\frac{1}{2} \rightarrow +\frac{3}{2}$ and $-\frac{1}{2} \rightarrow -\frac{3}{2}$ transitions (Fig. 1-12a). This then provides a convenient method for the determination of the sign of the quadrupole splitting since if the doublet lies to positive velocity, then the sign of Δ is positive and *vice versa.* As η approaches 1 the spectrum assumes a symmetrical triplet-triplet structure showing that the sign of Δ is indeterminate in this case (Fig. 1-12b).

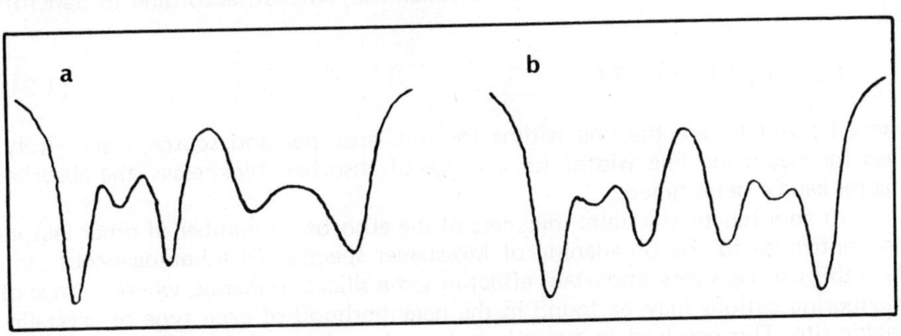

Figure 1-12. Simulated Mössbauer spectra for a polycrystalline sample with $\Delta = 2$ mm s^{-1} and $\Gamma = 0.35$ mm s^{-1} in an external magnetic field of 45 kOe parallel to the γ-beam (a) $\eta = 0$, (b) $\eta = 1$ (adapted from Collins and Travis, 1967).

1-2.5. Line Shapes

The absorption cross section, σ_o, required for γ-rays to produce a transition between the nuclear ground and excited states is given by

$$\sigma_o = \frac{1}{2\pi}\,\frac{h^2c^2}{E_o^2}\cdot\frac{2I_e+1}{2I_g+1}\cdot\frac{1}{1+\alpha} \qquad [1\text{-}20]$$

where h is Planck's constant, c is the velocity of light, E_o is the transition energy, I_e, I_g are the nuclear spins in the excited and ground states, respectively, and α is the internal conversion coefficient.

Over a range of energies the absorption cross section is

$$\sigma = \sigma_o\,[1 + 4\,(E-E_o)^2/\Gamma^2]^{-1}, \qquad [1\text{-}21]$$

where E_o is the energy of the incident γ-ray and $\Gamma = h/2\pi\tau$, the energy width of the nuclear excited state (the natural linewidth), where τ is the mean life of the excited state (i.e. $t_{1/2}/0.693$).

The magnitude of the resonance absorption is dependent upon the effective thickness, t, of the absorber

$$t = nf\sigma_o \qquad [1\text{-}22]$$

where n is the number of atoms of the Mössbauer isotope per unit area, and f is the recoil-free fraction.

The area under the absorption peak is given by

$$A = \frac{1}{2}\pi f_s \Gamma t, \qquad [1\text{-}23]$$

where f_s is the recoil-free fraction of the source.

The Lorentzian line shape described by equation [1-21] and the area described by equation [1-23] hold well for thin absorbers, but with thick samples these expressions are no longer valid. The experimental line width, Γ_{ex}, is systematically increased by increasing absorber thickness, so that, according to Bancroft (3),

$$\Gamma_{ex} = \Gamma_a + \Gamma_s + 0.27\,\Gamma t \qquad [1\text{-}24]$$

where Γ_a and Γ_s are the line widths for thin absorber and source, respectively. Thus by measuring line widths for a range of absorber thicknesses, the absorber f-factor can be determined.

In addition to the finite thickness of the absorber, a number of other factors can contribute to the broadening of Mössbauer spectra: (i) Inhomogeneous sample – this can be a very important effect in some silicate minerals, where a range of neighboring cations may be found in the neighborhood of each type of crystallographic site. This can lead to a number of quadrupole components which may not be resolved from one another. (ii) Unresolved quadrupole splittings – when a sample has quadrupole splittings less than the line width, the individual components are not resolved and a broadened line is observed. This is not likely to be an important effect in the study of soils and minerals since quite large splittings are usually obtained. (iii) External vibrations – a Mössbauer spectrometer should be sited with care (preferably in a basement) since coupling to the vibration of the building can occur with resulting line broadening. Another external source of vibration can come from vacuum pumps connected to the apparatus and this practice should be avoided whenever possible. When it is necessary to use a vacuum pump while running a spectrum great care should be taken to isolate the absorber from the vibrations of the pump. (iv) Relaxation effects – if the hyperfine parameters in a material fluctuate rapidly then the experimental spectrum will correspond to the mean value of the hyperfine field (Fig. 1-13a), whereas, if the fluctuation rate is slow, the individual hyperfine fields are observed (Fig. 1-13b). At intermediate rates a broadened spectrum is obtained. The variation with relaxation time of the spectrum of a sample with a large internal magnetic field is illustrated in Fig. 1-14.

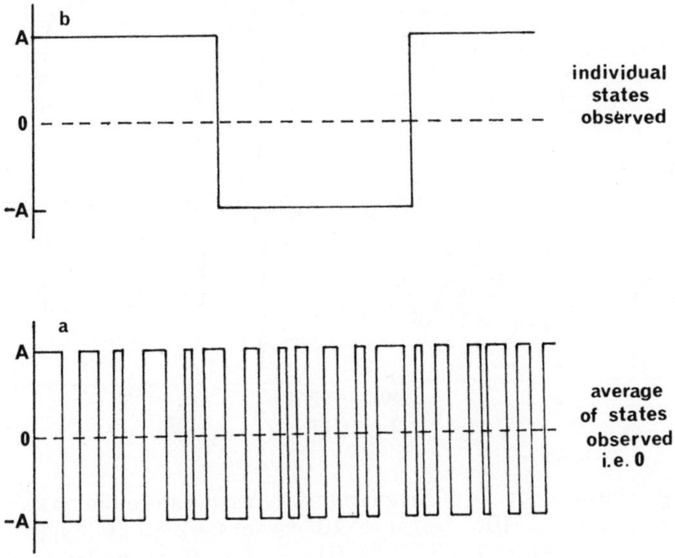

Figure 1-13. An illustration of (a) rapid and (b) slow fluctuation of a material between the two states A and –A.

This behavior results either from the inversion in direction of the magnetic hyperfine field in a paramagnetic material as a result of a spin-flip process or by the collective reorientation of the magnetic moment direction in fine particles. In the study of soils, extremely fine particles are often encountered and it is important to understand the influence of particle size on the Mössbauer spectrum, particularly when the technique is being used for the qualitative analysis of soil samples. When a material is cooled below its magnetic ordering temperature the spins of the magnetic ions tend to lock together producing the magnetic ordering. In extremely small particles (~ 10–10^3 magnetic ions) the spins may all be inverted simultaneously as a result of thermal excitation, with the result that magnetization of all sublattices is reserved. The energies of the spin states are equal and the mean time

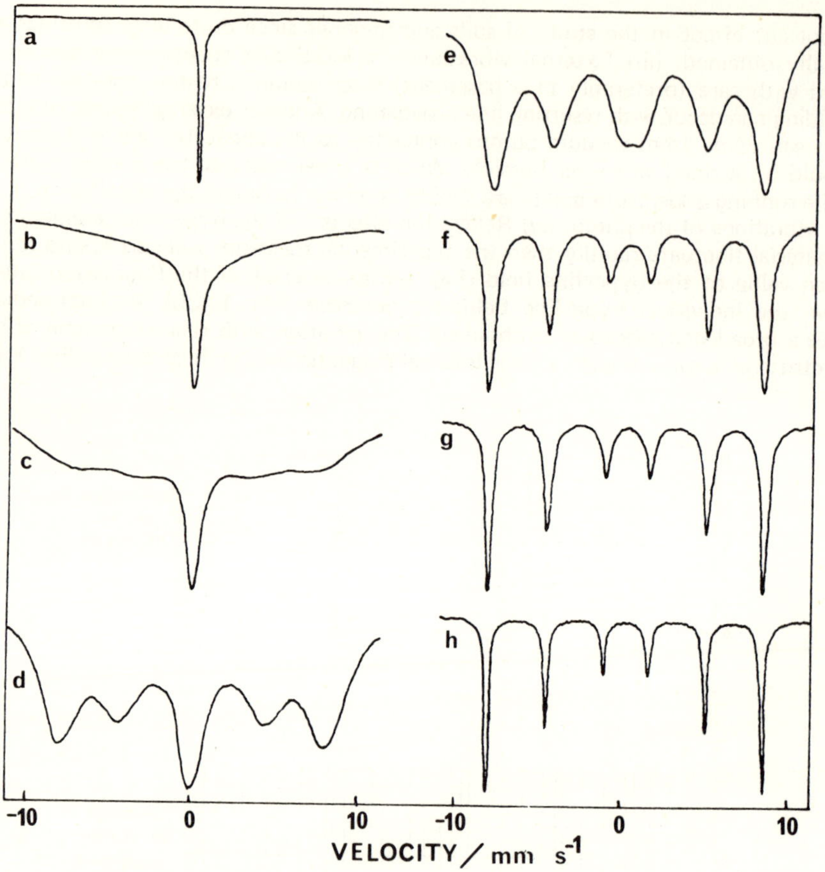

Figure 1-14. The dependence of the magnetic hyperfine structure on relaxation time (a) $t = 10^{-12}$s, (b) $t = 10^{-9}$s, (c) $t = 2.5 \times 10^{-9}$s, (d) $t = 5 \times 10^{-9}$s, (e) $t = 7.5 \times 10^{-9}$s, (f) $t = 2.5 \times 10^{-8}$s, (g) $t = 7.5 \times 10^{-8}$s, (h) $t = 10^{-6}$s (adapted from Wickman, 1966).

between spin flips is proportional to exp(KV/kT), where K is the anisotropy energy of the material, *V* is the particle volume, k is the Boltzmann constant and T is the temperature. Thus, with extremely small particles, magnetic hyperfine structure may be absent below the magnetic ordering temperature. If there is a range of particle sizes in the sample the magnetic hyperfine structure may appear over quite a large temperature range and will exhibit broadening analogous to that shown in Fig. 1-14 c-f. At low enough temperatures the hyperfine parameters will be identical to those shown by large particles.

1-2.6. Recoil-free fraction (f-factor)

It was mentioned in Section 1-2.5 that the area under a Mössbauer peak contains terms which include the source f-factor, f_s, and the absorber f-factor, f_a, the latter appearing as part of the effective thickness of the absorber, t. Since f-factors can vary appreciably from one sample to another, the relative areas of peaks from a mixture will not normally represent the relative proportions of the components, but the product of their concentrations and corresponding f-factors. The recoil-free fraction may be expressed as

$$f = \exp(-4\pi^2 \langle x^2 \rangle / \lambda^2), \tag{1-25}$$

where λ is the wavelength of the γ-ray and $\langle x^2 \rangle$ is the mean square displacement of the Mössbauer atom from its equilibrium position under thermal vibration. The f-factor, therefore, varies with temperature, decreasing rapidly at high temperatures. Because this temperature dependence is governed largely by the nature of the lattice, quantitative measurements should be performed at the lowest temperatures conveniently obtainable.

In some cases the amplitudes of the mean-square vibration displacements may vary along different directions in the crystal. This effect, known as the Goldanskii-Karyagin effect (19), causes the f-factor to vary with crystal orientation and gives rise to unequal peak heights in a quadrupole spectrum from a randomly oriented, polycrystalline sample. Decreasing the temperature tends to bring the f-factors closer together and the difference in intensity of the peaks decreases. Caution should be used, however, in interpreting this phenomenon as proving the existence of a Goldanskii-Karyagin effect, since by increasing the f-factors by lowering the temperature the effective thickness of the sample will also have been increased. This may then in turn increase the degree of saturation (which results in the area under a peak being less than that expected from the effective thickness) and affect the more intense peak to a greater extent than the weaker peak.

1-2.7. Second-order Döppler shift

This is the small decrease in energy of the γ-ray emission or absorption that results from relativistic effects of the thermal vibration velocity of the nuclei. It has the value $-1/2(\langle v^2 \rangle / c^2)\, E\gamma$, where $\langle v^2 \rangle$ is the mean of the square of the atom vibration velocity in the lattice, c is the velocity of light and $E\gamma$ is the γ-transition energy.

The second order Döppler shift is a component of the measured isomer shift and, being temperature dependent, must be taken into account if temperature dependence of the s-electron density is being studied.

Further treatment of the theoretical aspects of Mössbauer spectroscopy may be found in references 5, 17, 18, 29 and 42.

1-3. INSTRUMENTATION AND EXPERIMENTAL PROCEDURES

A Mössbauer spectrometer consists basically of a drive unit which moves the source, a γ-ray detector, and data storing device along with various amplifiers and some form of output device (Fig. 1-15).

The drive unit consists of a linear velocity transducer, which operates like a moving coil loudspeaker, consisting of a driving coil and an electromagnetically-isolated pick-up coil, and a function generator which controls the motion of the drive. The source is held at one (or both) end(s) of the transducer.

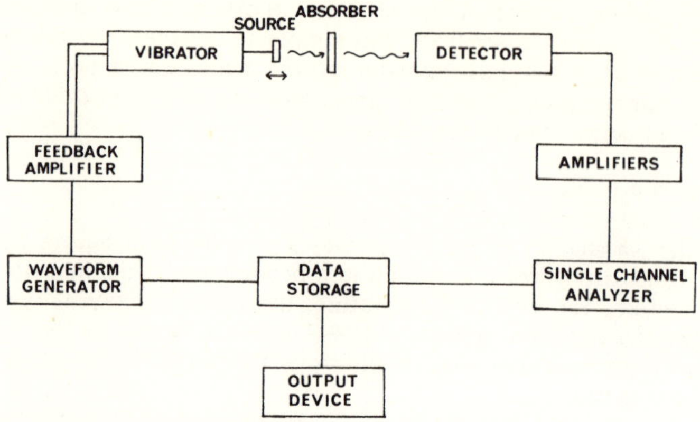

Figure 1-15. Simplified block diagram of a Mössbauer spectrometer.

Two basically-different systems may be used for generating the velocity of the drive unit, namely constant velocity or velocity sweep devices. In a constant velocity device the source is moved towards or away from the absorber at a constant velocity for a fixed period of time. This procedure is then repeated for a succession of different velocities. With a velocity sweep drive system the spectrum is scanned by varying the velocity of the source during a single sweep, usually at constant acceleration. The principal velocity modes provided by the function generator are either sawtooth or triangular, the former having the advantages of using all the channels of the analyzer, the latter having a smaller deadtime between sweeps (Fig. 1-16). In a typical system the data-logging device, usually a multichannel analyzer, and the wave function generator are synchronized so that the γ-rays from a given velocity are always fed into the same channels, each of which corresponds to a fixed velocity increment.

It should be pointed out that the background absorption in a spectrum is perfectly flat only if the source-detector distance remains constant, i.e. provided the Döppler motion is applied to the absorber rather than the source. However, it is far more convenient to move the source, so the amplitude of motion is kept to a minimum and a slightly curved (parabolic) baseline is obtained. This curvature can be essentially eliminated if a triangular waveform is used by combining the two mirror-image spectra that are obtained. Devices have also been produced which allow a preselected velocity range to be scanned normally while the unrequired velocities are swept quickly. This allows maximum resolution to be obtained in a particular area of interest, a facility which is sometimes useful when dealing with overlapping magnetic spectra. The constant velocity mode is also useful in this respect, permitting the isolated scanning of one peak with high statistical accuracy in a short period of time.

Several different sources, which combine a single line emission with a high f-factor and a small linewidth (close to the natural linewidth), are commercially available for ^{57}Fe. They are usually in the form of alloys, Ir and Pd being among the best matrices as far as f-factor and Γ are concerned. Ir has the advantage of remaining a single line emitter at very low temperatures, whereas Pd is slightly cheaper and easier to use in strong sources (>50 mCi). The source is prepared by evaporating or electroplating the ^{57}Co onto a metal foil, which is then heated *in vacuo*, and finally mounted in a suitable holder.

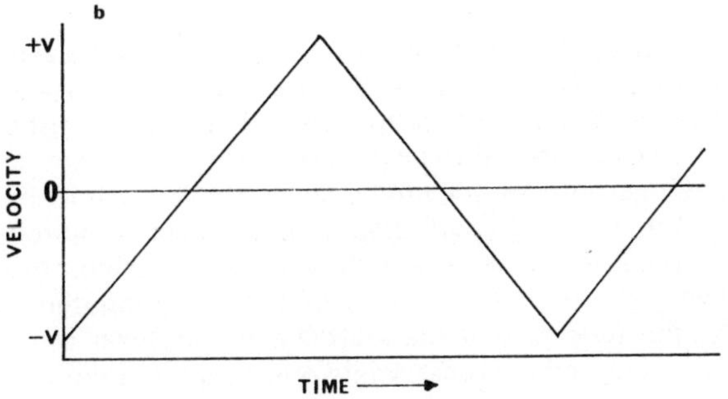

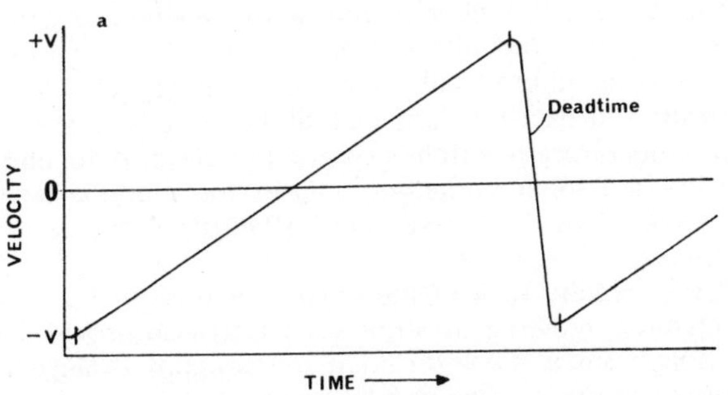

Figure 1-16. Variation of velocity with time for a constant acceleration drive system with (a) sawtooth and (b) triangular waveforms.

The nature of the detection system is governed by the type of experiment being performed, i.e. whether it is in the transmission or back-scatter mode. The function of the detector is to detect the Mössbauer γ-ray as efficiently as possible, while at the same time excluding any other γ- or x-rays. It consists of a γ-ray counter, a preamplifier, amplifier, a single channel analyzer and a multichannel analyzer (or computer or data logger). In the transmission mode, which is used in most experiments, conventional gas-filled proportional counters are commonly used. They are normally filled with one of the heavier inert gases, e.g. Ar, Kr or Xe, with nitrogen or methane as quenching gas. They are fairly cheap, provide good resolution of the 14.4 keV γ-ray, and thus have a high signal-to-noise ratio. Other detectors in common use include scintillation counters, which have poorer resolution than proportional counters, and Li-drifted Ge counters which have better resolution than proportional counters but are expensive. The feature of high resolution given by the Li-drifted Ge detector is of little advantage for ^{57}Fe, where radiation energies are well separated, and its much greater cost, combined with the necessity of running it at liquid nitrogen temperature, account for its lack of use in ^{57}Fe experiments. In back-scatter experiments the absorber is usually incorporated within a proportional counter which is set up to detect either back-scattered x-rays or conversion and Auger electrons. The back-scattered x-rays are thought to escape from a depth of $\leqslant 10^{-6}$ m, whereas the mean escape depth of conversion electrons is $\sim 3.5 \times 10^{-8}$ m.

Multichannel analyzers are normally used for data accumulation, combining this function with control of the sweep time of the drive unit in the velocity sweep mode. Spectra are continuously displayed and the experimental data are obtained through a teletype, floppy disk or other output device when sufficient counts have been accumulated for adequate signal-to-noise ratios. With the recent development of microcomputers and microprocessors a trend away from the use of multichannel analyzers has commenced. By using a dedicated mini- or micro-computer, computation of spectra can proceed simultaneously with data accumulation, thus increasing efficiency and removing the need for the use of remote computers in most instances. Alternatively, the functions of the multichannel analyzer can be replaced by a microprocessor at much reduced cost, while retaining the traditional modes of operation.

For the study of samples of soil and clay minerals, accessories are often required which allow the samples to be studied at a variety of temperatures, both above and below ambient. It is desirable, therefore, that a laboratory possess cryostats capable of operating at liquid nitrogen and liquid helium temperatures and preferably with variable temperature facilities for operating at intermediate temperatures. A high temperature furnace may also be useful especially if one is interested in the study of high temperature reactions occurring in clays. A further accessory which may be of value is a superconducting magnet for applying large external magnetic fields to samples. Figs. 1-17, 1-18, and 1-19 illustrate the designs of some accessories in common use.

Calibration of the velocity scan of the drive system is usually performed by use of a standard reference material as absorber, which also provides a standard reference point for isomer shifts, since different sources emit at slightly different energies. The two most commonly-used standards are iron metal and sodium nitroprusside. Iron metal is magnetically-ordered and gives 6 lines, the positions of which are known with a high degree of accuracy. Thus by comparing the peak channel numbers with the known velocities of these peaks, the velocity increment of the spectrometer can be readily calculated. Also, since there are 6 points for this velocity calibration for large scan ranges, any non-linearity in the drive waveform can be easily detected. Sodium nitroprusside gives only a 2-peak spectrum and hence does not give any check on the drive linearity, but it has often been used as a reference point for isomer shifts, having one of the lowest values observed for iron compounds. Absolute calibration of the velocity of the drive system can be obtained by using a laser interferometer (Fig. 1-20). The incoming laser beam is split into two parts by a beamsplitter, both parts being reflected by prisms. One of the prisms is fixed to the beam splitter, whereas the other is attached to the driving tube of the velocity transducer. A displacement of this latter prism generates a fringe pattern, which is detected by a photodiode. The fringe counts are transformed into pulses, each of which corresponds to a displacement of half a wavelength of the laser light (i.e. 3.164×10^{-4} mm for a He-Ne laser). The pulses are stored in the multichannel analyser memory, thus permitting the precise determination of the velocity corresponding to each channel.

In designing an experimental set-up the separation of the source from the detector has to be carefully selected. Thus, although it is desirable to have them close together in order to maximize the count rate, placing them too close leads to broadening of the spectra. This is because the emitted γ-ray makes an angle θ with the direction of motion of the source (Fig. 1-21), so that the energy shift due to the Döppler motion is $(v/c)E_\gamma \cos\theta$, where v, c, and E_γ have the same meanings as

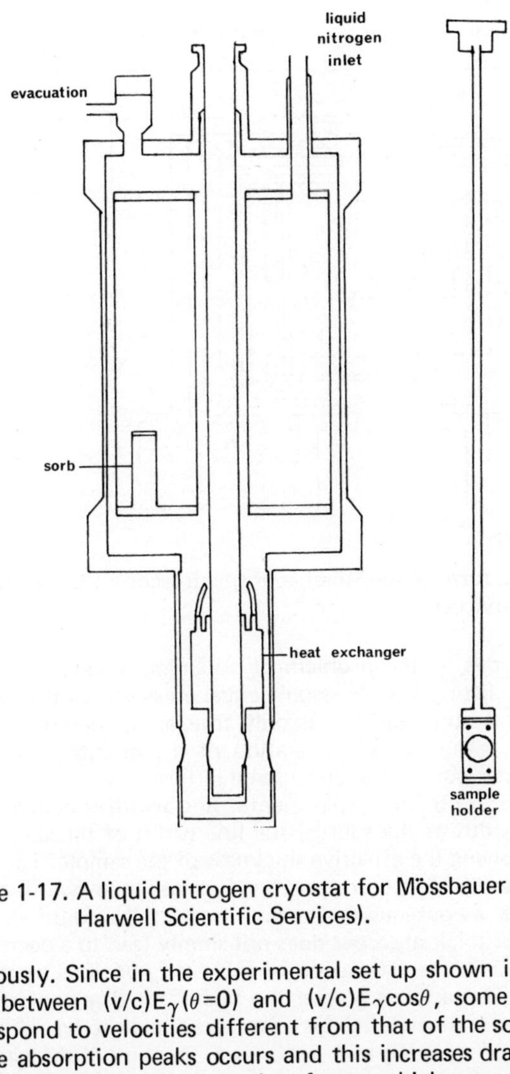

Figure 1-17. A liquid nitrogen cryostat for Mössbauer spectroscopy (design by Harwell Scientific Services).

previously. Since in the experimental set up shown in Fig. 1-21 the γ-ray energies vary between $(v/c)E_\gamma(\theta=0)$ and $(v/c)E_\gamma\cos\theta$, some of the counts detected will correspond to velocities different from that of the source. Hence some broadening of the absorption peaks occurs and this increases dramatically at very small source to detector distances. A further factor which may place constraints on the geometrical arrangement of the set-up concerns the overloading of the electronics of the detector at very high count rates. Since the count rates are fastest in the regions of no absorption (i.e. the baseline), these counts are lost to a greater extent than those at an absorption peak and may consequently affect the goodness-of-fit criteria applied to the analysis of the data.

The preparation of the absorber is of crucial importance in obtaining a good Mössbauer spectrum. If the concentration of ^{57}Fe per unit area is too low a large proportion of the γ-rays will pass straight through the absorber and there will be a high background (i.e. poor signal-to-noise), whereas if the concentration is too high there will be saturation of the most intense peaks and a consequent broadening and under-estimation of their intensities. A further item which is important, especially

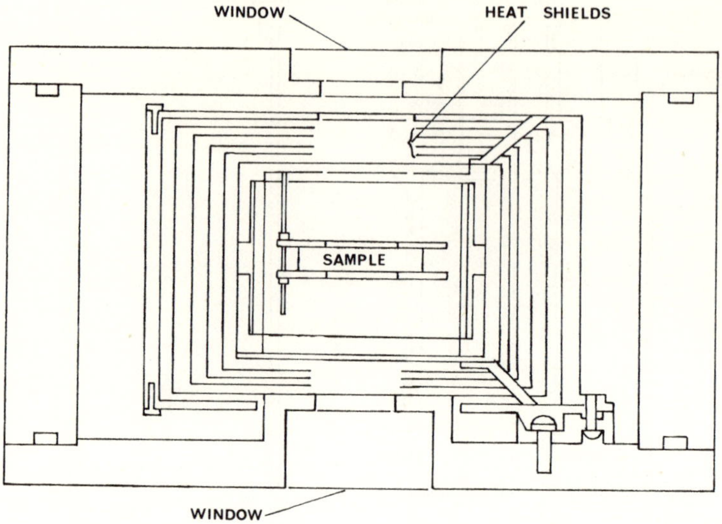

Figure 1-18. High temperature furnace for Mössbauer spectroscopy (design by Harwell Scientific Services).

when dealing with clay minerals, is the problem of obtaining randomly oriented powders since most computer-fitting models assume equal areas and widths of both components of a quadrupole doublet and this is only true, as was described in the previous section, if there is complete randomization of the orientations of the crystallites. These two problems will now be considered in turn.

As discussed in section 1-2.5, the experimental line width is composed of contributions from the line width of the source, the line width of the absorber at zero thickness and a term involving the effective thickness of the sample. Thus with increasing thickness of sample there is a corresponding increase in linewidth and, in the case of overlapping peaks, a consequent decrease in spectral resolution. However, with overlapping peaks a thick absorber does not simply lead to a decrease in resolution but also to an error in the estimation of the peak positions by normal computer-fitting procedures. This error is greater the higher the degree of overlap. For ^{57}Fe the optimum amounts of material for absorbers appear to be ~ 3 mg Fe per cm^2 if there is no magnetic ordering and ~ 10 mg Fe per cm^2 if the sample is magnetically ordered.

Absorbers need to be uniform since any gross local variations in thickness will lead to either saturation or high background counts or both on a local scale, with the result that a spectrum with less than optimum quality is obtained. Absorber holders are usually very simple. A cross-section of the type used at the Macaulay Institute is shown in Fig. 1-22, and is made of polymethylmethacrylate. While this holder is cheap to produce, some workers simply use a piece of lead with a hole in it and use cellotape for the windows. This type of cell can make it difficult to obtain a uniformly thin absorber from a sample with high iron content. In these cases it is usual to mix the sample with an inert matrix composed of atoms of low atomic number. Examples of materials that have been used include polyethylene powder, aluminium powder, sugar and alumina. It is important to use light elements since many heavy elements scatter or absorb the Mössbauer γ-ray with consequent decrease in spectral intensity. It is also for this reason that it is often difficult to obtain good quality spectra from samples with < 1% natural iron, and ~ 0.1% is an approximate lower limit for Mössbauer spectroscopy.

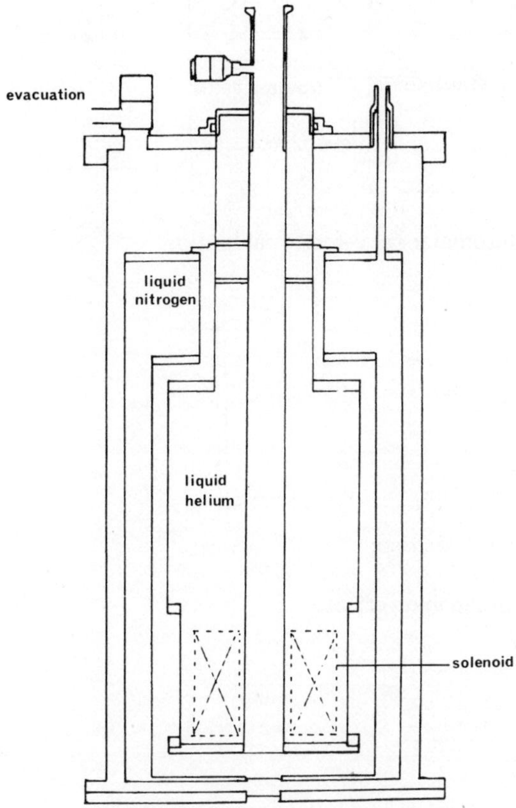

Figure 1-19. A basic liquid helium cryostat with super conducting magnet (design by The Oxford Instrument Company). In this design the source may either be driven vertically by mounting it in the central tube, or horizontally by using windows similar to those shown in Fig. 1-17.

With platey samples such as sheet silicates it is often difficult to obtain completely randomly oriented samples, since there exists a certain degree of preference for alignment of the sheets with the face of the holder (Fig. 1-23). The result of this is that the two components of a quadrupole doublet may not be equal in intensity, thus increasing the difficulty in subsequent computational procedures. Even worse, though, if the existence of preferred orientation in the sample is not appreciated then it can lead to incorrect analysis of the spectrum, e.g. the presence of an additional component centered around the more intense peak might be assumed or a Goldanskii-Karyagin effect (19) (anisotropy of the recoil-free fraction) might be postulated. Various groups of workers have different methods for

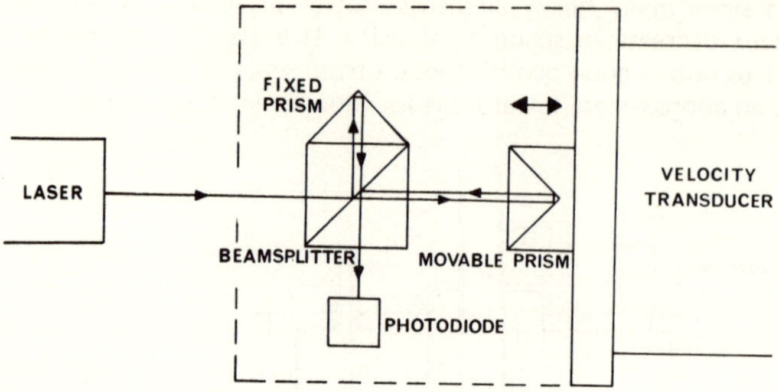

Figure 1-20. Laser interferometer for velocity calibration.

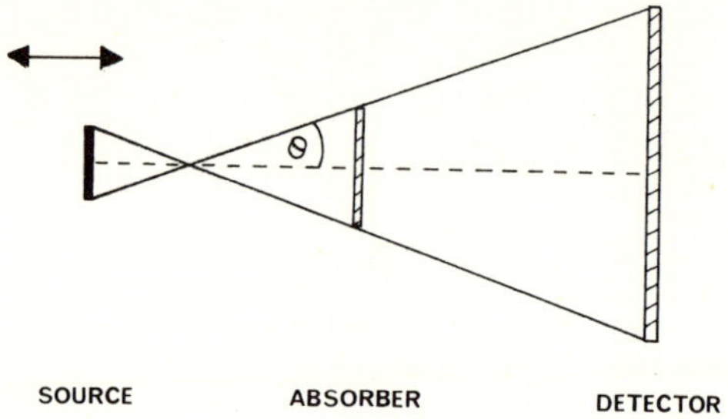

Figure 1-21. The cosine broadening effect.

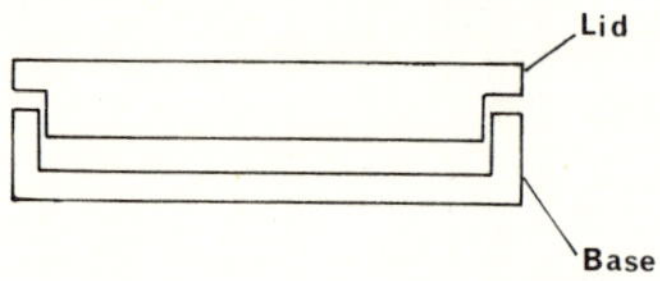

Figure 1-22. A simple absorber holder.

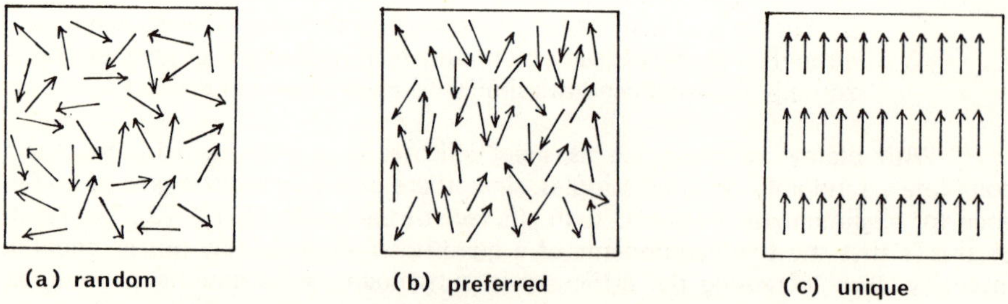

Figure 1-23. States of orientation for species within a solid; (a) random, as in a glass, (b) preferred, where in this example there is a preference for alignment in a vertical plane, and (c) unique as in a single crystal.

minimizing the effects of preferred orientation within their absorbers, one of the simplest being to shake the sample with at least five times its volume of polyethylene powder, which is in the form of small spheres. The plates are attracted to the surface of the polyethylene and random orientation is more or less assured. Care should, however, be exercised in packing the sample into the absorber holder. No obvious orientation effects have been observed at the Macaulay Institute when such samples were pressed into discs, although on such occasions much larger polyethylene to sample ratios were used.

A better method for eliminating texture effects from a Mössbauer spectrum involves orienting the absorber holder so that the γ-ray passes through it at an angle of 54.7° instead of at right angles (14). This works if we consider that the difference between an absorber with preferred orientation and a completely randomized absorber is a non-random distribution of the angles θ, which govern the orientation of the crystallites relative to the plane of the holder (Fig. 1-24); the distribution of ϕ will remain random. Thus the overall powder spectrum has a bigger contribution from those components corresponding to orientations of the crystallites in the plane of the holder than would be expected for a completely random absorber. However, at an orientation of 54.7° to the γ-ray direction the two quadrupole components from a single crystal have equal intensity. Thus by orienting the absorber holder at this angle to the γ-radiation, the effects of preferential orientation on the intensities are not observed.

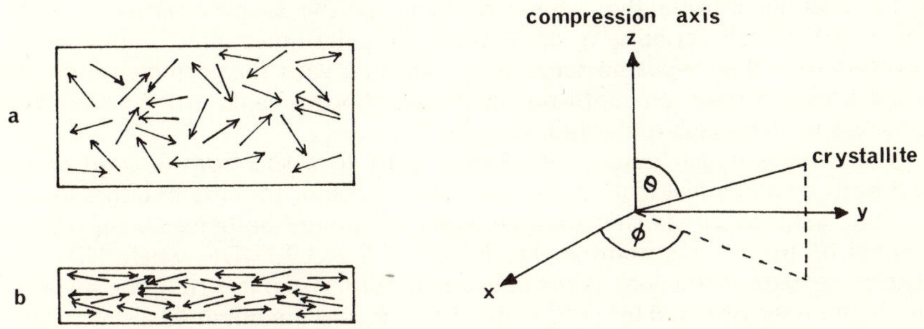

Figure 1-24. Conversion of random (a) into preferred (b) orientation by compression along an axis.

Analysis of the data comprising a Mössbauer spectrum is almost invariably carried out by computer and a number of programs for doing this are readily available. In most cases it is usual to assume that the thin absorber approximation holds, with each peak having Lorentzian shape, although it is easy to use any other lineshape function if there is good reason for doing so (e.g. to fit a convolution of Lorentzian and Gaussian functions for thick absorbers). For a randomly-oriented absorber the number of variables required to define each peak, i.e. position, width, and area or height, may be decreased for quadrupole split spectra by assuming equal areas and widths for the two components of a doublet. In magnetically-ordered samples further constraints can be introduced since the positions of all six lines are not independent. In the case of two or three overlapping components it is usually necessary to use such constraints in order to obtain a converging fit with the computer. The computer program involves the fitting of a function, Y(x), containing a number of variables to a set of experimental data points. The function, as already stated, usually consists of a set of peaks of Lorentzian shape, which is given by

$$Y_{(x)} = \frac{Y_{(o)}}{1 + \left(\frac{X - X_{(o)}}{\Gamma'}\right)^2} \qquad [1\text{-}26]$$

where $Y_{(o)}$ is the intensity at the maximum absorption position $X_{(o)}$ and Γ' is one half of the peak width at half height. Therefore, for each peak $Y_{(o)}$, $X_{(o)}$ and Γ' are the independent variables along with two parameters which specify the baseline position and curvature. In a fitting operation the objective is to minimize X^2, which is the sum of the squares of the deviation of each point in the fitted spectrum from the corresponding experimental point divided by the variance at a single point. Thus

$$\chi^2 = \sum_{i=1}^{NCH} W_i \, [Y_i - Y_{fi}]^2 \qquad [1\text{-}27]$$

where NCH is the number of elements in the spectrum, W_i is the inverse of the variance at channel i, Y_i is the observed count at channel i, and Y_{fi} is the computed value for channel i using the estimated values of the spectral variables. Using $d\chi^2/dq = 0$ for each variable, q, corrections are determined which minimize χ^2. The procedure is then repeated successively, starting with the corrected estimates from the previous iteration, until no significant improvements in the value of χ^2 are obtained from successive iterations.

The criteria which determine whether a fit to a spectrum is good or not depend both on statistical factors and one's knowledge of the system under investigation. For a fit to be statistically acceptable χ^2 should lie between the 1% and 99% limits of the χ^2 distribution, i.e. $NDF + 2.2 \pm 3.3\sqrt{NDF}$, where NDF, the number of degrees of freedom, is the number of points used in fitting the spectrum minus the number of variables used in the fit. Once having obtained a statistically-acceptable fit, it is necessary to ask oneself if the fitted parameters are meaningful: i.e. are the isomer shift values sensible; do the number of components correspond to the number of iron-containing sites in the sample; are all of the constraints used justifiable; are the line widths reasonable; etc.? It must always be remembered that the computer tests whether or not the model given to it can satisfy the experimental data, it never proves that the model is correct. One should never accept uncritically the values of parameters obtained from computer fitting a spectrum.

1-4. APPLICATION OF MÖSSBAUER SPECTROSCOPY TO THE STUDY OF SILICATE MINERALS

This section will be concerned with a brief general survey of some of the published work on the main groups of silicate minerals with the aim of illustrating the types of spectra that are obtained and the interpretations that have been made by various workers.

1-4.1. Chain Silicates

Pyroxenes. The basic structure of pyroxenes consists of SiO_4 tetrahedra linked to form chains of composition $(SiO_3)_n$ (Fig. 1-25). These chains are held together by cations bound to the non-bridging oxygen atoms (Fig. 1-26). There are two crystallographically-distinct positions, M_1 and M_2. The cations in the M_1 positions are coordinated to 6 oxygen atoms in a nearly regular octahedron, while the cations in the M_2 site are coordinated to between 6 and 8 oxygens in a distorted environment. The general chemical formula for pyroxenes can be expressed as $R^{2+}SiO_3$, with $R^{2+} = Ca^{2+}$, Mg^{2+}, Fe^{2+}, Mn^{2+} or Na^+ for the M_2 sites; and $R^{2+} = Mg^{2+}$, Fe^{2+}, Mn^{2+}, Al^{3+} or Fe^{3+} for the M_1 sites. In addition there is the possibility of substituting Al^{3+} or Fe^{3+} for some of the Si^{4+}.

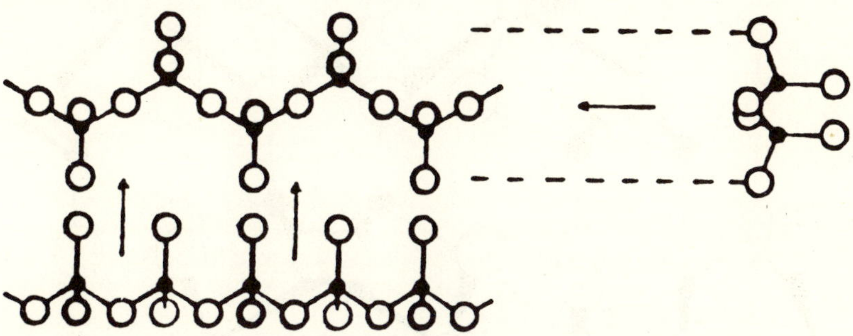

Figure 1-25. The configuration of $(SiO_3)_n$ chains in pyroxenes.

A typical low temperature spectrum from an orthopyroxene with approximate composition $(Mg, Fe)_2Si_2O_6$ is shown in Fig. 1-27. The inner doublet has been assigned to Fe^{2+} in the M_2 site and the outer doublet to Fe^{2+} in the M_1 site (50). It could be argued that these assignments were made because the smallest quadrupole splittings for Fe^{2+} arise from the sites with greatest distortion from cubic symmetry, but in the case of pyroxenes there is also XRD evidence that Fe^{2+} prefers the M_2 position in orthopyroxene. It thus appears that Fe^{2+} in the two types of site in pyroxenes can be distinguished by Mössbauer spectroscopy. Annealing the sample at 1000°C produced the spectrum shown in Fig. 1-28 (50), which shows that a partial redistribution of iron between the two sites has occurred. This type of observation has led some workers to suggest that Mössbauer spectroscopy has potential uses as a geothermometer, especially since changes with pressure can also be observed. With spectra run at room temperature there is a less complete separation of the peaks from the two types of site (Fig. 1-29) (13). It has also been found that for some clinopyroxenes, at least, anomalies in relative areas of the peaks arise if the spectrum is simply fitted to 2 doublets, there being an apparent overestimation of the peaks from the M_2 site compared to XRD results. Explanations offered have suggested (i) the presence of a domain structure in which the M_2 doublets for the 2 phases are more or less coincident but the M_1 doublets are further separated with one of these components overlapping appreciably with the M_2 peaks (54), or (ii) the effects of variation in composition at next-nearest-neighbor sites (13). In the hedenbergite-ferrosilite series, for example, the composition changes from $CaFeSi_2O_6$ to $Fe_2Si_2O_6$. For intermediate members it may be

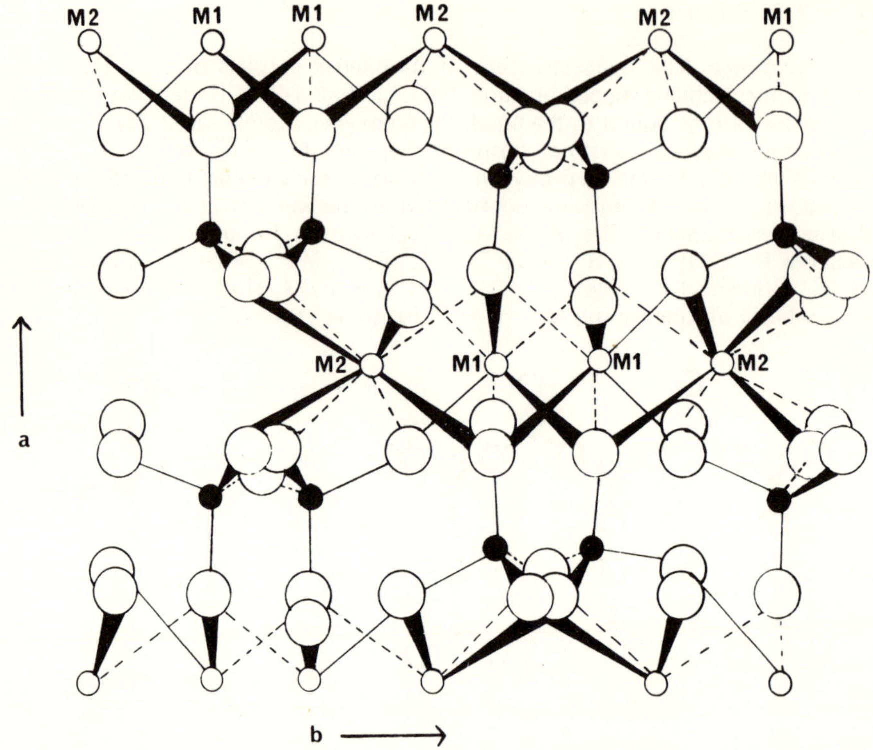

Figure 1-26. The crystal structure of diopside – a pyroxene.

considered that the composition of the M_2 sites is a mixture of Ca and Fe, with the M_1 sites being occupied by Fe. Thus since each M_1 octahedron shares edges with $3M_2$ polyhedra and $3M_1$ octahedra, there are 4 basically different types of next-nearest-neighbor configurations depending on whether the adjacent M_2 sites are occupied by 3Ca, 2Ca and 1Fe, 2Fe and 1Ca or 3Fe. The M_2 polyhedra share corners, but no edges, with other polyhedra, so the next-nearest-neighbor contribution may be smaller. With both of the above explanations more than one doublet needs to be fitted to the M_1 sites (Fig. 1-30).

In some other pyroxenes Fe^{3+} may also be present, although it is not usual to be able to separate components from the M_1 and M_2 sites. The presence of Fe^{3+} in tetrahedral sites in a synthetic ferridiopside have, however, been distinguished (Fig. 1-31) (31). In some other pyroxenes the number of inequivalent octahedral sites is increased so that, for example, in spodumene there are two M_1 and M_2 and in omphacities four M_1 and four M_2 structurally distinct positions. Complex spectra may, therefore, result and the assignment of computer-fitted components can be quite tentative.

Amphiboles. Whereas pyroxene structures are based on single chains of $(SiO_3)_n$ tetrahedra the amphibole structures are composed of double chains (Fig. 1-32) held together by octahedral cations (Fig. 1-33). In this case, though, there are 4 inequivalent octahedral sites. If $Ca_2Mg_5Si_8O_{22}(OH)_2$ is taken as a basic formula

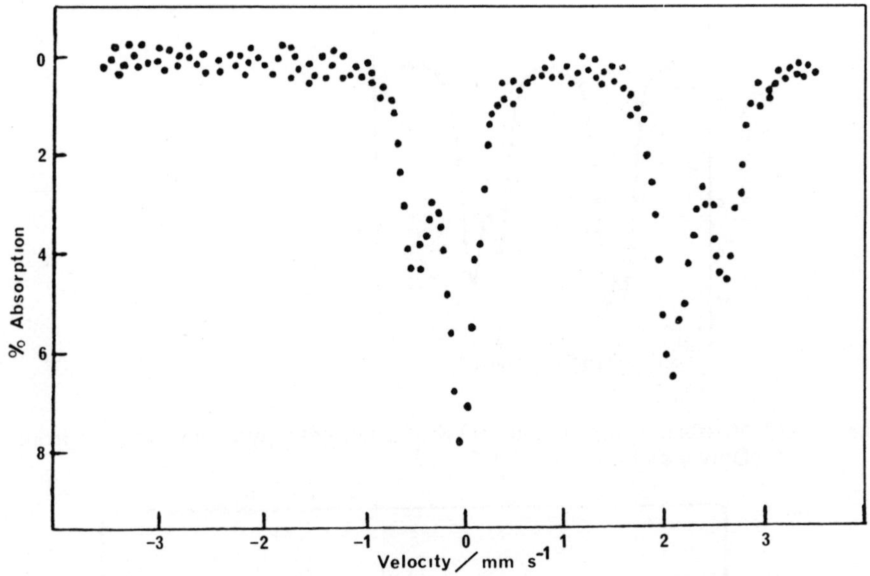

Figure 1-27. Mössbauer spectrum of an orthopyroxene at 77K (adapted from Virgo and Hafner, 1969).

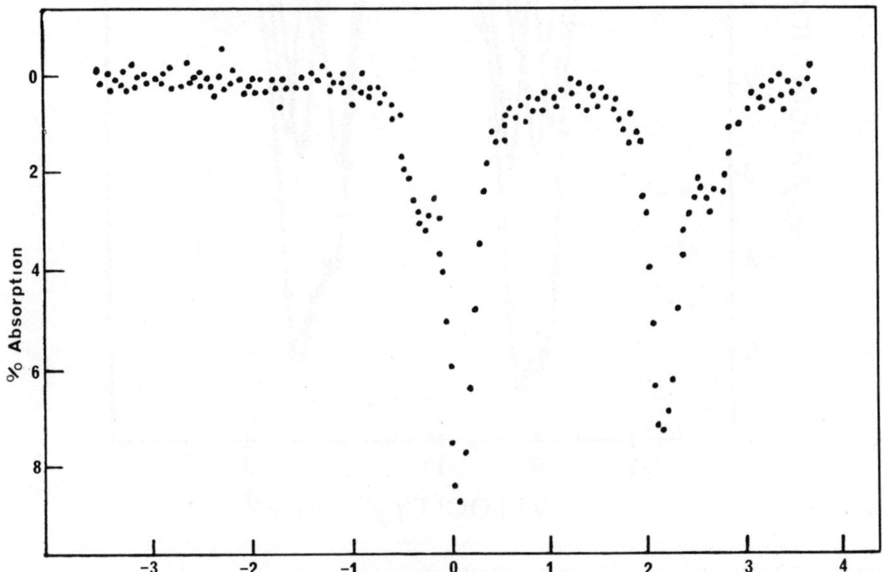

Figure 1-28. Mössbauer spectrum at 77K of the orthopyroxene used for Fig. 1-27 after it had been annealed at 1000° C (adapted from Virgo and Hafner, 1969).

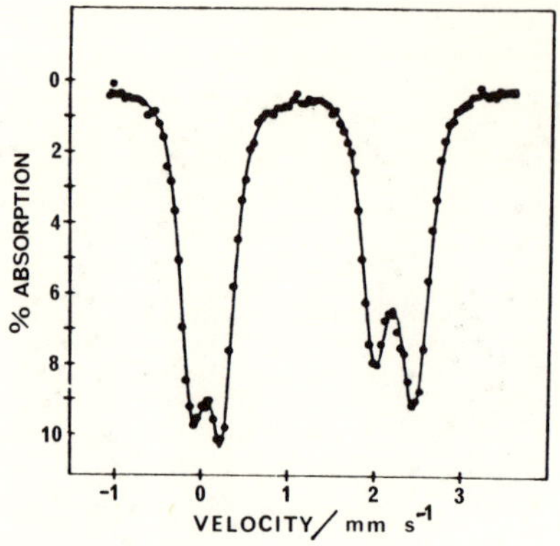

Figure 1-29. Mössbauer spectrum at 300 K of a synthetic pyroxene (adapted from Dowty and Lindsley, 1973).

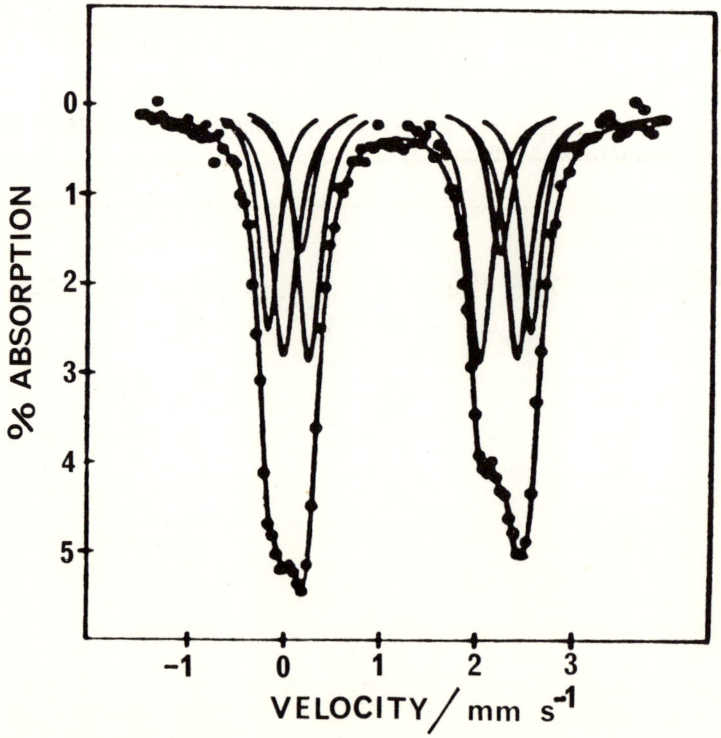

Figure 1-30. Mössbauer spectrum at 300K of a synthetic pyroxene fitted to 4 doublets (adapted from Dowty and Lindsley, 1973).

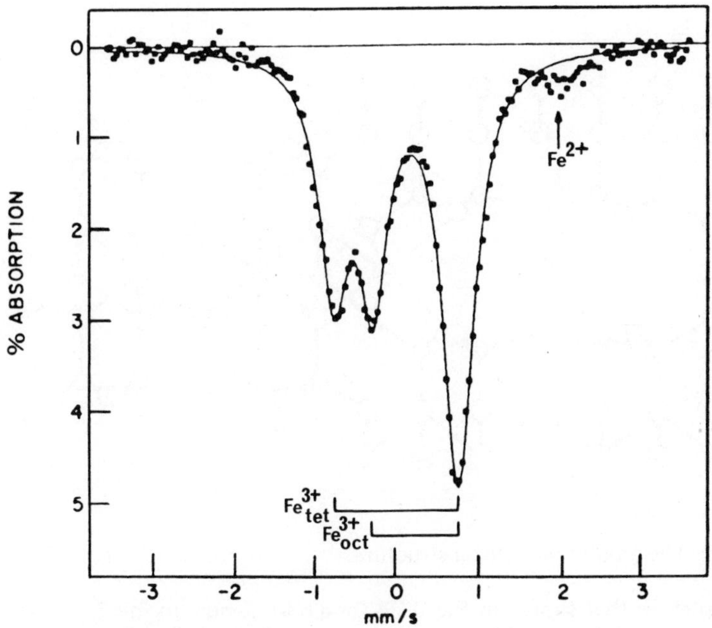

Figure 1-31. Mössbauer spectrum of ferridiopside at 300K (adapted from Hafner and Huckenholz, 1971).

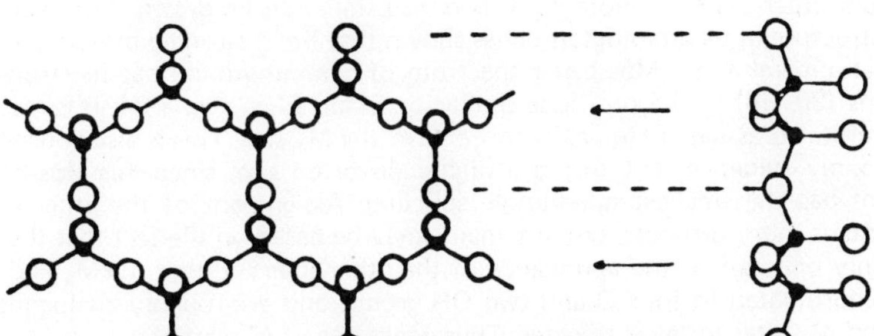

Figure 1-32. The configuration of $(SiO_3)_n$ chains in amphiboles.

for amphiboles, the various types of substitution that can occur are: (i) (Al, Fe^{3+}) may replace up to two of the eight Si atoms in the chains; (ii) Fe^{2+}, Mn, Mg are completely interchangeable; (iii) (Al, Fe^{3+}) may replace up to about one of the five Mg atoms in six-coordination positions; (iv) (Fe^{2+}, Mg) may replace Ca; (v) the total (Ca, Na, K) may rise from 2 to 3; and finally, (vi) OH^-, F^- are interchangeable and the maximum number of 2 may be decreased by replacement by O.

Thus there may be 4 different types of octahedra containing Fe^{2+}, which may also contain Fe^{3+} and, in addition, there is the possibility of some tetrahedral Fe^{3+}. For all but end-member compositions, complex Mössbauer spectra are

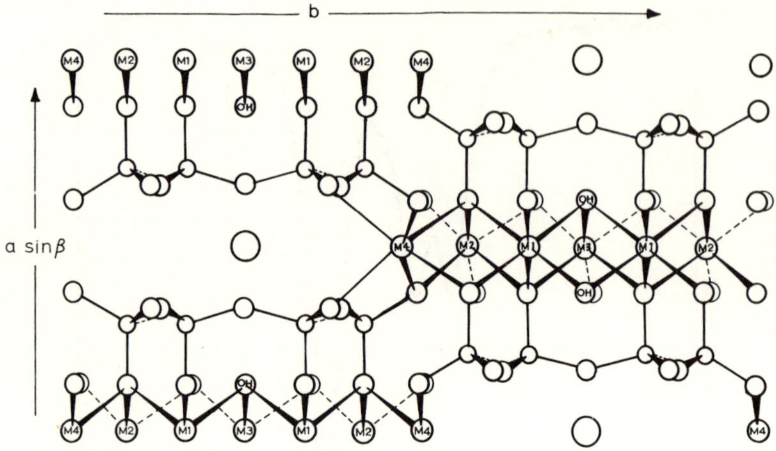

Figure 1-33. The amphibole crystal structure.

expected, such as that shown in Fig. 1-34 for a hornblende. In this figure attempts to obtain statistically-acceptable fits by increasing the number of Fe^{2+} components are illustrated (28). This spectrum also shows that there is, at most, a very small amount of tetrahedral Fe^{3+} and that, if there is more than one type of octahedral Fe^{3+}, they must all have similar parameters. An unambiguous assignment of the Fe^{2+} peaks from this spectrum alone must be tentative but, using information gained from other systems, more definite conclusions can be drawn. For example an x-ray structure of a cummingtonite has shown that Fe^{2+} strongly prefers the M_4 site in that mineral and a Mössbauer spectrum of cummingtonite has its strongest absorptions (30, 38) in the positions similar to E and E′ in Fig. 1-34. It is reasonable, therefore to assign these peaks to Fe^{2+} in the M_4 site. This is also consistent with the x-ray evidence that this is a highly distorted site, since this Mössbauer component has the smallest quadrupole splitting. Assignment of the other Fe^{2+} components is more difficult, but arguments may be based on the fact that the M_3 site has only one half of the abundance of the other sites and that the M_3 and M_1 sites are coordinated to four O and two OH groups and are roughly analogous to the M_1 and M_2 sites in layer silicates. Thus peaks A and A′ may be assigned to the M_1 site, which is the most symmetrical. One could argue about the relative distortions at sites M_2 and M_3 and then attempt to assign peaks B, B′ and C, C′ to them, but when it is recalled from the pyroxene work that there will be next-nearest-neighbor effects which will produce a multiplicity of peaks from each site, it will be seen that further analysis of this spectrum would be futile. Indeed, the relative contents of the sites to which peaks have already been assigned might be quite different to those indicated by the Mössbauer spectra.

1-4.2. Layer Silicates

The crystal structures are based upon a hexagonal network of linked silicon-oxygen tetrahedra (Fig. 1-35). The 1:1 family of minerals is formed when each

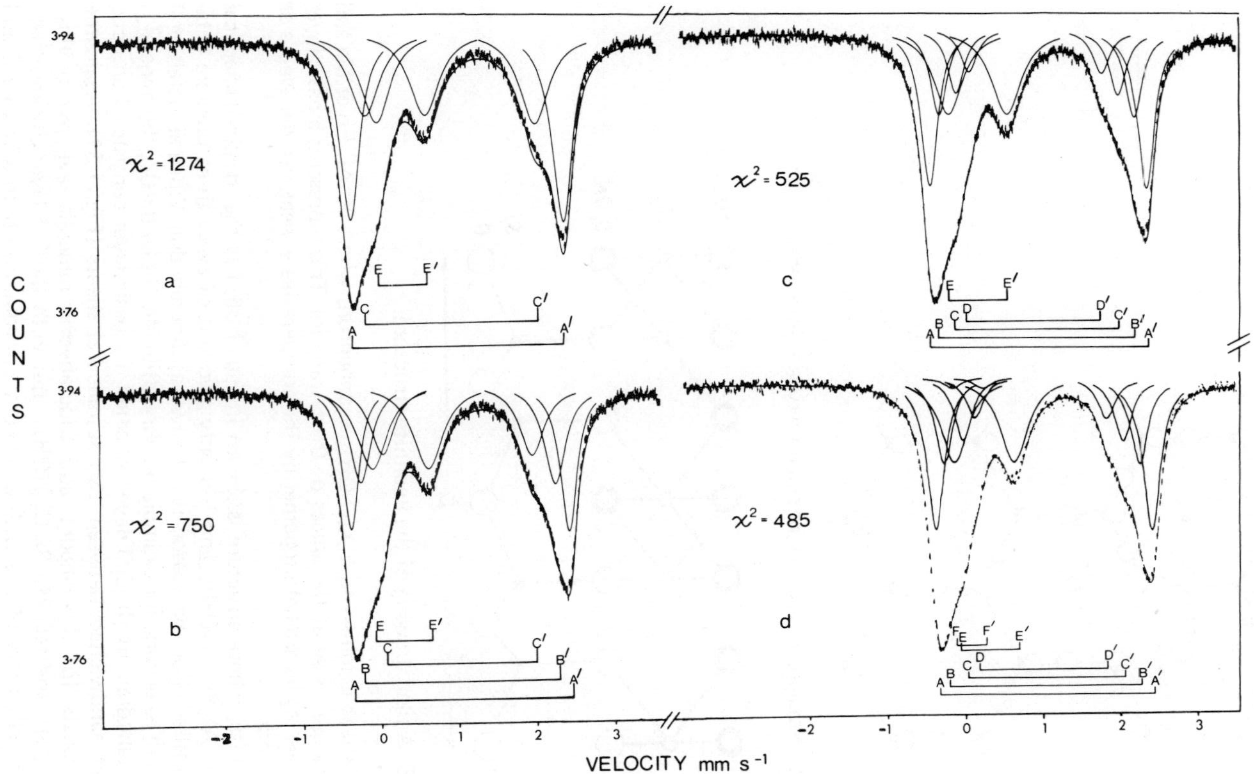

Figure 1-34. Various computer fits to the Mössbauer spectrum of a hornblende at 300K (from Goodman and Wilson, 1976). (a) 3-doublet fit, (b) 4-doublet fit, (c) 5-doublet fit and (d) 6-doublet fit. Peaks AA′, BB′, CC′, and DD′ correspond to Fe^{2+}, peaks EE′ to octahedral Fe^{3+} and peaks FF′ to tetrahedral Fe^{3+}.

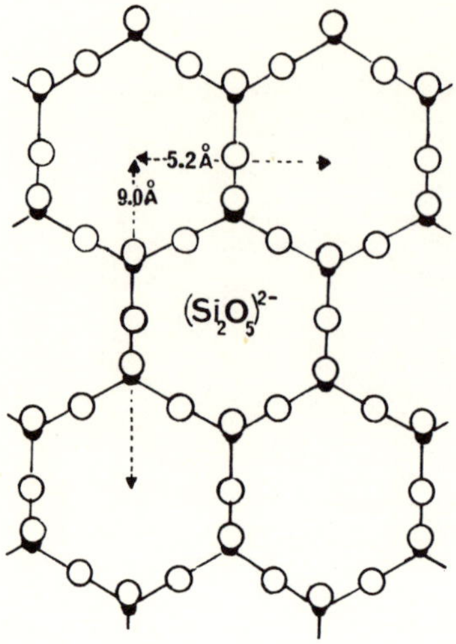

Figure 1-35. Hexagonal network of silicon-oxygen tetrahedra.

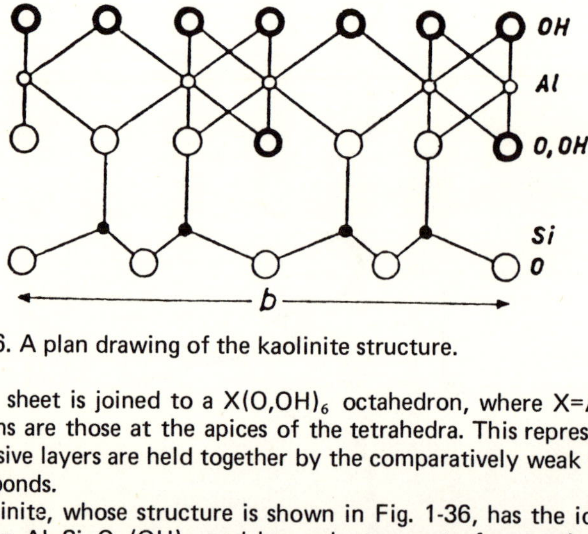

Figure 1-36. A plan drawing of the kaolinite structure.

tetrahedral sheet is joined to a $X(O,OH)_6$ octahedron, where X=Al, Fe or Mg and the O atoms are those at the apices of the tetrahedra. This represents a basic layer and successive layers are held together by the comparatively weak van der Waal's or hydrogen bonds.

Kaolinite, whose structure is shown in Fig. 1-36, has the idealized chemical composition $Al_2Si_2O_5(OH)_4$ and has only two out of every three possible octahedral sites filled. For this reason it is known as dioctahedral. When all octahedral sites are filled, as occurs, for example, in chrysotile $Mg_3Si_2O_5(OH)_4$, the mineral is called trioctahedral. In the 2:1 layer silicates the basic layer consists of an octahedral sheet sandwiched between two tetrahedral sheets (Fig. 1-37). Again the possibility exists for dioctahedral and trioctahedral minerals, e.g. pyrophyllite $Al_2Si_4O_{10}(OH)_2$ and talc $Mg_3Si_4O_{10}(OH)_2$, but with the 2:1 layer silicates there are two different types of site with octahedral coordination both with four O and two OH groups. The OH groups may be arranged either *cis* (M_2) or *trans* (M_1) to one another, the former type of site being twice as abundant as the latter. Substitution of Al for Si in the tetrahedral sheets produces a charge deficiency which is

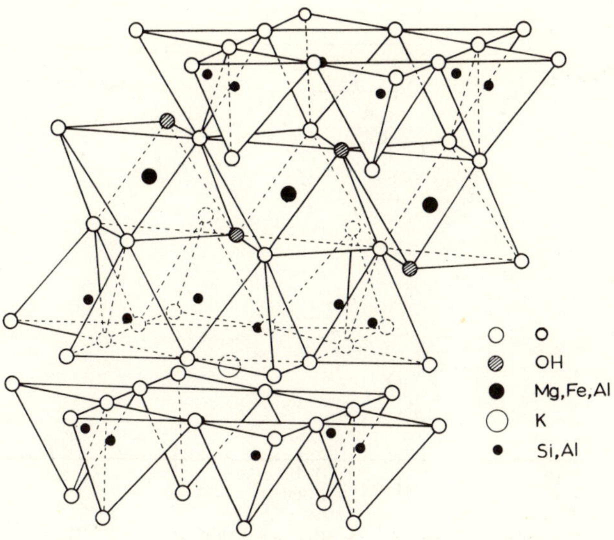

Figure 1-37. The 2:1 layer silicate structure.

compensated for by the presence of interlayer cations, hence muscovite $KAl_2(Si_3,Al)O_{10}(OH)_2$ and phlogopite $KMg(Si_3,Al)O_{10}(OH)_2$. As with the amphiboles there appears to be complete interchangeability within the groups of ions (Mg, Fe^{2+}, Mn) and (Al, Fe^{3+}) and in addition some divalent ions may substitute in the octahedral sites in dioctahedral minerals and trivalent ions in trioctahedral minerals.

A further group of layer silicates are the chlorites which have a 2:1:1 layer structure, being a combination of a 2:1 layer and a brucite-like interlayer.

1:1 layer silicates. Kaolinite has been extensively studied and, although there is some substitution of Fe^{3+} for Al^{3+}, most of the iron in natural samples is in associated impurity phases, such as hematite or goethite (33, 35, 36). These can be distinguished from the structural iron at room temperature only if they are well crystallized and produce magnetic hyperfine splitting, since the quadrupole doublet from small particles of goethite is almost coincident with that from the lattice Fe^{3+}. At lower temperatures, poorly-crystalline goethite is magnetically-ordered and can easily be distinguished from the lattice iron. With pure kaolinites the spectra are usually weak, arising solely from Fe^{3+} in the octahedral sheets (Fig. 1-38).

With the trioctahedral minerals the iron is usually present as Fe^{2+} when there is no tetrahedral substitution. However, when there is substitution within the tetrahedral sheet appreciable quantities of Fe^{3+} may be found both in the octahedral and tetrahedral sheets (Fig. 1-39).

Dioctahedral 2:1 layer silicates. This group consists of minerals such as muscovite, montmorillonite, nontronite, illite, glauconite and celadonite. In a muscovite that contains only ferric iron, there was no evidence for the occupancy of more than one type of site although the line widths indicated that there was a distribution of quadrupole splittings at that site (20) (Fig. 1-40). The results from

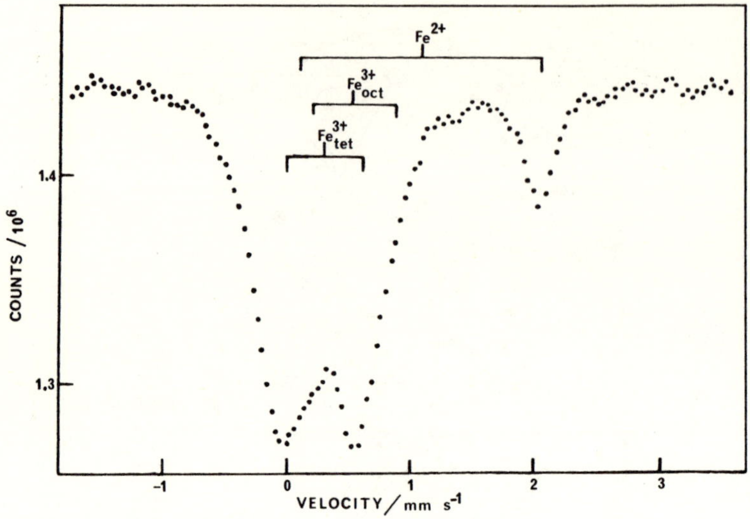

Figure 1-38. Mössbauer spectrum of a kaolinite at 300 K.

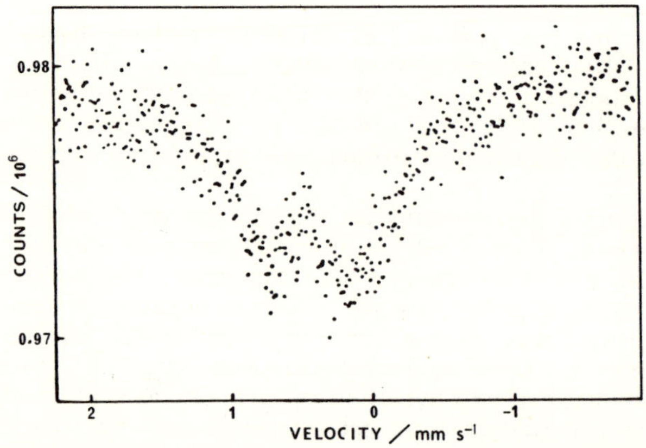

Figure 1-39. Mössbauer spectrum of a crostedite at 300 K (adapted from Coey, personal communication).

Fe^{2+} containing specimens, however, have been widely interpreted as showing clear evidence for the presence of more than one Fe^{2+}-containing site (34) (Fig. 1-41). There is disagreement in the literature concerning the assignment of these Fe^{2+} sites, although by analogy with the trioctahedral mica, biotite, the outer doublet may correspond to the M_2 site (20). The Fe^{3+} doublet must almost certainly correspond to the M_2 site also.

In order to illustrate the problems of fitting a spectrum from a fairly simple system, the spectrum of a nontronite will now be considered. This sample, which corresponds to sample CRO in reference 26, has a chemical analysis of

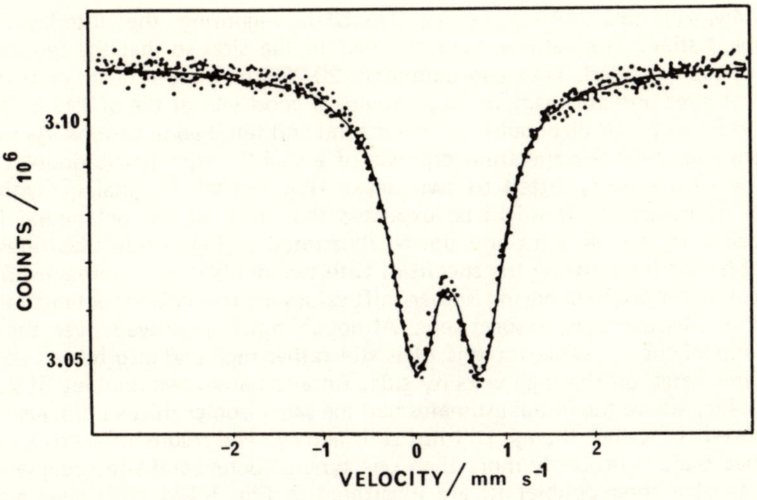

Figure 1-40. Mössbauer spectrum of a ferrian muscovite at 300 K (from Goodman, 1976).

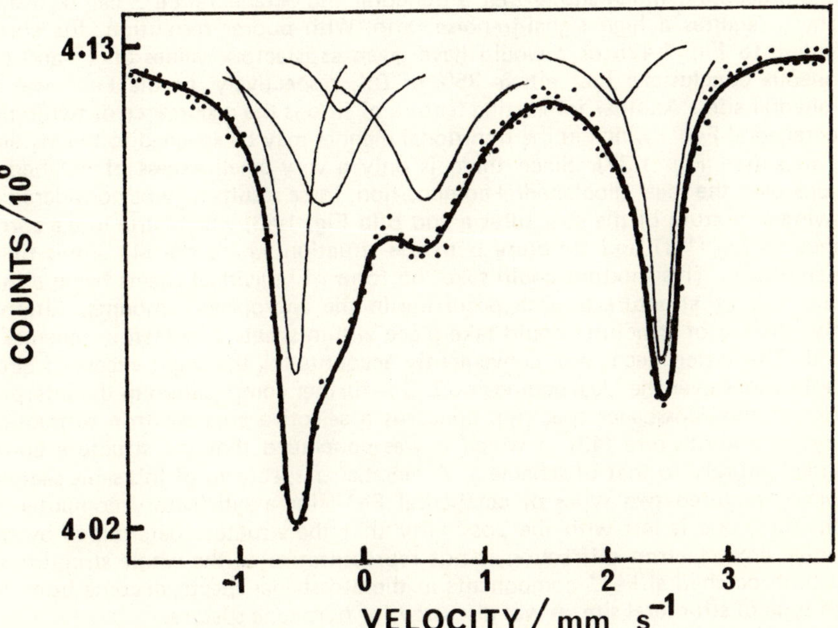

Figure 1-41. Mössbauer spectrum of a Fe^{2+}-containing muscovite at 300K (adapted from Hogg and Meads, 1970).

$(Fe_{3.90}Mg_{0.24})$ $(Si_{6.75}Al_{0.06}Fe_{1.19})O_{20}(OH)_4$, ignoring the interlayer exchangeable cation. The cations were assigned to the sites so that the tetrahedral sites were exactly filled. Thus approximately 20-25% of the Fe must occur in the tetrahedral sites and this sample may provide a good test of the ability of Mössbauer spectroscopy to distinguish the octahedral and tetrahedral cations. As can be seen from Fig. 1-42 the spectrum consists of a slightly asymmetric doublet that cannot be satisfactorily fitted to two peaks (Fig. 1-42a). By analogy with the structure of muscovite it might be expected that most of the octahedral Fe^{3+} would occur in the M_2 sites and the fit illustrated in Fig. 1-42b illustrates the results obtained from fitting the spectrum with two doublets with initial estimates for the computer program having isomer shift values appropriate for octahedral and tetrahedral environments, respectively. Although much improved over the one doublet model the χ^2 value for this fit is still rather high and also has an area of obvious mismatch on the high velocity side. An alternative two-doublet fit shown in Fig. 1-42c, where the initial estimates had the same isomer shift values, also gave an improved χ^2 value, though still not satisfactory. The results of these two fits imply that there is probably more than one type of octahedral site occupied and the results of a three-doublet fit are illustrated in Fig. 1-42d. This gives a satisfactory value for χ^2 and isomer shifts which are consistent with 1 type of tetrahedral and 2 types of octahedral environment for the Fe^{3+}. The areas of the peaks corresponding to the tetrahedral site represent $\sim$ 20% of the total absorption and are consistent with the chemical analyses. It must be stressed, however, that this fitting procedure, which shows that tetrahedral and octahedral Fe^{3+} can be distinguished, requires a high signal-to-noise ratio. With poorer resolution, fits corresponding to Fig. 1-42b or c would have given satisfactory values of χ^2 and the erroneous conclusions that either 35% or 0%, respectively, of the Fe^{3+} was in tetrahedral sites. Another interesting feature of fit d is the occurrence of two types of octahedral Fe^{3+}, which in the traditional manner may be assigned to the M_2 and M_1 sites (see Fig. 1-37). Since there is only a very small excess of octahedral cations over the ideal dioctahedral composition, these results may be considered as showing a mixture of the structures a and b in Fig. 1-43, where structure a corresponds to Fig. 1-37 and structure b to the situation where the M_1 site is filled preferentially. This mixture could take the form of individual sheets being either structure a or structure b, each occurring in the appropriate amounts. Alternatively, mixing of structures could take place within sheets as shown in schemes c and d. This latter mechanism conveniently accounts for the slight excess of octahedral cations over the ideal composition. One further complication in the interpretation of this Mössbauer spectrum concerns a selective area electron diffraction study of a nontronite (43) in which it was concluded that the structure corresponded entirely to that of scheme a. A Mössbauer spectrum of this same sample, however, required two types of octahedral Fe^{3+} for a satisfactory computer fit (22). Thus, one is left with the possibility that the structure determined by the selective area electron diffraction is not representative of the whole structure or that both octahedral Fe^{3+} components in the Mössbauer spectrum come from the same type of structural site, as was discussed for pyroxene spectra.

In the study of montmorillonites, no evidence has been found for the presence of Fe^{3+} in the tetrahedral sheets. The quadrupole splittings of the octahedral Fe^{3+} are appreciably greater than those of Fe^{3+} in nontronite and there is an apparent correlation between the b-axis parameter and Δ in these silicates (46) (Fig. 1-44). One difficulty encountered in the study of montmorillonites, particularly

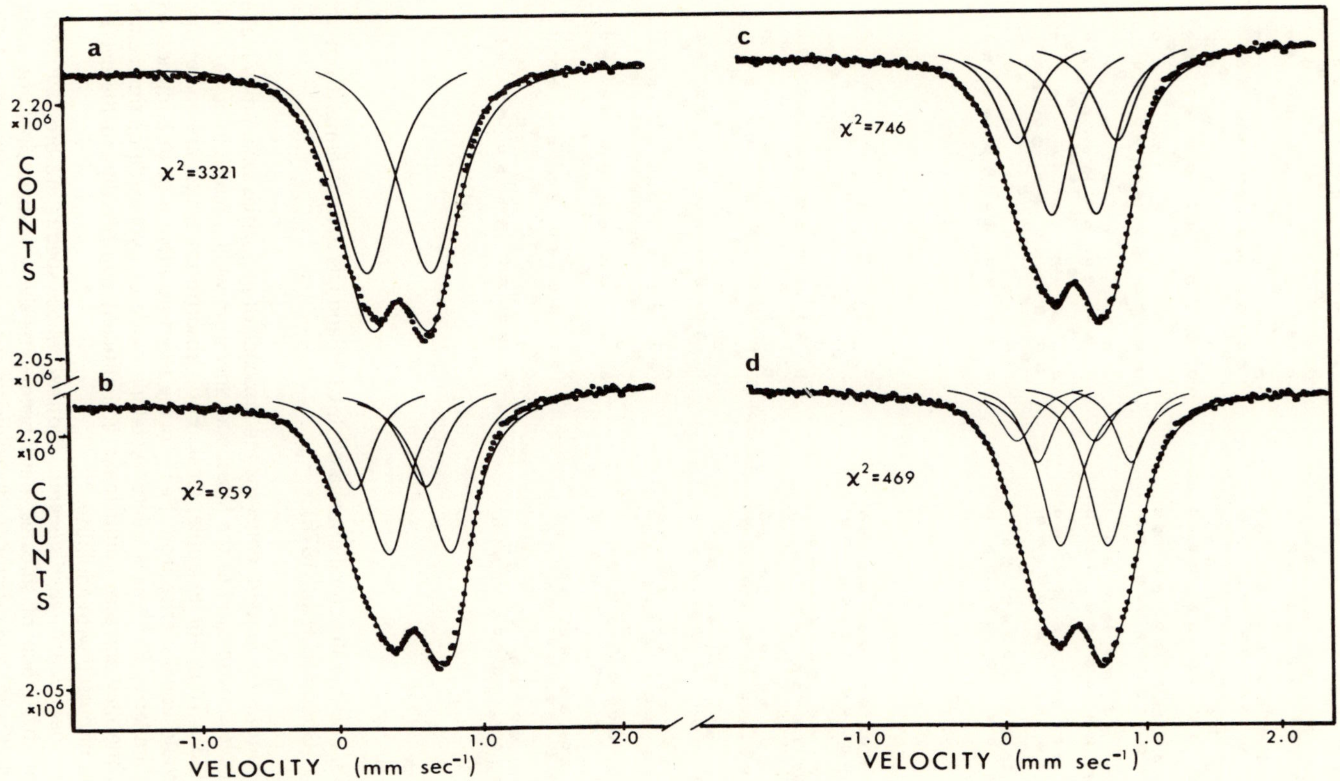

Figure 1-42. Various computer fits to the Mössbauer spectrum of a nontronite at 77K (a) one component, (b) one octahedral plus one tetrahedral component, (c) two octahedral components and (d) two octahedral plus one tetrahedral component (from Goodman *et al.*, 1976).

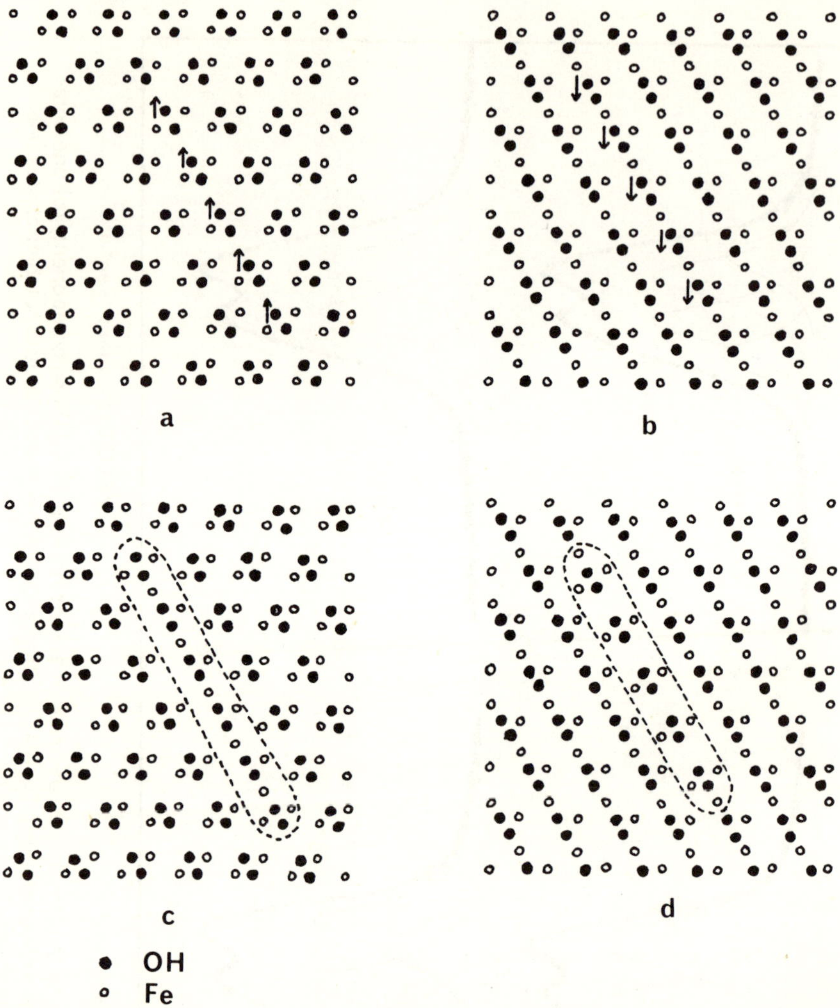

Figure 1-43. Possible structural arrangements of Fe and OH in nontronite (from Goodman *et al.,* 1976).

those of low total Fe content, concerns the presence of impurities which may make an appreciable contribution to the Mössbauer spectrum when present in quite small amounts (23). Thus, for example, weak Fe^{2+} absorptions which are sometimes observed could equally arise from Fe^{2+} in the montmorillonite or in a micaceous impurity. Similarly, Fe^{3+} as a microcrystalline oxide or hydroxide or simply adsorbed on the clay surfaces could make a significant contribution to the overall spectrum as discussed with kaolinite.

Trioctahedral 2:1 layer silicates. Biotites, whose spectra are broadly representative of this group of silicates, have been extensively studied since the early 1960s (1, 4, 32, 34) and usually produce spectra similar to that shown in Fig. 1-45. This spectrum has been fitted to two Fe^{2+} components and one Fe^{3+} component,

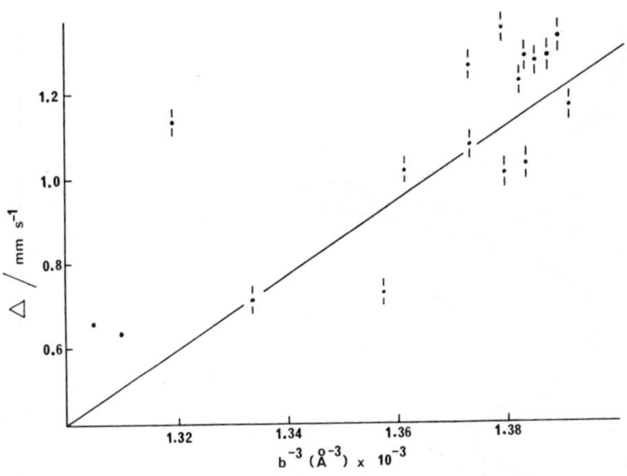

Figure 1-44. The dependence of Γ on b^{-3} from the M_1 site in smectites (adapted from Rozenson and Heller-Kallai, 1977).

all with isomer shifts consistent with octahedral coordination. Some workers have fitted spectra to two octahedral Fe^{3+} components, although there is usually no statistical evidence to support this, and there has also been an observation of tetrahedral Fe^{3+} in phlogophite (48). It is in this latter type of observation that Mössbauer spectroscopy is most valuable, since the quantitative assignment of the octahedral components to the two types of site is complicated by lattice effects. The range of quadrupole splittings associated with one crystallographic site in a mineral showing the usual level of isomorphous substitution has been calculated to be comparable to the difference between the values of Δ from the two sites when long range lattice effects are ignored (21, 44). Thus, although work with iron-rich biotites has shown that the component with the larger value of Δ must correspond to Fe^{2+} in the M_2 site (32), any conclusions about the relative distribution of iron between the two types of site must be tenuous.

2:1:1 layer silicates. Chlorites may be considered as having a 2:1:1 structure since they consist of alternate layers of talc-like and brucite-like structures (Fig. 1-46). One interesting structural question concerns the possible distribution of iron between these layers. The Mössbauer spectra have narrow lines and, even with very iron-rich samples, there is little difference between the values of Δ for the Fe^{2+} components (24) (Fig. 1-47). The major component has Δ and δ values similar to the major component in biotite and it is, therefore, tempting to assign it to the M_2 site in the talc-like layer. However, there is no evidence as to the expected value for Δ for the brucite-like layer, although in brucite itself Δ is ~ 2.9 mm/sec, compared to the value of ~ 2.7 mm/sec obtained for the major component in chlorites. The

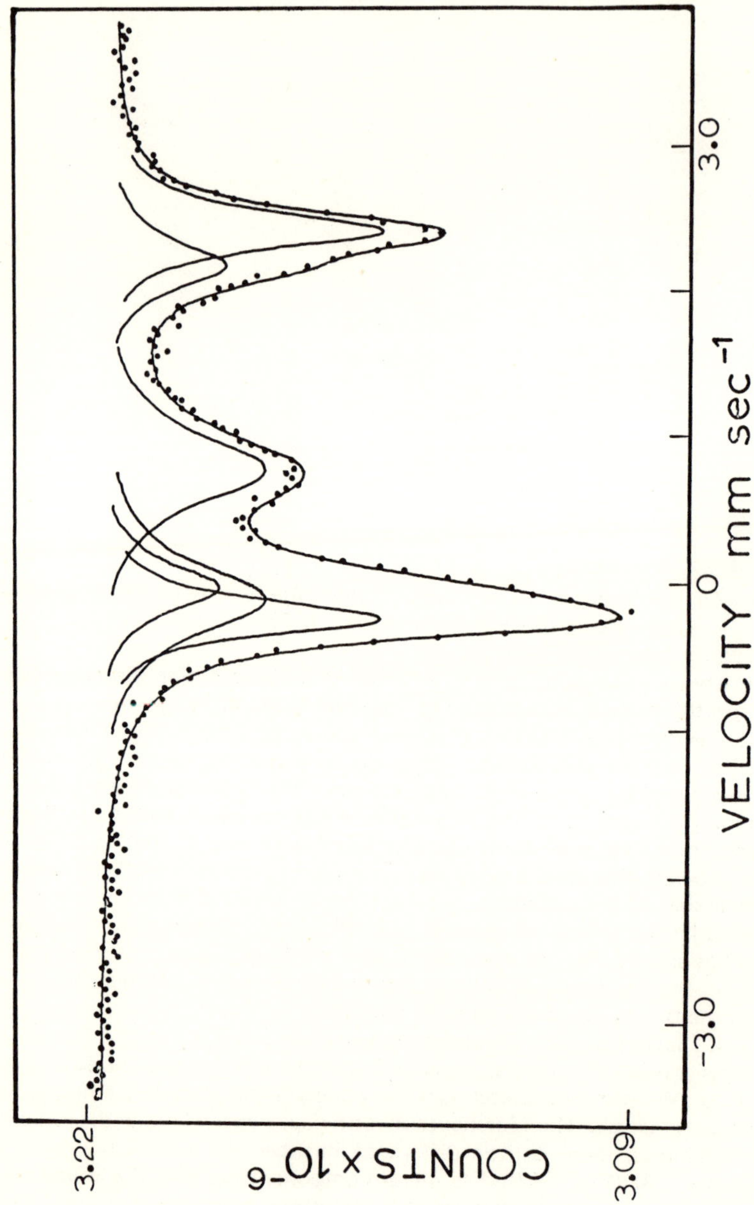

Figure 1-45. Mössbauer spectrum of a biotite at 300 K (from Goodman and Wilson, 1973).

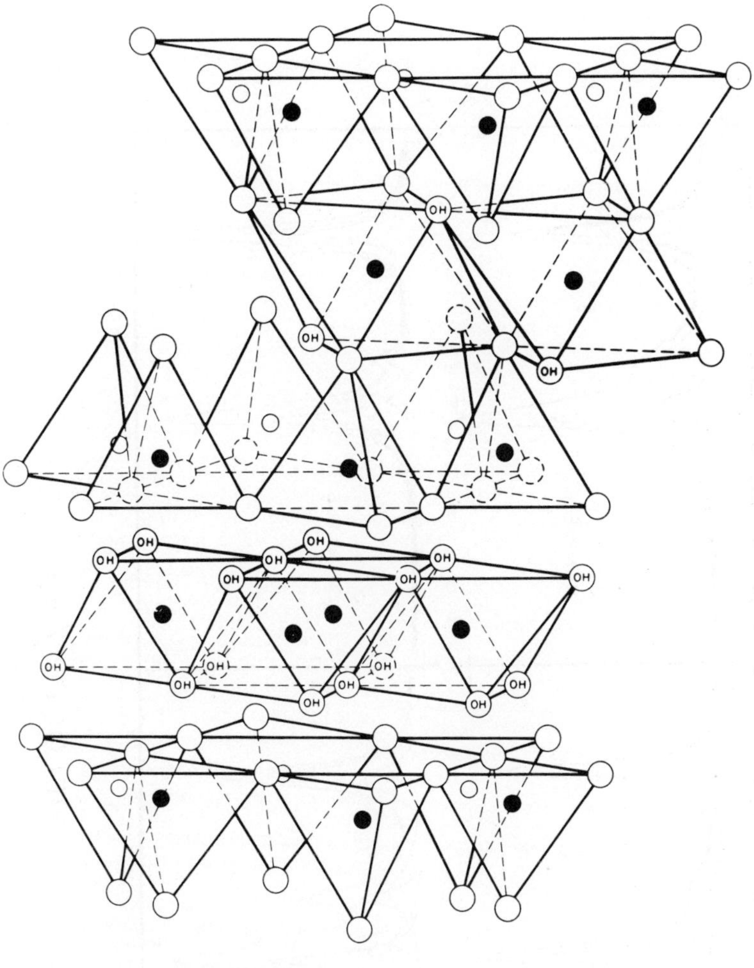

Figure 1-46. The structure of chlorite.

Δ value of ~ 2.4 mm/sec for the minor Fe^{2+} component in the chlorites should be contrasted with the value of ~ 2.2 mm/sec obtained with biotite, so no great significance should be attached to variations of a few tenths of a mm/sec in Δ from one mineral to another. In contrast to the uniformity of Fe^{2+} parameters shown in Fig. 1-47, there is appreciable variation in value for the Fe^{3+} components from one sample to another. In some specimens evidence for the presence of tetrahedral Fe^{3+} is clearly seen (Fig. 1-47 a and c), even though analyses indicate more than enough Al available to fill those tetrahedral sites not occupied by Si.

1-5. THE STUDY OF MINERAL ALTERATION REACTIONS

This section is concerned with the study, in a little more detail than that on

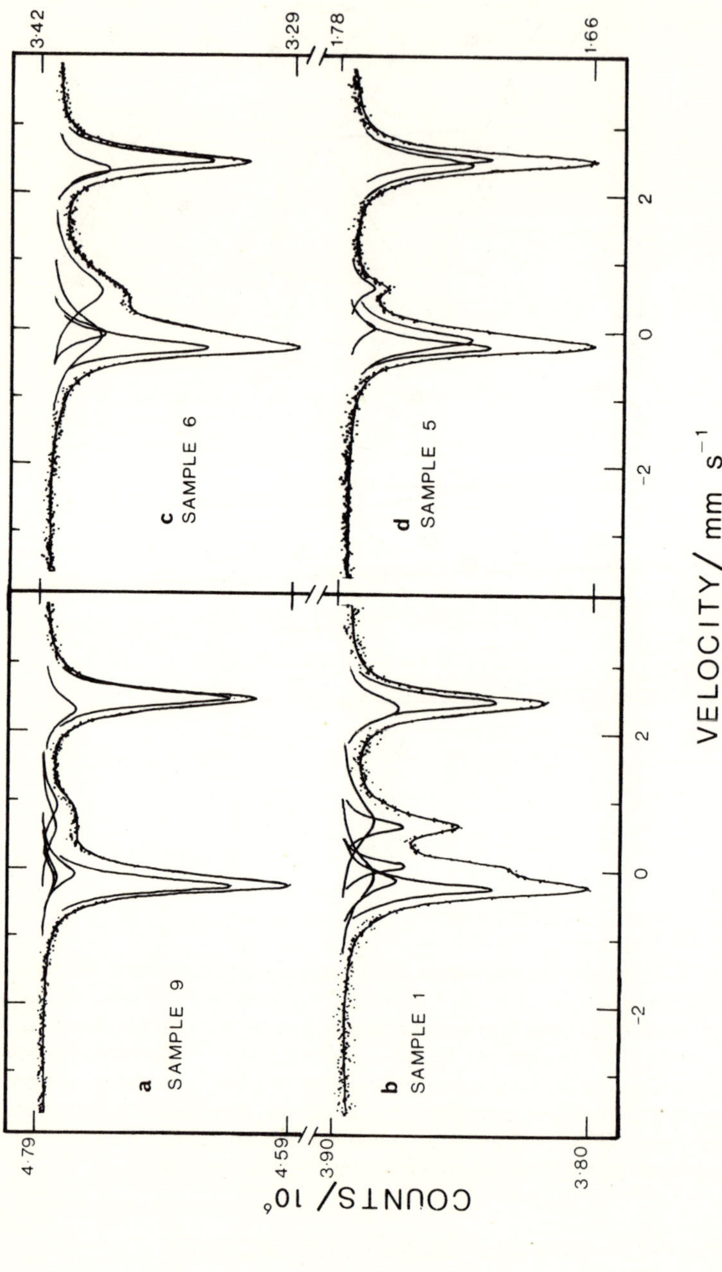

Figure 1-47. Mössbauer spectra of four chlorite specimens at 300K (from Goodman and Bain, 1978) in which sample details are given).

the silicate minerals, of some examples of the use of Mössbauer spectroscopy in the study of mineral alteration reactions. It has for convenience been broken down into three main groups, namely (i) the investigation of natural weathering processes occurring in the soil, (ii) the study of chemical alteration of minerals carried out in the laboratory and (iii) the study of changes that occur as a result of heat treatment and thermal decomposition.

1-5.1. Application to the study of natural weathering processes

As an example of natural weathering processes the transformation of biotite and hornblende, coexisting within the same soil profile (Fig. 1-48), will be compared (27, 28). The biotite in the bedrock is transformed into hydrobiotite in the C horizon, and then into an interstratified vermiculite-chlorite in the B horizon. In the more acidic A horizon the brucite layers of the chlorite tend to break down to yield a more vermiculitic product. Chemical analyses showed that the initial transition of biotite to hydrobiotite is characterized by loss of iron and extensive oxidation (55). This is readily confirmed by the Mössbauer spectra a, b, and c of Fig. 1-49 (see Table 1-5 for computed parameters), where the transition is accompanied by a substantial decrease in the intensities of the Fe^{2+} peaks, AA′ and BB′.

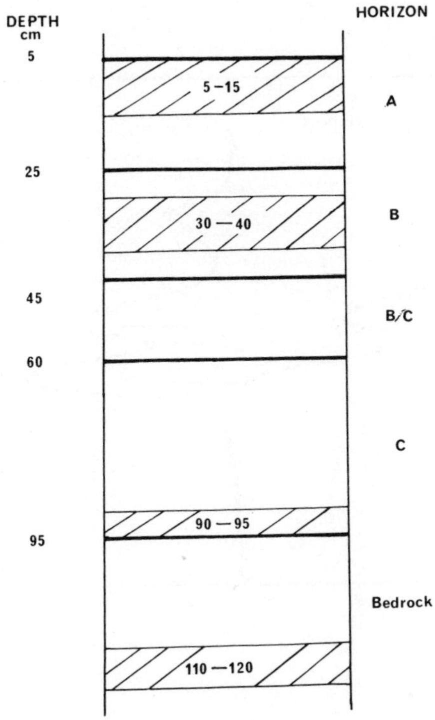

Figure 1-48. Soil horizons and sampling positions for the biotites and hornblende used in the weathering studies (from Goodman and Wilson, 1973).

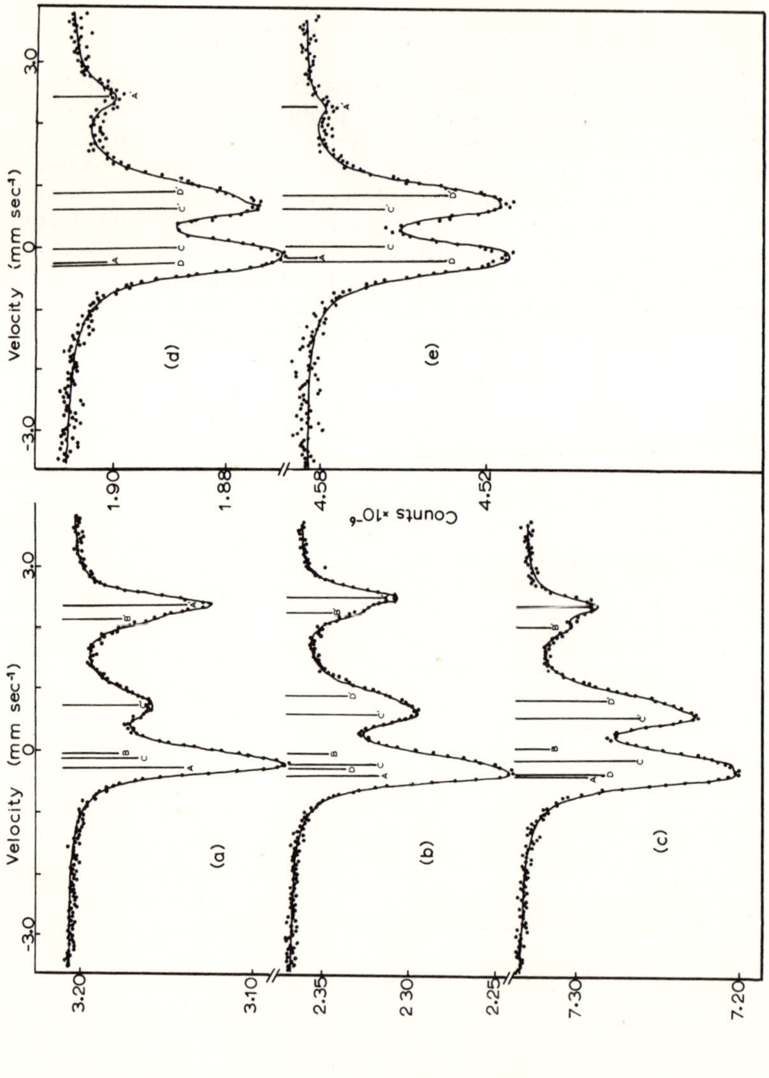

Figure 1-49. Mössbauer spectra at 300 K of biotite (a) and its weathering products, (b) partly oxidized bedrock, (c) hydrobiotite from C horizon, (d) and (e) interstratified vermiculite-chlorite from B and A horizons, respectively (from Goodman and Wilson, 1973).

Table 1-5. Computer parameters for biotite and its weathering products (from Goodman and Wilson, 1973).

Sample	Chem. Fe^{2+}	Fe^{2+} M2 Δ†	M2 δ†*	M2 Γ†	M2 %‡	M1 Δ†	M1 δ†*	M1 Γ†	M1 %‡
Fresh biotite	7.47	2.64 (0.02)	1.13 (0.01)	0.30 (0.01)	36 (3)	2.19 (0.04)	1.11 (0.01)	0.40 (0.03)	21 (3)
Partly oxidized biotite (bedrock)		2.66 (0.01)	1.13 (0.01)	0.29$^{\chi}$ (0.01)	26 (1)	2.11 (0.03)	1.15 (0.01)	0.29$^{\chi}$ (0.01)	11 (1)
Hydrobiotite (C horizon)	2.37	2.75 (0.04)	1.12 (0.02)	0.33 (0.02)	18 (1)	1.98 (0.04)	1.15 (0.02)	0.33$^{\chi}$ (0.02)	8 (1)
Interstratified vermiculite-chlorite (B horizon)	1.15	2.71 (0.18)	1.13 (0.08)	0.50 (0.06)	15 (2)				
Interstratified vermiculite-chlorite (A horizon)	1.21	2.51 (0.45)	1.15 (0.22)	0.34 (0.03)	5 (1)				

† all values in mm sec^{-1}.
* isomer shifts quoted relative to Fe metal.
‡ % refers to the percentage of the total Fe.
χ half widths constrained to be equal.
The errors quoted in brackets include standard deviations and covariance contributions.

Table 1—5 continued.

Sample	Chem. Fe^{3+}	Fe^{3+} M2 Δ†	δ†*	Γ†	%‡	Fe^{3+} M1 Δ†	δ†*	Γ†	%‡	χ^2
Fresh biotite	4.23	0.89 (0.03)	0.38 (0.02)	0.60 (0.02)	43 (2)					253
Partly oxidized biotite (bedrock)		0.76 (0.05)	0.34 (0.03)	0.42 (0.02)	37 (5)	1.08 (0.12)	0.41 (0.06)	0.46 (0.04)	26 (5)	245
Hydrobiotite (C horizon)	6.68	0.72 (0.02)	0.35 (0.01)	0.41 (0.02)	41 (5)	1.23 (0.11)	0.37 (0.03)	0.46 (0.03)	33 (6)	185
Interstratified vermiculite-chlorite (B horizon)	7.06	0.64 (0.03)	0.37 (0.01)	0.41 (0.03)	35 (9)	1.17 (0.10)	0.38 (0.03)	0.57 (0.05)	49 (11)	237
Interstratified vermiculite-chlorite (A horizon)	6.93	0.63 (0.03)	0.37 (0.01)	0.34 (0.03)	28 (8)	1.13 (0.06)	0.36 (0.02)	0.51 (0.02)	68 (10)	241

† all values in $mm.sec^{-1}$.
* isomer shifts quoted relative to Fe metal.
‡ % refers to the percentage of the total Fe.
χ half widths constrained to be equal.
The errors quoted in brackets include standard deviations and covariance contributions.

Analytical results show that the other stages in the weathering sequence are also characterized by further oxidation and loss of iron (Table 1-5). In Fig. 1-49 spectrum a, unaltered biotite, was fitted to two Fe^{2+} doublets and one Fe^{3+} doublet, as was described in the previous section. However, in the spectra of partly oxidized biotite in the bedrock (spectrum b) and the hydrobiotite (spectrum c) the presence of an additional Fe^{3+} component was clearly seen, and indeed was required for a satisfactory value of χ^2. Two Fe^{3+} components were also required in fitting the interstratified vermiculite-chlorite specimens. A closer examination of the computed values in Table 1-5 reveals that the single Fe^{3+} component computed for the unaltered biotite has a Δ intermediate between those values of Δ obtained for the Fe^{3+} components in these oxidized samples. It is likely, therefore, that the unaltered biotite contains two Fe^{3+} components even though statistically only one is required for a satisfactory fit. Also in passing from a biotite to hydrobiotite the two computed Fe^{2+} components seem to disappear at roughly the same rate. In the spectra of the interstratified vermiculite-chlorites the Fe^{2+} contents are too low for more than one Fe^{2+} component to be fitted meaningfully. Although the Fe^{2+} components are lost at similar rates the ratios of the Fe^{3+} components change appreciably. In fact there is no great change in the intensity of the component with the smaller Δ throughout the weathering sequence, whereas the component with the larger Δ increases dramatically and accounts for about 2/3 of the total iron in the sample taken from the A-horizon.

The difficulty now is in the interpretation of such results. In the conventional scheme the Fe^{2+} peaks AA′ would be assigned to the M_2 site and BB′ to the M_1 site. However, as stated in the last section, lattice effects prevent such a clear distinction from being made. Nevertheless, even if there is no accurate quantitative correlation between peak areas and site contents, an approximate qualitative relationship probably holds. Thus these results indicate that oxidation is not taking place preferentially at one type of site during the weathering sequence. An explanation of the behavior of the Fe^{3+} components is more difficult. If the arguments for assigning the Fe^{2+} sites are applied here, then the inner doublets CC′, would correspond to the M_2 site and the outer doublets DD′, to the M_1 site. The results would then indicate that the percentage of the total iron as Fe^{3+} in the M_2 site increases on going from biotite to hydrobiotite, then decreases with further weathering, whereas the Fe^{3+} in the M_1 site increases throughout. Thus, there is an implication that iron is lost preferentially from M_2 sites. There are, however, a number of difficulties in accepting this conclusion. First, since iron is being lost from the structure, vacancies will be created within the structure and consequently will affect the electric field gradients at neighboring octahedral sites. Hence, some of the newly-formed Fe^{3+} ions may experience quite different lattice contributions to the electric field gradient than either existed originally or exists at other octahedral sites. Peaks DD′ may, therefore, contain a contribution from Fe^{3+} in M_2 sites. Second, in the weathering sequence the Fe^{2+} content drops from 7.47% to 1.21%, whereas the Fe^{3+} content only increases from 4.23% to 6.93%. If the composition of the major ions is assumed to remain the same then the increase in Fe^{3+} content accompanying the weathering is appreciably less than that required to retain charge balance after oxidation and expulsion of 1/3 of the resulting Fe^{3+} ions. Thus, since the total iron content in the most highly-weathered samples is lower than expected for a simple alteration process, there is the implication that regions of the biotite contain higher levels of iron than the bulk and that these regions are more completely broken down on weathering. Thus the iron remaining in the structure in the most highly weathered samples may be representative of phases least abundant in iron in the original biotite (it should be noted that in the biotite → hydrobiotite transition the increase in Fe^{3+} content is only slightly less than two-thirds the decrease in Fe^{2+} content). The question of regions of higher iron content being selectively destroyed raises the possibility that part of component DD′ is associated with iron not present in the structure at all, possibly as an amorphous phase or as adsorbed surface ions. (Treatment of sample with citrate-dithionite solutions to remove surface coatings was inconclusive since noticeable breakdown of the structure of the slightly weathered samples, e.g. hydrobiotite, was observed which sheds doubt on the integrity of highly weathered samples treated similarly). The presence of such components in some montmorillonites has recently been demonstrated by EPR spectroscopy (23) and have Δ very similar to component DD′.

With the hornblende samples much less dramatic mineralogical changes take place during the weathering sequence. However, it has been shown that the original hornblende contains lamellar intergrowths of another amphibole phase, which is richer in iron, and that this iron-rich phase weathers preferentially to yield a swelling mineral in the C-horizon (56). The major hornblende phase appears by conventional techniques to remain unaltered throughout the soil profile. The computer fitting of the unweathered hornblende was discussed in the previous section and was found to contain four Fe^{2+} components and an octahedral Fe^{3+} component. The tetrahedral Fe^{3+} content was very small and the possible presence of such a component has been ignored in these studies of the weathering sequence. Representative spectra of the weathered hornblendes are shown in Fig. 1-50 (28). Large particles (> 150 μm) vary only slightly from one horizon to another or from the unweathered bedrock sample, but changes are observed in the spectra of smaller particles. These show up as a decrease in the relative intensities of peaks BB′ and a corresponding increase in intensity of peaks CC′ with increasing weathering for any of the samples (Table 1-6). In section 1-4.1 peaks AA′ and DD′ were assigned to the M_1 and M_4 sites in the amphibole structure (Fig. 1-33). Conclusive assignments of peaks BB′ and CC′ to the M_2 and M_3 sites were impossible, so no differentiation could be made between enriched versus depleted sites relative to Fe^{2+}. However, on structural grounds it seems unlikely that weathering would result in the preferential removal of Fe^{2+} from the M_2 sites, since the M_2 and M_4 sites are located at the edges of the talc-like strips (Fig. 1-51) and provide the binding forces that link the chains parallel to the a and b crystallographic axes. The nature of the occupancy of these positions markedly affects the b parameter of the unit cell (12) and removal of ions from such sites would lead to the disintegration of the amphibole structure. On the other hand, occupancy of M_1 and M_3 sites in hornblende has only a minor influence on unit cell dimensions, and it may be significant that unit cell parameters calculated from x-ray powder patterns show no significant differences. It is, therefore, reasonable to suppose that M_1 and M_3 cations would be more easily displaced than M_2 and M_4 cations. Thus peaks BB′ are unlikely to correspond to Fe^{2+} in the M_2 sites. Consequently, peaks AA′, BB′, CC′ and DD′ can be assigned to Fe^{2+} in the M_1, M_3, M_2 and M_4 sites, respectively. The intensities of peaks EE′, which correspond to Fe^{3+} in the structure, remain remarkably constant throughout the weathering sequence – a behavior in contrast to that of biotite, where oxidation of Fe^{2+} to Fe^{3+} during weathering is the most obvious phenomenon. In the hornblende samples both the quadrupole splitting and linewidth increase in the more highly weathered samples, indicating that the

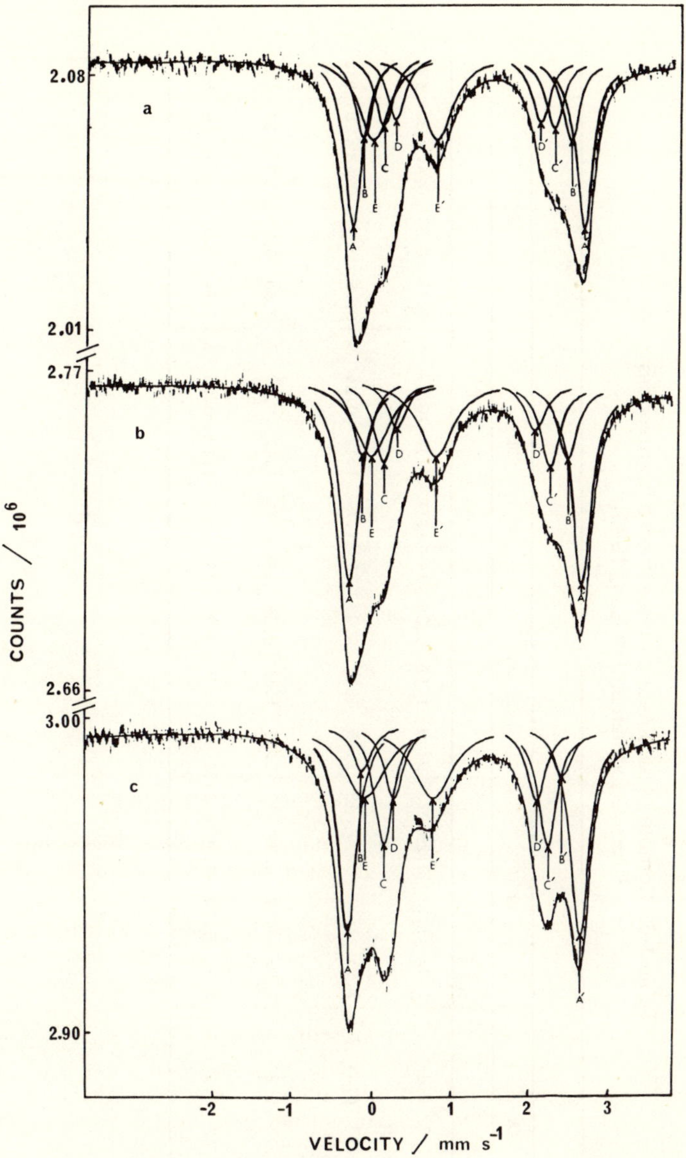

Figure 1-50. Mössbauer spectra at 300 K of weathered hornblendes from (a) C horizon, > 150 m, (b) C horizon, < 150 m and (c) B horizon, < 150 m (adapted from Goodman and Wilson, 1976).

environment of Fe^{3+} is becoming more distorted – a situation that could arise either because Fe^{3+} is formed at more distorted sites at the same rate as it is lost from the original sites or because weathering leads to lattice distortions that affect the electric field gradient at the Fe^{3+} ions. The former could be the case only if

Table 1-6. Computed parameters for weathered hornblendes (from Goodman and Wilson, 1976).

	Fe^{2+}																Fe^{3+}				χ^2
	AA'				BB'				CC'				DD'				EE'				
	Δ	δ	Γ	%	Δ	δ	Γ	%	Δ	δ	Γ	%	Δ	δ	Γ	%	Δ	δ	Γ	%	
Bedrock	2.86	1.13	0.26	27	2.57	1.11	0.29	21	2.10	1.12	0.29	14	1.71	1.10	0.27	8	0.76	0.34	0.47	30	534
A, B, C Horizons Coarse fractions	2.84	1.13	0.26	29	2.55	1.11	0.30	19	2.05	1.13	0.29	15	1.74	1.12	0.26	9	0.76	0.35	0.48	28	532
C Horizon <150μ	2.88	2.13	0.27	30	2.59	1.12	0.30	17	2.05	1.14	0.36	21	1.71	1.12	0.30	7	0.80	0.35	0.54	25	461
B Horizon <150μ	2.88	1.13	0.26	28	2.58	1.12	0.30	12	2.02	1.15	0.31	26	1.75	1.13	0.25	8	0.84	0.32	0.59	26	509

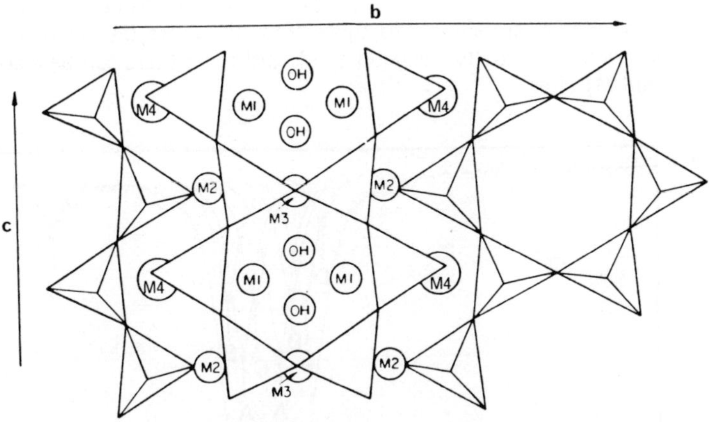

Figure 1-51. Amphibole crystal structure showing positions of octahedral cations relative to the talc-like strips (from Goodman and Wilson, 1976).

Fe^{3+} is lost preferentially from the M_1 sites which are the least distorted, but the Mössbauer results for the unweathered hornblende provide no information on the distribution of Fe^{3+} between the octahedral sites. Also, structural studies indicate that some concentration of Fe^{3+} in the M_2 sites might be expected. One is, therefore, led to the conclusion that the average electric field gradient at the Fe^{3+} sites increases as a result of weathering, similar to the situation occurring for Fe^{2+}. As a consequence the detailed interpretation of the Fe^{2+} results might be questioned further.

1-5.2. The study of mineral alteration in the laboratory

Mössbauer spectroscopy can be used not only in the study of stable phases produced as a result of chemical treatment of minerals, but also to study the nature of iron in unstable intermediate phases by rapid cooling of specimens to liquid nitrogen temperatures. The study of structural changes brought about in nontronites by chemical reducing agents (47) will be given as an example of the use of Mössbauer spectroscopy in this type of work. The Mössbauer spectra of nontronites were discussed in section 1-4.2 and it was shown that two octahedral and one tetrahedral Fe^{3+} components were usually required for an acceptable fit. This section will be concerned with the influence of the composition of nontronites on the nature of the reaction products formed as a result of treatment with the reducing agents hydrazine and dithionite.

The spectra obtained from the reduction of several nontronite specimens with hydrazine have been found to be similar to one another (47) and a typical spectrum is shown in Fig. 1-52a. Peaks AA′ and BB′ arise from the Fe^{3+} in octahedral sites and peak CC′ from tetrahedral sites as discussed for the original nontronites. Peaks DD′ are assigned to Fe^{2+} and correspond to ~ 10% of the total iron. The widths of these Fe^{2+} peaks are quite small indicating that reduction could be occurring selectively at one type of site. By comparing these results with those obtained from untreated specimens it appears that the reduction takes place at the M_2 sites (from peaks AA′) but, because of large errors in computation of the relative areas of the Fe^{3+} components, definite conclusions cannot be made. Also these Fe^{2+} parameters are similar to those from the more distorted site in dioctahedral micas, which is usually assigned to M_1 although there can be a contribution from distorted M_2 sites.

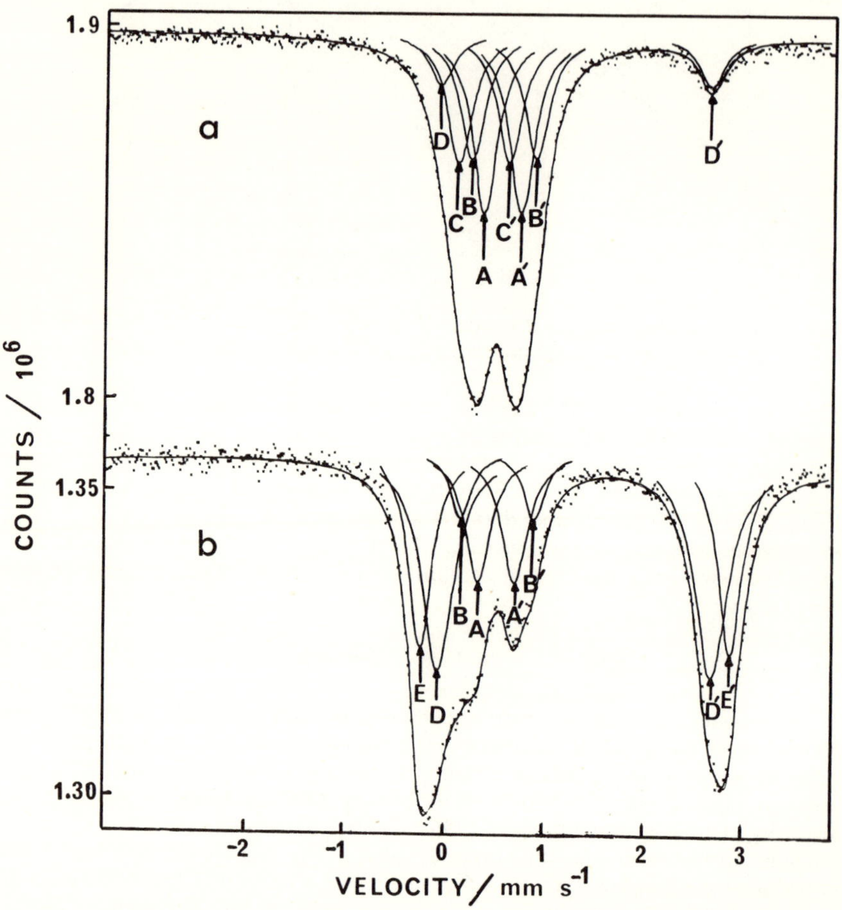

Figure 1-52. Mössbauer spectra at 77 K of a nontronite with high tetrahedral Fe^{2+} content reduced (a) with hydrazine and (b) with dithionite solution (from Russell *et al.,* 1979).
Doublets AA′ and BB′ correspond to Fe^{3+} in octahedral sites, CC′ to Fe^{3+} in tetrahedral sites and DD′ and EE′ to Fe^{2+} in octahedral sites.

The behavior of the same series of nontronites on dithionite reduction showed a marked dependence on structural composition (47). Samples that contain little or no tetrahedral Fe^{3+} behave with dithionite as they do with hydrazine, but those with higher tetrahedral Fe^{3+} contents exhibit rapid and extensive reduction accounting for 60-80% of the iron (Fig. 1-52b). The spectra were all fitted to four doublets (two Fe^{2+} and two Fe^{3+} components) although this must be an oversimplification since there were three Fe^{3+} components in each of the untreated samples. All computed results, for both hydrazine and dithionite reduction, are shown in Table 1-7, where the sample descriptions are as referred to in the original work (47). Because they contain a contribution from tetrahedral Fe^{3+}, the δ values for the dithionite-reduced samples in Table 1-7 are slightly lower than those computed for octahedral Fe^{3+} in the untreated samples. By assuming that the isomer shifts for both octahedral and tetrahedral components are unchanged on reduction it was possible to calculate approximate tetrahedral Fe^{3+} contents of the dithionite-reduced samples on the basis that the δ values in Table 1-7 represent a weighted average of the contributions from the two types of site. The calculated tetrahedral Fe^{3+} contents for these samples that had undergone extensive reduction were much smaller than in the original nontronites. Thus, since the Fe^{2+} values are typical of octahedral Fe^{2+}, considerable loss of tetrahedral iron must have occurred. Assignment of the Fe^{2+} components to sites within the structure cannot realistically be made because substantial structural rearrangement must have occurred after removal of the tetrahedral Fe^{3+}.

In order to investigate further the extent of any structural decomposition, samples were treated with unbuffered 1% w/v sodium dithionite solutions for various times, after which they were allowed to oxidize by exposure to the air. The resulting spectra showed no evidence for the presence of any Fe^{2+} and representative spectra are shown in Fig. 1-53 along with the spectrum of the untreated specimen. A 5-minute treatment with the dithionite solution resulted in a spectrum that bears a strong resemblance to that of the original nontronite, except that there is a decrease in intensity of peaks CC′ from tetrahedral Fe^{3+}. In contrast, the spectrum from the sample which had received two 20-minute treatments shows little overall resemblance to that of the original nontronite. The weak central peaks have similar parameters to the main peaks (AA′) of nontronite and probably correspond to this phase, but the spectrum is dominated by a pair of broad peaks which must correspond to a range of environments for the iron. The parameters of this broad doublet are similar to those obtained from a hydrous ferric oxide gel prepared by aging ferric nitrate solutions, and indicate the near complete decomposition of the nontronite. Hence at least one of the Fe^{2+} components in the reduced nontronite samples most probably corresponds to a separate phase and not to Fe^{2+} in the nontronite structure.

1-5.3. The study of thermal alteration

In this section two problems will be discussed. First, a structural problem where attempts have been made to locate the iron in chlorites by a combination of limited thermal decomposition and chemical treatment (24). Second, a more general application to the study of archaeological samples will be discussed with particular reference to the problem of identification of firing conditions (6).

Representative spectra of some chlorites were presented in section 1-4 (Fig. 1-47) but the question of the extent of substitution of iron in the brucite-like sheet was left unanswered. It has been reported that heating a chlorite beyond the temperature of its major DTA endotherm followed by digestion with HCl, leads to the loss of the brucite-like hydroxide sheet and the formation of vermiculite (45). The work (24) in which Mössbauer spectroscopy has been used in the study of this reaction will now be reported in some detail. Chlorites, with iron contents ranging

Table 1-7. Computed results for Mössbauer spectra of reduced nontronites (from Russell *et al.*, 1979).

Nontronite sample and treatment	Fe^{3+} AA'				Fe^{3+} BB'				Fe^{3+} CC'				Fe^{2+} DD'				Fe^{2+} EE'				
	Δ	δ	Γ	%	Δ	δ	Γ	%	Δ	δ	Γ	%	Δ	δ	Γ	%	Δ	δ	Γ	%	χ^2
WAS + hydrazine[1]	0.37	0.45	0.38	61[4]	0.73	0.43	0.31	21									2.96	1.26	0.42	18	502
WAS + dithionite[1]	0.37	0.46	0.36	60[4]	0.72[3]	0.46	0.29[3]	14[4]									2.98	1.27	0.44	26	455
GAR + hydrazine[1]	0.36	0.47	0.34	59[4]	0.68	0.47	0.28	29[4]					2.58	1.22	0.33	13					533
GAR + dithionite[1]	0.36	0.47	0.34	61[4]	0.67	0.46	0.27	26[4]					2.65	1.23	0.39	13					466
CLA + hydrazine[2]	0.37	0.49	0.30	37	0.63	0.50	0.30	26	0.50	0.30	0.30	27	2.71	1.22	0.30	11					472
CLA + dithionite	0.37	0.46	0.32	20	0.72[3]	0.47	0.25[3]	8					2.74	1.26	0.41	44[4]	3.10	1.27	0.29	28	499
CRO + hydrazine[2]	0.34	0.49	0.29	34	0.61	0.49	0.29	29	0.48	0.30	0.29	28	2.72	1.25	0.29	9					481
CRO + dithionite	0.27[3]	0.46	0.33[3]	14[4]	0.57[3]	0.47	0.30[3]	10[4]					2.73[3]	1.27	0.40	46[4]	3.09	1.27	0.27	31[4]	459
KOE + hydrazine[2]	0.33	0.49	0.29	31	0.57[3]	0.49	0.29	32	0.50	0.30	0.29	25	2.77	1.24	0.29	11					400
KOE + dithionite	0.33[3]	0.46	0.31[3]	20[4]	0.62[3]	0.46	0.32[3]	17[4]					2.71	1.26	0.39	32	3.08	1.28	0.29	30	604
AMO + hydrazine[2]	0.34	0.49	0.30	34	0.60	0.49	0.30	29	0.51	0.31	0.30	27	2.77	1.24	0.30	10					456
AMO + dithionite	0.26[3]	0.45	0.29[3]	7	0.59[3]	0.47	0.36[3]	12					2.69	1.28	0.42	41	3.08	1.28	0.32	40	671
CAL + hydrazine[2]	0.33	0.49	0.29	35	0.61	0.49	0.29	31	0.46	0.30	0.29	29[4]	2.74[3]	1.29[3]	0.29	5					507
CAL + dithionite	0.33	0.45	0.33[3]	20	0.68[3]	0.46	0.25[3]	8					2.73	1.27	0.40	40[4]	3.07	1.27	0.28	33	505

All values are in mm s^{-1} with the isomer shift, δ, relative to iron metal.

1 These spectra were fitted to three doublets because of the low amounts of tetrahedral iron in the untreated specimens.

2 The fits to these spectra assume that all components have equal values for the peak width, Γ.

3 The standard deviations for the quadrupole splitting, Δ, isomer shift, δ, and peak width, Γ, are $\leqslant$0.02 mm s^{-1} except for those marked [3], where the standard deviations are in the range 0.03-0.07 mm s^{-1}.

4 The standard deviations for the amounts of each component are $\leqslant$4%, except for those marked [4] where the standard deviations are in the range 5-10%.

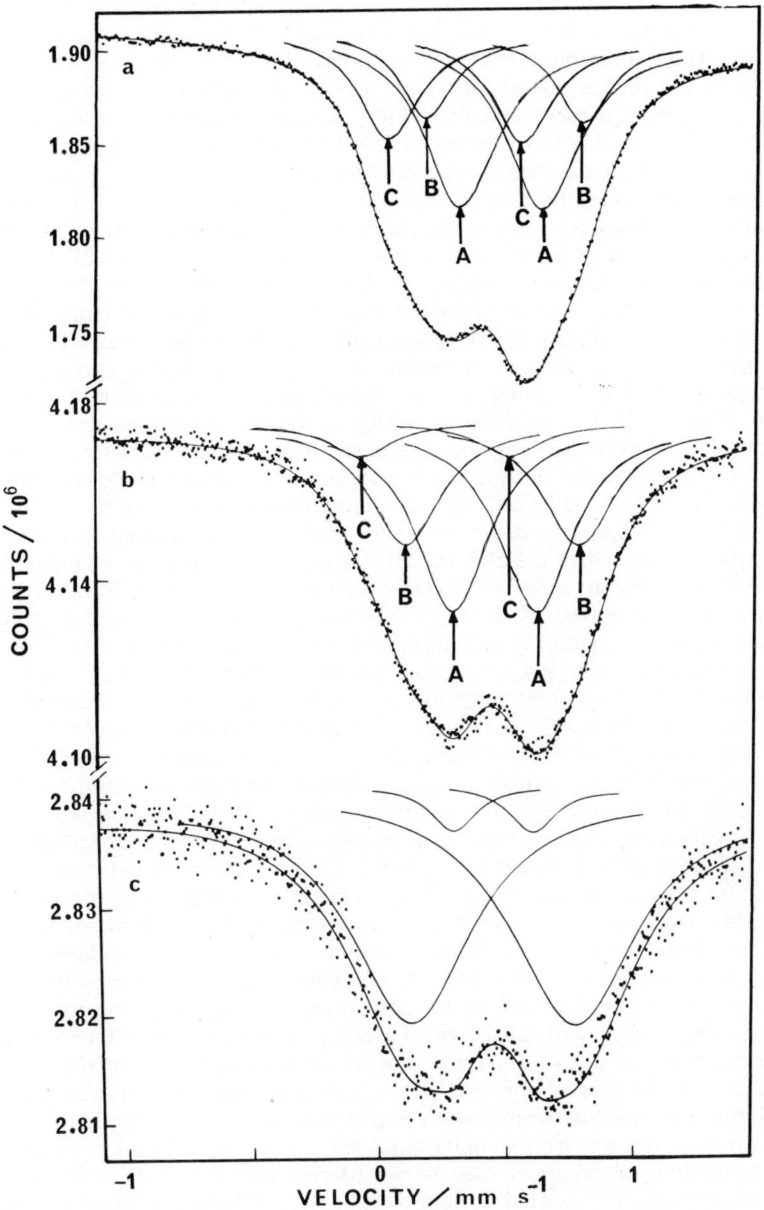

Figure 1-53. Mössbauer spectra at 77 K of a nontronite (a) and after reoxidation following dithionite treatment for (b) 5 min. (c) 2 x 20 min. (from Russell *et al.*, 1979).

from ~ 2% to 30% from five locations were investigated. Three of these samples could be considered to be low-iron specimens and their spectra are a, b and c in Fig. 1-47. (The sample numbers are those referred to in the original work). They were chosen for further investigation because of the differences in their original spectra; e.g. sample 5 contained little Fe^{3+}, whereas samples 1 and 6 had about 40% of their iron in this form. However, the isomer shifts indicated that the Fe^{3+} in the sample 1 had octahedral coordination but that in sample 6 was largely tetrahedral. In each case heating the sample in N_2 to a temperature just below the DTA endotherm resulted in some oxidation of the iron, although this had little effect on the Mössbauer parameters of the remaining Fe^{2+} (Table 1-8). There was also little change in the parameters for the Fe^{3+} in sample 1, but for the other two samples, the Fe^{3+} produced as a result of heat treatment had much larger values of Δ than any of the original Fe^{3+} components. Heating to the peak of the DTA endotherm led in each case to complete oxidation (Fig. 1-54 a and b), a large percentage of the Fe^{3+} having a much larger value of Δ than in the original chlorites (Table 1-8). Isomer shifts were slightly lower than the typical values for 6-coordination, thus indicating that there might be a small contribution to the spectra from components with a lower coordination number. Acid treatment of those chlorites which had been heated to just below their principal endotherm produced little change, but when the same treatment was applied to the sample that had been heated beyond this endotherm some conversion to vermiculite occurred (indicated by an x-ray diffraction peak at 14.9Å). The Mössbauer results indicated that at least part of the Fe^{3+} with the larger Δ was removed in each case by acid treatment, suggesting that this component could represent the iron that had been expelled from the lattice by the heat treatment. With the high-iron chlorite (sample 9, Table 1-8) there was also little change as a result of heating in N_2 to just below the major endotherm. On further heating, however, there was only a slight increase in Fe^{3+} content (Fig. 1-55a), in complete contrast to the behavior of the low-iron samples. The parameters for both Fe^{2+} and Fe^{3+} were now quite different to those in the original specimens (Table 1-8). The envelopes of the Fe^{2+} peaks were much broader and in each case contained at least 2 doublets, with δ decreasing with decreasing Δ, which is the behavior expected if decreases in coordination number are accompanied by increases in distortion at the Fe^{2+}-containing sites. The broad Fe^{3+} peaks are also indicative of a range of environments for this ion. Treatment of these heated samples with HCl did not induce vermiculitization, and produced no changes in the Mössbauer spectra. Heating these same chlorites in O_2 to the peak of, or beyond, the DTA endotherm led to complete oxidation (Fig. 1-55b) with all of the Fe being found in sites of 6-coordination with various levels of distortion. Treatment with HCl produced no further changes.

Chemical analyses of the HCl extracts showed that on average about one-third of the iron was lost from those samples that had been heated beyond their endotherm. For the low-iron samples this represents only a small proportion of their total cations, although it may be of interest that sample 6, which had the highest tetrahedral Fe^{3+} content before treatment, lost the largest percentage of its iron. With the high-iron specimens a considerable proportion of the total cations was removed, which evidently led to complete disruption of the structure since the samples became poorly diffracting to x-rays. The question now is: Can such experiments provide further information of the site populations of iron in the original chlorites? The iron-rich samples underwent only partial oxidation on heating in N_2 to the temperature of dehydroxylation of the brucite-like sheet, whereas the low-

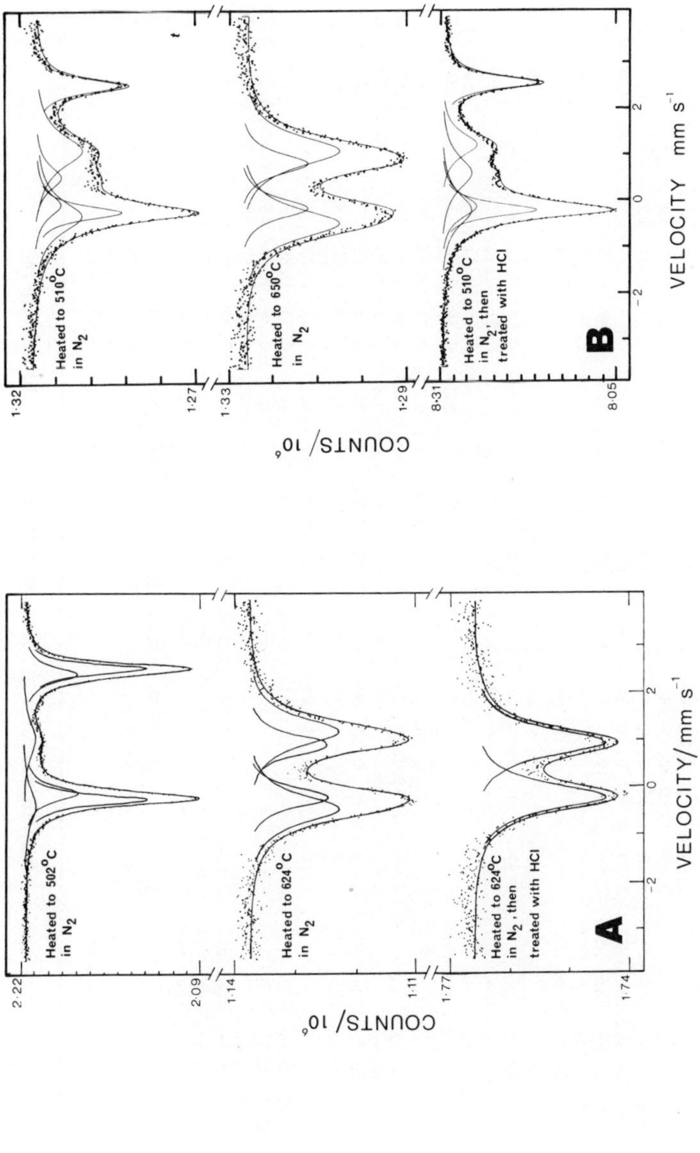

Figure 1-54. Mössbauer spectra at 300 K of 2 low iron chlorites, (a) sample 5 and (b) sample 6 from Goodman and Bain, 1978, subjected to various treatments.

Table 1-8. Mössbauer parameters for some chlorites and their thermally altered products. (Adapted from Goodman and Bain, 1978).

Sample Number	Treatment	Fe(II) Δ	δ	Γ	%	Δ	δ	Γ	%	Fe(III) Δ	δ	Γ	%	Δ	δ	Γ	%	χ^2
9		2.69	1.13	0.28	63	2.35	1.12	0.42	17	0.72	0.26	0.61	11	1.06	0.37	0.52	9	617
	415C in N_2	2.69	1.14	0.36	81	-	-	-	-	0.79	0.35	0.62	19	-	-	-	-	437
	585C in N_2	2.46	1.06	0.50	29	1.96	0.98	0.67	46	0.94	0.53	0.81	25	-	-	-	-	522
	415C in N_2, then HCl	2.59	1.16	0.38	34	-	-	-	-	-	-	-	-	1.21	0.33	0.70	66	555
	585C in N_2, then HCl	2.46	1.07	0.46	26	1.97	0.99	0.66	48	0.92	0.49	0.85	26	-	-	-	-	541
	610C in O_2	-	-	-	-	-	-	-	-	0.84	0.34	0.40	31	1.39	0.34	0.67	69	922
	610C in O_2, then HCl	-	-	-	-	-	-	-	-	0.78	0.34	0.39	22	1.31	0.35	0.76	78	504
7		2.67	1.14	0.35	44	2.39	1.15	0.55	33	-	-	-	-	1.10	0.36	0.85	23	410
	610C in N_2	2.57	1.08	0.41	21	2.07	1.00	0.64	39	1.08	0.40	0.72	39	-	-	-	-	541
	610C in N_2, then HCl	2.52	1.11	0.46	18	2.07	1.01	0.77	41	1.10	0.37	0.80	42	-	-	-	-	429
	610C in O_2	-	-	-	-	-	-	-	-	1.24	0.35	0.73	100	-	-	-	-	649
	610C in O_2 , then HCl	-	-	-	-	-	-	-	-	0.94	0.36	0.54	42	1.53	0.34	0.75	58	354
6		2.71	1.13	0.28	43	2.37	1.14	0.34	19	0.66	0.24	0.75	39	-	-	-	-	515
	510C in N_2	2.64	1.13	0.37	36	-	-	-	-	0.57	0.24	0.63	19	1.37	0.40	0.85	45	474
	650C in N_2	-	-	-	-	-	-	-	-	0.92	0.32	0.55	32	1.51	0.29	0.81	68	540
	510C in N_2, then HCl	2.69	1.13	0.27	30	2.33	1.14	0.29	8	0.71	0.15	0.69	26	1.36	0.42	0.76	36	933
	650C in N_2, then HCl	-	-	-	-	-	-	-	-	1.08	0.33	0.82	100	-	-	-	-	389
5		2.70	1.13	0.25	41	2.51	1.13	0.36	51	0.58	0.32	0.27	8	-	-	-	-	591
	502C in N_2	2.68	1.13	0.26	51	2.43	1.14	0.39	33	1.36	0.35	0.86	17	-	-	-	-	492
	624C in N_2	-	-	-	-	-	-	-	-	1.11	0.34	0.54	43	1.69	0.33	0.77	57	804
	725C in N_2	-	-	-	-	-	-	-	-	1.08	0.36	0.50	38	1.57	0.36	0.73	62	486
	502C in N_2, then HCl	2.65	1.13	0.27	63	2.28	1.08	0.30	12	1.05	0.32	0.89	25	-	-	-	-	706
	624C in N_2, then HCl	-	-	-	-	-	-	-	-	1.19	0.35	0.71	100	-	-	-	-	519
	725C in N_2, then HCl	-	-	-	-	-	-	-	-	1.19	0.36	0.68	100	-	-	-	-	442
1		2.69	1.13	0.28	40	2.46	1.13	0.41	21	0.57	0.39	0.24	15	0.82	0.34	0.76	24	413
	510C in N_2	2.67	1.11	0.33	51	-	-	-	-	0.70	0.37	0.50	49	-	-	-	-	612
	655C in N_2	-	-	-	-	-	-	-	-	0.60	0.38	0.27	20	1.27	0.30	0.97	80	401
	655C in N_2, then HCl	-	-	-	-	-	-	-	-	0.59	0.39	0.24	24	0.99	0.35	0.83	76	400

All values in mm s^{-1}. Isomer shifts relative to iron metal.

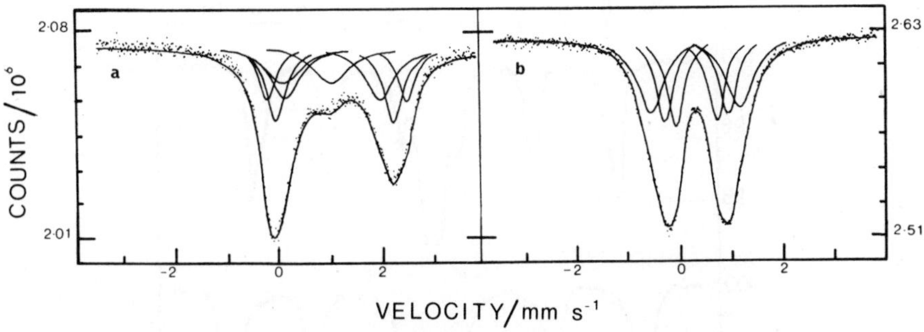

Figure 1-55. Mössbauer spectra at 300 K of an iron-rich chlorite that had been heated in (a) N_2 and (b) O_2 (from Goodman and Bain, 1978).

iron specimens were completely oxidized. Since the former specimens must have contained Fe^{2+} in both the brucite and talc-like sheets, it seems reasonable to conclude that the unoxidized iron is in the talc-like sheet. Extension of this conclusion to indicate that the oxidized iron must be in the brucite-like sheet is not supported by the analytical results since sample 5, which was almost completely converted to vermiculite, lost only 27% of its iron on HCl extraction. If allowance is made for HCl attack on the talc-like layer, then appreciably less than 50% of the Fe^{2+} can be in the brucite-like sheet in the original chlorite. Thus, some oxidation of Fe^{2+} in the talc-like layer occurred, probably via an internal dehydrogenation reaction, vis,

$$2Fe^{2+} + 2OH^{-} \rightarrow 2Fe^{3+} + 2O^{2-} + H_2 \quad [1\text{-}28]$$

although this does not occur extensively in iron rich specimens.

As a final conclusion no components in the Mössbauer spectra could be specifically assigned to the brucite-like sheet, and any Fe^{2+} occurring there must have parameters similar to that in the talc-like sheet.

The study of potter's clays, the changes occurring on firing under various conditions, and the attempts to relate their Mössbauer spectra to those of ancient pottery provides a natural link between this section dealing with alteration of clays and the next section, in which the study of whole soil samples will be considered. The potter's clays consist of a mixture of minerals, some of which are expanding lattice silicates, others being the common soil forming minerals such as quartz, feldspar, etc. A considerable proportion of the iron, however, may be in the form of oxide or hydrous oxide phases associated with these minerals, so that the Mössbauer spectrum of a typical unfired clay may have a considerable contribution from magnetically ordered components (Fig. 1-56). The variation with temperature of intensity of the magnetically-ordered phases relative to the central doublet will be considered more fully in the next section. The transformations that occur in the clay during the firing processes are determined partly by the composition of the clay itself and partly by the firing conditions, the latter depending on the firing temperature and whether oxidizing or reducing conditions were used. Examples of spectra taken at room temperature and 4.2K are shown in Fig. 1-57. As an example of the application of Mössbauer spectroscopy to an archaeological problem a brief summary will be made of the work of Bouchez *et al.* (6) on the origin and forma-

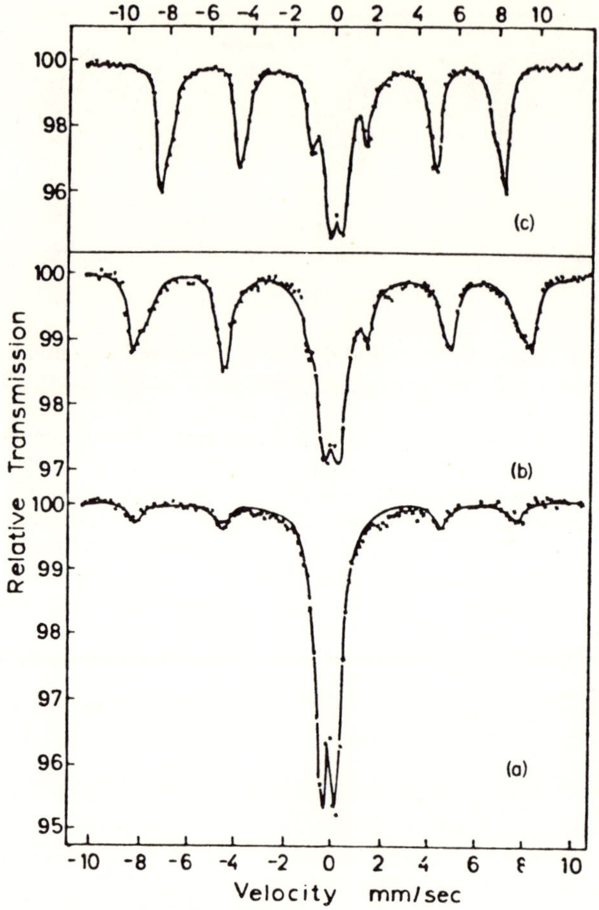

Figure 1-56. Mössbauer spectra of an unfired clay at (a) 300 K, (b) 77 K and (c) 4.2 K (from Kostikas *et al.*, 1976).

tion of two types of pottery from Turkestan dating from the third millenium, B.C. The two types of pottery, one red and one grey, were found to coexist over archaeological levels covering several centuries. It is of interest to know whether the differences are related to the manufacturing techniques or whether one type appeared as a result of immigration of people into the area. Analytical techniques showed no significant differences in the elemental composition of the two types of pottery but Mössbauer spectroscopy indicated that the grey form was predominantly Fe^{2+} and the red form largely Fe^{3+} This suggested that the difference between the two ceramics was the firing conditions under which they were produced. By studying local clays over a range of temperatures under both oxidizing and reducing conditions the authors were able to conclude that the red pottery was fired at 1050 ± 50° C under oxidizing conditions and the grey form some 50-100° C lower and under reducing conditions.

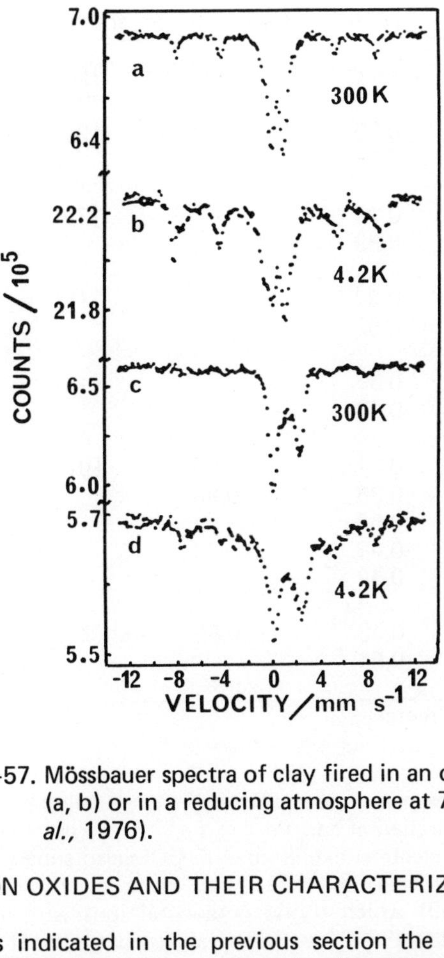

Figure 1-57. Mössbauer spectra of clay fired in an oxidizing atmosphere at 925° C (a, b) or in a reducing atmosphere at 750° C (c, d) (from Chevalier *et al.*, 1976).

1-6. IRON OXIDES AND THEIR CHARACTERIZATION IN SOILS

As indicated in the previous section the various silicate minerals are not readily distinguished from one another in mixtures on the basis of the magnitudes of δ and Δ for their Fe^{2+} and Fe^{3+} components. Indeed, identification of such minerals by Mössbauer spectroscopy is unnecessary since x-ray diffraction (XRD) of powder samples is able to identify the major crystalline components both quickly and easily. Difficulties arise with XRD when dealing with poorly crystalline components since they are usually only weakly diffracting and in some instances may not give a diffraction pattern at all. In soils, especially in the upper horizons, there are considerable amounts of secondary minerals that are either microcrystalline or amorphous. This section will deal with examples of the application of Mössbauer spectroscopy to the study of poorly crystalline iron oxides or their precursors in soils.

The minerals of most interest in this work are the oxides, hematite (α-Fe_2O_3), maghemite (γ-Fe_2O_3) and magnetite (Fe_3O_4) and the oxyhydroxides, goethite (α-FeOOH), akaganéite (β-FeOOH) and lepidocrocite (γ-FeOOH). At low temperatures all are magnetically ordered, but at room temperature the last two are paramagnetic. Their parameters are summarized in Table 1-9. The spectrum of

Table 1-9. Mössbauer parameters for iron oxides and hydroxides

Sample	T(k)	(Hkoe)	δ^{+} (mm s^{-1})	Δ (mm s^{-1})	Ref.
α-Fe_2O_3	300	516	0.36		[2]
	77	527	0.48		
γ-Fe_2O_3	300	502	0.25		[39]
		503	0.39		
Fe_3O_4	300	493	0.27		[15]
		460	0.52		
α-FeOOH	300	384	0.37		[16]
	77	504	0.48		
β-FeOOH	300	–	0.37	0.55	[9]
		–	0.38	0.95	
	77	473	0.52		
		463	0.48		
		437	0.48		
γ-FeOOH	77	–	0.48	0.55	[37]
	4.2	460	0.51		

+ Isomer shifts are relative to Fe metal

Fe_3O_4 (Fig. 1-58) shows the presence of two different magnetically-ordered sites, corresponding to Fe^{3+} in the tetrahedral and Fe^{2+} + Fe^{3+} in the octahedral sites, the latter ions undergoing rapid electron exchange. β-FeOOH also shows the presence of more than one type of site (Figs. 1-59, 1-60) as indicated by the room temperature spectrum (Fig. 1-59) which shows octahedral ions with at least 2 different quadrupole splittings, and the low temperature spectrum (Fig. 1-60) which shows that there are at least 3 different magnetic fields present. Since there is only one type of structural site in the β-FeOOH lattice (Fig. 1-61a), these components have been interpreted as arising from the different arrangements of halide ions in the interlayer spaces (Fig. 1-61b).

It was mentioned in section 1-2 that the temperatures at which magnetic ordering is observed in Mössbauer spectra vary with crystal size for microcrystalline samples. An alternative way of looking at this situation is that, at a temperature below the magnetic ordering temperature in a microcrystal, the ratio of magnetically-ordered to paramagnetic components will be related to the mean of the particle size distribution. This is illustrated in Fig. 1-62 for α-FeOOH. The decrease in ordering temperature is even more dramatic if there is isomorphous substitution of aluminum for iron in the mineral structure as almost certainly occurs in the soil (Fig. 1-63). Thus it can be seen that in the small particle size fractions of the oxide minerals that are likely to be found in the upper horizons of the soil, the magnetic ordering temperature may be appreciably lower than that recorded for well-crystallized synthetic samples. The magnetic ordering temperature should, therefore, be

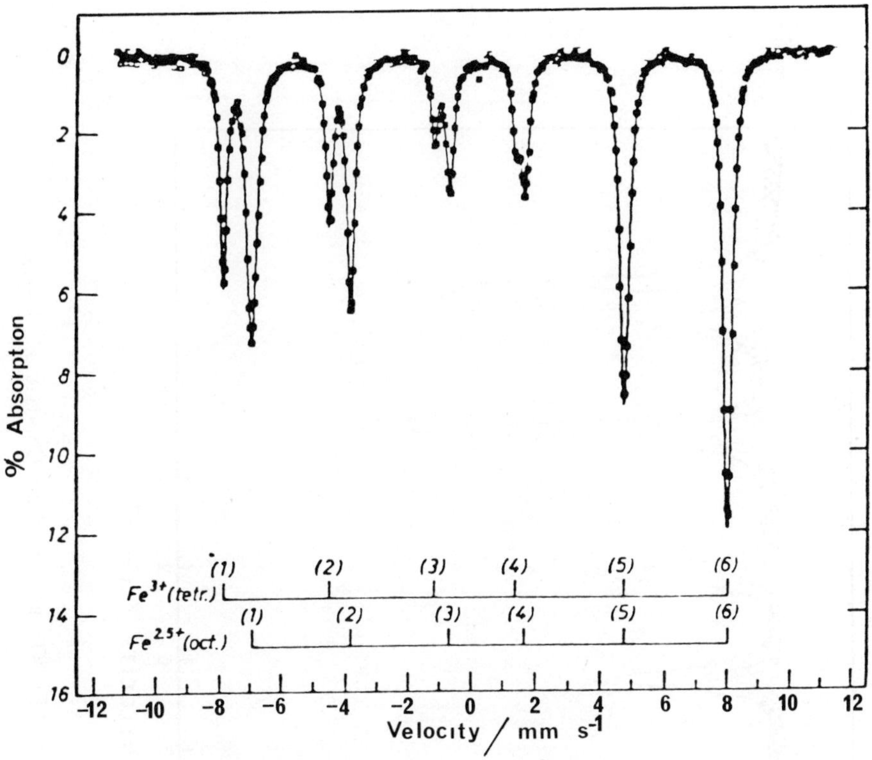

Figure 1-58. Mössbauer spectrum of Fe_3O_4 at 300 K (from Weber and Hafner, 1971).

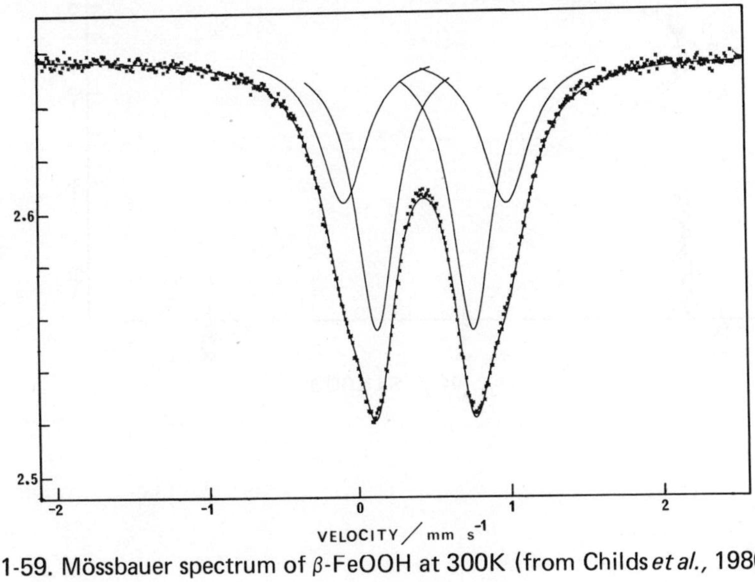

Figure 1-59. Mössbauer spectrum of β-FeOOH at 300K (from Childs *et al.*, 1980).

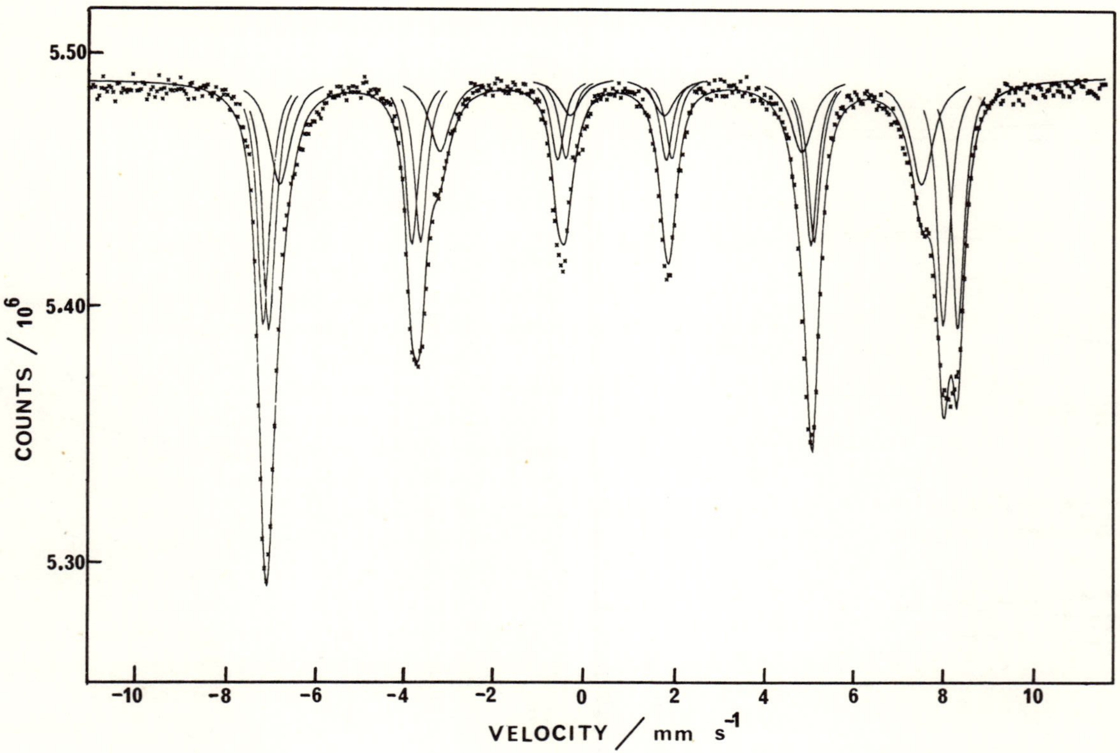

Figure 1-60. Mössbauer spectrum of β-FeOOH at 77K (from Childs *et al.*, 1980).

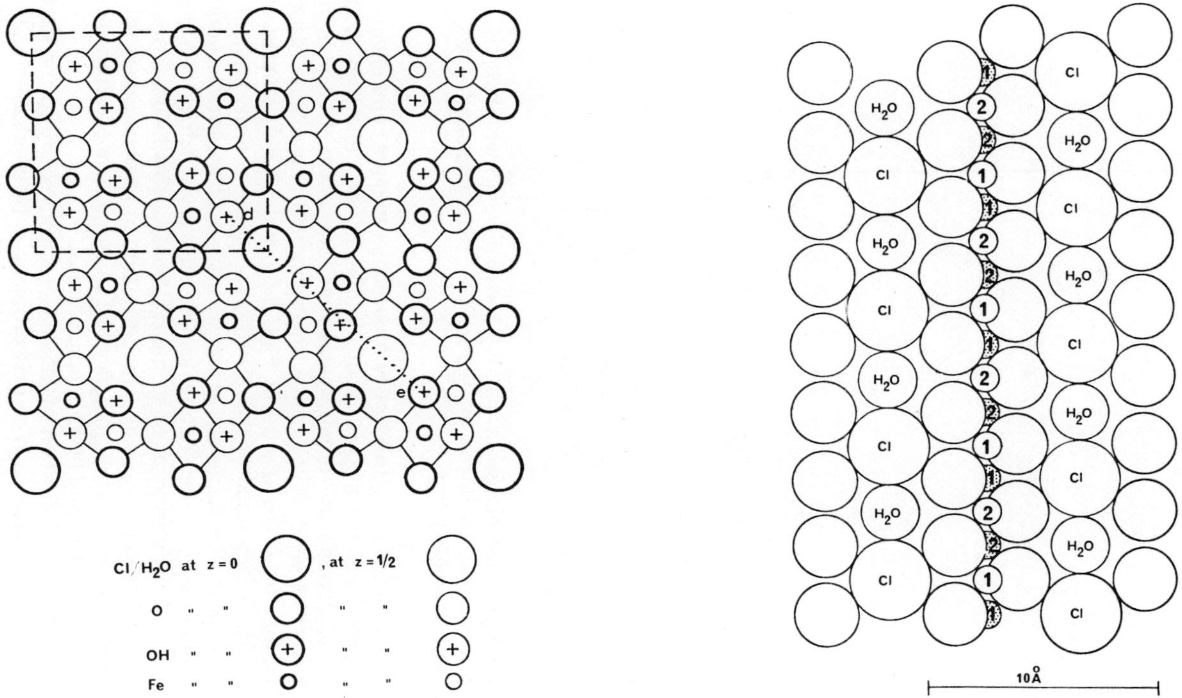

Figure 1-61. (a) Basic structure of β-FeOOH, (b) arrangement of atoms along d-e in plane perpendicular to the paper, which illustrates two different types of Fe atoms (from Childs *et al.*, 1980).

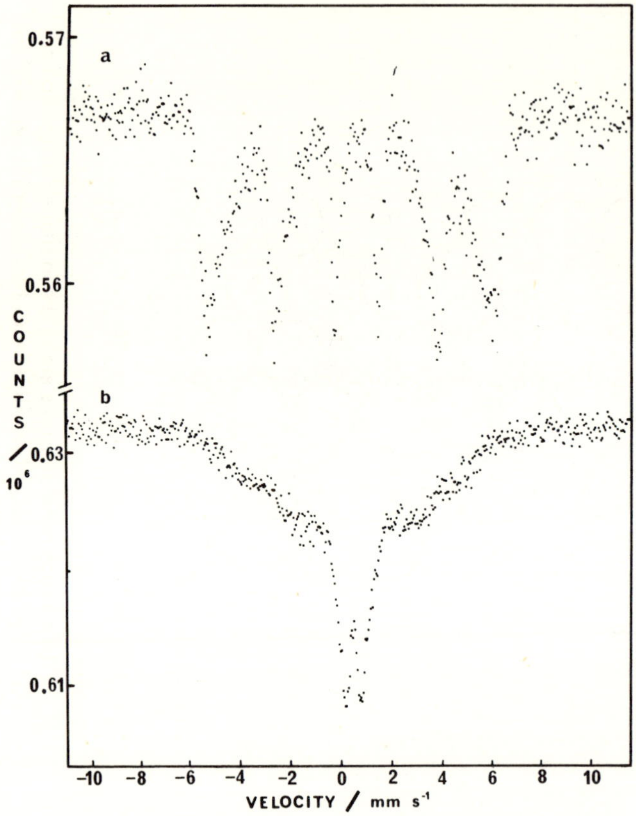

Figure 1-62. Variation of Mössbauer spectra at 300 K of α-FeOOH with particle size. Surface area (a) 75 and (b) 107 m^2g^{-1}. (Goodman and Lewis, unpublished results).

used with caution when attempting to identify a particular mineral component. A further difficulty arises from the magnitude of the internal magnetic field. Computed values for microcrystals are usually somewhat lower than the values for macrocrystals even up to several degrees below the temperature at which ordering occurs. As can be seen from the spectra in Figs. 1-62 and 1-63, the peaks are broadened asymmetrically as a result of the relaxation processes and any computer fit that assumes Lorentzian line shapes will obviously underestimate the magnitude of the internal magnetic field, H. In a natural sample, such asymmetric peaks may lead to the erroneous identification of other mineral components having a somewhat smaller H than the main component. The only hope of resolving these difficulties is to record spectra well below the magnetic ordering temperature which, for many samples, means using liquid helium temperatures.

A few examples of the application of Mössbauer spectroscopy to the study of soil samples will now be presented. The first problem to be considered will be that of characterising the secondary iron in pans formed in podzolic soils (25).

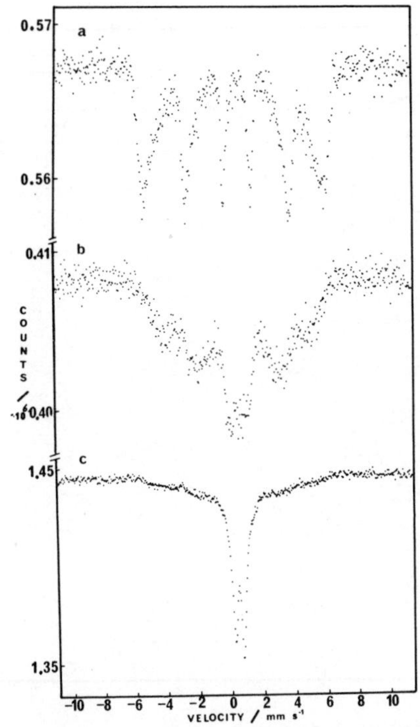

Figure 1-63. Variation of Mössbauer spectra at 300 K of α-(Fe, Al)OOH with Al content (a) 0%, (b) 5.1% and (c) 7.7% Al (Goodman and Lewis, unpublished results).

Podzols are found extensively in cool-temperate to temperate-humid climates, and are usually characterized by a highly-leached, whitish-grey A horizon, directly beneath which is a red-brown layer rich in iron and organic matter. This layer may sometimes include a thin iron pan which is relatively impervious to water, leading to impeded drainage in the soil. For a comparison the spectra obtained from a ground water gley are shown in Fig. 1-64. This sample was shown by XRD to consist largely of goethite (α-FeOOH). The Mössbauer spectra at room temperature show a partially-collapsed magnetic structure and at 77K a well resolved magnetic structure with H = 486 kOe, where 1 Oersted (Oe) $\equiv 10^{-4}$ Tesla (T). This is somewhat lower than the value for pure macrocrystalline α-FeOOH of 504 kOe (16) but is similar to that computed for a microcrystalline sample. In contrast to this sample the podzol iron pans gave only a broad doublet at both room temperature and 77K (Fig. 1-65), and, although other weak peaks were obtained in some samples, they could be assigned to minerals present throughout the profile. Two doublets were required for statistically-acceptable computer fits to the spectra. On cooling the samples below 77K magnetic ordering was observed to take place over a range of temperatures (Fig. 1-66). The peaks in the magnetically-ordered spectra are broad and spectra were fitted to two components, having values of the magnetic field, H, of ~ 500 and ~ 450 kOe, respectively. Since the value of the internal magnetic field in α-FeOOH is 504 kOe and, since the blocking temperature for

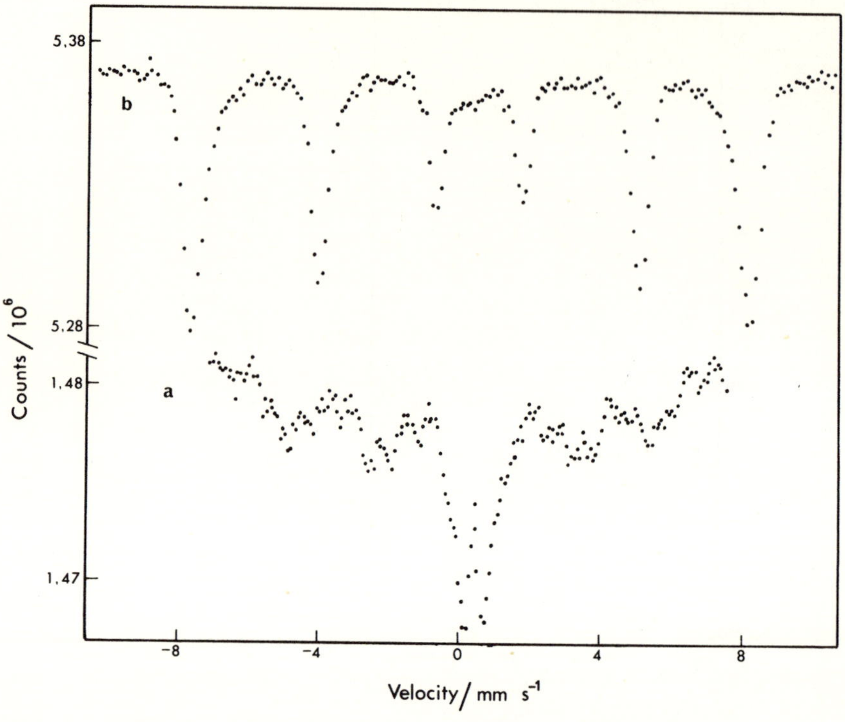

Figure 1-64. Mössbauer spectra of an iron pan from a ground water gley (a) at 300 K at (b) 77 K (from Goodman and Berrow, 1976).

magnetic ordering decreases with decreasing particle size, the possibility exists that these spectra correspond to extremely small particles of goethite (probably with considerable substitution of Al for Fe) with the component with the smaller field arising as a result of fitting the asymmetric relaxation spectra to two components with Lorentzian line shapes. Alternatively, either β- or γ-FeOOH, or both, could also be present.

The next problem to be discussed concerns the identification of secondary iron oxides in some red and yellow-brown soils (8). These samples originated in New Zealand and were taken in pairs, one of each color, from sites very close to one another. The Mössbauer spectra at room temperature of three pairs of samples are shown in Fig. 1-67. There is a clear distinction between the red and yellow-brown samples. Each red sample has a 6-line, magnetic hyperfine component (a) and a 2-line paramagnetic component from Fe^{3+} (b). In contrast there is no magnetic component in the yellow-brown samples. In some samples a further doublet (c′) from Fe^{2+} was also observed. Extraction with dithionite removed the 6-line components and also reduced appreciably the intensities of the Fe^{3+} doublets. The parameters for the 6-line components are consistent with those of hematite (α-Fe_2O_3). The magnetic field is much greater than that of antiferromagnetic goethite (α-FeOOH) at room temperature. There is no evidence for the presence of two magnetic components as would be required for magnetite (Fe_3O_4) and,

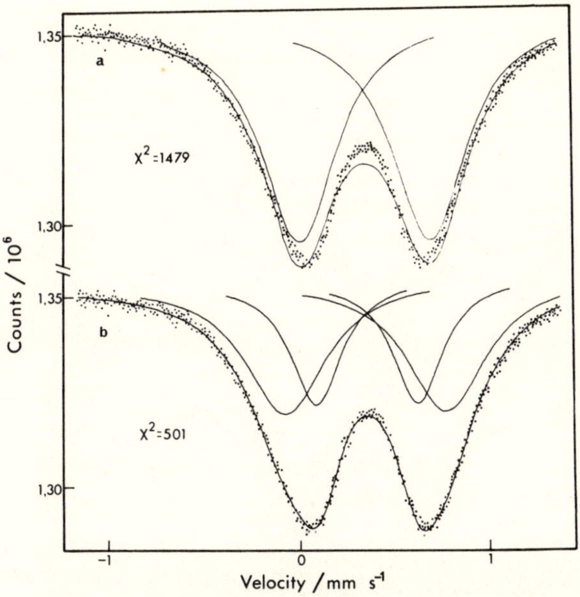

Figure 1-65. Mössbauer spectrum of a podzol iron pan at 300 K, fitted to (a) 1 doublet and (b) 2 doublets (from Goodman and Berrow, 1976).

although maghemite (γ-Fe_2O_3) has similar parameters to those observed (Table 1-9), magnetite was the only mineral that could be detected by XRD in magnetic extracts of the soils. The amounts of magnetite were too small to have been evident in the whole soil spectra. The Fe^{3+}-doublets (b, b′) may arise from a number of sources: (i) Fe^{3+} in aluminosilicate minerals; (ii) Fe^{3+} in the oxyhydroxides, akaganéite, lepidocrocite, or superparamagnetic goethite (i.e. goethite below its Néel temperature, but undergoing relaxation so that its magnetic field is averaged to zero); (iii) Fe^{3+} in a poorly-ordered precursor of the oxyhydroxide, amorphous $(FeOH)_3$; or, (iv) Fe^{3+} in superparamagnetic hematite. The doublet c′ is attributable to Fe^{2+} in aluminosilicates.

At 77K (Fig. 1-68) both red and yellow-brown samples have 6-line components in their spectra. For some red samples two 6-line components (d and e) are evident, with e being similar to e′ for the yellow-brown samples. Central Fe^{3+} doublets (f, f′) were also seen in most spectra and Fe^{2+} doublets (g′) in some of them. Component d in the red samples has parameters attributable to hematite and corresponds to component a in Fig. 1-67. Components e, e′ are probably either goethite or akaganéite. Although these two components can be distinguished when present as single minerals, the poorer signal-to-noise ratio with soil samples does not permit unambiguous identification in these spectra. An attempt was also made to distinguish these two minerals on the basis of their room temperature spectra. Fig. 1-69 shows the spectra of two yellow-brown samples compared with those obtained from β-FeOOH and Al-substituted α-FeOOH. It can be seen that one sample gives a spectrum almost identical to that of superparamagnetic goethite, whereas the other more closely resembles akaganéite. However, other Fe^{3+} forms may contribute to these central doublets as mentioned earlier, so any assignments must be tentative.

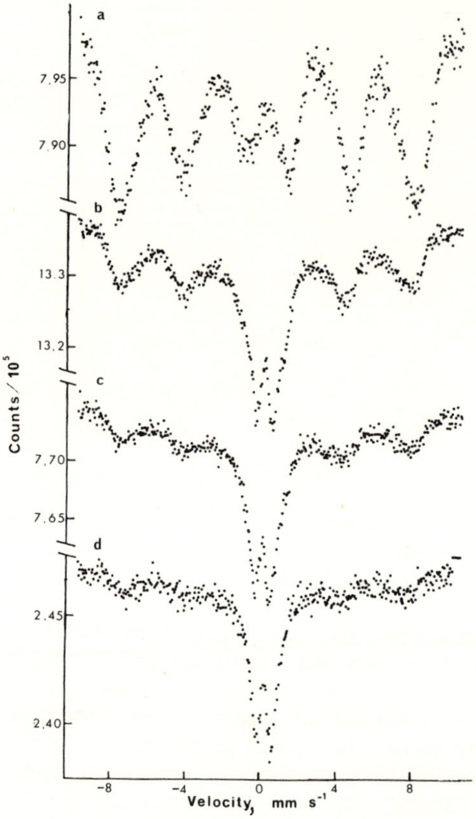

Figure 1-66. Mössbauer of a podzol iron pan (a) at 5.4° K, (b) at 25° K, (c) at 30° K and (d) at 35° K (from Goodman and Berrow, 1976).

Approximate concentrations of hematite could be calculated from the areas under the spectra by assuming a constant f-factor for all components, and the concentration of hematite correlated well with the color. It was also concluded that hematite was present in all of the red samples but in none of the yellow-brown samples, the oxyhydroxides goethite or akaganéite were present in all of the yellow-brown samples and most of the red ones. Thus, the main distinction between the red and yellow-brown samples is the presence and absence of hematite, respectively.

In another study of soil samples (41) magnetic components have been separated and studied in detail in addition to the complete soil samples. The results for the original samples at room temperatures are shown in Fig. 1-70 and for magnetically-separated components in Fig. 1-71. The samples labelled TB3, TB7 and Caldy Hill came from soils that had been subjected to recent burning and it can be seen that the magnetic separation technique is a very effective way of increasing the Mössbauer signal strength. The Annecy soil samples had also had a history of burning and it can be seen that its 6-line magnetic component was also increased in intensity by magnetic separation; but the spectrum of sample TB1, which was from

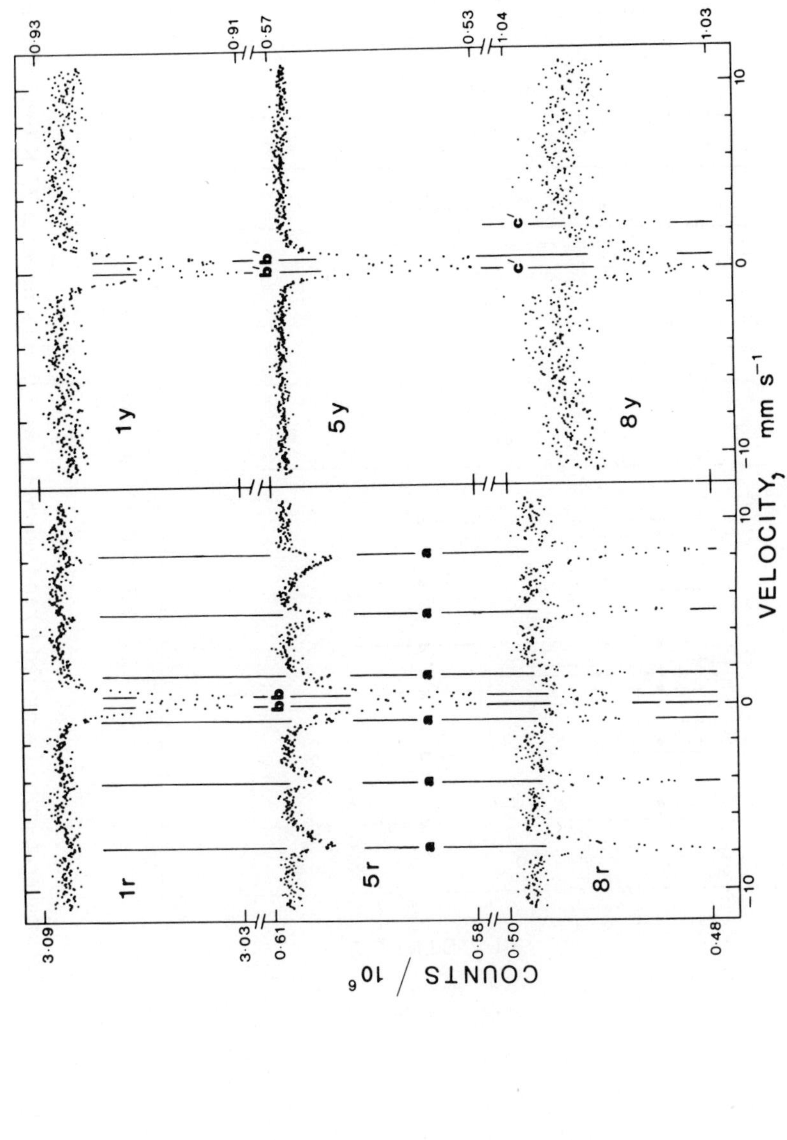

Figure 1-67. Mössbauer spectra of some red (r) and yellow-brown (y) soil samples at 300°K (from Childs *et al.*, 1978).

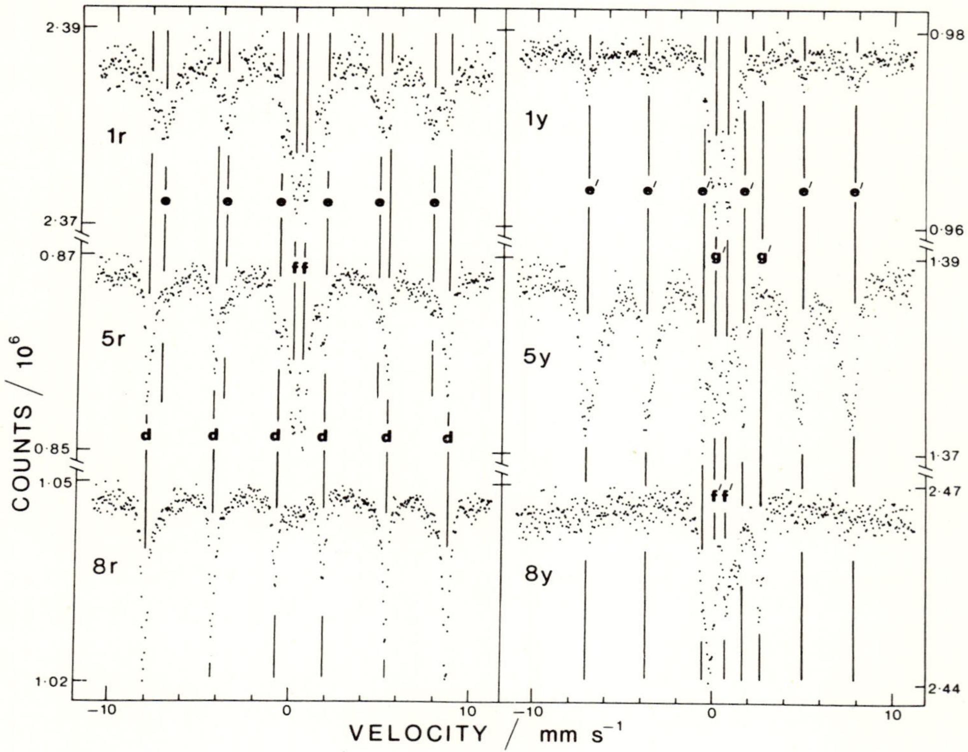

Figure 1-68. Mössbauer spectra of some red (r) and yellow-brown (y) soil samples at 77° K (from Childs *et al.*, 1978).

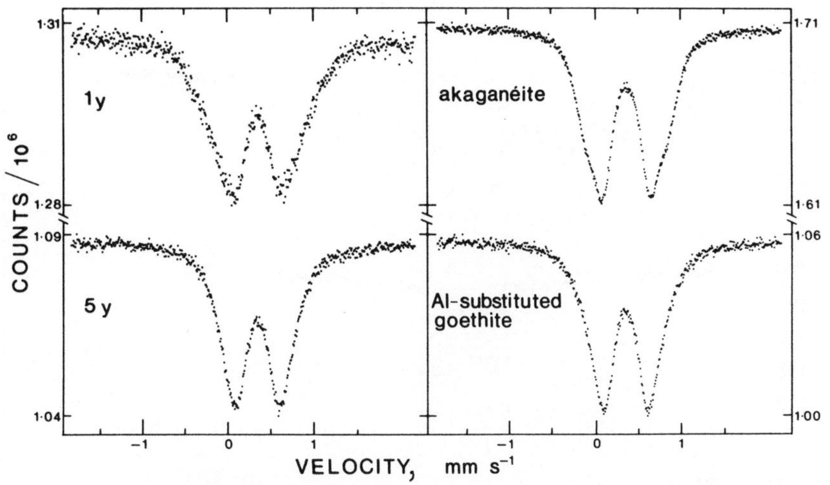

Figure 1-69. Mössbauer spectra of 2 yellow-brown soil samples compared to α- and β-FeOOH (from Childs *et al.,* 1978).

a stream bed in an unburnt soil catchment, was not greatly changed by magnetic separation. The magnetic structure at 300° K could be fitted to three sets of 6-line components for TB3, TB7 and Caldy Hill (see Fig. 1-72) and one 6-line component for Annecy. The parameters for the magnetic components in Fig. 1-72 are consistent with the presence of hematite (α-Fe_2O_3) and magnetite (Fe_3O_4). Other components seen in these spectra were (i) a central Fe^{3+} doublet which decreased in intensity on cooling (Fig. 1-73), thus indicating the presence of a superparamagnetic component as well as a paramagnetic component from Fe^{3+} in silicate minerals; and, (ii) a Fe^{2+} component, seen particularly well in sample TB1. It is extremely difficult to distinguish the magnetic components in the Mössbauer spectra at 4.2° K because of the large degree of overlap between magnetite, hematite, maghemite and the super-paramagnetic component, and also the complexity of the magnetite spectrum at this temperature. By applying large external magnetic fields to the samples, it was possible to obtain much greater information from these low temperature spectra. Thus, whereas for a completely random distribution of hyperfine field directions the line intensities are 3:2:1:1:2:3, in large external magnetic fields the internal field in a ferrimagnet, such as magnetite or maghemite, becomes aligned with the external field and, for H parallel to the γ-direction, the line intensities are 3:0:1:1:0:3. For antiferromagnets, such as hematite or goethite, the internal field direction is not affected by the external magnetic field. The spectra of the oxide minerals and magnetically-separated soil samples are shown in Fig. 1-74. In interpreting such spectra it should be borne in mind that if the ferrimagnetic component is microcrystalline the alignment of spins may not be complete and hence the intensities of peaks 2 and 5 may not always be reduced to zero. Nevertheless, in samples TB3 and TB7, where magnetite is the major magnetic component, the decrease in intensities of these peaks is appreciable. It can also be seen that the spectrum of the component in TB1, which was superparamagnetic at room temperature, has the appearance of an antiferromagnetic material. Its hyperfine field at 4.2° K was 483KOe, suggesting that it might be goethite, although

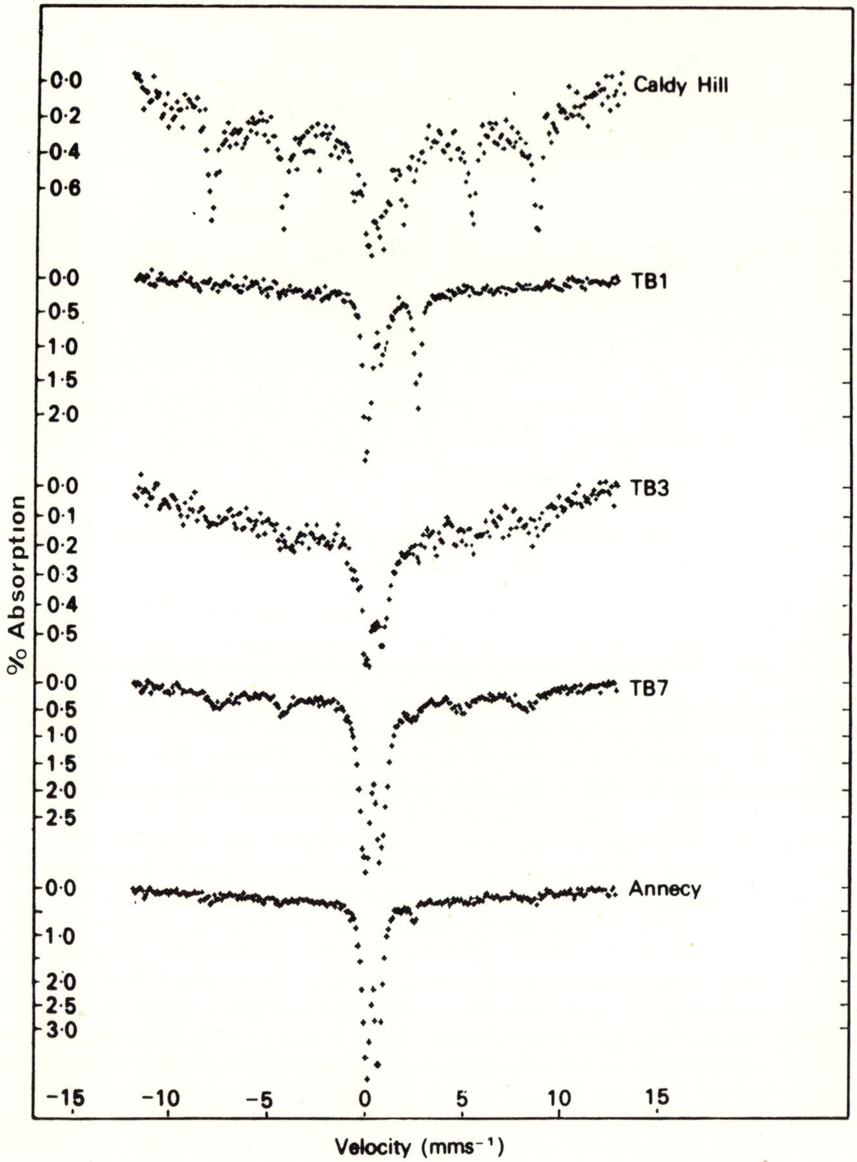

Figure 1-70. Mössbauer spectra of some whole soil samples at 300° K (from Longworth *et al.*, 1979).

Longworth *et al.* (41) did not rule out the possibility of microcrystalline hematite.

This work shows that differentiation between hematite and maghemite, which have very similar Mössbauer parameters, becomes possible in the presence of a large external magnetic field. The two hyperfine fields in magnetite and their appreciably lower values allow it to be distinguished from the other two oxides in

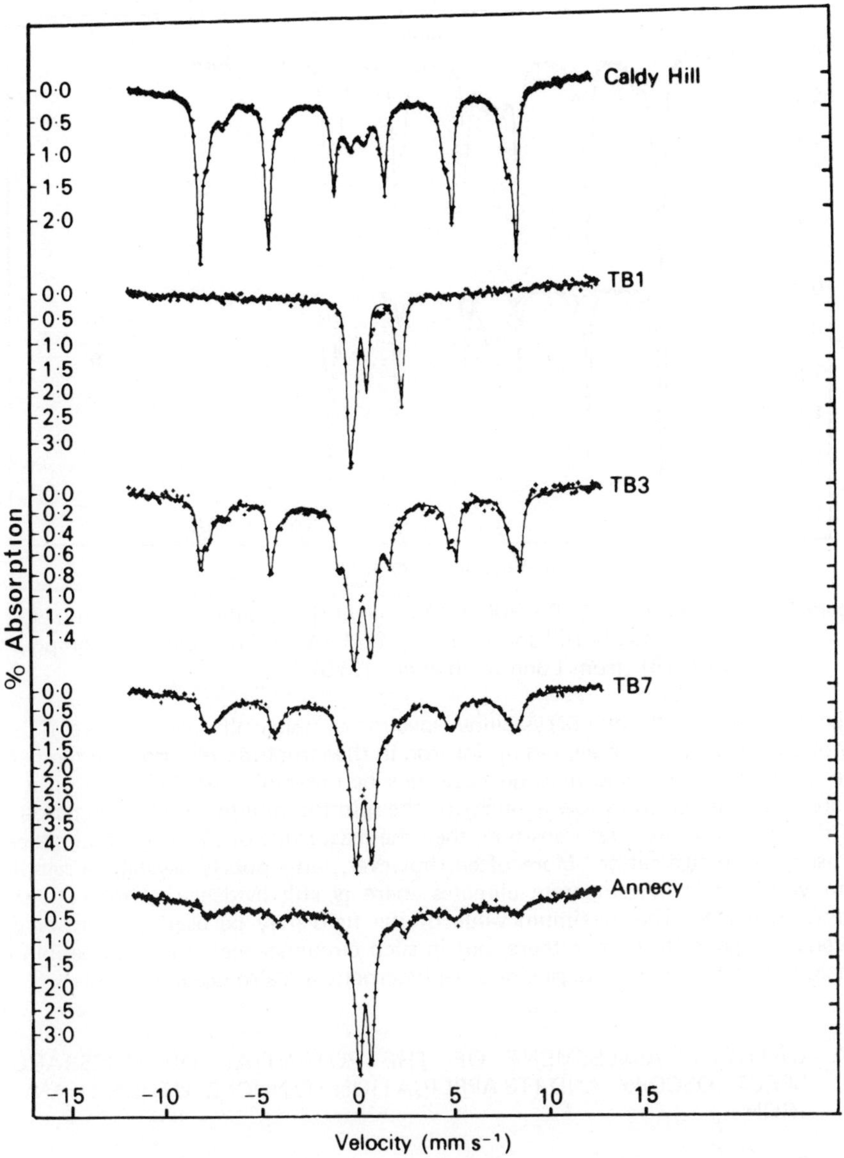

Figure 1·71. Mössbauer spectra of magnetically-separated soil samples at 300° K (from Longworth *et al.*, 1979).

room temperature spectra, although of course, an applied field at low temperature could be used to confirm the distinction between it and hematite. Unambiguous distinction between the various oxyhydroxides is not always possible in soil samples, although well-crystallized samples can be identified by taking account of both the temperature at which ordering occurs and the magnitude of the internal

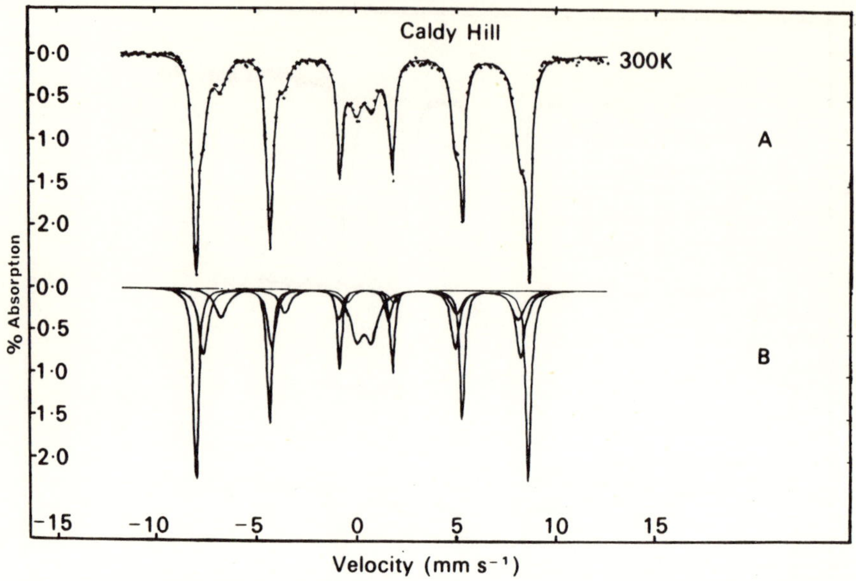

Figure 1-72. Computer fit to the 300° K Mössbauer spectrum of the magnetic extract of Caldy Hill soil from Fig. 1-71 (A) and its individual components (B) (from Longworth *et al.*, 1979).

magnetic field. With microcrystalline specimens, especially when there is isomorphous substitution of aluminum for iron in the structure, the ordering temperature is of little value and may on occasions be misleading. If the spectra can be recorded at temperatures low enough for the spin-flip rate to be slow compared to the time of the Mössbauer transition, then the magnitude of the magnetic field can be used for identification. More often, however, with poorly crystalline samples from young soils in temperate climates there is still evidence of relaxation in spectra at 4.2° K. The maximum value of the field may be used to identify the presence of goethite if it is there, but in such circumstances it is not possible to conclude whether or not akaganéite or lepidocrocite are also present.

1-7. CRITICAL ASSESSMENT OF THE POTENTIAL OF MÖSSBAUER SPECTROSCOPY, AND ITS APPLICATION TO NUCLEI OTHER THAN IRON

1-7.1. Summary of applications of ^{57}Fe Mössbauer spectroscopy

In the study of silicate minerals, Mössbauer spectroscopy is able to distinguish clearly between the high spin ions Fe^{2+} and Fe^{3+}. The isomer shift is the most important parameter here since there is typically a difference of about 0.7 mm sec^{-1} between the two ions (Fig. 1-8). The quadrupole splitting is also often used for distinguishing Fe^{2+} from Fe^{3+}, but there are instances where lattice sites may be extremely distorted, e.g. as a result of dehydroxylation processes, and on such occasions the quadrupole splittings for the two ions may be similar. Also it

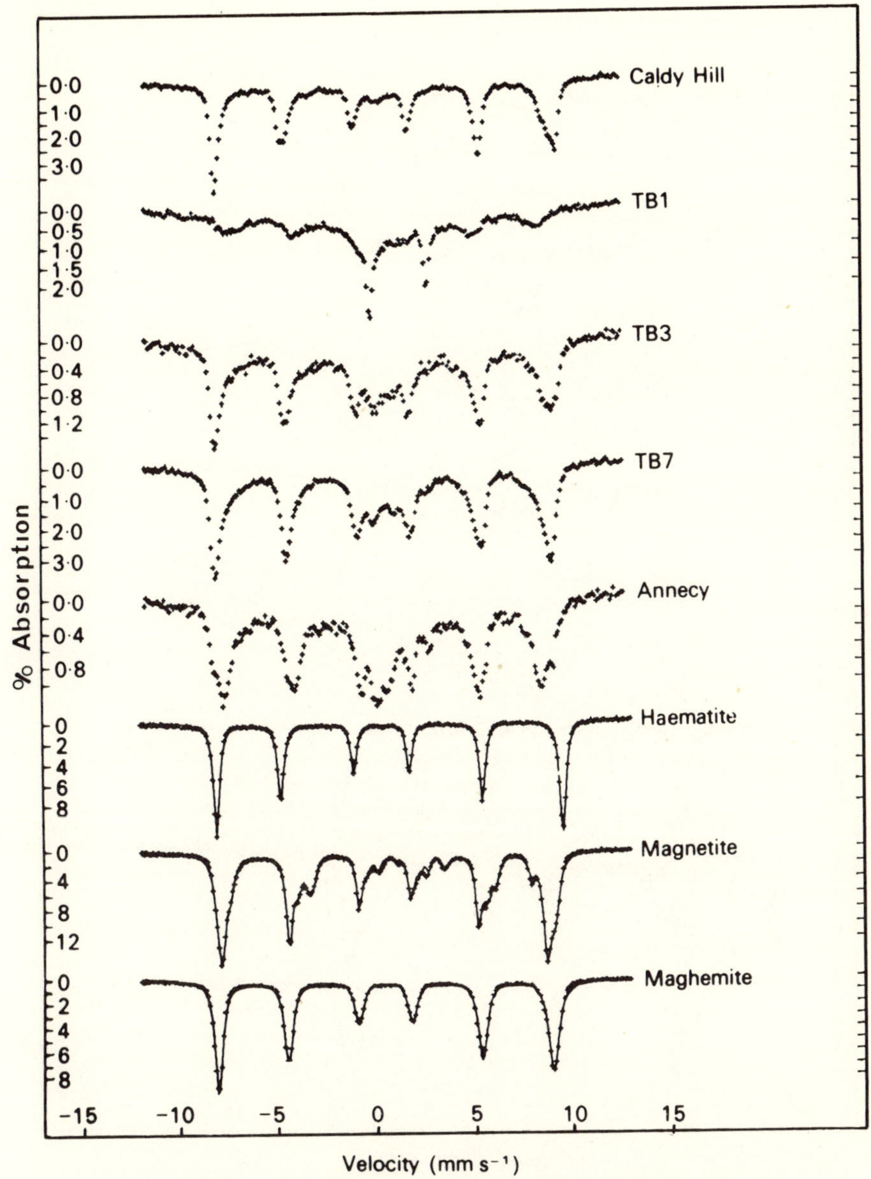

Figure 1-73. Mössbauer spectra of magnetically-separated soil samples and oxide standards at 4.2° K (from Longworth *et al.,* 1979).

should be remembered that the quadrupole splitting for high spin Fe^{2+} is temperature dependent, since thermal population of electronic excited states can readily occur. Thus, in a room temperature spectrum, there may be an appreciable contribution to Δ from excited states, which will have the opposite sign to the ground state contribution (d_{xz}, d_{yz} as compared to d_{xy}). This situation is illustrated in

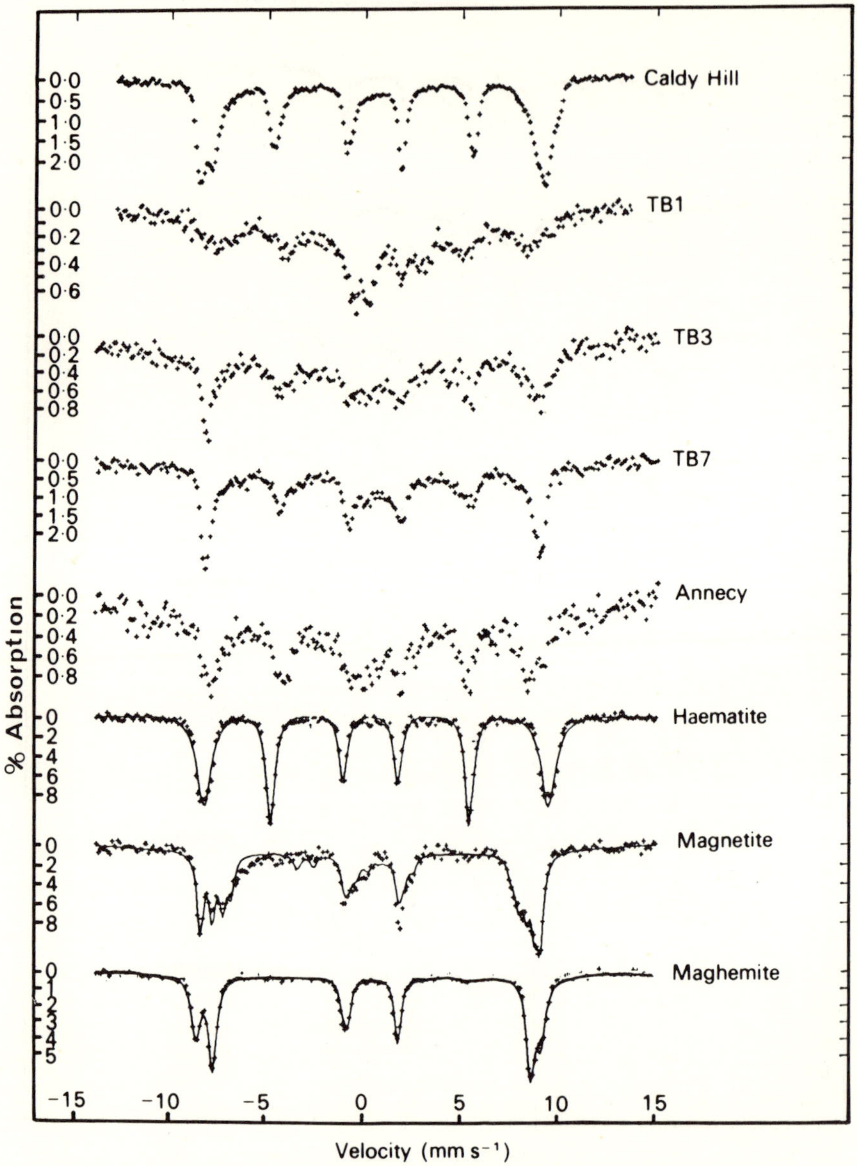

Figure 1-74. Mössbauer spectra of samples in Fig. 1-73 in an applied magnetic field at 30 kOe parallel to the γ-beam (from Longworth *et al.*, 1979).

Fig. 1-75, where the valence terms at 77° K and 300° K have been expressed relative to the distortion from cubic symmetry and then combined with the lattice term to illustrate the resultant quadrupole splitting. (Note that for perfectly cubic symmetry the d_{xz}, d_{yz} and d_{xy} orbitals are degenerate and both valence and lattice

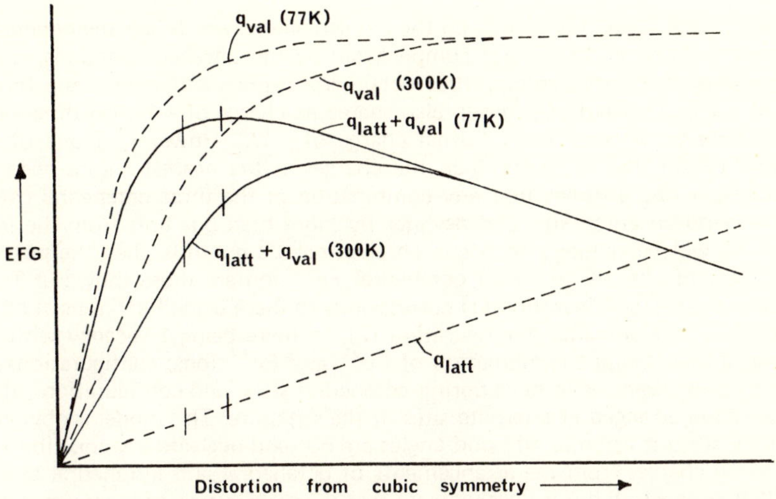

Figure 1-75. Diagrammatic representation of the variation of the total electric field gradient for Fe^{2+} with lattice distortion.

terms are zero). Also if there is a small range of lattice contributions to the electric field gradient at a site in a mineral, then, for fairly small distortions from cubic symmetry, i.e. near the bars marked in Fig. 1-75, Δ will not only be smaller at room temperature than at 77° K, but may also have a greater range of values.

The isomer shift may also give information on the coordination number of the ion. As can be seen from Fig. 1-8 the isomer shift increases with increasing coordination number. In this figure, however, there is considerable overlap between the isomer shifts from the various coordination numbers, but if the element to which the iron is bound is held constant there is much less spread in isomer shift values. Thus for coordination to oxygen atoms Fe^{3+} isomer shifts are usually in the range 0.15-0.25 mm sec^{-1} (relative to iron metal) for 4-coordination and 0.30-0.40 mm sec^{-1} for 6-coordination, both sets of values applying to measurements at room temperature. Similar relationships also apply to Fe^{2+}, and there is usually no difficulty in assigning the coordination number to a particular spectral component. The biggest problem lies in separating the peaks from components with different coordination numbers in the same sample as was mentioned in earlier sections. Uncritical use of computer programs can lead to the derivation of incorrect parameters for the spectral components. Incorrect conclusions may, therefore, be drawn. By recognizing the likely ranges for isomer shifts for the various coordination numbers, such errors may sometimes be avoided.

One of the major uses to which Mössbauer spectroscopy has been put in mineralogy has been in computing the distribution of Fe^{2+} and Fe^{3+} ions over the various sites in a crystal lattice. As has been stressed in earlier sections, this type of application is full of pitfalls for the unwary, especially when attempts are being made to distinguish between sites with the same coordination number on the basis of quadrupole splittings. As an example, the lattice contribution to the electric field gradients at the crystallographic sites in a biotite will be considered in terms of the effective charge distribution on the coordinated anions. Considering the composition $K(Si_3Al)(R^{2+}_{3-y}R^{3+}_{2y/3}\square_{y/3})O_{10}(OH)_2$, where $\square$ represents vacancies in the octahedral layer, the charge on the oxygen anions which link the octahedral and tetrahedral sheets, that is not compensated by the tetrahedral cations, is −1. This charge is distributed among three octahedral cations and so carries a formal charge of −1/3. The hydroxyl group also shares its charge of −1 with three octahedral cations so it also has a formal charge of −1/3. However, if one of the octahedral cation sites is empty then the charge on the nearest anions becomes −1/2. Thus, if one assumes that any combination of the three octahedral cation charges around an anion site that deviates by more than one unit from the ideal value of +6 will be unlikely to occur on electrostatic grounds, then the possible arrangements of charges about an octahedral Fe^{3+} ion are those shown in Table 1-10. In this table, configuration (1) corresponds to there being Fe^{2+} ions at all six neighboring octahedral sites; configuration (2), to there being 1 vacancy with the remaining 5 sites being a combination of Fe^{2+} and Fe^{3+} ions; configurations (3) and (4) to two vacancies in neighboring octahedral sites; and configuration (5), to three vacancies arranged at alternate sites in the structure. This model is obviously very crude, since it assumes all bond angles are 90° and neglects any contributions to the anion charges from the arrangements of octahedral and tetrahedral cations (other than vacancies), but it does indicate that the distribution of vacancies within the octahedral sheet may well contribute substantially to the lattice electric field gradient, q_{latt}. Further support for this conclusion has been obtained by carrying out calculations of the electric field gradients at the sites in micas using atomic coordinates determined by single crystal XRD measurements. Such calculations indicate that the q_{latt} values are not greatly different for the two crystallographic sites and that the spread of values that may be obtained at each type of site is comparable to, if not greater than, these differences. Hence this may provide the explanation as to why the Fe^{3+} components from the two types of site are not usually resolved in the Mössbauer spectra of trioctahedral micas, whereas the Fe^{2+} components are resolved at room temperature because of their position on the region of steep slope in Fig. 1-75. At 77°K the Fe^{2+} absorption in biotite is much less asymmetrical than at room temperature, indicating that there is a much narrower distribution of electric field gradients at this temperature.

In the study of soils there is little possibility of using Mössbauer spectroscopy for the identification of specific silicate minerals; but, as indicated in section 1-6 there is considerable scope for its use in the identification of secondary oxide components. A considerable amount of progress can be made in studies between room temperature and 77°K on reasonably well-defined components, but with top-soils or other horizons rich in organic matter the minerals are often very poorly defined. The use of very low temperature facilities (liquid helium), preferably also incorporating a superconducting magnet, is then necessary. However, by combining Mössbauer spectroscopy with various physical separation techniques it may sometimes be possible to characterize the various oxide components.

With regard to future lines of research, little has been said about the use of spectrometers in backscattering geometry. Although less sensitive than transmission geometry, backscattering experiments have the advantage of being able to look selectively at the surface regions of particles. The mean escape depths for conversion electrons are of the order of a few hundred Å (i.e. 3-4 x 10^{-8} m) and, by comparing such spectra with those obtained in transmission geometry, it is possible to get information on ions close to the surface, especially if the iron in that surface can be enriched with ^{57}Fe. By careful preparative methods it might be possible to study the surface regions of large crystals of oxides or other synthetic minerals as distinct from microcrystalline samples that are normally used. There may also be applications to the study of mineral alteration especially since electron microscopic studies of resistant minerals indicate a selective erosion of surface regions.

In conventional experiments the study of silicate minerals is bound to continue, but this author expects to see more calculations of electric field gradients for sites in typical mineral structures in attempts to authenticate conclusions that have been drawn from Mössbauer spectra. There is plenty of scope for the study of alteration reactions; the effects of various chemical reagents commonly used for cleaning up minerals prior to investigation of the minerals themselves urgently need investigating, e.g. in my own laboratory, unbuffered sodium dithionite has been shown, not only to destroy nontronites with tetrahedral Fe^{3+}, but also to increase the Fe^{3+} content of biotite. This latter observation is surprising and presumably occurs because the dithionite causes appreciable mineral breakdown and this results in oxidation of the newly exposed Fe^{2+} on exposure to air. The study of samples at elevated temperatures might be extremely useful in the identification of reactions occurring at DTA peaks, provided the f-factor remains high enough to give a satisfactory signal-to-noise ratio. In the study of soils it should be possible to help with the explanation of soil forming processes. This can be achieved, not only by the identification of secondary iron oxides as described in the last section but also, by careful sampling, through the elucidation of reactions such as gleying occurring in the soil.

1-7.2. Elements other than ^{57}Fe

There is little possibility of using Mössbauer spectroscopy for the study of elements other than iron in soils and clay minerals, because of the generally low natural abundance of suitable isotopes. The elements other than iron that can readily be studied are tin, antimony, iodine, europium and dysprosium. The last two of these are rare earths and the other three are elements that are generally classed as trace elements. Even in soils formed in areas in which these elements are mined there is little possibility of obtaining good Mössbauer spectra although there have been applications to specific minerals. These will not be reviewed here, but a brief description will be given on the nuclear transitions involved with each isotope and the type of spectrum that can be expected.

The Mössbauer isotope for tin is ^{119}Sn, which has a natural abundance of about 8.6%. The source, metastable ^{119}Sn produced by the (n, γ) reaction on ^{118}Sn in the thermal neutron flux of a nuclear reactor, has a half life of 245 days, and the Mössbauer transition, at 23.9 KeV, occurs between a spin 1/2 ground state and a spin 3/2 excited state. Spectra are, therefore, similar to those observed with ^{57}Fe with two exceptions; the radius of the excited state nucleus with ^{119}Sn is larger than the radius of the ground state and line widths are much broader. Thus, with tin there is a positive relationship between isomer shift and s-electron density. The larger linewidth for ^{119}Sn combined with a smaller quadrupole moment than for ^{57}Fe makes the detection of small electric field gradients more difficult (these may all be considered as lattice-derived since both Sn^{2+} and Sn^{4+} have symmetrical electron distributions). Nevertheless, many tin compounds give quadrupole splittings and in chemistry a considerable amount of success has been achieved in assigning oxidation states and coordination number to tin ions.

With ^{121}Sb the Mössbauer transition is between a ground state with spin 5/2 and an excited state with spin 7/2 and with ^{129}I the ground and excited state spins

Table 1-10. Relative quadrupole splittings for the various anion arrangements around a Fe^{3+} ion in biotite.

	Charge arrangement	Δ (relative values)
1.	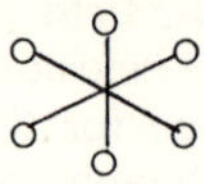	0
2.	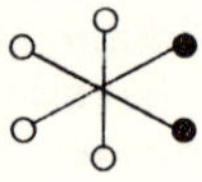	1
3.	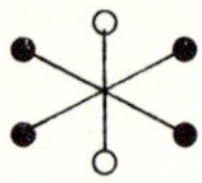	2
4.	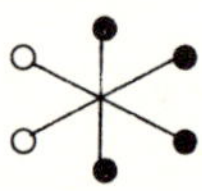	1
5.	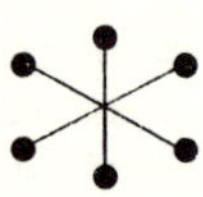	0

○ = – 1/3; ● = – 1/2.

are 7/2 and 5/2, respectively. Spectra are, therefore, considerably more complex than for ^{57}Fe or ^{119}Sn, since an electric field gradient splits these states into three and four sublevels (Fig. 1-76). The γ-ray energies for the transitions and source half-lives for ^{121}Sb and ^{129}I are 37.2 KeV, 76 years and 27.8 KeV, 33 days, respectively. With ^{121}Sb the radius of the nuclear excited state is smaller than that of the ground state but with ^{129}I the excited state is larger. Thus with ^{121}Sb, like ^{57}Fe, there is an inverse relationship and with ^{129}I, like ^{119}Sn, there is a direct relationship between isomer shift and s-electron density.

1-7.3. Energy Units

The energy unit commonly used in Mössbauer spectroscopy is one of convenience, the mm sec^{-1}. For conversion to more familiar units one uses the relationship,

$$\Delta E = \frac{v}{c} E_{\gamma} \qquad [1\text{-}29]$$

where ΔE is the energy change measured in the Mössbauer experiment, v is the source velocity, c is the velocity of light, and E_{γ} is the energy of the Mössbauer transition. Thus for ^{57}Fe, E_γ = 14.39 KeV, so that for v = 1 mm sec^{-1}, ΔE = 4.80 x 10^{-8} eV. Conversion factors for other energy units for ^{57}Fe are given in Table 1-11.

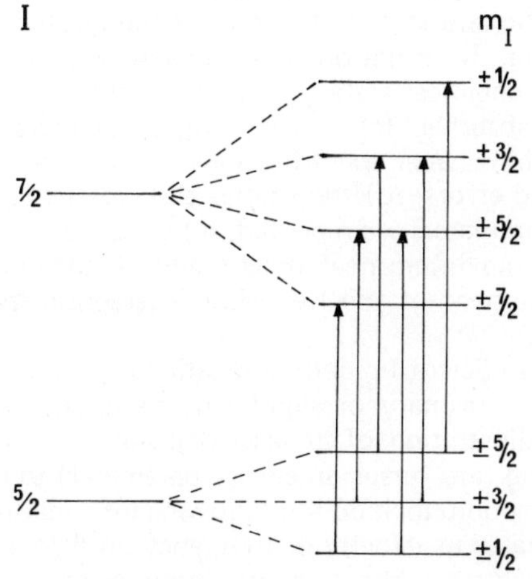

Figure 1-76. The splitting of the nuclear energy levels of ^{121}Sb (ground state I = 5/2, excited state I = 7/2) by an electric field gradient.

Table 1-11. Energy conversion table for ^{57}Fe Mössbauer spectroscopy

	mm s^{-1}	cm^{-1}	joule	eV	Mc/s
1 mm s^{-1}	1	3.87×10^{-4}	7.69×10^{-27}	4.80×10^{-8}	11.61
1 cm^{-1}	2.58×10^{3}	1	1.986×10^{-23}	1.240×10^{-4}	2.998×10^{4}
1 joule	1.30×10^{26}	5.035×10^{22}	1	6.242×10^{18}	1.509×10^{27}
1 eV	2.08×10^{7}	8.066×10^{3}	1.602×10^{-19}	1	2.418×10^{8}
1 Mc/sc	8.61×10^{-2}	3.336×10^{-5}	6.625×10^{-28}	4.136×10^{-9}	1

1-7.4. Conventions for the reporting of Mössbauer data

This final section will be concerned with reporting experimental results. So many papers appear with insufficient specification of details, e.g. isomer shift reference compound omitted, that a guideline, based on recommendations of IUPAC and the Mössbauer Effect Data Center, for reporting Mössbauer results has been produced.

Text. The text should include information about: (a) the method of sample mounting, sample thickness, sample confinement, and appropriate composition data; (b) the form of the absorber (single crystal, polycrystalline powder, inert matrix if used, isotopic enrichment, etc.); (c) the apparatus and detector used and comments about the associated electronics and data acquisition time if unusual; (d) the geometry of the experiment (transmission, scattering, angular dependence, etc.); (e) any critical absorbers or filters, if used; (f) the method of data reduction and cruve-fitting procedure (See Notes 1 and 2); (g) the isomer shift convention used or the isomer shift of a standard (reference) absorber. Positive velocities are defined as source approaching absorber. Sufficient details concerning the isomer shift standard should be included to facilitate interlaboratory comparison of data; and, (h) an estimate of systematic and statistical errors of the quoted parameters.

Numerical or tabulated data. Information collected and summarized in tabular form should include: (a) the chemical state of source matrix and absorber; (b) the temperature of source and absorber; (c) values of the parameters required to characterize the features in the Mössbauer spectrum (given in mm sec^{-1} or other appropriate units) with estimated errors; (d) the isomer shift reference point with respect to which the position parameters are reported; (e) the observed line-widths defined as the full-width at half maximum peak-height; and, (f) the line intensities or (relative) areas of each component of the hyperfine interaction spectrum observed, when pertinent.

Figures Illustrating Spectra. Scientific communications in which Mössbauer effect measurements constitute a primary or significant source of experimental information should include an illustration of at least one spectrum (i.e. % transmission or absorption or counting rate versus an energy parameter) to indicate the quality of the data. Such figures should include the following information features: (a) a horizontal axis normally scaled in velocity or frequency units (e.g. mm sec^{-1} or MHz. Channel number or analyzer address values should not be used for this purpose; see Note 3); (b) a vertical axis normally scaled in counts per channel or related units (see Note 4); (c) an indication, for at least one data point, of the statistical counting error limits (See Note 5); and (d) individual data points (rather than a smoothed curve alone) should be shown. Computed fits should be indicated in such a way that they are clearly distinguishable from the experimental points.

NOTES

1. If data are analyzed by computer, a brief description of the program should be given to identify the algorithm used. The number of constraints should be specified (e.g. equal line-widths or intensities, etc.), and a measure of the goodness of fit should be indicated.
2. If measurements of very high accuracy are reported and the discussion of the reality of small effects is an important part of the work, then the following items should be included:

    (i) the functional form and all parameters used in fitting (i.e. the constraints should be clearly stated);

    (ii) the treatment of the background (e.g. assumed energy independent, experimentally subtracted, etc.);

    (iii) the relative weighting of abscissa and ordinate (e.g. equal weighting);

    (iv) a measure of the statistical reliability;

    (v) the number of replications and the agreement between these if applicable;

    (vi) an estimate of systematic errors as primary results.
3. Constant acceleration spectrometers to be used for work in the mm sec^{-1} range can be calibrated with respect to velocity using either metallic iron foil of at least 99.99% purity or an optical method based on interferometric or Moiré pattern techniques.

4. It has become customary to display data obtained in transmission geometry with the resonance maximum 'down' and scattering data with the resonance maximum 'up'. In either case, sufficient data should be shown far enough from the resonance peaks to establish the non-resonant base-line.

5. In most instances (where the data are uncorrected counting results), the standard deviation (i.e. the square root of the second moment of the distribution) is given by $N^{1/2}$, where N is the number of counts scaled per velocity point. For corrected data (i.e. when background or other non-resonant effects are subtracted from the raw data), the error propagated should be computed by normal statistical methods which are briefly described in the text or figure legend. Fiducial marks bracketing the data point to show the magnitude of the standard deviation are often useful in indicating the spread of the data.

REFERENCES

1. Annersten, H. 1974. Mössbauer studies of natural biotites. *Am Mineral.* **59**: 143-151.
2. Artman, J.O., A.H. Muir, and H. Weidersich. 1968. Determination of the nuclear quadrupole moment of iron-57 from α-ferric oxide data. *Phys. Rev.* **173**: 337-343.
3. Bancroft, G.M. 1973. Mössbauer spectroscopy: an introduction for inorganic chemists and geochemists. McGraw-Hill, London.
4. Bancroft, G.M. and J.R. Brown. 1975. A Mössbauer study of coexisting hornblendes and biotites: quantitative Fe^{3+}/Fe^{2+} ratios. *Am. Mineral.* **60**: 265-272.
5. Bhide, V.G. 1973. Mössbauer effect. Tata McGraw-Hill.
6. Bouchez, R., J.M.D. Coey, R. Coussement, K.P. Schmidt, M. Van Rossum, J. Aprahamian, and J. Deshayes. 1974. Mössbauer study of firing conditions used in the manufacture of the grey and red ware of Tureng-Tepe. *J. Phys. Colloq. C6,* **35**: 541-546.
7. Chevalier, R., J.M.D. Coey, and R. Bouchez. 1976. A study of iron in fired clay: Mössbauer effect and magnetic measurements. *J. Phys. Colloq. C6,* **37**: 861-865.
8. Childs, C.W., B.A. Goodman, and G.J. Churchman. 1978. Application of Mössbauer spectroscopy to the study of iron oxides in some red and yellow/brown soil samples from New Zealand. *Proc. Inter. Clay Conf.* 1978 (Pub. 1979): 555-565.
9. Childs, C.W., B.A. Goodman, E. Paterson, and F.W.D. Woodhams. 1980. A Mössbauer spectroscopic investigation of the nature of iron in akaganéite (β-FeOOH). *Aust. J. Chem.* (In Press).
10. Coey, J.M.D., personal communication.
11. Collins, R.L. and J.C. Travis. 1967. The electric field gradient tensor. *In* I.J. Gruverman, Ed. Mössbauer Effect Methodology. Vol. 3. Plenum, New York. pp. 123-161.
12. Colville, P.A., W.G. Ernst, and M.C. Gilbert. 1966. Relationships between cell parameters and chemical compositions of monoclinic amphiboles. *Am. Mineral.* **51**: 1727-1754.
13. Dowty, E. and D.H. Lindsley. 1973. Mössbauer spectra of synthetic hedenbergite-ferrosilite pyroxenes. *Am. Mineral.* **58**: 850-868.
14. Ericsson, T. and R. Wäppling. 1976. Texture effects in 3/2-1/2 Mössbauer spectra. *J. Phys. Colloq. C6,* **37**: 719-723.
15. Evans, B.J. and S.S. Hafner. 1969. Iron-57 hyperfine fields in magnetite (Fe_3O_4). *J. Appl. Phys.* **40**: 1411-1413.
16. Forsyth, J.B., I.G. Hedley, and C.E. Johnson. 1968. The magnetic structure and hyperfine field of goethite (α-FeOOH). *J. Phys. C. Ser. 2,* **1**: 179-188.
17. Gibb, T.C. and N.N. Greenwood. 1971. Mössbauer spectroscopy. Chapman and Hall, London.
18. Gol'danskii, V.I. and R.H. Herber. 1968. Chemical applications of Mössbauer spectroscopy. Academic Press, New York.
19. Gol'danskii, V.I., E.F. Makarov, and V.V. Khrapov. 1963. Difference in two peaks of quadrupole splitting in Mössbauer spectra. *Phys. Letts.* **3**: 344-346.
20. Goodman, B.A. 1976. The Mössbauer spectrum of a ferrian muscovite and its implications in the assignments of sites in dioctahedral micas. *Miner. Mag.* **40**: 513-517.

21. Goodman, B.A. 1976. The effect of lattice substitutions on the derivation of quantitative site populations from the Mössbauer spectra of 2:1 layer lattice silicates. *J. Phys. Colloq. C6,* **37**: 819-823.
22. Goodman, B.A. 1978. The Mössbauer spectra of nontronites: consideration of an alternative assignment. *Clays Clay Miner.* **26**: 176-177.
23. Goodman, B.A. 1978. An investigation by Mössbauer and EPR spectroscopy of the possible presence of iron-rich impurity phases in some montmorillonites. *Clay Miner.* **13**: 351-356.
24. Goodman, B.A. and D.C. Bain. 1978. Mössbauer spectra of chlorites and their decomposition products. *Proc. Inter. Clay Conf.* 1978 (Pub. 1979): 65-74.
25. Goodman, B.A. and M.L. Berrow. 1976. The characterization by Mössbauer spectroscopy of the secondary iron in pans formed in Scottish podzolic soils. *J. Phys. Colloq. C6,* **37**: 849-855.
26. Goodman, B.A., J.D. Russel, A.R. Fraser, and F.W.D. Woodhams. 1976. A Mössbauer and I.R. spectroscopic study of the structure of nontronite. *Clays Clay Miner.* **24**: 53-59.
27. Goodman, B.A. and M.J. Wilson. 1973. A study of the weathering of a biotite using the Mössbauer effect. *Miner. Mag.* **39**: 448-454.
28. Goodman, B.A. and M.J. Wilson. 1976. A Mössbauer study of the weathering of hornblende. *Clay Miner.* **11**: 153-163.
29. Gruverman, I.J., Ed. Mössbauer effect methodology (series). Plenum Press, New York.
30. Hafner, S.S. and S. Ghose. 1971. Iron and magnesium distribution in cummingtonites. *Z. Krist.* **133**: 301-326.
31. Hafner, S.S. and H.G. Huckenholz. 1971. Mössbauer spectrum of synthetic ferri-diopside. Nature (London), *Phys. Sci.* **233**: 9-11.
32. Häggstrom, L., R. Wäppling, and H. Annersten. 1969. Mössbauer study of iron-rich biotites. *Chem. Phys. Letts.* **4**: 107-108.
33. Hogg, C.S., P.J. Malden, and R.E. Meads. 1975. Identification of iron-containing impurities in natural kaolinites using the Mössbauer effect. *Miner. Mag.* **40**: 89-96.
34. Hogg, C.S. and R.E. Meads. 1970. The Mössbauer spectra of several micas and related minerals. *Miner. Mag.* **37**: 606-614.
35. Janot, C., H. Gibert, and C. Tobias. 1973. Charactérisation de kaolinites ferrifères par spectrometrie Mössbauer. *Bull. Soc. Fr. Mineral Cristallogr.* **96**: 281-291.
36. Jefferson, D.A., M.J. Tricker, and A.P. Winterbottom. 1975. Electron-microscopic and Mössbauer spectroscopic studies of iron-stained kaolinite minerals. *Clays Clay Miner.* **23**: 355-360.
37. Johnson, C.E. 1969. Antiferromagnetism of γ-FeOOH: a Mössbauer effect study. *J. Phys. C Ser. 2,* **2**: 1996-2002.
38. Kamineni, D.C. 1973. X-ray and Mössbauer characteristics of a cummingtonite from Yellowknife, District of Mackenzie. *Can. Mineral.* **12**: 230-232.
39. Khalafalla, D. and A.H. Morrish. 1972. Ferrimagnetic cobalt-doped gamma-ferric oxide micropowders. *J. Appl. Phys.* **43**: 624-631.
40. Kostikas, A., A. Simopoulos, and N.H. Gangas. 1976. Analysis of archaeological artifacts. *In* R.L. Cohen, Ed. Applications of Mössbauer spectroscopy. Vol. **1**, Academic Press, New York. pp. 241-261.
41. Longworth, G., L.W. Becker, R. Thompson, F. Oldfield, J.A. Dearing, and T.A. Rummery. 1979. Mössbauer effect and magnetic studies of secondary iron oxides in soils. *J. Soil Sci.* **30**: 93-110.

42. May, L., Ed. 1971. An introduction to Mössbauer spectroscopy. Plenum Press, New York.
43. Mering, J. and A. Oberlin. 1967. Electron-optical study of smectites. *Clays Clay Miner.* **15**: 3-25.
44. Mineeva, R.M. 1978. Relationship between Mössbauer spectra and defect structure in biotites from electric field gradient calculations. *Phys. Chem. Minerals* **2**: 267-277.
45. Ross, G.J. and H. Kodama. 1974. Experimental transformation of a chlorite into vermiculite. *Clays Clay Miner.* **22**: 205-211.
46. Rozenson, I. and L. Heller-Kallai. 1977. Mössbauer spectra of dioctahedral smectites. *Clays Clay Miner.* **25**: 94-101.
47. Russell, J.D., B.A. Goodman, and A.R. Fraser. 1979. Infrared and Mössbauer studies of reduced nontronites. *Clays Clay Miner.* **27**: 63-71.
48. Sanz, J., J. Meyers, L. Vielvoye, and W.E.E. Stone. 1978. The location and content of iron in natural biotites and phlogopites: a comparison of several methods. *Clay Miner.* **13**: 45-52.
49. Sternheimer, R.M. 1963. Shielding and antishielding effects for various ions and atomic systems. *Phys. Rev.* **146**: 140-160.
50. Virgo, D. and S.S. Hafner. 1969. Fe^{2+}, Mg order-disorder in heated orthopyroxenes. *Mineral Soc. Amer. Spec. Pap.* **2**: 67-81.
51. Walker, L.R., G.K. Wertheim, and V. Jaccarino. 1961. Interpretation of the ^{57}Fe isomer shift. *Phys. Rev. Letts.* **6**: 98-101.
52. Weber, H.P., and S.S. Hafner. 1971. Vacancy distribution in nonstoichiometric magnetites. *Z. Krist.* **133**: 327-340.
53. Wickman, H.H. 1966. Mössbauer paramagnetic hyperfine structure. *In* I.J. Gruverman, Ed. Mössbauer Effect Methodology, Vol. **2**, Plenum Press, New York. pp. 39-66.
54. Williams, P.G.L., G.M. Bancroft, M.G. Bown, and A.C. Turnock. 1971. Anomalous Mössbauer spectra of C2/c clinopyroxenes. *Nature (London), Phys. Sci.* **230**: 149-151.
55. Wilson, M.J. 1970. A study of weathering in a soil derived from a biotite-hornblende rock. Pt. 1. The weathering of biotite. *Clay Miner.* **8**: 291-303.
56. Wilson, M.J. and V.C. Farmer. 1970. A study of weathering in soil derived from a biotite-hornblende rock. Pt. II. The weathering of hornblende. *Clay Miner.* **8**: 435-444.

Chapter 2

NEUTRON SCATTERING METHODS OF INVESTIGATING CLAY SYSTEMS

D. Keith Ross
Department of Physics

Peter L. Hall
Department of Chemistry

The University of Birmingham
Birmingham B 15 2TT
United Kingdom

2-1. INTRODUCTION

2-1.1. Historical Survey

Of the techniques surveyed in the present proceedings, neutron scattering is probably the least familiar to clay scientists because the number of publications on the application of the technique to clay systems is still relatively small. We therefore start this survey with a brief account of the development of the technique, followed by a discussion of the various types of measurement that can be made and of the potential advantages of such measurements in comparison with other techniques. We conclude the introduction with some recent examples of the application of the technique in a variety of areas of physics, chemistry and biology to illustrate the significant contributions that have been made in fields ranging from the study of phase transitions to the mechanisms of cell division. The second section gives a rather simple account of the theory of neutron scattering, while the third section describes the experimental techniques that have been developed to obtain data for comparison with theoretical predictions. Finally, the fourth section describes the applications of neutron scattering methods to studies of clay minerals.

Neutron scattering as a practical technique has its origins in the mid 1950's when reasonably copious beams of thermal neutrons (i.e. those with a kinetic energy distribution in equilibrium with that of the moderator) became available from a new generation of nuclear research reactors. These reactors had been constructed at national nuclear laboratories (e.g. Harwell and Brookhaven). The first type of instrument to be developed was the neutron diffractometer, which was

J. W. Stucki and W. L. Banwart (eds.), Advanced Chemical Methods for Soil and Clay Minerals Research, 93–168.

directly parallel to the equivalent x-ray instruments; the theory developed rapidly making use of the existing x-ray analysis (12). The early inelastic neutron scattering spectrometers, however, were developed not for solid state research as such but to evaluate the energy transfer cross sections of reactor moderators (graphite, heavy water, water, zirconium hydride and certain organic liquids), data that was considered necessary for the design of nuclear reactors. Although this need was soon satisfied, it provided the impetus for the development of the first generation of inelastic scattering spectrometers which were able to demonstrate the usefulness of the technique for investigating the microscopic properties of a wide range of materials. The technique thereafter developed rapidly within the national reactor establishments. In the United Kingdom, however, university scientists were involved at an early stage and on a large scale through the Science Research Council and this encouraged the rapid dissemination of information about the technique first among physicists and chemists and later among materials scientists and biologists.

The next stage of development was the construction of purpose built high flux beam reactors (HFBR's), at Brookhaven and Oak Ridge in the USA and at the Institut Laue-Langevin at Grenoble, France. These reactors used a highly enriched reactor core both cooled and moderated by D_2O. In order to maximize the flux available for extraction down beam tubes, however, the core was made as compact as possible so that it was considerably undermoderated, i.e. most of the neutrons leaving the core had energies greater than thermal energies and their moderation was completed in the D_2O that surrounds the core. Hence the beam tubes were designed to extract neutrons from the point in the D_2O where the thermal flux peaked. A diagram of the Grenoble reactor is given in Fig. 2-1 and the calculated flux distribution is given in Fig. 2-2 (57).

These HFBR's, having a peak thermal neutron flux of $\sim 10^{15}$ n/cm^2/sec represent about the realistic maximum that can be achieved from a continuous fission source because the heat generated/unit volume of core is near the maximum that can be dissipated using water cooling. Higher fluxes must therefore be achieved by changing to a reaction that generates less heat/neutron produced and/or employs the pulsed neutron source principle so that high instantaneous fluxes are available for lower average heat generation rates (19). By coincidence, both Argonne National Laboratory (USA) and the Rutherford Laboratory (UK) recently proposed the conversion of outdated $\sim$ 10 GeV accelerators to produce intense proton pulses of GeV energies and then using the spallation reaction to generate pulses of neutrons in a heavy metal target. A similar installation is under construction at Los Alamos. In a spallation reaction a heavy nucleus is broken up into a range of smaller nuclei plus up to 10 neutrons. If the design predictions are fulfilled these facilities will be producing intensities in the 1980s that will be for many purposes higher by an order of magnitude than HFBR intensities (100, 111). All these intense neutron sources, both steady state and pulsed, support an array of instruments which require a large body of scientists to keep them fully occupied. It is therefore an appropriate time to be introducing the technique to a wider audience as there will be expanded opportunity for new users to enter the field. Moreover, as fully engineered instruments and appropriate technical support are generally available, it is possible for scientists whose main expertise lies in other techniques to do neutron scattering experiments on systems in which they are interested. Thus it is the aim of this chapter to provide clay scientists with sufficient background information to propose their own neutron experiments.

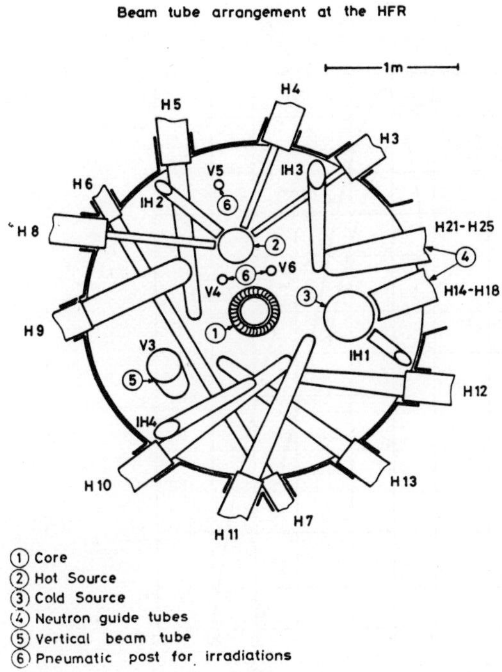

Figure 2-1. Plan of the Grenoble High Flux Beam Reactor.

2-1.2. Classification of Neutron Scattering Measurements

At this point it would seem to be appropriate to specify the four distinct areas of neutron scattering and the general kinds of information each produces.

Neutron Diffraction. The elastic 'coherent' scattering of neutrons from periodic lattices, either polycrystalline or single crystal, is exactly analogous to x-ray diffraction as the wavelength of thermal neutrons is similar to that of x-rays, both being comparable with average interatomic separations in materials. Measurement of the positions of the Bragg reflections defines the size and symmetry of the unit cell while measurement of the intensity of each reflection yields the actual location of the atoms within the unit cell.

Because of the easy availability of intense monochromatic x-ray sources, one would not use neutron diffraction unless one were seeking information not available from x-rays. Situations where this applies include: (a) the location of hydrogen or deuterium atoms; (b) the differentiation between elements of similar atomic number; (c) the determination of magnetic structures; (d) determination of the structures of liquids and amorphous solids; and (e) structural determinations in-

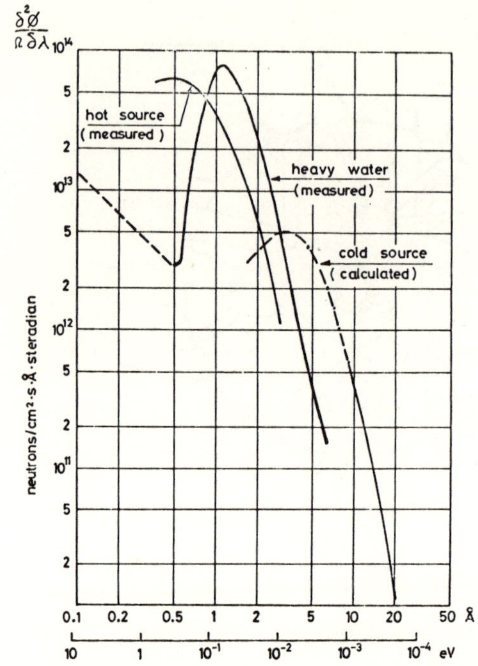

Figure 2-2. Neutron flux distributions in the Grenoble reactor.

volving large unit cells where it is convenient to use long wavelength radiation. The first and last are clearly relevant to clays and have provided the motivation for the work carried out so far.

Small Angle Neutron Scattering. The measurement of coherent elastic diffraction at small angles yields information of density fluctuations in materials over distances typically in the range 50-5000 Å. The main advantage of neutrons over x-rays in this case is that one may select a sufficiently long wavelength to avoid Bragg reflections completely. The main areas of application include lattice distortions around defects, precipitation phenomena in alloys, voids in ceramics and irradiated materials, and particle sizes in catalysts, colloids and biological molecules. Here our interest is in clay/water colloids. One important aspect of this technique when applied to colloidal suspensions or solutions of macromolecules is the method of 'contrast variation' by hydrogen-deuterium substitution. This method is based on the different neutron scattering properties of these nuclei. Thus, for example, adjustment of the H_2O-D_2O ratio in the solvent phase enables the small angle scattering from particular components of the system to be selectively highlighted (58).

Quasi-Elastic Neutron Scattering. We turn now to non-elastic scattering. The important basic factor is that the energies of thermal neutrons are comparable to those of atoms undergoing thermally activated motion in a solid or liquid. Since the masses are comparable, their velocities are also comparable. Thus, whether we regard an interaction as being a collision between two point masses or a Doppler effect in the scattering of a wave packet by a moving scattering center, the result is that the gain or loss of a significant and easily measured proportion of the neutron's initial energy will occur during the scattering event. If the sample is a rigid solid, the neutron will either have an elastic collision or will transfer one or more quanta of energy to or from any of the modes of vibration of the solid. If, however, the sample contains atoms free to diffuse, the elastic peak becomes broadened in energy, an effect that is known as quasi-elastic neutron scattering. As will be shown in section 2-2, measurement of this broadening can provide information on the nature of the diffusive motion involved e.g. jump diffusion in a periodic lattice, rotational diffusion of a molecule in a crystal or random diffusion within a restricted volume. Here, the primary application in clays is to the behavior of water intercalated between the clay layers. This topic will be covered in some detail in section 2-4.

Inelastic Neutron Scattering. As mentioned above, inelastic neutron scattering occurs when the neutron exchanges one or more quanta of energy with the solid. The most important example, where the information produced is absolutely unique, is in the measurement of phonon dispersion curves, i.e. the determination of the relationship between the wave vector and frequency of phonons in symmetry directions, which determines in principle the forces acting between the atoms in the crystal (see section 2-2). These measurements depend almost exclusively on rather specialized instruments known as Triple Axis Spectrometers which use a computer control system to measure the frequency associated with a predetermined set of wave vectors. On the other hand, for a wide range of energy transfers, particularly for samples containing hydrogen, inelastic scattering can be used to determine the frequencies and relative intensities of the various possible modes of vibration. Here, particularly where the dispersion is small, the probability of neutron scattering can be determined by very simple expressions uninfluenced by the selection rules involved in infrared absorption and Raman scattering.

2-1.3. Examples of the Scope of Neutron Scattering Measurements

A number of examples of recent experiments in rather diverse and specialized fields are given below to illustrate the range of measurements currently being performed with neutrons.

Superlattice Formation in βPd/D. For many years it has been known that there is an anomaly in the specific heat of both βPd/H and βPd/D in the vicinity of 50K. In the absence of any measureable effect in x-ray diffraction, the field was open to speculation until a neutron superlattice reflection of very low intensity was observed at the (1, ½, 0) position in reciprocal space (Fig. 2-3; 8,9). This has been interpreted in terms of an ordering among the D atoms on the octahedral interstitial sites in the Pd lattice.

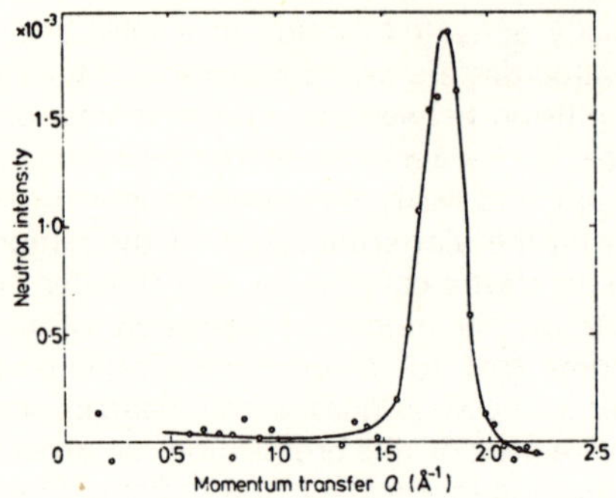

Figure 2-3. Superlattice reflection from βPdD at 50 K indicating ordering of D atoms on octahedral interstitial sites.

Small Angle Scattering from Insect Flight Muscle. Fig. 2-4 is obtained by using the contrast between H_2O and D_2O (84).

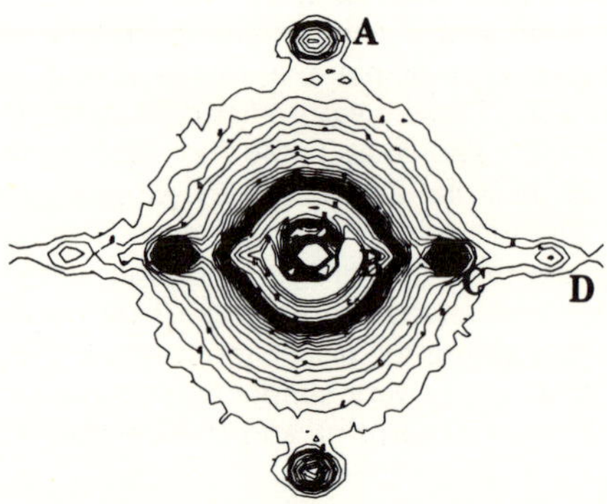

Figure 2-4. Contour plot of small angle neutron diffraction from relaxed insect flight muscle in D_2O.

Quasi-elastic Neutron Scattering for the Determination of the Mechanism of Self Diffusion in Sodium. The diagram (Fig. 2-5) shows the variation of the quasi-elastic broadening with angle of scatter of the neutrons for different crystal orientations. The solid lines represent a model that assumes certain nearest neighbor jumps by monovacancies. The data indicate that the process is somewhat complex (35).

Inelastic Neutron Scattering Measurement of the Vibrational Modes of Adsorbed Hydrogen Atoms on Platinum Black. The spectrum of hydrogen vibrations is shown in Fig. 2-6. See Howard *et al.* (53) for further details.

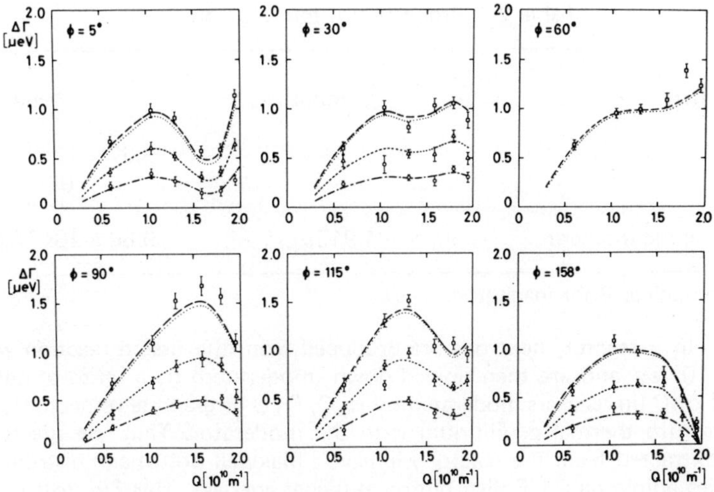

Figure 2-5. Quasi-elastic broadening curves for self diffusion in a single crystal of sodium at three temperatures. (□ = 96° C; △ = 85° C; ○ = 70° C.)

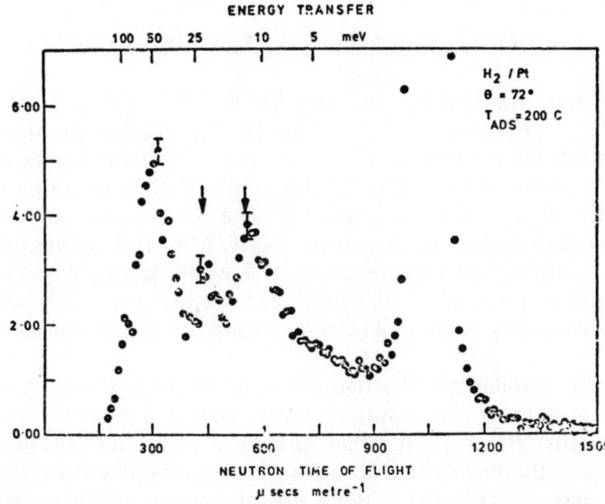

Figure 2-6. Inelastic time-of-flight spectrum for hydrogen adsorbed on Pt black at 200° C.

2-2. ELEMENTARY NEUTRON SCATTERING THEORY

2-2.1. The Neutron-Nucleus Interaction

The Properties of the Neutron. The neutron is a neutral particle of mass almost equal to that of the proton (Table 2.1). It has spin ½ and a magnetic moment of $\mu_n = \gamma\mu_N$ where γ is a constant (−1.93) and μ_N is the nuclear Bohr

Table 2-1. Properties of the Neutron

Quantity	Symbol	Value
Mass	m	1.67×10^{-27} kg
Charge	—	0
Spin	s	½
Magnetic dipole moment	$\mu_n = -1.913\mu_N$*	9.66×10^{-27} JT^{-1}

*μ_N is the nuclear Bohr magneton.

magneton. In a reactor, neutrons are produced from the fission reaction with an energy $\sim 10^6$ eV and are then slowed down (moderated) by a series of collisions with H, D or C (in reactors moderated by H_2O, D_2O or graphite respectively) until they come into thermal equilibrium with the moderator. Thus the spectrum of neutrons extracted from the reactor will have a Maxwell-Boltzmann distribution at low energies adjoining a 1/E distribution at higher energies. This 1/E distribution is characteristic of neutrons slowing down in a moderator. It is convenient to define the neutron spectrum in terms of the neutron flux, ϕ (E), where ϕ (E) = v n(E) where n(E) is the neutron density per unit energy and v is the neutron velocity. In these terms the spectrum becomes

$$\phi(E) = \Phi \ \{ [E/(k_B T)^2] \exp(-E/k_B T) + H(E\text{-}E_L)\, C/E \ \}$$

where $H(E\text{-}E_L)$ is zero for $E < E_L$ and one for $E > E_L$. ϕ is the total thermal neutron flux ($\sim 10^{19}$ neutrons m^{-2} S^{-1} in a HFR), $\Phi C/E$ is the slowing down flux and k_B is Boltzmann's constant. E_L is the lower limit for the slowing down spectra and can be taken to be $\sim 5k_B T$. The value of C is characteristic of the moderator (large for H_2O, small for graphite). The mean neutron energy in the Maxwell-Boltzmann distribution is, of course, $3k_B T/2$ but it is conventional to say that a neutron of energy E has a temperature of T where $E = k_B T$ (meV). For T = 293 K, the corresponding energy is 25.3 meV and the corresponding velocity, 2.2 km/s. This velocity is usually taken as a reference value for thermal neutrons.

The de Broglie wavelength of a neutron is given by $\lambda = h/mv$ Å where h is Planck's constant and m is the neutron mass. Hence a neutron with energy 25.3meV has a wavelength of 1.8 Å which is typical of atomic spacings in solids and the interactions of thermal neutrons with solids must be described by quantum mechanics. In this case, it is normal to write the momentum of the neutron $\underline{p} = \hbar\underline{k}$ where $\hbar = h/2\pi$ and $|\underline{k}| = (2\pi/\lambda)(\text{Å}^{-1})$.

On the other hand, the energy of the neutron is often determined by measuring its time-of-flight over a fixed distance and therefore it is often convenient to define its reciprocal velocity, $\tau\,(\mu s m^{-1})$.

Two other measures of neutron energy may be found in the literature. When comparing energy transfers, ΔE, with infra-red spectra, one often finds units of cm^{-1}, which refers to the inverse of the wavelength of the equivalent electromagnetic quanta of energy (λ_{em}) i.e. $1/\lambda_{em} = \Delta E/hc$. Alternatively, where the energy is transferred to a vibrational mode it is often given in terms of the angular frequency of the mode (in terahertz) i.e. $\Delta E = \hbar\omega$. For convenience in conversion between these units we can write $E = 0.08617T = 5.227v^2 = 81.81/\lambda^2 = (5.227 \times 10^6)/\tau^2 = 0.122/\lambda_{em} = 23.8\ \omega$ using the units as defined above.

In most research reactors the moderator is at ambient temperature and the resulting spectrum is such that useful quantities of neutrons can be obtained over an energy range from 5-100 meV. If the low energy range is of particular interest it can be enhanced by introducing a volume of cold moderator, typically liquid H_2 or D_2 with a temperature $\sim$ 20 K, and this will yield useful flux in the range 0.1-10 meV. If higher energies are required a volume of hot moderator can be used, typically graphite at a temperature of about 3000°C yielding useful neutrons with energies up to $\sim$ 500 meV. The actual spectra from such sources at the ILL are shown in Fig. 2-2.

Neutron Cross Sections. The theoretical treatment of neutron scattering in the following sections is intentionally simplified with emphasis on the physical principles involved rather than on mathematical rigor. In particular, we shall exclude treatment of systems having unpaired electron spins which interact significantly with the magnetic moment of the neutron. As interactions with the nuclear magnetic moments are negligibly small this means that we can safely ignore the neutron's magnetic moment. The probability of direct neutron-nucleus interactions occurring can be measured experimentally and expressed in terms of a cross section/nucleus, usually denoted σ and measured in barns i.e. 10^{-28} m^2. In a medium having n nuclei per unit volume, the probability of a neutron interacting in an element of thickness dx will be $n\sigma dx$. If we denote the number of neutrons in a parallel beam/unit time/unit area incident on a sample to be ϕ_o and the number penetrating to depth x without interaction to be $\phi(x)$, then $N\sigma\phi(x)dx$ will be the number of interactions in the thickness dx which is the reduction in ϕ_o in dx, therefore, $o\phi = -N\sigma\phi(x)dx$ or on integrating $\phi(x) = \phi_o\ e^{-N\sigma x}$.

The product $N\sigma$ is normally written Σ and called the macroscopic cross section. It may be noted that it is equivalent to the linear attenuation coefficient in x-ray terminology.

Various kinds of neutron-nucleus interactions can occur. We need only concern ourselves with $\sigma_a(E)$, $\sigma_f(E)$ and $\sigma_s(E)$—absorption (followed by emission of one or more λ-ray quanta), fission and scattering cross sections, respectively. The first two are straightforward in that they involve the removal of the interacting neutron. The last is more complex in that the neutron has a certain probability of scattering into energy interval dE′ about a final energy E′ and into a solid angle dΩ at an angle of scattering of θ.

This is written $\dfrac{d^2\sigma}{d\Omega dE'}(\theta\ E\rightarrow E')$ such that

$$\frac{d\sigma}{d\Omega}(\theta,E) = \int\frac{d^2\sigma}{d\Omega dE'}(\theta,E\rightarrow E')dE' \text{ and } \sigma_s(E) = \int\frac{d\sigma}{d\Omega}(\theta,E)d\Omega \qquad [2\text{-}1]$$

where $\dfrac{d\sigma}{d\Omega}(\theta,E)$ is the angular scattering cross section. Neutron inelastic scattering involves the measurement of $\dfrac{d^2\sigma}{d\Omega dE'}(\theta E\rightarrow E')$ and neutron diffraction, of $\dfrac{d\sigma}{d\Omega}(\theta,E)$.

Scattering From a Fixed Nucleus. In the previous section we defined cross sections in terms of particles. As the scattering must be described by quantum mechanics, however, it is essential to translate these ideas into concepts of quantum mechanics.

First, we will show that an incident parallel beam of neutrons can be represented by the wave function $\psi = Ce^{i\mathbf{k_o}.\mathbf{r}}$ where $\mathbf{k_o}$ is the wave vector parallel to the neutron propagation direction. This is a solution of the Schrödinger Equation for $V(\mathbf{r}) = 0$ as can be seen by substitution:

$$\frac{\hbar^2}{2m}\nabla^2\psi + (E - V(r))\psi = 0 \qquad [2\text{-}2]$$

that is

$$\left(-\frac{\hbar^2}{2m}.\ k_o^2 + E\right)\psi = 0$$

This means that $E = \dfrac{\hbar^2 k_o{}^2}{2m}$ is an eigenvalue. [2-3]

It is also a plane wave because it will have a constant phase at all points such that $\mathbf{k_o}\ .\ \mathbf{r}$ is constant. Further, planes separated by a distance λ will have a phase difference of $k_o\lambda = 2\pi$. The neutron density in the incoming beam is $\psi\psi^* = C^2$, so that the incident neutron flux is

$$\phi = C^2 v = \frac{C^2\hbar k_o}{m} \qquad [2\text{-}4]$$

Consider the scattering from a fixed point nucleus at r = 0. In these circumstances the scattering must be elastic in the frame of reference of the fixed nucleus and the scattered wave can in general be written

$$\psi'(\mathbf{r}) = C'\frac{f(\theta)}{r}e^{ik'r} \qquad [2\text{-}5]$$

where k′ is taken parallel to **r** and θ is the angle between k′ and k_o. The wave patterns for incident and scattered waves are shown in Fig. 2-7. This general expression can be expanded in terms of partial waves, each term in the expansion corresponding to a particular value of the angular quantum number (ℓ) which quantizes the angular momentum of the neutron relative to the nucleus. However, where the wavelength of the incident neutron ($\sim 10^{-10}$ m) is much greater than the diameter of the scattering potential ($\sim 10^{-14}$ m) only the zeroth ($\ell = 0$) term can exist. As this 's-wave' scattering is entirely isotropic we can take $f(\theta) = 1$.

We will now confirm that this form of $\psi'(\mathbf{r})$ satisfies Schrödinger's equation using the spherically symmetric version of ∇^2 i.e.

$$\frac{1}{r^2}\frac{d}{dr}\left(r^2\frac{d\psi}{dr}\right) + \frac{2m}{\hbar^2}(E - V(r))\,\psi'(r) = 0 \qquad [2\text{-}6]$$

where for r greater than the nuclear diameter we can, as before, take V(r) = 0. On substitution we now have

$$\frac{1}{r^2}\left(-ik'e^{ik'r} + r(ik')^2 e^{ik'r} + ik'e^{ik'r}\right) + \frac{2m}{\hbar^2}E\frac{1}{r}e^{ik'r} = 0 \qquad [2\text{-}7]$$

so that $E' = \frac{\hbar^2 k'^2}{2m}$ is the eigenvalue as before. [2-8]

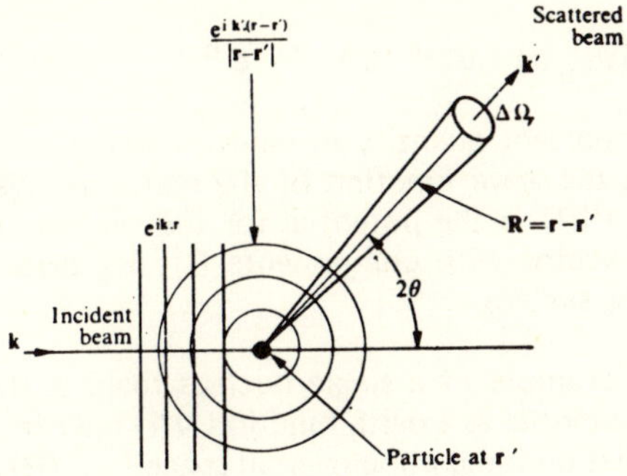

Figure 2-7. Illustration of incident plane waves and scattered spherical waves for a fixed scattering nucleus.

The scattered neutron probability density in solid angle $d\Omega$ between r and r + dr is $r^2\ \psi'\psi'^*\ dr d\Omega$ so that the number of neutrons scattered into $d\Omega$/unit time is $vr^2\ (C'^2/r^2)d\Omega$.

Remembering that the incident flux is vC^2 we have, by definition

$$\frac{d\sigma}{d\Omega} = \frac{vC'^2\, d\Omega}{vC^2\ d\Omega} = \frac{C'^2}{C^2} = b^2 \qquad [2\text{-}9]$$

where b is called the nuclear scattering amplitude or scattering length of the nucleus. The corresponding value of σ_s is $4\pi b^2$.

This result can also be derived from the general quantum mechanical analysis of neutron scattering. The detail of this approach can be found elsewhere (98, 104) but some of the important equations will be given here to illustrate the general outline of the arguments involved. Using standard scattering theory and the first Born Approximation i.e. the assumption that the incident wave function is not significantly changed by the presence of the scattering center (only strictly applicable outside the nuclear potential V(r)) one obtains the result

$$\left(\frac{d^2\sigma}{d\Omega dE'}\right)_{\nu \to \nu'} = \frac{k'}{k_o}\left(\frac{m}{2\pi\hbar^2}\right)^2 \langle k'\nu'|V\ |k_o\nu\rangle^2\ \delta\ (E_\nu - E_\nu' + E_o - E'). \qquad [2\text{-}10]$$

Here ν and ν' refer to the initial and final eigenstates of the scattering system having energies E_ν and E_ν', respectively and the δ function therefore ensures energy conservation. The term $(\mathbf{k}'\nu'|\ \mathbf{V}\ |\mathbf{k_o}\nu)$, known as the matrix element for scattering from direction $\mathbf{k_o}$ to $\mathbf{k}'$, may be written in full

$$(\mathbf{k}'\nu'|\mathbf{V}|\mathbf{k_o}\nu) = \int\int \psi^*_{k'}(\mathbf{r})\chi^*_{\nu'}(\mathbf{R})\ \mathbf{V}(\mathbf{R})\psi_{k_o}(\mathbf{r})\chi_\nu(\mathbf{R})d\mathbf{R}d\mathbf{r} \qquad [2\text{-}11]$$

The neutron wave functions are written for incident and scattered plane waves each normalized to unit neutron density so that

$$\psi^*_{k'}(\mathbf{r})\psi_{k_o}(\mathbf{r}) = e^{-ik'.r}e^{ik_o.r} = e^{iQ.r} \qquad [2\text{-}12]$$

where $\mathbf{Q} = \mathbf{k}_o - \mathbf{k}'$, a most important quantity in neutron scattering known as the wave vector transfer, $\chi_\nu(\mathbf{R})$ is the wave function of the scattering system normalized to unity over all space, $V(\mathbf{R})$ is the potential for the neutron-nucleus interaction and $\mathbf{R}$ is a composite vector with components $\mathbf{R}_1$, $\mathbf{R}_2$ describing the coordinates of all the nuclei in the system.

Now, returning to our example of a single nucleus fixed at the origin (i.e. $V(\mathbf{R}) = V(\mathbf{r})$) let us write the potential as a delta function $V(\mathbf{r}) = a\,\delta(\mathbf{r})$. Because the scattering is elastic $\nu = \nu'$ so that on integrating over all space $\int \chi^*_\nu(\mathbf{R})\chi_\nu(\mathbf{R})\, d\mathbf{R} = 1$ and have, after integrating over E′ (equation [2-1])

$$\frac{d\sigma}{d\Omega} = \left(\frac{m}{2\pi\hbar^2}\right)^2 a^2 \left[\int \exp(i\mathbf{Q}.\mathbf{r})\,\delta(\mathbf{r})\, d\mathbf{r}\right]^2 \qquad [2\text{-}13]$$

But $\int \exp(i\mathbf{Q}.\mathbf{r})\,\delta(\mathbf{r})d\mathbf{r} = 1$, therefore

$$\frac{d\sigma}{d\Omega} = \left(\frac{m}{2\pi\hbar^2}\right)^2 a^2 \qquad [2\text{-}14]$$

Thus, by comparison with the definition of b, the (nuclear) scattering length (equation [2-9]), we have

$$V(\mathbf{r}) = \frac{2\pi\hbar^2}{m} b\,\delta(\mathbf{r}). \qquad [2\text{-}15]$$

This potential is known as the Fermi pseudopotential. It is defined so as to give the correct scattered wave function outside the nucleus. Since we are not interested in what happened inside the nucleus it is satisfactory for our purposes. It may be noted that although it has been defined in a situation where $\nu = \nu'$ it is equally valid for $\nu \neq \nu'$. Also the sign of b has been arbitrarily chosen so that it is positive for the majority of isotopes and this in fact corresponds to the case of a repulsive hard sphere potential. In a number of nuclei, however, especially where the compound nucleus formed by the neutron and the target nucleus is near to a resonance level, the scattering length turns out to be negative.

The mass used in the expression for the Fermi pseudopotential is strictly speaking the reduced mass of the neutron/nucleus system i.e. $\mu = mM(m + M)$.

Thus for a fixed nucleus as described here (i.e. $M \to \infty$), $\mu = m$. However, for a free nucleus, we can define a new scattering length f (free nucleus), such that given $V(\mathbf{r})$ is the same for both cases, equation [2-15] yields

$$(b/m) = (f/\mu) \text{ i.e. } f = (m/m + M)b. \qquad [2\text{-}16]$$

Scattering From a Set of Fixed Nuclei. Consider a set of N nuclei fixed at positions $\mathbf{R}_i$ each having nuclear scattering lengths b_i. The potential becomes

$$V(\mathbf{r}) = \frac{2\pi\hbar^2}{m} \sum_i b_i\, \delta(\mathbf{r}-\mathbf{R}_i) \qquad [2\text{-}17]$$

Now equation [2-13] can be written generally

$$\frac{d\sigma}{d\Omega} = \frac{m}{2\pi\hbar^2}\, [\int \exp(i\,\mathbf{Q}.\mathbf{r})\, V(\mathbf{r})d\mathbf{r}]^2. \qquad [2\text{-}18]$$

When defined per nucleus this case becomes

$$\frac{d\sigma}{d\Omega} = \frac{1}{N} \,|\sum_i b_i \exp(i\mathbf{Q}.\mathbf{r}_i)|^2 \qquad [2\text{-}19]$$

It may be noted that this expression is also obtained by adding scattered wave functions as in equation [2-5]. On expansion, it becomes

$$\frac{d\sigma}{d\Omega} = \frac{1}{N} \sum_{ij} b_i b_j \exp(i\,\mathbf{Q}.(\mathbf{r}_i-\mathbf{r}_j)) \qquad [2\text{-}20]$$

Here the diagonal terms have no phase factor and can be written $\frac{1}{N}\sum_i b_i^2 = < b^2 >$ where the brackets indicate the average value. Therefore

$$\frac{d\sigma}{d\Omega} = < b^2 > + \frac{1}{N} \sum_{i,j\neq i} b_i b_j \exp(i\mathbf{Q}.(\mathbf{r}_i-\mathbf{r}_j)) \qquad [2\text{-}21]$$

There is no correlation between the value of b_i, b_j or the corresponding phase angle, $\exp(i\,\mathbf{Q}.(\mathbf{r}_i-\mathbf{r}_j))$, so each term can be averaged separately in a process known as the Random Phase Approximation. The reason for this result is seen, if one selects from the large population of N nuclei a subset which have exactly the same value of b_j and $(\mathbf{r}_i-\mathbf{r}_j)$. Because of the lack of any correlation, the values of b_i in this subset will be typical of the whole population and can therefore be replaced in the summation by $< b >$. Doing this for all such subsets and summing over all subsets we have

$$\sum_{i,j\neq i} b_i\, b_j \exp(i\,\mathbf{Q}.(\mathbf{r}_i - \mathbf{r}_j)) = <b> \sum_{i,j\neq i} b_j \exp(i\,\mathbf{Q}.(\mathbf{r}_i - \mathbf{r}_j)). \qquad [2\text{-}22]$$

Repeating the argument we have finally

$$\frac{d\sigma}{d\Omega} = < b^2 > + \frac{<b^2>}{N} \sum_{i,j\neq i} \exp(i\,\mathbf{Q}.(\mathbf{r}_i-\mathbf{r}_j)) \qquad [2\text{-}23]$$

and on replacing the diagonal terms

$$\frac{d\sigma}{d\Omega} = \underbrace{< b^2 > - <b>^2}_{\text{incoherent}} + \underbrace{\frac{<b>^2}{N} \sum_i \sum_j \exp[i\,\mathbf{Q}.(\mathbf{r}_i-\mathbf{r}_j)]}_{\text{coherent}}. \qquad [2\text{-}24]$$

The first part of this expression, $<b^2> - <b>^2$ describes an isotropic component known as the incoherent part of the scattering. The corresponding cross section $\sigma^{inc} = 4\pi(<b^2> - <b>^2)$. The second part of the expression contains all the interference effects and is thus known as coherent scattering. The corresponding cross section, σ^{coh},is defined as $4\pi\langle b\rangle^2$. It should be emphasized that the incoherence arises not from any loss of the phase relationship in the scattering process but from fluctuations (both positive and negative) in the amplitude of the scattered wave from each nucleus.

The reasons for the fluctuations in the scattering amplitude from nucleus to nucleus are spin effects and isotope effects. The first arises because the form of the neutron-nucleus interaction for a given isotope, k, depends, for non zero spin nuclei, on the relative directions of the spins. Thus if the nuclear spin is I, the total spin of the compound nucleus will be $J = I + ½$. Quantum statistics dictate that the probability of getting each value of J depends on the number of quantum states associated with that value. Let p^+ and p^- be the probability of having parallel and antiparallel interactions and b^+ and b^- be the corresponding scattering lengths, then

$$p^+ = \frac{2(I + ½) + 1}{2(I + ½) + 1 + 2(I - ½) + 1} = \frac{I + 1}{2I + 1}$$

$$p^- = \frac{I}{2I + 1} \qquad [2\text{-}25]$$

and the mean scattering length for the kth isotope is $\langle b_k\rangle = p_k^+ b_k^+ + p_k^- b_k^-$. We can write the overall mean as $\langle b\rangle = \sum_k \omega_k (p_k^+ b_k^+ + p_k^- b_k^-)$ where ω_k is the concentration of the kth isotope (i.e. $\sum_k \omega_k = 1$). The corresponding value of $\langle b^2\rangle$ is

$$\langle b^2\rangle = \sum_k \omega_k (p_k^+ b_k^{+2} + p_k^- b_k^{-2}). \qquad [2\text{-}26]$$

Values of Scattering Cross Sections. Values of σ^{inc} and σ^{coh} are known for all elements and many individual isotopes (10, 11, 12, 59, 86). It should be noted that these values are not directly related to the total cross sections measured by low energy neutrons. The bound cross section, $\sigma_b = 4\pi\langle b^2\rangle = \sigma^{inc} + \sigma^{coh}$, is usually obtained from the high energy free atom cross section, σ_f, using equation [2-16] where $\sigma_f = 4\pi f^2$ and σ^{coh} is obtained from measurements of $\langle b\rangle$, for instance by analysis of the intensity of different reflections from crystals with well known structures. In general measured values vary in a fairly random fashion from isotope to isotope, behavior that is in strong contrast to x-ray scattering powers which rise strongly with the number of electrons/atom, Z, being identical for different isotopes of the same element. Two particular cases will be mentioned, namely hydrogen and vanadium. The neutron-proton interaction is unusual because of the fact that the anti-parallel orientation forms a compound nucleus that is close to a virtual state of the deuteron with zero spin making the scattering amplitude b^- unusually strong and negative. At the same time b^+ is slightly less than one-third as large and positive. Hence $\langle b\rangle = 3/4b^+ + 1/4b^-$ is small and negative, and σ^{coh} is only about 2% of σ^{inc}. This means that in structural work, it is often advantageous to replace H with D, for which σ^{coh} predominates, to avoid the strong incoherent (isotropic) scattering from H. There are cases, however, where the negative scattering length of hydrogen is an advantage.

The other notable case of incoherence is vanadium which is only 7% coherent, also due to spin effects ($V^{51} > 99\%$ abundant). Being much heavier than hydrogen, thus minimizing recoil effects, and being easily available in a pure form, vanadium is normally used as a standard isotropic scatterer for intercalibration of neutron detectors. Scattering lengths and cross sections of interest in clay work are given in table 2-2.

Table 2-2

Element or Isotope	Spin	Scattering Lengths $\times 10^{14}$ m			Cross Sections $\times 10^{28}$ m²			
		b^+	b^-	<b>	σ^{tot}	σ^{coh}	σ^{inc}	σ^{abs}
$_1H^1$	½	1.085	4.75	−0.374	81.5	1.76	79.7	0.19
$_1H^2$	1	0.953	0.098	0.67	7.6	5.6	2.0	0.0003
$_6C^{12}$	0			0.6626	5.517	5.517	0	0.003
$_7N^{14}$	1			0.940	11.4	11.1	0.3	1.1
$_8O^{16}$	0			0.577	4.24	4.24	0	0.0001
$_{12}Mg$	0			0.516	3.70	3.34	0.36	0.04
$_{13}Al^{21}$	5/2			0.345	1.5	1.49	0.01	0.13
$_{14}Si$	0			0.4151	2.2	2.165	0.03	0.06
$_{23}V^{51}$	7/2			−0.051	5.1	0.033	4.8	2.8
$_{26}Fe$	0			0.951	11.8	11.36	0.44	1.4

*Quoted at 1.08Å – and normally proportional to neutron wavelength.

2-2.2. Neutron Diffraction

Diffraction Intensities and Structure Factors. The measurement of coherent diffraction of neutrons by a crystalline material involves the cross section $(d\sigma/d\Omega)^{coh}$, which is proportional to the number of neutrons coherently scattered into unit solid angle in real space. This quantity is given by

$$(d\sigma/d\Omega)^{coh} = |\, \Sigma <b_i> \exp(i\mathbf{Q}.\mathbf{R}_i)\,|^2 \qquad [2\text{-}27]$$

where $<b_i>$ is the average scattering amplitude of an atom at position vector $\mathbf{R}_i$, the summation being carried out over the entire crystal. Let us decompose each vector $\mathbf{R}_i$ into two components $\mathbf{R}_i = \mathbf{n} + \underline{\rho}$ where $\mathbf{n}$ is the lattice vector of any given unit cell and ρ is the position within the unit cell. Remembering that the value of the scattering amplitudes $<b_i>$ depends only on the nature of the atom but not on which unit cell the atom occupies, then one may separate equation [2-27] into two terms

$$(d\sigma/d\Omega)^{coh} = |\sum_n \exp\,(i\mathbf{Q}.\mathbf{n})\,|^2 \times |\sum_\rho <b_\rho>\exp(i\mathbf{Q}.\underline{\rho})\,|^2 \qquad [2\text{-}28]$$

Here n refers to the summation over all unit cells and ρ to the summation over all atoms within the unit cell.

The derivation of most useful quantities for diffraction experiments is done in terms of reciprocal space, since the wave-vector transfer, $\mathbf{Q}$, has dimensions of reciprocal length, i.e. is proportional to $1/\lambda$ where λ is the neutron wavelength. This concept is analogous to a spatial Fourier transformation, as will be shown.

If the basis vectors of the unit cell be a_1, a_2 and a_3 the points of the lattice are defined by

$$\mathbf{n} = n_1 \mathbf{a}_1 + n_2 \mathbf{a}_2 + n_3 \mathbf{a}_3 \qquad [2\text{-}29]$$

where n_1, n_2 and n_3 are integers. The reciprocal unit cell may be defined in a similar way to have basis vectors b_1, b_2 and b_3, so that its points are

$$\underline{\tau} = m_1 \mathbf{b}_1 + m_2 \mathbf{b}_2 + m_3 \mathbf{b}_3 \qquad [2\text{-}30]$$

where m_1, m_2 and m_3 are integers. The reciprocal lattice vectors are perpendicular to planes containing the direct lattice vectors, and may be defined as

$$\mathbf{b}_1 = \frac{2\pi\, \mathbf{a}_2 \times \mathbf{a}_3}{V},\quad \mathbf{b}_2 = \frac{2\pi\, \mathbf{a}_3 \times \mathbf{a}_1}{V},\quad \mathbf{b}_3 = \frac{2\pi\, \mathbf{a}_1 \times \mathbf{a}_2}{V} \qquad [2\text{-}31]$$

where $V = a_1 . (\mathbf{a}_2 \times a_3)$ is the volume of the unit cell. Now $b_i.a_j = 2\pi\delta_{ij}$ where δ_{ij} is the Kronecker delta, i.e.

$$\begin{aligned} \mathbf{b}_i.\mathbf{a}_j &= 2\pi \text{ if } i = j \\ &= 0 \text{ if } i \neq j \end{aligned} \qquad [2\text{-}32]$$

Thus $\mathbf{n}\underline{\tau} = 2\pi(n_1 m_1 + n_2 m_2 + n_3 m_3) = 2\pi N$ where N is an integer.

It can be shown that the first term in equation [2-28] has sharp maxima at the points $\underline{\tau}$, i.e. where the wave vector transfer $\mathbf{Q}$ corresponds to a reciprocal lattice vector $\underline{\tau}$. Integrating over the region of space surrounding each lattice point, it can be shown that the magnitude of this term is $8\pi^3 N_o/V$ where N_o is the total number of unit cells in the crystal (110). Though the magnitude of the first term in equation [2-28] is $4\pi^2 N_o^2$ for $\mathbf{Q} = \underline{\tau}$, it can be shown that the peak width is proportional to $N_o^{-1/3}$, so that the integrated intensity of each reflection is proportional to the number of unit cells, as might be expected. The coherent cross-section therefore becomes

$$(d\sigma/d\Omega)^{coh} = \frac{8\pi^3 N_o}{V} \sum_{\underline{\tau}} |F_{\underline{\tau}}|^2\, \delta(\mathbf{Q} - \underline{\tau}) \qquad [2\text{-}33]$$

where the Dirac delta function expresses the condition that the wave vector transfer is equal to the reciprocal lattice vector. The expression

$$F_{\underline{\tau}} = \sum_{\underline{\rho}} \langle b_{\underline{\rho}} \rangle \exp(i\, \underline{\tau}.\underline{\rho}) \qquad [2\text{-}34]$$

is known as the structure factor of the unit cell for the reciprocal lattice vector $\underline{\tau}$.

Crystallographers, as opposed to solid state physicists, employ a somewhat different notation in which the factor 2π is omitted from the definition of the reciprocal lattice vectors. In this notation, equation [2-34] becomes

$$F_{hkl} = \sum_{\rho} \langle b_\rho \rangle \exp\left[2\pi i\left(h\bar{x}_\rho + k\bar{y}_\rho + l\bar{z}_\rho\right)\right] \qquad [2\text{-}35]$$

where $\bar{x}_\rho$, $\bar{y}_\rho$ and $\bar{z}_\rho$ are the fractional atomic coordinates, i.e. $\bar{x}_\rho = \frac{x_\rho}{a}$, $\bar{y}_\rho = \frac{y_\rho}{b}$ and $\bar{z}_\rho = \frac{z_\rho}{c}$ measured along $\mathbf{a}_1$, $\mathbf{a}_2$ and $\mathbf{a}_3$ respectively.

Here m_1, m_2 and m_3 and $|\mathbf{a}_1|$, $|\mathbf{a}_2|$ and $|\mathbf{a}_3|$ in equations [2-29] and [2-30] are replaced by the more familiar h, k and l (Miller Indices) and a, b and c respectively (unit cell dimensions). As was mentioned previously, the wave vector transfer is $\mathbf{Q} = \mathbf{k}_0 - \mathbf{k}'$. For elastic scattering, where there is no change of energy (and hence wavelength) as the magnitudes of the incident and scattered wave vectors are identical, i.e. $|\mathbf{k}'| = |\mathbf{k}_0|$. From Fig. 2-7 and elementary trigonometry it can be shown that the magnitude of Q is given by $Q = 2\,k_0 \sin\theta_B = 4\pi\sin\theta_B/\lambda$ where θ_B is the Bragg angle. Now for adjacent hkl planes separated by a distance d, one has $|\underline{\tau}| = 2\pi/d$, and since $|\mathbf{Q}| = |\underline{\tau}|$ for elastic scattering this implies that $1/d = 2\sin\theta/\lambda$ satisfying Bragg's Law (104).

With neutron diffraction, as for x-ray diffraction, the size and shape of the unit cell is derived from the positions of the observed reflections, while unit cell structure determination involves measurement of the intensities of as many reflections as possible to obtain the structure factors F_{hkl}. The latter may be Fourier transformed (i.e. converted back from reciprocal to real space) giving a series expansion of the nuclear scattering density distribution in the crystal. We give here the outline of the theory for the simplest case only, i.e. a one dimensional Fourier projection for a centrosymmetric crystal. This has direct relevance to studies on clay intercalates, as described in Section 2-4. For a more complete discussion of neutron diffraction see Bacon (12), Willis (109) and Turchin (104).

One-Dimensional Structure Determination. Let us consider a measurement involving one series of reflections only, e.g. the basal or ($00l$) reflections, as is frequently true in the case of clay minerals. In this case equation [2-29] reduces to

$$F_{00l} = \sum_{\rho} \langle b_\rho \rangle \cos\left(\frac{2\pi l z_\rho}{d}\right) \qquad [2\text{-}36]$$

The integrated intensities of the reflections are proportional to the squares of the structure factors. However it is necessary to correct the intensities observed experimentally for three effects: (a) the finite detector solid angle, i.e. the proportion of the intensity which the detector actually receives, which depends on the diffraction angle, detector geometry and the mosaic spread of the crystal; (b) attenuation of the beam in the sample; and (c) the effect of the vibrational modes of the atoms, which reduces the overall observed intensity.

The first effect is known as the Lorentz Factor, denoted here by L(Q). For single crystals this is given by $1/\sin 2\theta$ and for a random powder by $\cos\theta/\sin^2(2\theta)$. For the intermediate case of preferred orientation, common in clay specimens, the expression for L(Q) is more complex as discussed by Reynolds (80) and Hawkins (48). The attenuation factors, A(Q), depend on the sample thickness and geometry. The general problem is to calculate the fractional attenuation $d\ell = -\Sigma d\ell i$ occurring within any length element $d\ell$ and to integrate this over the whole crystal. For slab samples equations have been published by Carlile (17). A more complete treatment of attenuation effects, involving multiple scattering corrections is given by Sears (91), though the formal expansion in terms of the number of scatters, the first term of which corresponds to the basic attenuation formula, was first derived by Vineyard (106). The final correction, D(Q), is the Debye-Waller or temperature factor discussed more fully in section 2-2.4 in connection with incoherent elastic and quasi-elastic scattering. For an atom whose vibrational amplitude in the z-direction is u_z, this takes the form

$$D(Q) = \exp(-Q^2 \langle u_z{}^2 \rangle) \qquad [2\text{-}37]$$

where < > denotes the root mean square vibrational amplitude. We have tacitly assumed here that all atoms have identical Debye-Waller factors, which can be sufficiently accurate if the quantity of data is limited. Full crystallographic studies, however, strictly require independent Debye-Waller factors for each atom to be fitted during data refinement.

It should be noted here that the polarization factor, which depends on the relationship between the direction of the polarization of the x-ray beam and the angle of diffraction, does not apply in the case of neutron diffraction. The observed intensities are therefore given by

$$I^{obs}_{00l} = \frac{8\pi^3 N_o}{V} \mid F_{00l} \mid^2 A(Q)\, D(Q)\, L(Q). \qquad [2\text{-}38]$$

The determination of atomic positions thus requires applying the above corrections to the observed intensities to obtain experimental values of F_{00l} which may be directly compared with theoretical structure factors calculated on the basis of a hypothetical model structure. Alternatively, since we know that the nuclear scattering amplitude function $\rho(z)$ is centrosymmetric and of periodicity d, then it may be written as the Fourier series

$$\rho(z) = \sum_{l=\infty}^{\infty} \phi(l)\, A(l) \cos\left(\frac{2\pi l z}{d}\right) \qquad [2\text{-}39]$$

where the $\phi(l)$ are phase factors (+1 for the centrosymmetric case) and the $A(l)$ are the amplitudes of each term. From the theory of the Fourier transform, and recalling that cosine is an even function, it can be shown that

$$\rho(z) = \frac{2}{d} \sum_{l=1}^{\infty} \phi(l)\, F_{00l} \cos(2\pi l z/d) \qquad [2\text{-}40]$$

neglecting the forward scattering. Two remaining problems are to determine the phases $\phi(l)$ and to allow for the fact that only a finite number of terms can be obtained. The former can be overcome if the majority of the atomic co-ordinates are already known from x-ray data. The latter problem causes spurious 'wiggles' in the final curve for $\rho(z)$ that can be partly eliminated by the use of an appropriate smoothing function (65).

2-2.3. Small Angle Neutron Scattering

This aspect of neutron scattering is rapidly developing into an important technique for studies on colloids, macromolecules, and porous materials. It is concerned with coherent scattering of neutrons at very low angles (sometimes only a few minutes of arc). It differs from Bragg diffraction in that the interference terms producing the observed scattering arise in this case not from regular crystal planes, but from inhomogeneities such as particle edges (or conversely the boundaries of pores). The **Q**-dependence of the small-angle scattering yields information concerning the size distribution of these inhomogeneities or zones in a range from 10 - 5000 Å.

The basic theory of small angle neutron scattering (SANS) corresponds directly to the case of small angle x-ray scattering which has been treated in detail in the book by Guinier and Fournet (38). A brief summary of the important results is given here, though our notation differs somewhat for reasons of consistency. A useful review of the SANS technique, with particular reference to studies of biological molecules, has been published by Jacrot (58).

SANS results from fluctuations of the scattering amplitude density $\rho(R)$ where $\rho(R)dR = \sum_{dR} \langle b_i \rangle$. As was shown in the previous section, the structure factor F(Q) is the spatial Fourier transform of $\rho(R)$, and the intensity of scattering, denoted here as I (Q), is proportional to the square of the modulus of F(Q). We may therefore write

$$I(\mathbf{Q}) = \langle | F(\mathbf{Q}) |^2 \rangle = \langle | \int \rho(\mathbf{R}) e^{i\mathbf{Q}.\mathbf{R}} d\mathbf{R} |^2 \rangle \qquad [2\text{-}41]$$

where the integral is performed over the entire volume of the scattering system, and the brackets denote an ensemble average.

In general $\rho(\mathbf{R})$ will be constant only in a perfectly homogeneous medium. As stated previously, and in common with Bragg diffraction, the SANS intensity is determined by the extent of the fluctuations in $\rho(\mathbf{R})$ from point to point. An important concept, which we will return to later is the contrast between the average scattering amplitude density in the zones of interest (e.g. particles or voids) and that of the surrounding medium.

Interpretation of SANS data depends on being able to make approximations to the integral in equation [2-41]. The appropriate form of this approximation depends on the shape and relative orientation of the scattering zones.

Consider an assembly of widely separated centrosymmetric particles in a vacuum. In this case the scattering can only arise from the particles themselves. Since they are centrosymmetric, the phase relationships between the scattered waves are such that the exponential terms in equation [2-41] are purely real and equal to cos(**Q**.**R**). If all particle orientations are equally probable the spatial average of this term can be evaluated analytically and is equal to sin(QR)/QR. This may be expanded as a series, the first three terms of which are

$$\frac{\sin(QR)}{QR} = 1 - \frac{(QR)^2}{6} + \frac{(QR)^4}{120} + \ldots\ldots \qquad [2\text{-}42]$$

The structure factor can then be written as

$$\langle F(Q)\rangle = \int_0^\infty \rho(R)\frac{\sin(QR)}{QR}4\pi R^2 dR \qquad [2\text{-}43]$$

Let us now define two important quantities, firstly a 'neutronic' radius of gyration, R_g, where

$$R_g^2 = \frac{\int \rho(\mathbf{R})\, R^2\, d\mathbf{R}}{\int \rho(\mathbf{R})\, d\mathbf{R}} \qquad [2\text{-}44]$$

which is directly analogous to the expression for the radius of gyration in classical mechanics. Secondly, we may define a mean scattering length density, ρ_m, given by

$$\rho_m = (1/V) \int \rho(\mathbf{R})\, d\mathbf{R} \qquad [2\text{-}45]$$

where V is the volume of the sample.

If Q = 0 in equation [2-43] and equation [2-45] is integrated, the result is $F(0) = \rho_m V$ so that the intensity of scattering at Q = 0 is given by

$$I(0) = \rho_m{}^2 V^2. \qquad [2\text{-}46]$$

Utilizing equations [2-44] to [2-46], and expanding $|F(Q)|^2$ as a power series it can be shown that a good approximation to equation [2-41] is

$$I(Q) = I(0) \exp(-Q^2 R_g{}^2/3) \qquad [2\text{-}47]$$

which is known as the Guinier Approximation (38). This approximation is valid for Q values which do not greatly exceed the reciprocal mean particle dimensions.

Up to this point we have considered particles in a vacuum, i.e. a situation where there is a large contrast between the scattering amplitude density of the particles and that of the surrounding medium. In cases where one has a two-phase system, such as a colloidal suspension, the important quantity becomes $\overline{\rho}$, defined as $\overline{\rho} = \rho_m - \rho_s$ where ρ_s is the mean scattering length density of the solvent. In this case equation [2-46] yields $I(0) = (\overline{\rho}V)^2$, implying that the intensity at Q = 0 vanishes when $\rho_m = \rho_s$. An important experimental technique is that of contrast variation, which depends upon the fact that hydrogen and deuterium have scattering lengths of opposite sign (see section 2-2.1). Thus the value of ρ_s for an H_2O/D_2O mixture can readily be calculated and depends markedly on composition.

Variations in the Q-dependence of the SANS intensity with relative contrast can yield information regarding the microstructure of the colloidal system. Thus for homogeneous particles the value of R_g is equivalent to the mechanical radius of gyrations, whereas for non-homogeneous particles (e.g. clay aggregates in soils with entrained water) the observed value of both R_g and the contrast match point ($\rho_m = \rho_s$) will depend on the fluctuations of $\rho(R)$ within the inhomogeneous particles due to variations in composition (22).

For finite Q the most appropriate form of the approximation to equation [2-41] depends on the shape of the individual particles. Various forms of equation [2-47] for anisotropic particle shapes are given by Guinier and Fournet (38). Most importantly, for the case of thin disc-shaped particles of height R and thickness H, where R>>H, it can be shown (60) that

$$I(Q) \propto \frac{1}{Q^2 R^2} \exp(-Q^2 H^2/12) \qquad [2\text{-}48]$$

This equation implies that a plot of log Q^2 I(Q) against Q^2 will be linear in this approximation, and will enable the mean value of H to be estimated.

At higher Q values it can be shown that $I(Q) \propto Q^{-4}$ which is known as Porod's Law, and is valid where Q is significantly greater than 1/R (or 1/H in the case of thin discs).

The studies of montmorillonite sols by SANS are discussed in section 2-4.3 (20, 22).

2-2.4. Quasi-elastic Neutron Scattering

The Van Hove Correlation Functions. The analysis discussed in the earlier sections was concerned with rigidly bound nuclei. The thermal motions of bound atoms, which must be present, were only considered in so far as they affect the elastic scattering through the Debye-Waller factor. In this section, we discuss the effect on the elastic part of the scattering of the atoms being allowed to diffuse away from their permanent site. Inelastic scattering, which is due to the exchange of quanta with the vibrations of a bound system, will be discussed in the next section.

The treatment of scattering from diffusive motions is based on the Van Hove space-time correlation functions. The derivation of the relationship between these functions and the scattering cross sections is beyond the scope of the present treatment and the reader is referred to the standard books (98, 104, 109). This section discusses why the results obtained are physically reasonable.

First let us consider a random arrangement of identical rigidly bound atoms as in an amorphous solid. The coherent angular elastic scattering cross section as given in equation [2-24] may be rewritten,

$$\left(\frac{d\sigma}{d\Omega}\right)^{coh} = \langle b\rangle^2 \int e^{i\mathbf{Q}.\mathbf{r}} \frac{1}{N} \sum_{i,j} \delta(\mathbf{r}-\mathbf{R}_i+\mathbf{R}_j)d\mathbf{r} \qquad [2\text{-}49]$$

and we can define a new function g(r), the static pair distribution function, given by

$$g(\mathbf{r}) = \frac{1}{N}\sum_{ij} \delta(\mathbf{r} - \mathbf{R}_i + \mathbf{R}_j) - \delta(\mathbf{r}) \qquad [2\text{-}50]$$

$$= \frac{1}{N} \sum_{i,j\neq i} \delta(\mathbf{r} - \mathbf{R}_i + \mathbf{R}_j).$$

This function, which for periodic structures is known as the Patterson Function, describes the probability distribution of atoms about any atom taken as origin and averaged by taking each atom as origin in turn. A typical shape for this function is shown in Fig. 2-8a. It is zero going out from the origin for a distance equal to twice the hard sphere radius of the atom followed by a more or less well defined peak at the nearest neighbor distance and then by further peaks of decreasing amplitude tending asymptotically to the number density of the system, n = N/V where V is the volume containing the N atoms.

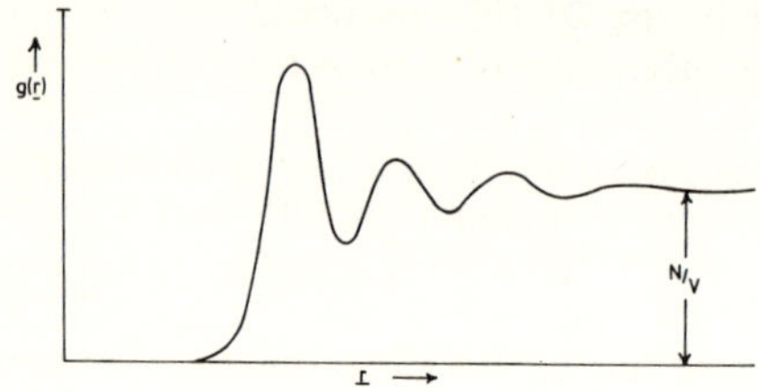

Figure 2-8. (a) Typical static pair distribution function for a liquid.

We can now define a function S(Q), known as the structure factor by

$$S(\mathbf{Q}) = \int \exp(i\,\mathbf{Q}.\mathbf{r})\,(g(\mathbf{r}) + \delta(\mathbf{r}))\,d\mathbf{r} = 1 + \int \exp(i\,\mathbf{Q}.\mathbf{r})\,g(\mathbf{r})\,d\mathbf{r} \qquad [2\text{-}51]$$

such that the coherent angular cross section is

$$\left(\frac{d\sigma}{d\Omega}\right)^{coh} = \langle b\rangle^2\, S(\mathbf{Q}). \qquad [2\text{-}52]$$

It will be noted that S(Q) is a unique function for a given scattering system, whereas the shape of $(d\sigma/d\Omega)$ will be scaled by the wave vector of the neutrons in use. The meaning of this function in the presence of thermal vibrations will follow. The main point to be made here is that the variables **r** and **Q** are related through the Fourier transformation. Van Hove (105) generalized this relationship to include time dependence through its conjugate variable $\omega = (E_o - E')/\hbar$ (here ω is taken as being positive for energy transfer from the neutron). By analogy with equation [2-51] we can define the coherent scattering function

$$S^{coh}(\mathbf{Q},\omega) = \frac{1}{2\pi}\int e^{i(\mathbf{Q}.\mathbf{r} - \omega t)}\, G(\mathbf{r},t)\,d\mathbf{r}\,dt. \qquad [2\text{-}53]$$

and also a similar relationship for the incoherent scattering function

$$S^{inc}(\mathbf{Q},\omega) = \frac{1}{2\pi}\int e^{i(\mathbf{Q}.\mathbf{r} - \omega t)}\, G_s(\mathbf{r},t)\,d\mathbf{r}\,dt. \qquad [2\text{-}54]$$

The functions $G(\mathbf{r},t)$ and $G_s(\mathbf{r},t)$ are the Van Hove correlation functions. The former is the pair function and the latter the self function. They are defined in terms of quantum mechanical operators, $R(\mathbf{r})$, describing the position of the nucleus at time t, but the usefulness of the analysis is often restricted to situations where diffusion is sufficiently slow for $\mathbf{R}(t)$ to be considered as a position vector in the classical sense. When this is true $G^{Cl}(\mathbf{r},t)$ can be defined as the probability that there will be a nucleus at position **r** at time t given that *any* nucleus was at the origin at time t = 0, and similarly $G_s^{Cl}(\mathbf{r},t)$ is the probability of finding a nucleus at position **r** at time t given that *that* nucleus was at the origin at t = 0.

The scattering functions defined in this way are related to the corresponding inelastic scattering cross sections by the expressions

$$(d^2\sigma/dEd\Omega)^{coh}\ (E_o \rightarrow E', \theta) = (k'/k_o) <a>^2\ S^{coh}(Q,\omega) \qquad [2\text{-}55]$$

and

$$(d^2\sigma/dEd\Omega)^{inc}\ (E_o \rightarrow E', \theta) = (k'/k_o)\ (<a^2> - <a>^2)\ S^{inc}(Q,\omega) \qquad [2\text{-}56]$$

These expressions are analogous to equation [2-52] given that $k' = k_o$ for elastic scattering. The k'/k_o term arises from the ratio of the densities of neutron states in the incident and scattered beam as in the basic equation [2-10]. Again as in [2-50] we can write, in the classical limit,

$$G^{Cl}(\mathbf{r}, t) = \frac{1}{N} \sum_{i,j} <\delta(\mathbf{r} - \mathbf{R}_i(t) + \mathbf{R}_j(0))>. \qquad [2\text{-}57]$$

Here the averaging signs<......> have been introduced to indicate that the δ function has been averaged over all starting times for a system in thermal equilibrium at a fixed temperature. Given that all the atoms are identical, the distribution about any atom j will be the same as for any other j so that summation over j will yield N time the distribution for a given j, say j = 0. Thus

$$G^{Cl}(\mathbf{r}, t) = \sum_i < \delta(\mathbf{r} - R_i(t) + R_o(0))> \qquad [2\text{-}58]$$

and by analogy

$$G_s^{Cl}(\mathbf{r}, t) = < \delta(\mathbf{r} - R_o(t) + R_o(0))> \qquad [2\text{-}59]$$

The forms of $G_s^{Cl}(\mathbf{r}, t)$ and $G^{Cl}(\mathbf{r}, t)$ for a typical liquid are shown in Fig. 2-8b and 2-8c, respectively. $G_s(\mathbf{r},0)$ consists of a δ function at **r** = 0 which spreads out with time, typically with a Gaussian shape, until the probability of finding the atom at any point approaches zero. G(**r**, t) starts at time zero as G(**r**, 0) = δ(**r**) + G(**r**) which follows from comparison of [2-50] and [2-57]. As time progresses the central δ function spreads out, as for $G_s(\mathbf{r}, t)$ and the modulation of g(r) gradually disappears until at long times it is flat everywhere having a value of n, the number density of the liquid.

Now a number of properties of these correlation functions will be examined. First, consider integrals of the correlation functions over space. The integrals over space can be obtained from equation [2-58] and [2-59] as

$$\begin{aligned} &\int G(\mathbf{r}, t)\, d\mathbf{r} = N \\ &\int G_s(\mathbf{r}, t)\, d\mathbf{r} = 1. \end{aligned} \qquad [2\text{-}60]$$

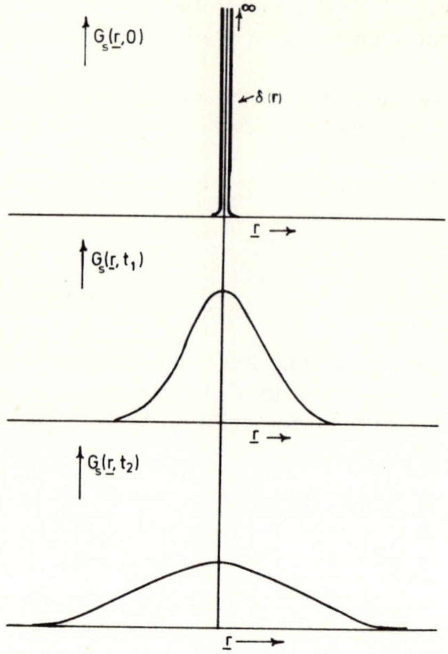

Figure 2-8. (b) Self correlation function for a liquid for three sequential times.

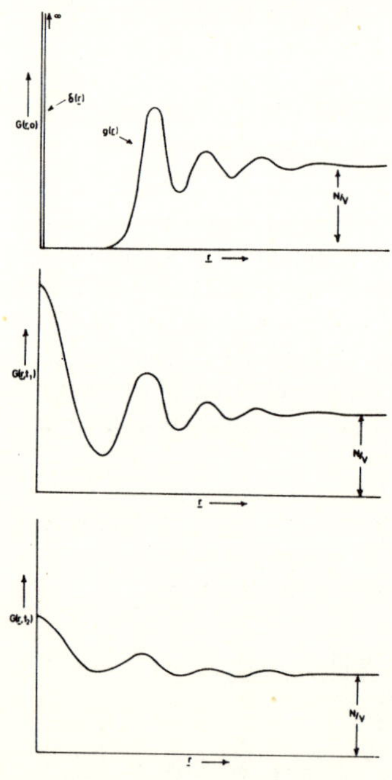

Figure 2-8. (c) Total correlation function for a liquid for three sequential times.

Second, consider the correlation functions at zero time. The various energy moments of the scattering functions can be obtained by integrating equation [2-53] over $d(h\omega)$. It is particularly useful to derive the zeroth moment $S_o(Q)$ from

$$S_o(Q) = \int S^{coh}(Q,\omega)\, d(\hbar\omega) = \frac{1}{2\pi\hbar} \int\int\int \exp(i\,(\mathbf{Q}.\mathbf{r} - \omega t)\, G(\mathbf{r}, t)\, d\mathbf{r} dt d(\hbar\omega) \qquad [2\text{-}61]$$

Performing the integral over ω first and remembering that

$$1/2\pi \int \exp(-i\omega t)\, d\omega = \delta\ (t),$$

$$S_o(\mathbf{Q}) = \int\int \exp(i\,\mathbf{Q}.\mathbf{r})\, G(\mathbf{r}, t)\delta(t)\, d\mathbf{r}\, dt = \int \exp(i\,\mathbf{Q}.\mathbf{r})\, G(\mathbf{r},0) d\mathbf{r} \qquad [2\text{-}62]$$

Remembering that $G(\mathbf{r}, 0) = \delta(\mathbf{r}) + g(\mathbf{r})$ we can identify $S_o(\mathbf{Q})$ with the structure factor defined in equation [2-51]. It is clear, therefore, that to obtain $g(\mathbf{r})$ in a real system having thermal motion one must measure the zeroth moment of the scattering function. This causes no difficulty in x-ray scattering because the possible energy transfers with the system are negligible compared with the energy of the quanta involved ($\sim$ keV) and therefore for fixed angle of scatter the value of Q remains constant regardless of energy transfer and the detector performs the integration over ω. In the case of neutrons, however, the values of $|\mathbf{k}'|$ and therefore of $\mathbf{Q}$ vary significantly at constant angle causing the integrated counts in the detector not to yield $S(\mathbf{Q})$ directly. The assumption that they do is known as the static approximation and improvements on this are a major problem in the measurement of liquid structure factors by neutrons (73).

Third, consider the correlation function at long time. Having discussed the properties of $G(\mathbf{r}, 0)$ let us consider the behavior of the function at long time, $G(\mathbf{r},\infty)$, when it is presumed to have reached an asymptotic shape, i.e.

$$G(\mathbf{r}, t) = G^1(\mathbf{r},t) + G(\mathbf{r},\infty), \qquad [2\text{-}63]$$

where $G^1(\mathbf{r}, t)$ is a function that tends to zero for $t \to \infty$. By substitution into equation [2-53], $G(\mathbf{r},\infty)$ gives rise to a singularity at $\omega = 0$ or

$$S(Q,0) = (1/2\pi) \int e^{i\mathbf{Q}.\mathbf{r}} G(\mathbf{r},\infty)\, d\mathbf{r}. \qquad [2\text{-}64]$$

In a liquid $G(\mathbf{r},\infty)$ conveys no useful information being a constant. It corresponds to $S(\mathbf{Q},0) = N\delta(\mathbf{Q})$ or is indistinguishable from the unscattered beam. In a solid, $G(\mathbf{r},\infty)$ consists of the fully developed thermal clouds, and the corresponding $S(\mathbf{Q}, 0)$ is Bragg scattering modified by the Debye-Waller factor.

In a similar way we can define $S^{inc}(\mathbf{Q},\infty)$. In a solid, this yields the Debye-Waller factor

$$S^{inc}(\mathbf{Q}, 0) = e^{-\langle u^2\rangle Q^2} \qquad [2\text{-}65]$$

where $\langle u^2 \rangle$ is the mean square deviation in the thermal cloud. An important case is where one has diffusion within a limited volume of space, such as in the rotational diffusion of a large molecule when $G(\mathbf{r},\infty)$ could become the surface of a sphere. In this case, the corresponding $S(Q,0)$ is known as the Elastic Incoherent Structure Factor (EISF).

Fourth, consider detailed balance. Returning to the important question of when one can use the classical definition of $G(\mathbf{r},t)$, let us first describe an important property of neutron cross sections known as "Detailed Balance". This states that when neutrons are in complete equilibrium with a medium (this situation is in fact unobtainable in practice, as it would require an infinite non-absorbing medium), then the rate at which neutrons are scattered from energy E_o to energy E' and through angle θ is equal to the rate at which they are scattered back. There are a number of ways of proving this statement but we accept it to be the only reasonable way of retaining a neutron energy distribution that is independent of time. Thus from equation [2-1]

$$E_o/(kT)^2 \exp(-E_o/kT)\,(d^2\sigma/dE'd\Omega)\,(E_o \rightarrow E',\theta) =$$
$$(E'/(kT)^2)\exp(-E'/kT) \times (d^2\sigma/dE_o d\Omega)\,(E' \rightarrow E_o,\theta). \qquad [2\text{-}66]$$

On replacing the cross sections with the appropriate scattering functions we have $S(\mathbf{Q},\omega) = \exp\,(t\omega/kT)S(-\mathbf{Q}, -\omega)$. Also, if, as for all systems apart from non-centrosymmetric crystals, $G(\mathbf{r},t) = G(-\mathbf{r}, t)$, then it follows that

$$S(-\mathbf{Q}, -\omega) = S(\mathbf{Q}, -\omega) = \exp\,(-t\omega/kT)\,S(\mathbf{Q},\omega). \qquad [2\text{-}67]$$

The important conclusion here is that $S(\mathbf{Q},\omega)$ is not symmetric in ω, and therefore, the corresponding $G(\mathbf{r}, t)$ obtained by inverse Fourier transformation must be complex. This is another way of saying that the Classical Approximation is not valid and therefore, as mentioned above, the position vectors of the atoms must be regarded as quantum mechanical operators. In the limit of small ω the difficulty does not arise because $e^{-t\omega/kT} \sim 1$. At larger values, however, we may use an approximation by Schofield (88). Thus let us define a symmetric version of $S(\mathbf{Q}, \omega)$, written $\bar{S}(\mathbf{Q},\omega)$, where

$\bar{S}(\mathbf{Q},\omega) = \exp(-\hbar\omega/2kT)\,S(\mathbf{Q}, \omega)$.Hence

$\bar{S}(\mathbf{Q},-\omega) = \exp(\hbar\omega/2kT)\,S(\mathbf{Q},-\omega)$ and from equation [2-67] $\bar{S}(\mathbf{Q},\omega) = \bar{S}(\mathbf{Q}, -\omega)$.

On an inverse Fourier transformation of this function we obtain a real correlation function $\bar{G}(\mathbf{r}, t)$ and Schofield's approximation is to take $G^{Cl}(\mathbf{r},t) \approx \bar{G}(\mathbf{r}, t)$. Reversing this process we can obtain from $G^{Cl}(\mathbf{r}, t)$ a scattering function that obeys detailed balance.

It may be noted here that the requirement that $\bar{S}(Q,\omega)$ is real also means that $G^{Cl}(\mathbf{r}, t)$ is even in time. From the definition of $G^{Cl}(\mathbf{r}, t)$ it is not immediately obvious how the function should behave at negative time but it is quite reasonable to assume that the atom will diffuse outwards from the δ function going backwards in time just as it does going forward in time. We can ensure this behavior by using $|t|$ for t in expressions for $G(\mathbf{r}, t)$ obtained for positive time.

Fifthly, we may define useful functions known as the intermediate scattering functions. In calculations of the scattering function it is often convenient to start with the intermediate function $I(\mathbf{Q}, t)$ or the intermediate self function $I_s(\mathbf{Q},t)$, which is obtained by Fourier transforming $G(\mathbf{r}, t)$ and $G_s(\mathbf{r}, t)$ in space i.e.

$$I(\mathbf{Q}, t) = \int \exp(i\mathbf{Q}.\mathbf{r})\, G(\mathbf{r}, t)\, d\mathbf{r} \quad \text{and} \tag{2-68}$$

$$I_s(\mathbf{Q}, t) = \int \exp(i\mathbf{Q}.\mathbf{r})\, G_s(\mathbf{r}, t)\, d\mathbf{r} \tag{2-69}$$

Substituting for $G(\mathbf{r}, t)$ and $G_s(\mathbf{r}, t)$ from equations [2-58] and [2-59] we obtain

$$I(\mathbf{Q}, t) = \langle \sum_i \exp(i\mathbf{Q}.(\mathbf{R}_i(t) - \mathbf{R}_o(0)))\rangle \tag{2-70}$$

$$I_s(\mathbf{Q}, t) = \langle \exp(i\mathbf{Q}.(\mathbf{R}_o(t) - \mathbf{R}_o(0)))\rangle \tag{2-71}$$

One general property of $I^s(\mathbf{Q}, t)$ is particularly useful in building up models of diffusive motion. Let us consider a proton in a molecule that is undergoing rotational, vibrational and translational motion, and let us assume that these motions are entirely uncorrelated.

Thus $\mathbf{R}(t) = \mathbf{l}(t) + \mathbf{r}(t) + \mathbf{u}(t)$ where $\mathbf{l}(t)$ is the translation of the molecular centre of mass, $\mathbf{r}(t)$ is the distance rotated about the centre of mass and $\mathbf{u}(t)$ the amplitude of the vibrations. In performing the thermal average of $\mathbf{R}_o(t) - \mathbf{R}_o(0)$ we use the RPA discussed in section 2-2.1 to obtain,

$$I_s(\mathbf{Q}, t) = \langle\exp(i\mathbf{Q}.(\mathbf{l}(t) - \mathbf{l}(0)))\rangle \langle\exp(i\mathbf{Q}.(\mathbf{r}(t) - \mathbf{r}(0)))\rangle \times$$

$$\times \langle\exp(i\mathbf{Q}.(\mathbf{u}(t) - \mathbf{u}(0)))\rangle \tag{2-72}$$

$$= I_{trans}(\mathbf{Q}, t)\, I_{rot}(\mathbf{Q},t)\, I_{vib}(\mathbf{Q},t) \tag{2-73}$$

Now, on transformation to the scattering function, $S(\mathbf{Q},\omega)$ is the convolution in ω (indicated by $*$) of the transforms of the individual functions.

$$S^{inc}(\mathbf{Q},\omega) = S_{trans}(\mathbf{Q},\omega) * S_{rot}(\mathbf{Q},\omega) * S_{vib}(\mathbf{Q},\omega) \tag{2-74}$$

where

$$f(\omega) * g(\omega) \equiv \int f(\omega - \omega')\, g(\omega')\, d\omega'. \tag{2-75}$$

Here we have used the well known result that the Fourier transform of a product is the convolution of the Fourier transforms of the constituent functions. It should be noted that the use of the Random Phase Approximation in arriving at equation [2-73] is really a particular application of this result in that if a correlation function is made up of two independent correlation functions for two uncorrelated motions taking place simultaneously, the overall function at time t will be the convolution in $\mathbf{r}$ of the separate functions each taken at time t. Thus, the Intermediate Scattering Function will often be the easiest of the functions to calculate in that it does not involve convolutions in either $\mathbf{r}$ or ω. As an example of equation

[2-73] we will examine the influence of the vibrational motions of the proton on the quasi-elastic scattering. As shown earlier, let us write the vibrational correlation function

$$G_s(\mathbf{u},t) = G_s(\mathbf{u},\infty) + G^1(\mathbf{u},t). \qquad [2\text{-}76]$$

Thus from equation [2-65]

$$S_{vib}(\mathbf{Q},\omega) = e^{-<u^2>Q^2}\,\delta(\omega) + f(\mathbf{Q},\omega) \qquad [2\text{-}77]$$

where $f(\mathbf{Q},\omega)$ is the inelastic part of the spectra which does not normally contribute significantly at ω values within the quasi-elastic peak. When we convolute a delta function with another function we have

$$\int C\delta(\omega_1)\, f(\omega - \omega_1)d\omega_1 = Cf(\omega). \qquad [2\text{-}78]$$

Hence we can write the quasi-elastic part of $S(\mathbf{Q},\omega)$ as:

$$S^{inc}(\mathbf{Q},\omega) = e^{-<u^2>Q^2}\, S_{trans}(\mathbf{Q},\omega) * S_{rot}(\mathbf{Q},\omega). \qquad [2\text{-}79]$$

Thus the effect of the vibrational motions is just to attenuate the quasi-elastic peak intensity with increasing Q.

<u>The Correlation Function for Translational Diffusion: (1) Classical Diffusion in Three Dimensions.</u> The simplest formulation of the translational diffusion correlation function is obtained by solving the classical diffusion equation in an infinite medium with a delta function source term.

$$(\partial G_s/\partial t)\,(\mathbf{r},t) = D\nabla^2\, G_s(\mathbf{r},t) \text{ with } G_s(\mathbf{r},0) = \delta(\mathbf{r}) \qquad [2\text{-}80]$$

where D is the macroscopic diffusion coefficient. Here, because all origin points are equivalent, there is no need to average over initial configurations. The solution of this equation is a Gaussian:

$$G_s(\mathbf{r},t) = (4\pi D|t|)^{-3/2} \exp(-\mathbf{r}^2/4D|t|) \qquad [2\text{-}81]$$

This distribution has width $(2Dt)^{1/2}$, which implies a mean square value of r given by $<r^2> = 6Dt$. This equation may be directly transformed to yield (107)

$$S(\mathbf{Q},\omega) = \frac{1}{\pi}\,\frac{DQ^2}{(DQ^2)^2 + \omega^2}. \qquad [2\text{-}82]$$

This is a Lorentzian function with full width at half height (FWHH) of $2DQ^2$. For monatomic systems it is indeed found experimentally that, at low Q, the width of the quasi-elastic peak is proportional to Q^2, the constant being twice the macroscopic diffusion coefficient. At higher Q, however, the experimental curve always drops below this form. The reason for this discrepancy is that equation [2-81] implies a mean velocity of

$$\frac{<r^2>^{1/2}}{t} = \sqrt{\frac{6D}{t}} \quad [2\text{-}83]$$

which increases as t and $<r^2>^{1/2}$ get smaller. Eventually this velocity predicted from the differential equation becomes greater than the actual atomic velocity at temperature T, $\mathbf{v} \sim (2kT/m)^{1/2}$, and thus it is clear that macroscopic diffusion theory is bound to break down. This upper limit to the mean velocity will have the effect of reducing the broadening at large Q below the $2DQ^2$ curve.

(2) Classical Diffusion in Two Dimensions. In studying diffusion in lamellar systems, scattering functions for diffusion in two dimensions are of interest. It can be shown that in this case

$$S(Q,\omega) = \frac{1}{\pi} \frac{D_{\parallel} \cos^2\theta Q^2}{(D_{\parallel} \cos^2 \theta\ Q^2)^2 + \omega^2} \quad [2\text{-}84]$$

where $D_{\parallel}$ is the diffusion coefficient parallel to the plane ($\theta = 0$). This expression in the polycrystalline average gives rise to a cusp-shaped function (27, 101). On the other hand, if the diffusion equation is solved for diffusion between two parallel impermeable infinite plane boundaries separated by a distance d, it has been shown (41) that

$$S_{inc}(Q,\omega) = \sum_{n=0}^{\infty} A_n (Qd \sin\theta)\ L\ (D[Q^2 \cos^2\theta + (n\pi/d)^2]) \quad [2\text{-}85]$$

where θ is the angle between Q and the interlamellar plane and L(x) is a normalized Lorentzian of FWHM = 2x. Here

$$\sum_{n=0}^{\infty} A_n (x) = \frac{2x^2}{[x^2 - (n\pi)^2]^2} (1 - (-1)^n \cos x). \quad [2\text{-}86]$$

(3) Jump Diffusion. One important improvement on macroscopic diffusion theory, is to consider diffusion as a Markovian random walk (24). This assumption implies that diffusion takes place as a result of a series of uncorrelated instantaneous jumps or phases of motion each of which can be described by the same function $\rho(\mathbf{r})$ and that the jumps have a Poisson distribution in time, i.e. a fixed probability of jumping/unit time, Γ. In these circumstances, the spatial probability distribution after *n* jumps is given by the *n*-fold convolution of $\rho(\mathbf{r})$ with itself. We can therefore write

$$I(Q,t) = \sum_{n=0}^{\infty} (F(Q))^n T_n (t) \quad [2\text{-}87]$$

$$\text{where } F(Q) = \int \rho(\mathbf{r})\, e^{i\mathbf{Q}.\mathbf{r}}\, d\mathbf{r} \quad [2\text{-}88]$$

and T_n (t) is the probability of having jumped *n* times at time t, i.e.

$$T_n (t) = \frac{(\Gamma t)^n}{n!} e^{-\Gamma t}. \quad [2\text{-}89]$$

This series can be summed yielding

$$F(Q,t) = e^{-[1 - F(Q)]\Gamma t} \qquad [2\text{-}90]$$

which in turn can be Fourier transformed to yield

$$S(Q,\omega) = \frac{1}{\pi} \frac{\Gamma(1 - F(Q))}{\omega^2 + \Gamma^2 (1 - F(Q))^2}. \qquad [2\text{-}91]$$

It is of interest to examine the behavior of this general result at low Q. Expanding the exponential in [2-88] one obtains

$$F(Q) = \int \rho(\mathbf{r}) (1 + i\mathbf{Q}.\mathbf{r} - \tfrac{1}{2} (\mathbf{Q}.\mathbf{r})^2 \ldots)d\mathbf{r} \qquad [2\text{-}92]$$

For small Q and $\rho(\mathbf{r})$ symmetric in **r**, the first term is unity, the second is zero and the third yields the mean square deviation of $\rho(\mathbf{r})$ in the x, y and z directions. Hence $F(Q) = 1 - \tfrac{1}{2} (Q_x^2\langle x^2\rangle + Q_y{}^2\langle y^2\rangle + Q_z^2\langle z^2\rangle)$. For cubic symmetry, the mean square jump length $\langle r^2\rangle$ can be written $\langle r^2\rangle = 3\langle x^2\rangle = 3\langle y^2\rangle = 3\langle z^2\rangle$. Hence $\Gamma(1 - F(Q)) = Q^2\langle r^2\rangle\Gamma/6)$ so that equation [2-91] above becomes identical with the macroscopic theory result as expected.

A number of particular models described in the literature can now be obtained. The Chudley-Elliott model (25) was derived for jump diffusion on a periodic lattice such that $\rho(\mathbf{r}) = (1/m) \sum_{\ell} \delta(\mathbf{r} - \underline{\ell})$ where $\underline{\ell}$ is one of a set of m equally likely jump vectors. Here

$$F(Q) = (1/m) \sum_{\underline{\ell}} e^{iQ.\underline{\ell}} \qquad [2\text{-}93]$$

and the scattering is again Lorentzian, with a width that rises from zero and falls to zero again at the reciprocal lattice points. This model has been very successful in the description of quasi-elastic neutron scattering from hydrogen diffusing in a lattice (96).

A second example is the Singwi-Sjolander model (95). Their model involved a generalization of the above derivation to describe the actual time dependence in the jump phase. In its most familiar form, the mean time for the jump phase is reduced to zero and the model coincides with the assumptions of equation [2-91] but with a spherically symmetric $\rho(r)$ given by

$$\rho(r) = (r/r_0{}^2)e^{-r/r_0} \qquad [2\text{-}94]$$

for which distribution $\langle r^2\rangle = 6r_0{}^2$ and one obtains:

$$F(Q) = \frac{e^{-\langle u^2\rangle Q^2}}{1 + Q^2\langle r^2\rangle/6.} \qquad [2\text{-}95]$$

Again this yields a width of $2\,DQ^2$ at small Q and tends asymptotically to 2Γ at large Q. In their original formulation, Singwi and Sjolander assumed that the oscillatory behaviour was convoluted with $\rho(r)$ and that the next jump originated from this combined probability distribution as opposed to $\rho(r)$ itself yielding

$$F(Q) = \frac{e^{-<u^2>Q^2}}{1 + Q^2<r_1^2>/6} \qquad [2\text{-}96]$$

where the overall mean square jump length, $<r_1^2> = <r^2> + <u^2>$.

There are a number of other models for G(**r**, t) which are discussed in the literature. Apart from some minor variations on the above models, the main interest has been in developing models for liquids which provide a combined treatment of diffusional and vibrational motions (29, 66, 97). While such an approach is essential for monatomic liquids, in clay-water and clay-organic systems with high clay fractions, the diffusion is considerably hindered in various ways so that the separation into quasi-elastic and inelastic parts is quite clear. Therefore these unified models will not be considered further here.

The Correlation Function for Localized Diffusion. If the scattering atom is diffusing within a limited region of space, then G(**r**,∞) will not be constant throughout space and therefore the scattering will have a Q dependent elastic component, the Elastic Incoherent Structure Factor (EISF) defined earlier, given by

$$S_{el}^{inc}(\mathbf{Q},\omega) = \delta(\omega) \int e^{i\mathbf{Q}.\mathbf{r}} G(\mathbf{r},\infty) d\mathbf{r}. \qquad [2\text{-}97]$$

In general this function decreases from unity at **Q** = 0 and approaches 0 for large Q, normally with some increasingly damped oscillations.

An example of this behavior that has been covered is the existence of an elastic peak for diffusion between two parallel planar boundaries (equation [2-85]). Here for n = 0 and θ = 90° (i.e. Q perpendicular to the interlamellar plane) the Lorentzian becomes a delta function whose amplitude is given by

$$A_o(Qd) = \frac{2[1 - \cos(Qd)]}{(Qd)^2} = \frac{\sin^2(Qd/2)}{(Qd/2)^2} \qquad [2\text{-}98]$$

This kind of analysis can be applied for any system in which there is a restricted space available for the diffusing atom.

A similar situation also exists for molecules undergoing rotational diffusion. This can occur for rather spherical molecules, the centers of which are located in a well-defined lattice while the molecules themselves undergo rotational diffusion. For a single molecular species this normally happens for a restricted range of temperatures, which is known as the plastic phase of the crystal. Molecules can also reorientate in a rigid lattice consisting of a different species e.g. NH_4 reorientations in the NH_4Br lattice (63) or one can have reorientations of CH_3 groups within a larger molecule. Alternatively, in a molecular liquid, if center of mass diffusion is entirely uncorrelated with rotational diffusion (possibly not a good approximation for water because if diffusion takes place by the breaking and remaking of two of the hydrogen bonds, the center of mass will move for each rotation) one can use the separation of the intermediate scattering function and calculate $S_{rot}(\mathbf{Q},\omega)$ separately before convoluting in ω with S_{trans} and S_{vib}. This type of analysis has been most important in the understanding of motions in liquid crystals (27). We would note in passing that where macroscopic diffusion exists, the convolution with $S_{rot}(Q,\omega)$ means that the EISF becomes broadened. However, if the rates of the two processes are sufficiently different, a separation can still be made experimentally.

As an example of localized diffusion we shall first derive $S^{inc}(Q,\omega)$ for the simplest case of localized diffusion, namely the process of jump diffusion between two fixed sites. We shall then quote results that have been obtained for a number of more complex situations.

Let the diffusion be between two sites separated by the vector **r** and let Γ be the jump probability/unit time. Now for even numbers of jumps, the atom returns to its origin while for odd numbers of jumps it is displaced by $\pm\, \mathbf{r}_o$ (after averaging over the initial sites). Then using the expansion described above for Markovian random walk diffusion (equation [2-87]), we have

$$I(Q,t) = e^{-\Gamma t}[1 + F(Q)\Gamma t + \frac{\Gamma t^2}{2!} + F(Q)\,\frac{(\Gamma t)^3}{3!} + ...] \qquad [2\text{-}99]$$

$$= \tfrac{1}{2}\,\{\,[1 + F(Q)] + [1 - F(Q)]\; e^{-2\Gamma t}\}$$

where $F(Q) = \tfrac{1}{2}(e^{iQ.r_0} + e^{-iQ.r_0}) = \cos \mathbf{Q.r_o}$.

Hence

$$\mathbf{S(Q,\omega) = \tfrac{1}{2}\,[1 + \cos(Q.r_0)]}\; \delta(\omega) + + [1 - \cos(\mathbf{Q.r_0})]\,\frac{1}{\pi}\,\frac{2\Gamma}{(2\Gamma)^2 + \omega^2} \qquad [2\text{-}100]$$

where the coefficient of the δ function is the EISF (in this case there is no damping of the oscillations) and where the Lorentzian term has a Q dependent amplitude and a constant width in contrast to the unrestricted diffusion case.

The simplest extension of this model, due to Barnes (14), was to the case of uniaxial jump rotation, for jumps between multiple sites on a circle, a model that can be averaged analytically over all orientations of the axis for the polycrystalline case.

Uniaxial continuous diffusion has also been analyzed by Favro (31) who obtained a series expansion in terms of Lorentzians of increasing width. Richardson (81), however, has pointed out that the Barnes model for six or more sites on the circle gives results which are indistinguishable from the Favro expansion and is more convenient to use in practice. A further model for uniaxial diffusion in a cosine potential has been given by Dianoux and Volino (28).

Turning to the spherical case, a general formalism for diffusion on the surface of a sphere can be given in terms of the Sears expansion (90)

$$S^{inc}_{rot}(Q,\omega) = \delta(\omega) j_0^2\,(QR_o) + \sum_{\ell=1}^{\infty} (2\ell + 1)\, j_\ell^2\,(Q\,R_0)\tilde{F}_\ell(\omega) \qquad [2\text{-}101]$$

where the j_ℓ are spherical Bessel functions, R_o is the distance of the proton from the center of the molecule, and the functions $\tilde{F}$ are the Fourier transformed orientational self correlation functions, namely:

$$\tilde{F}_\ell(\omega) = \frac{1}{2\pi} \int e^{i\omega t} \langle P_\ell(\cos\beta(t))\rangle.$$

Here, $\beta(t)$ is the angle through which the vector to a given proton rotates within a time interval t and P_ℓ is the ℓth Legendre Polynomial. The forms of $\langle P_\ell(\cos\beta(t))\rangle$ can be calculated for specific models as continuous diffusion or isotropic large angle jumps (89, 90).

A complete discussion of these models can be found in Lovesey and Springer (66). Their main use of restricted diffusion models to date has been to interpret the shape of the EISF as it has proven difficult to separate the widths of a series of superimposed Lorentzians (27).

2-2.5. Inelastic Neutron Scattering

General Theory. As has been mentioned above, inelastic neutron scattering occurs when one or more quanta of energy are transferred to or from the vibrational modes of the scattering system. The measurement of the cross sections for such events has provided very valuable information in many areas of physics and chemistry. While the potential exists, there have been few such measurements on clay systems to date, and the treatment given here is rather brief. We shall first consider scattering from an isolated simple harmonic oscillator (SHO) and extend this treatment to an incoherent scattering solid, and then briefly consider coherent inelastic scattering.

The advantage of starting with a simple harmonic oscillator is that the cross sections can be derived in a fairly simple manner from equation [2-10] and the basic quantum mechanics of the SHO (87). Let us first write the wave function of the SHO $\phi(\mathbf{R}) = \phi_{n1}(x)\,\phi_{n2}(y)\,\phi_{n3}(z)$ where $\phi_n(x)$ is the wave function of the nth state of a linear oscillator. Introducing the characteristic length, $g = (\hbar/M\omega)^{1/2}$, where M is the mass of the oscillator and ω is its natural frequency, we can write the wave function

$$\phi_n(x) = \pi^{-1/4} b^{-1/2} \exp(-x^2/2g^2) \frac{H_n(x/g)}{(2^n n!)^{1/2}} \qquad [2\text{-}102]$$

where $H_n(\xi)$ are the Hermite polynomials which can be generated using the equation

$$H_n(\xi) = (-1)^n \exp(\xi^2) \frac{d^n}{d\xi^n} \exp(-\xi^2) \qquad [2\text{-}103]$$

and are normalized by the relation

$$\int_{-\infty}^{\infty} \exp(-\xi^2)\, H_n(\xi)\, H_m(\xi)\, d\xi = 2^n n! \sqrt{\pi \delta_{nm}}. \qquad [2\text{-}104]$$

It is well known that these wave functions have eigenvalues of $\hbar\omega(n + 1/2)$. Denoting the initial states ν by the quantum numbers n_1, n_2 and n_3 and the final states ν' by

n_1', n_2', n_3' and taking Q parallel to z (for an isotropic oscillator) we can write the matrix element in equation [2-11] after introducing the pseudopotential (equation [2-15])

$$\langle \mathbf{k}'\nu'|\mathbf{V}|\mathbf{k}_o\nu\rangle = \int \phi^*_{n_1'}(x)\phi^*_{n_2'}(y)\,\phi^*_{n_3'}(z)\left(\frac{2\pi h^2}{m}b\right) e^{iQz}\phi_{n_1}(x)\phi_{n_2}(y)\,\phi_{n_3}(z)\,dxdydz \qquad [2\text{-}105]$$

Using the normalization condition (equation [2-104]) to do the integration over x and y, we see that the cross section is only finite for $n_1 = n_1'$ and $n_2 = n_2'$ and we have

$$\langle \mathbf{k}'\nu'|\mathbf{V}|\mathbf{k}_o\nu\rangle = \frac{2\pi h^2}{m}\; b \int \phi^*_{n'}(z)e^{-iQz}\phi_n(z)\,dz. \qquad [2\text{-}106]$$

This integral can be evaluated quite straightforwardly (104) to yield the following expression by substituting the matrix element into equation [2-106], for a set of identical incoherently scattering oscillators and summing over all possible final states :

$$\left(\frac{d^2\sigma}{d\Omega dE'}\right)_\nu = \langle b^2\rangle \frac{k}{k_o}\sum_{n'=0}^{\infty}\frac{n!\,n'!}{2^{|n-n'|}}\exp(-Q^2g^2/2)\,(Q^2g^2)^{|n'-n|}$$

$$\times\, |F_{nn'}(Q^2g^2)|^2\,\delta(E'-E_o+h\omega(n'-n)) \qquad [2\text{-}107]$$

where

$$F_{nn'}(Q^2g^2) = \sum_{m=0}^{n}\frac{(-Q^2g^2)^m}{2^m\,m!(n-m)!(n'-n+m)!} \quad \text{for } n' \geqslant n$$

or $\qquad\qquad\qquad\qquad\qquad\qquad\qquad\qquad\qquad\qquad$ [2-108]

$$= \sum_{m=0}^{n'}\frac{(-Q^2g^2)^m}{2^m\,m!(n'-m)!(n-n'+m)!} \quad \text{for } n \geqslant n'$$

This equation still has to be averaged over a thermal equilibrium of initial states n but before doing this, it is useful to consider the low temperature limit when the system is initially in the ground state n = 0. For elastic scattering, n' = 0.

$$(d\sigma/d\Omega)^{el}\,(0\to 0) = \langle b^2\rangle\,\exp(-Q^2g^2/2) \qquad [2\text{-}109]$$

This is exactly analogous to equation [2-65] and arises directly by Fourier transforming the probability distribution $\phi^*0(z)\,\phi 0(z)$ which is in fact a Gaussian with mean square deviation $\langle z^2\rangle = g^2/2$. Since the distribution is spatially isotropic one may write $\langle u^2\rangle$ in place of $\langle z^2\rangle$.

In addition to elastic scattering, the neutron can also raise the oscillator to excited states which are given for n' = 1, for example, by

$$(d^2\sigma/dE'd\Omega)\,(0\to 1) = \langle b^2\rangle\,(k/k_o)\,(Q^2g^2/2)\,e^{-Q^2g^2/2}. \qquad [2\text{-}110]$$

The important point to note here is that the one quantum transfer is proportional to $Q^2\langle u^2\rangle\ e^{-Q^2\langle u^2\rangle}$. Higher energy transfer terms are proportional to $(Q^2\langle u^2\rangle)^n\ e^{-Q^2\langle u^2\rangle}$. If these terms are all integrated over angle, one obtains the total cross section of a SHO in its ground state. This is a situation typical of hydrogen in a metal, e.g. zirconium dihydride, where $\hbar\omega$ is usually $\sim$ 100 meV and the assumption of ground state occupation is well justified at room temperature. The experimental data for ZrH_2 is compared with the theory in Fig. 2-9. As shown, each hump has low energy thresholds as E_o exceeds $\hbar\omega$, $2\hbar\omega$ etc. and higher values of n′ are successively introduced into the equation.

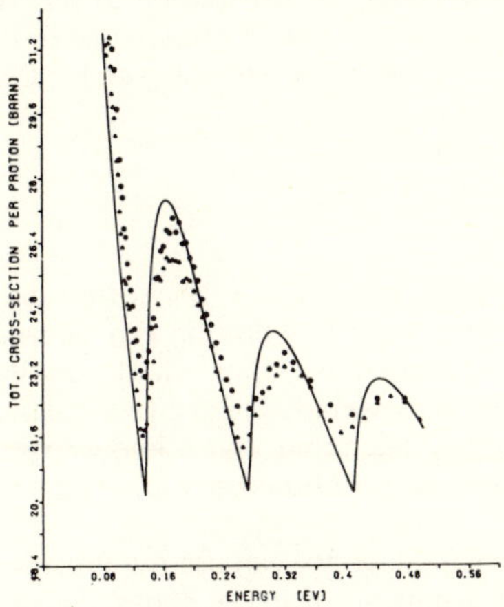

Figure 2-9. Comparison of total cross-section data for $ZrH_{1.92}$ at 293 K from calculations based on simple harmonic oscillator theory (74).

Now, consider the situation where the oscillator is in thermal equilibrium at a temperature T such that a number of states are excited. An assembly of such oscillators will have energies distributed according to Bose-Einstein statistics. For such a system, the probability of the nth state being occupied, p_n, is

$$p_n = (1 - e^{-\hbar\omega/kT})e^{-n\hbar\omega/kT}. \qquad [2\text{-}111]$$

It is easy to show that this expression satisfies the two necessary conditions

$$\text{(i)}\ \sum_{n=0}^{\infty} p_n = 1 \text{ and (ii)} \frac{p_{n+1}}{p_n} = e^{-\hbar\omega/kT}.$$

It is also easily shown that the average value of n, written $\langle n\rangle$, is

$$\langle n\rangle = \sum_{n=0}^{\infty} n\, p_n = \frac{e^{-\hbar\omega/kT}}{1-e^{-\hbar\omega/kT}}. \qquad [2\text{-}112]$$

We can now define the thermal equilibrium cross section for elastic scattering at temperature T

$$(d\sigma/d\Omega)^{el} = \langle b^2\rangle \sum_n p_n (n!)\, F_{nn'}(Q^2 g^2) e^{-Q^2 g^2/2}\, \delta(E' - E_o) \qquad [2\text{-}113]$$

This appears to be a complicated expression but can be simplified greatly using a result known as the Bloch Identity, the proof of which is described by Squires (98). Applied here this identity yields

$$\left(\frac{d\sigma}{d\Omega}\right)^{el}_T = \langle b^2\rangle\, e^{-(\langle n\rangle + ½)\, Q^2 g^2} \qquad [2\text{-}114]$$

This corresponds to saying that the probability distribution in real space is still a Gaussian but now of width $\langle u^2\rangle = (\langle n\rangle + ½)\ Q^2 g^2$. Reference to the expression for $\langle n\rangle$ shows that this increases from the zero point value at T = 0 and gradually slopes upward to reach

$$\langle u^2\rangle_{T\to\infty} = \frac{kT}{\hbar\omega} g^2 = \frac{kT}{M\omega^2} \qquad [2\text{-}115]$$

at high temperatures. It may be noted that $\langle n\rangle + ½$ is often written ½ coth $(\hbar\omega/2kT)$. The potential in which the atom is vibrating can be written $V(R) = (M\omega^2/2)\ (x^2 + y^2 + z^2)$ and $\langle u^2\rangle$ is proportional to $g^2 = \hbar/M\omega$ so that $\langle u^2\rangle$ gets smaller as the restoring force increases as expected. $\langle u^2\rangle$ also increases as the mass of the vibrating atom decreases. It may be noted that a knowledge of ω can be used to fix temperature factors in least-squares refinements of crystal structures.

Turning to the one-phonon transfer case, we can use the Bloch Identity again to simplify the averaging over the distribution of initial states. This process yields

$$(d^2\sigma/dEd\Omega)\ (E'=E_o \pm \hbar\omega) = \langle b^2\rangle\ (k/k_o)\ (\langle n\rangle + ½ \pm ½)\ (Q^2 b^2/2) \times$$
$$\times\, e^{-(\langle n\rangle + ½)Q^2 b^2}\ \delta(E_o - E' \pm \hbar\omega). \qquad [2\text{-}116]$$

It will be noted that this expression at low temperature ($\langle n\rangle \to 0$) predicts zero chance of neutron energy gain but clearly a finite chance of neutron energy loss as expected. Also $(\langle n\rangle + 1)/(\langle n\rangle) = e^{\hbar\omega/kT}$ and the cross sections obey detailed balance. They are also proportional to Q^2/ω as before. Similar results hold for higher terms.

Incoherent Scattering Solids. The analysis of the vibrational modes of solids is too large a topic to discuss in detail here but a few simple results will be given to illustrate how it follows naturally from the simple harmonic oscillator case. In a purely classical sense the vibrations of a solid can be viewed as the normal modes of a set of coupled harmonic oscillators. It may be shown that in a periodic lattice these modes can be taken as a set of plane waves of wave vector q and corresponding frequency $\omega_s(q)$. For a monatomic lattice there are three frequencies for each q value ($s=1\to 3$), normally two transverse and one longitudinal mode. By applying periodic boundary conditions to a finite crystal, it can be shown that there are 3N such q vectors uniformly distributed throughout the first Brillouin Zone of the crystal. The first Brillouin Zone of a crystal is the volume of reciprocal space around the origin bounded by planes perpendicularly bisecting the shortest reciprocal lattice vectors from the origin. From quantum mechanics the quanta of energy associated with these modes are known as phonons. A phonon has energy $\hbar\omega$ and can be regarded as having momentum $\hbar q$. The average number of phonons in a particular mode is $<n> +½$ as for the simple harmonic oscillator.

For small **q** vectors one can assume that $\omega(\mathbf{q}) = c\mathbf{q}$ where c is the velocity of sound in the crystal. The assumption that this relationship holds right out to the zone boundary yields the Debye model of specific heats. A more sophisticated approach (16) may be used to determine the actual values of $\omega^s(\mathbf{q})$ in terms of the force constants linking the atoms, assuming these interactions to be harmonic. Using a computer it is a simple matter to calculate the $\omega^s(\mathbf{q})$ corresponding to a large number of **q**'s uniformly distributed through the Brillouin Zone. Arranging these ω values as a histogram and normalizing to unit area, one obtains the frequency distribution, $f(\omega)$, of the solid. Thus $f(\omega)$ can be written

$$f(\omega) = \sum_{i,\ \omega<\omega^s(q_i)<\omega+\Delta\omega} \left(\frac{1}{3N\Delta\omega}\right) \qquad [2\text{-}117]$$

where the summation is over all values of ω which lie between ω and $\omega + \Delta\omega$

Using $f(\omega)$ we can immediately calculate the lattice specific heat in an improved version of the Debye Model. Thus the total energy in the crystal at temperature T is

$$E(T) = \int <n> \hbar\omega\, f(\omega)\, d\omega. \text{ Therefore}$$

$$C_v = \frac{dE(T)}{dT} = \int \frac{d<n>\hbar\omega}{dT} f(\omega)d\omega. \qquad [2\text{-}118]$$

where $<n> = \dfrac{e^{-\hbar\omega/kT}}{1 - e^{-\hbar\omega/kT}}$

This approach can be used to describe the incoherent scattering from a cubic monatomic lattice if we consider each normal mode to be an independent simple harmonic oscillator. Thus the incoherent elastic cross section is

$$(d\sigma/d\Omega)^{el} = <b^2>\, e^{-<u^2>Q^2}. \qquad [2\text{-}119]$$

By comparison with equation [2-114], writing $g^2 = \hbar/M\omega$

$$<u^2> = \int (<n> + ½)\, f(\omega)\, (\hbar/M\omega)\, d\omega$$

and by analogy the one phonon incoherent cross section

$$(d^2\sigma/d\Omega dE)^{1\ phonon} = <b^2>\, (k/k_o)\, (<n> + ½ \pm ½)\, (\hbar Q^2/2M\omega)\, f(\omega)\, e^{-<u^2>Q^2}. \qquad [2\text{-}120]$$

Higher phonon terms can now give a particular energy transfer as a result of any appropriate series of phonon transfer, $\Delta E = \hbar(\pm\omega_1 \pm\omega_2 \pm \omega_3 .....)$. Thus the two phonon term will be proportional to the convolution of the frequency distribution with itself

$$f_2(\omega) = \int f(\omega - \omega_1)\, f(\omega_1)\, d\omega_1$$

where ω_1 and $\omega_2 = (\omega - \omega_1)$ are the two contributing phonons. Higher terms are obtained by further convolutions. These terms, however, are also proportional to $(Q^2)^{N_p}$ where N_p is the number of phonons involved. Thus, if one makes the measurements at sufficiently low Q the one-phonon term will dominate. This then provides a powerful method of measuring $f(\omega)$ which is of fundamental importance in the interpretation of many solid state properties.

For more than one atom per unit cell, the analysis is somewhat more complex, in that the amplitudes of vibration of each atom in the unit cell are different for each mode of vibration $\omega^s(\mathbf{q})$. These amplitudes can be obtained from the Born-Von Karman analysis. Stated formally, if the $\omega^s(\mathbf{q})$ value is an eigenstate of the dynamical matrix, the amplitude, $\underline{\xi}^s(\mathbf{q})$, is the eigenvector associated with that eigenvalue. It consists of the real space components of the amplitudes of vibration of each atom in the unit cell suitably normalized. Now the neutron scattering is proportional to $|\mathbf{Q}.\underline{\xi}^s(\mathbf{q})|^2$ and instead of using $f(\omega)$, we must now use the amplitude weighted version (54). Although in the general case this is somewhat complicated, it reduces to a fairly simple problem for hydrogen vibrations in a rigid lattice. In principle one could use this method to determine the forces holding hydrogen in clay lattices.

Coherent Inelastic Scattering. Coherent inelastic scattering is an extremely important technique which has not as yet been applied to clay lattices and will be very briefly discussed. As would be expected, interference effects dominate the scattering. Just as in elastic coherent scattering one must satisfy the condition $\delta(E'-E_o)\,\delta(\mathbf{Q} - \underline{\tau})$, so in elastic coherent one phonon scattering the phonon energy and momentum are introduced into conservation conditions, i.e.

$$\delta(E' - E_o + \hbar\omega^s(\mathbf{q}))\,\delta(\mathbf{Q} - \mathbf{q} + \underline{\tau})$$

where the second δ function can be regarded as momentum conservation condition if we regard $\hbar\mathbf{q}$ and $\hbar\underline{\tau}$ as the momentum of the phonon and that given to the crystal respectively. The vectors involved are illustrated in Fig. 2-10. When neutrons scattered from a crystal satisfy both these conditions separately, there is a sharp peak in the cross section and by knowing $(E' - E_o)$ and $\mathbf{Q}$ a value of $\omega^s(\mathbf{q})$ can be obtained. This is the only practicable method of measuring the variation of ω with $\mathbf{q}$. Values are normally measured along symmetry directions in the crystal yielding dispersion curves. A typical set of such data is given in Fig. 2-11. Further details of this technique can be found elsewhere (66).

2-3. NEUTRON SCATTERING INSTRUMENTATION AND METHODS.

2-3.1. General Introduction

This section contains a brief description of current neutron scattering instruments, together with some general information on data analysis methods, particularly in connection with quasi-elastic scattering measurements. The information regarding current instrumentation is divided into four sections covering the topics of diffraction, small angle scattering, quasi-elastic and inelastic scattering in accor-

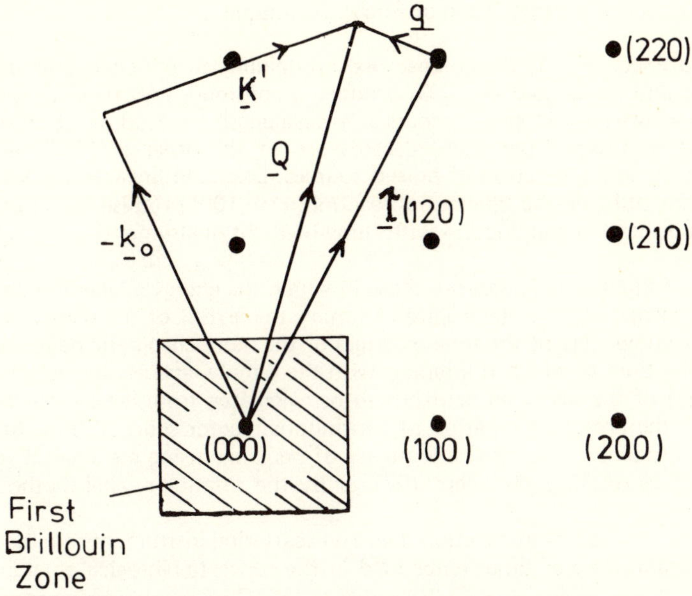

Figure 2-10. Illustration of momentum conservation in neutron-phonon scattering.

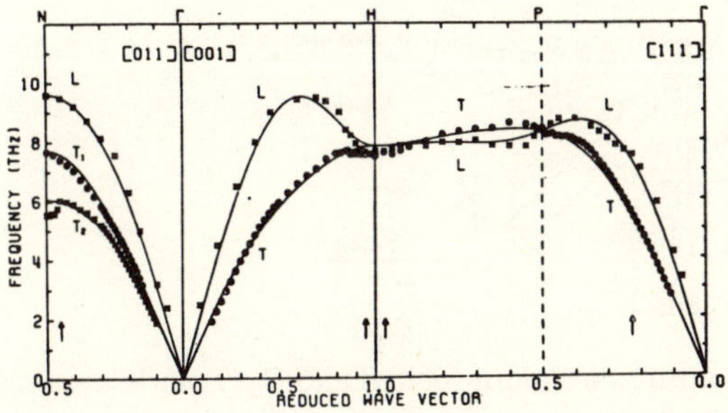

Figure 2-11. Typical phonon dispersion measured by neutron coherent inelastic scattering in a single crystal (69).

dance with the separation made previously. General information is given next regarding neutron sources and detectors while more detailed information is to be found elsewhere, e.g. Stirling (99).

2-3.2. Sources and Detectors; Time-of-Flight Technique

As outlined in 2-1, fission reactors, containing an enriched uranium core both cooled and moderated by D_2O, produce a continuous Maxwellian spectrum of neutrons whose peak intensity occurs at a wavelength $\lambda = 1.08$ Å. At present the highest neutron fluxes from such reactors are of the order of 10^{15} neutrons/ cm^2/second. A new generation of pulsed sources, based on linear accelerators(111) or sychrotrons utilizing the 'spallation' reaction, (19; 100) promise higher *effective* fluxes when utilized in conjunction with 'time-of-flight' instruments.

Time-of-flight instruments are those in which the energy analysis of scattered neutrons is carried out by electronically timing the arrival of the neutrons at the detectors. A knowledge of the time of origin of the monochromatic neutron pulse, and hence the time at which it interacts with the sample, enables the velocity (and hence energy) of the scattered neutrons to be calculated from the sample-detector distances. In the present generation of continuous reactor sources, 'time-of-flight' instruments produce monochromatic pulses of neutrons using mechanical velocity selectors such as rotating 'choppers' (97), or by the rotating crystal method (18).

At a fission reactor installation, neutron scattering instruments are located at the end of beam holes or tubes embedded in the concrete biological shielding (see Fig. 2-12).

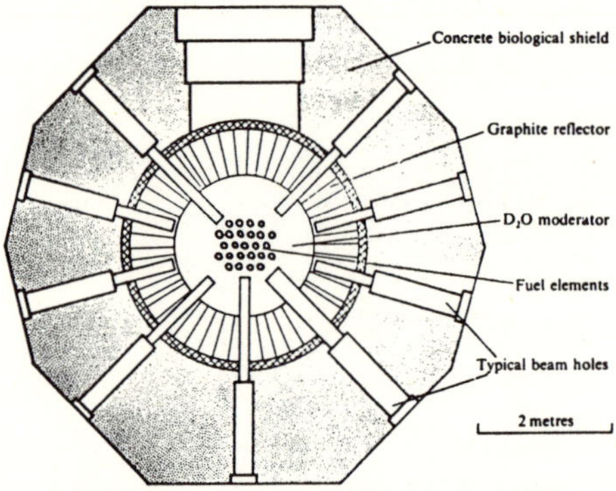

Figure 2-12. Typical layout of a uranium-fuelled and D_2O moderated reactor, showing beam holes.

Neutron detectors are based on the fact that, although neutrons themselves are non-ionizing, their collisions with certain light nuclei produce secondary charged particles which may be detected by ionization or scintillation methods. The two main types of detectors currently used are BF_3 proportional counters utilizing the (n,α) reaction and ^3He counters [(n,p) reaction]. Another type of detector, the position sensitive detector, is described briefly in section 2-3.4 in conjunction with small angle neutron scattering.

2-3.3. Neutron Diffractometers

General Description. Neutron diffractometers are directly analogous to x-ray instruments in that one requires a collimated, monochromatic source of radiation which interacts with the sample, the intensity of the diffracted radiation being then detected as a function of the angle of diffraction. The detector can usually be rotated independently of the sample, giving the possibility of various types of scan (see section 2-4.2). Diffractometers and samples are considerably larger than corresponding x-ray instruments in order to obtain sufficient counting rates as well as angular resolution.

One may distinguish two types of instruments: (a) powder, or two circle, diffractometers, and (b) single crystal, or four circle diffractometers. Both types of instrument can operate at thermal wavelengths ($\sim$ 1 Å) or at longer wavelengths via an intermediate cold source (see section 2-1). The advantage of the shorter wavelength instruments is that a larger range of momentum transfer, $\hbar Q$, may be obtained. The longer wavelength instruments enable materials with larger unit cells to be studied with greater resolution at low diffraction angles.

Powder Diffractometers. An example of a long wavelength powder diffractometer is the Guide Tube Small Angle Diffractometer in the Pluto reactor at AERE, Harwell, England (15, 50). This instrument operates at a wavelength of 4.7 Å, using beryllium polycrystals to filter out neutrons of less than 4 Å. After diffraction, a graphite crystal reflects neutrons of 4.7 Å into a BF_3 counter. The schematic layout of this instrument is illustrated in Fig. 2-13.

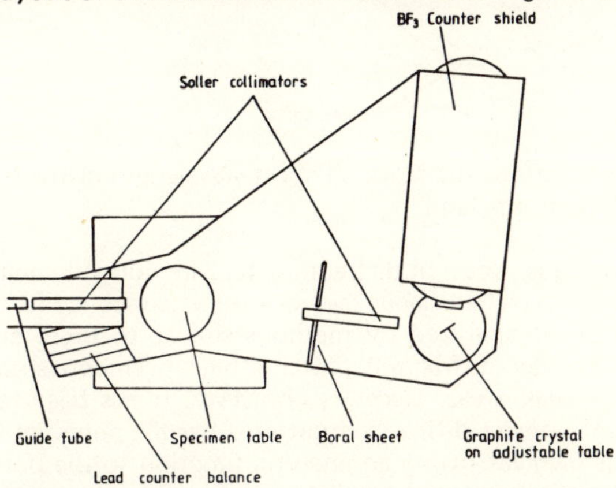

Figure 2-13. Schematic diagram of the Guide Tube Diffractometer at AERE, Harwell, England.

Single Crystal Diffractometers. An example of a single crystal diffractometer, which has been utilized in work on clays, is the D16 diffractometer at ILL, Grenoble, France. This is a conventional 4-circle diffractometer located at a cold source at the ILL reactor, which is able to operate at wavelengths between 3 and 6 Å. Sample environment chambers are available for instruments at controlled humidities (57).

A shorter wavelength single crystal diffractometer is the Mark VI four-circle diffractometer at AERE, Harwell (15, 99). Using a copper monochromator, an incident neutron wavelength of 1.18 Å is obtained. The instrument is located adjacent to a two-circle diffractometer whose wavelength can vary between 0.83 and 1.31 Å. The schematic layout of these instruments is illustrated in Fig. 2-14.

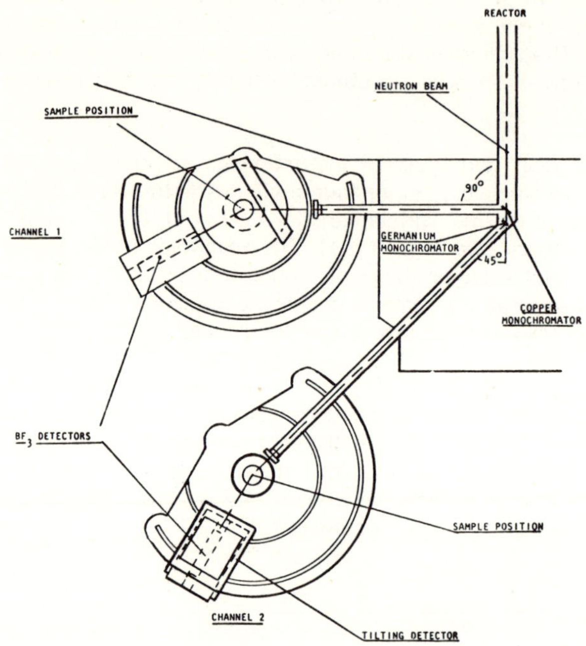

Figure 2-14. Schematic layout of the Mark VI short-wavelength diffractometers at AERE, Harwell, England.

Data Analysis Methods. Neutron diffractometers are normally computer controlled, data output being punched paper tape or magnetic tape for further processing. Single crystal data are analyzed by methods similar to those employed in x-ray crystallography. Powder diffraction patterns have been conventionally analyzed by integration of peak areas. Recently, however, it has been realized that direct profile analysis of powder diffraction patterns has the potential for yielding more information. This involves fitting an analytic function to the point-by-point experimental data by least squares methods (81).

2-3.4. Small Angle Scattering Instruments

One of the most versatile instruments for measurements of small angle neutron scattering is the D11 diffractometer at ILL (Grenoble, France).

The general layout of the instrument is illustrated in Fig. 2-15. D is a multidetector or position-sensitive detector containing 3808 actual elements of area 1 cm^3 filled with BF_3. The monochromator consists of a slotted mechanical velocity

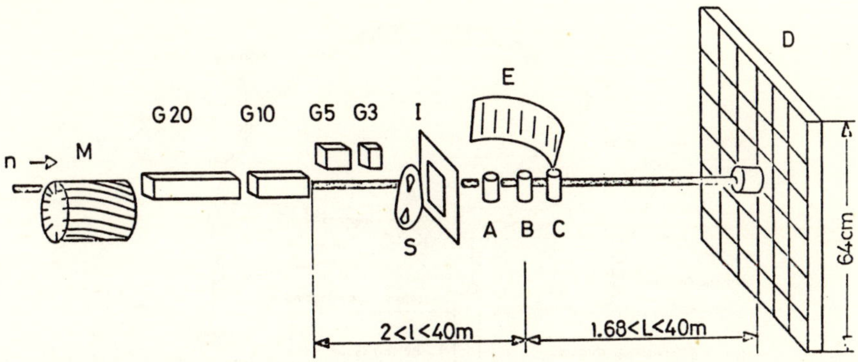

Figure 2-15. The D11 small angle scattering instrument at ILL, Grenoble.

selector. The multi-detector can be positioned at distances of up to 40 meters from the monochromator, giving measurements at small scattering angles corresponding to a Q-range of 10^{-3}–$0.5Å^{-1}$. A wide range of wavelengths is also possible depending on the rotary speed of the monochromator drum (56, 57). A similar instrument has recently been constructed at AERE, Harwell, the specifications of which are given in a recent report by Baston and Harris (15).

2-3.5. Quasi-Elastic Scattering Instruments

Introduction. Following is a discussion of two distinct instruments at ILL, Grenoble which have been used for measurements of translational and rotational diffusion in a wide range of liquids and molecular crystals. They are based on different principles, but often are used in parallel to yield complementary information. Differences arise from the fact that the two instruments are of different energy resolutions, and can therefore potentially separate diffusion processes taking place at different rates. Their application to studies of water diffusion in clays is discussed in section 2-4.4.

IN5 Multichopper Time-of-Flight Spectrometer. The IN5 spectrometer, whose schematic layout is illustrated in Fig. 2-16, is a 'time-of-flight' spectrometer, the wavelength of which can vary over a wide range. Four mechanical 'choppers' or rotors can be driven at speeds of up to 15,000 r.p.m. Their speeds and relative phases, or delay times, control both the incident wavelength and the energy resolution, as well as eliminating troublesome higher-order wavelengths. Elastic energy resolutions down to ca. 20 μeV can be achieved. The instrument is equipped with a large number of cylindrical ^{3}He detectors which can be positioned at up to 50 scattering angles between 8° and 135°. The sample is placed in a cylindrical box which can be filled with helium gas to minimize air-scattering effects.

The data collection is fully computer-controlled and can be recorded on magnetic tape.

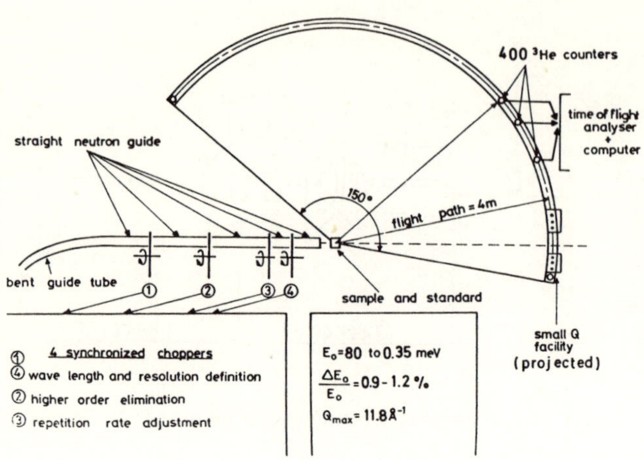

Figure 2-16. The IN5 multichopper time-of-flight spectrometer at ILL, Grenoble. (Quasi-elastic scattering).

Being a time-of-flight instrument, the cross section which is measured is not the standard double differential cross section $(d^2\sigma/d\Omega d\omega)$, but the equivalent time-of-flight cross section:

$$\frac{d^2\sigma}{d\Omega d\tau} = \frac{m}{\hbar\tau^3}\,\frac{d^2\sigma}{d\Omega d\omega} = \frac{m\sigma}{4\pi\hbar}\,\frac{\tau_0}{\tau^4}\,\text{Sinc}\,(\mathbf{Q},\omega) \qquad [2\text{-}121]$$

where m is the neutron mass. Moreover, the spectra are not precisely at constant Q in the quasi-elastic region, since in this case

$$Q = \frac{m}{\hbar}\left(\frac{1}{\tau^2} + \frac{1}{\tau_0{}^2} - \frac{2}{\tau\tau_0}\cos\theta\right)^{1/2} \qquad [2\text{-}122]$$

where θ is the scattering angle. In the above equation τ_0 and τ are the reciprocal velocities of the incident and scattered neutron respectively. The factor τ^3 in the denominator of the expression for $(d^2\sigma/d\Omega d\tau)$ implies that the 'time-of-flight' scale represented by equal intervals of τ is rather non-linear in energy. Data analysis has conventionally involved conversion of data to an energy scale, which can introduce errors. A direct simulation of the 'time-of-flight' spectrum profile, including inelastic and other background effects, has distinct advantages. The formalism of this approach, together with details of a FORTRAN computer program which performs least-squares fitting of data to a model 'time-of-flight' spectrum, is given in a recent paper by Hall *et al.* (45).

IN10 Backscattering Spectrometer. This instrument is based on the backscattering principle, which is designed to achieve very high elastic energy resolutions. This principle is based on a consideration of Bragg's Law, as discussed below. The instrument design is illustrated in Fig. 2-17. The wavelength is selected using a monochromator crystal (usually silicon). The energy range of the spectrum is selected by the frequency of the Doppler drive which oscillates the monochromator. After scattering from the sample, the neutrons are backscattered from silicon analyser crystals and detected by one of up to eight ^{3}He-counters. Energy resolutions down to ca. 1 μeV are obtainable. The instrument is fully automated, being controlled by a PDP 11 computer.

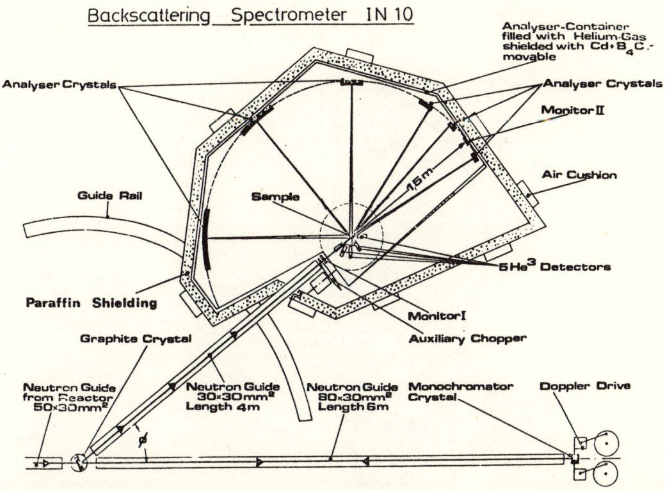

Figure 2-17. The IN10 backscattering spectrometer at ILL Grenoble. (High-resolution quasi-elastic scattering).

To discuss the backscattering principle consider Bragg's Law for first-order scattering, $\lambda = 2d \sin \theta$. Differentiating, one obtains $d\lambda = 2d \cos \theta \, d\theta$, whence $d\lambda/\lambda = \cot \theta \, d\theta$. Thus for a given angular range of scattering at the detector the wavelength (and hence energy) resolution is proportional to $1/\tan \theta$. This is a minimum when $\tan \theta$ tends to infinity at 90°. Thus backscattering (i.e. a Bragg angle of close to 90°) gives the optimum energy resolution for this type of instrument. More details are given in the paper of Alefeld *et al.* (7) and by Springer (97).

2-3.6. Instruments for Inelastic Incoherent Measurements

Generally the most useful instrument for inelastic scattering measurements on incoherently scattering samples is the beryllium filter spectrometer. Here one uses neutron energy loss scattering from a range of incident energies selected by a monochromating crystal to a fixed final energy. The scattered neutrons fall on a volume of polycrystalline beryllium, at liquid nitrogen temperature, where neutrons that can satisfy Bragg's Law are scattered out of the beam. For sufficiently large wavelengths, however, $\lambda = 2d \sin \theta$ cannot be satisfied, the cut-off coming at twice the maximum plane spacing in the crystal ($\theta = 90°$) which for Be is at 4 Å (.005 eV). Thus all scattered neutrons $.005 \text{ eV} > E' > 0$ are transmitted and detected by BF_3 counters behind the filter. Instruments of this type exist on the DIDO reactor at Harwell and at the ILL. The instruments are controlled by a computer which can set up the required incident energies. Useful energy ranges from $\sim$ 40 meV to several hundred meV can be covered by these instruments by selecting appropriate planes of the monochromator crystal. The Harwell instrument is illustrated in Fig. 2-18, and further details may be found elsewhere (15).

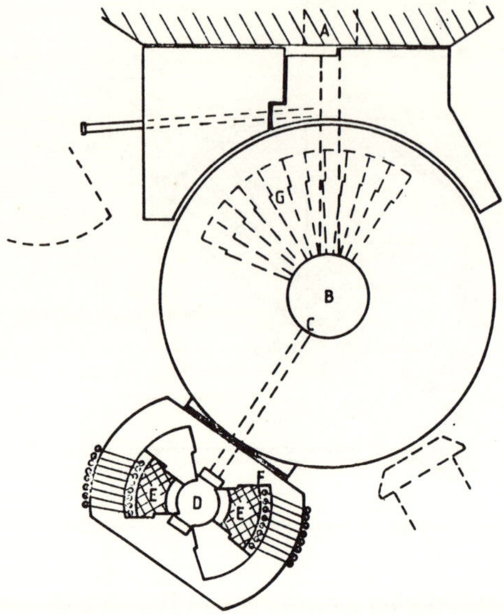

Figure 2-18. The Harwell Beryllium Filter Spectrometer (inelastic scattering).

2-4. APPLICATIONS OF NEUTRON SPECTROSCOPY TO STUDIES OF CLAY MINERALS.

2-4.1. Introduction

In this section the present status of the application of neutron spectroscopic techniques to the study of various aspects of the structure and interactions of clay minerals will be reviewed, all significant published articles which have come to our attention to date being covered. The experiments which have been performed up to the present time can be divided into four categories, the theoretical background of each which has been given in section 2-2. These are: (a) coherent Bragg diffraction measurements of structural aspects of clay minerals and clay intercalates; (b) small angle scattering measurements, chiefly of the colloidal properties of clay sols; (c) incoherent quasi-elastic scattering measurements of the motions of adsorbed and intercalated hydrogenous molecules; and (d) inelastic scattering studies of the vibrational modes of hydroxyl groups in clay minerals.

Other relevant published work on related materials of interest, such as nonsilicate layer structures (e.g. graphite and chalcogenides), oxides and zeolites are covered in a recent review of the applications of neutron scattering techniques to surface chemistry in general (40), in conjunction with this text, should give a broad overall perspective on the current activities in this challenging and rapidly expanding field.

2-4.2. Neutron Diffraction Studies

Structural Studies on Crystalline Micas. As discussed in section 2-2, one principal advantage of neutron diffraction (ND) over x-ray diffraction (XRD) is the fact that the coherent neutron scattering cross section of hydrogen is of comparable magnitude to those of most other elements, in particular for all elements commonly occurring in aluminosilicate structures. Thus the ND technique can readily locate hydrogen atom positions. Furthermore, if the crystal structure of the material has already been determined by XRD, as is usually the case, then the analysis of the neutron data is greatly simplified since the positions of all nonhydrogen atoms are already established. Alternatively, a full analysis of the data from an ND experiment can serve as an independent check on an x-ray crystal structure.

ND structural determinations of two micas, muscovite and phlogopite, have been achieved. Rothbauer (85) determined the structure of a $2M_1$ muscovite using a four circle diffractometer (wavelength 1.19 Å). From 625 observed reflections the monoclinic unit cell was found to have parameters a = 5.192 Å, b = 9.015 Å, c = 20.046 Å, β = 95.735°. The hydrogen positions calculated indicated that the OH bonds were orientated at an angle of approximately 78° to the c*-axis, in good agreement with previous infrared data (103) and with electrostatic energy calculations by Giese (34).

Rayner (79) determined the structure of phlogopite using a 4-circle diffractometer (wavelength 1.17 Å). From 293 observed reflections, the unit cell parameters were found to be a = 5.322 Å, b = 9.206 Å, c = 10.24 Å and β = 100.03°. In this case, the results indicated that the hydroxyl groups were orientated approximately parallel to the c*-axis, again a result in agreement with infrared measurements (92). The difference in the OH group orientations between these two micas is attributable to the differing charge distribution in the octahedral sheet between dioctahedral and trioctahedral micas. In the former case, the protons tend to point towards adjacent vacant octahedral cation sites. In the latter case, where all cation sites are occupied, the protons are located as far as possible from the neighboring Mg^{2+} ions. This implies that the protons lie much closer to the interlayer potassium ions in trioctahedral structures, a factor which may contribute to the lower stability of trioctahedral structures with respect to potassium exchange.

Neutron Diffraction Studies of Clay-Water Systems. In this section we will discuss the results of ND measurements on expanding lattice clays containing interlamellar water sheets, i.e. the three-layer minerals like smectite and vermiculite. In particular, our attention is focused on the information potentially available regarding the structural organization of the interlamellar water in these systems.

With the exception of orientated single flakes of vermiculite, experiments are necessarily restricted to polycrystalline samples in which the number of observable reflections has been too small to permit a full three-dimensional structural analysis. Nevertheless these samples can be prepared in such a manner as to exploit the plate-like morphology of the microcrystals to produce considerable preferred orientation in one dimension (with respect to the orientation of the c*-axes). As in the case of XRD, reflection measurements on such partially orientated samples will effectively enhance the intensities of the (00ℓ) reflections, providing the possibility of performing a one-dimensional Fourier synthesis from which the z-components of the atomic positions (their displacements with respect to the c*-axis) can be derived.

For ND measurements on polycrystalline preferentially orientated clay specimens, slab samples are required, of dimensions somewhat larger than for XRD measurements. The sample thickness necessary is determined by the need to obtain good statistics with reasonable counting times, while simultaneously minimizing extinction or multiple scattering effects, which would necessitate complex corrections to the experimental data. The latter requirement can only be fulfilled if the transmission factors of the samples do not fall below 90-95%. As for X-rays, the attenuation of a cold neutron beam on passing through any material is given approximately by a relationship of the type: $I = I_0 \exp(-\Sigma\ell)$ where I is the intensity remaining after penetrating a distance x into the sample, I_0 is the incident beam intensity and Σ is called a linear absorption coefficient or macroscopic cross-section.

From the neutron-scattering cross sections of the elements comprising clay structures, the values of Σ can be calculated. For a montmorillonite-water system containing two interlamellar water layers, Σ is approximately 0.2mm^{-1} (Appendix 1). Making the approximation that, for small values of Σx, $\exp(-\Sigma x) \simeq 1-\Sigma x$ one finds that I/I_0 values in the range 0.90-0.95 require thickness of approximately 0.25-0.50 mm. The attenuation of neutron beams in silicate materials is thus a factor of several hundred times less than for X-rays of comparable wavelengths. The other dimensions of the samples are not critical, but analysis is simplified if the sample remains completely illuminated by the incident beam at all sample orientations. This will normally limit these dimensions to a few centimeters.

Among the techniques which have been utilized to prepare slab samples of suitable dimensions from properly pre-treated homionic montmorillonite suspensions (75) are: (a) sedimentation, accompanied by suction of the excess water through a micropore filter (42); (b) compression, via a pressure plate technique (13, 21); and (c) heat-assisted evaporation (48). After drying, it is necessary to equilibrate the samples at a suitable relative humidity over standard solutions prior to the ND experiment.

For experiments on slab samples, all that is required is a diffractometer having a detector which can be scanned in a plane containing the normal to the slab, the detector angle being $2\theta_B$ where θ_B is the Bragg angle. Three typesof scans may be made: (a) $\omega = \theta_B, 2\theta_B$ scan. This is the normal 'half-angle' diffraction scan, in which the sample is rotated in steps of ω which are half the size of the steps made by the detector in 2θ. Thus the ideal reflection geometry for Bragg scattering is always maintained (Fig. 2-19), the resulting spectrum being the neutron equivalent of the conventional XRD pattern; (b) ω-scan with $2\theta_B$ constant. Having located the detector position $2\theta_B$ corresponding to the maximum for any given intense (00ℓ) reflection, the detector can be held fixed at this position while the sample is rotated in small increments of ω. This gives a 'rocking curve' from which the mosaic spread of a crystal, or particle orientation distribution function for a less well-ordered sample, can be measured (12); (c) Powder diffraction scan. In this case the sample is placed normal to the incident beam and the detector is rotated through the appropriate range of $2\theta_B$; and (d) transmission measurements. The sample transmission can be measured experimentally, with the sample at any angle to the beam, by one of two methods. In the first method, a low efficiency neutron monitor counts the flux of neutrons before and after penetrating the sample. In the second method neutron counts are taken with and without the sample in position with the detector located in the straight-through beam position. In this case it is necessary to create an extremely fine aperture by means of cadmium shielding in order not to exceed the counting rate capacity of the detector (21). Among the instruments which have been used in the studies discussed later are the Guide Tube Diffractometer at AERE Harwell U.K. (50) and the D16 diffractometer at ILL Grenoble, France. These instruments have incident wavelengths of approximately 4.7Å and 4.6Å respectively.

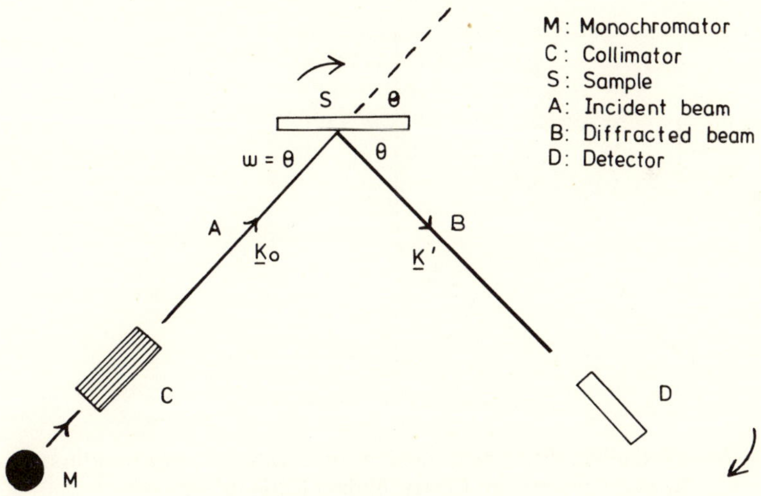

Figure 2-19. Geometry for neutron Bragg diffraction from a slab sample.

For most slab clay-water systems the ω-scans (rocking curves) are of the form illustrated in Fig. 2-20a (42). These data were from an Mg^{2+} montmorillonite (API No. 26, Clay Spur, Wyoming, U.S.A.) equilibrated with H_2O at 75% R.H. to produce a two layer hydrate having d_{001} = 15.6Å. The rocking curve was measured with the detector positioned at the maximum of the 001 reflection (Fig. 2-20b). The curve is characterized by two distinct minima, the size and positions of which are found to depend on (a) the neutron wavelength used and (b) the sample dimensions. In all cases it has been found that angular separation between the minima is approximately equal to $2\theta_B$, where θ_B is the Bragg angle for the reflection. They may therefore be attributed to attenuation maxima occurring when the path length within the sample of either the incident or diffracted beam reaches a maximum (i.e. lies in the plane of the slab sample). Cebula *et al.* (21) have reported similar data for Li+ montmorillonite hydrates.

With shorter wavelength diffractometers the Bragg angle for any reflection will be correspondingly smaller, and the two attenuation minima may merge into a single minimum, as in the data presented by Hawkins (48) and by Hawkins and Egelstaff (49).

These pronounced attenuation features occur only in relatively disordered samples where the spread in particle orientations is comparable to, or greater than the angle $2\theta_B$. The effect could be reduced by choosing a higher order (00*l*) peak about which to measure the rocking curve; unfortunately it is found that the higher order reflections in montmorillonite-H_2O systems are considerably weaker than the 001 peak. One has therefore to resort to making attenuation corrections (see Wignall (108) for the calculation of attenuation factors for slab geometries). The intensity distribution of the corrected rocking curve $I(\omega)$, is directly proportional to the number of platelets orientated correctly to give Bragg diffraction, and thus gives directly the 'mosaic spread' or orientation distribution function of the platelets.

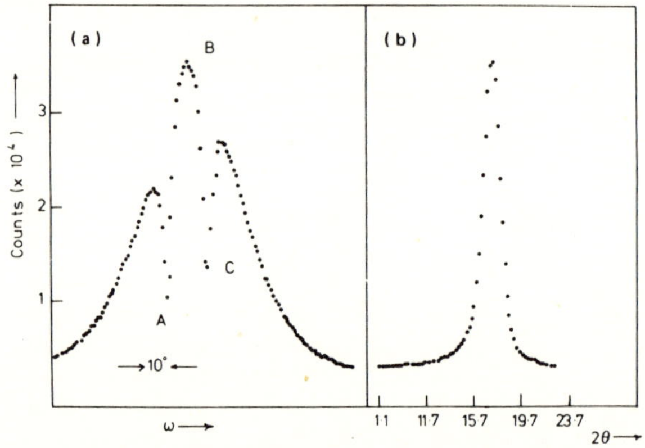

Figure 2-20. (a) Neutron diffraction rocking curve for Mg^{2+}-montmorillonite (d_{001} = 15.6 Å). (b) Corresponding (001) reflection.

Two pieces of information can be obtained from these curves: (a) the full width at half maximum (FWHM) $\Delta\omega$ of the orientation distribution function; and (b) the height of the background. For both monovalent and divalent cation-exchanged montmorillonites containing one or more interlamellar layers of water, $\Delta\omega$ values in the range 40° - 50° have been found (21, 42), i.e. there is a considerable degree of disorder. Moreover the background level in the wings of the rocking curves ($\omega - \omega_0 \sim 90°$) is slightly higher than the incoherent background beneath the (001) Bragg peak, indicating that a small but detectable fraction of the diffraction arises from platelets which are highly tilted from the plane of the specimen. Cebula *et al.* (21) have suggested that this may be attributed to a randomly distributed fraction of fine particle size, though this conclusion would seem somewhat at variance with results that follow. The results discussed so far were obtained for clays which were not fractionated with respect to particle size, having 0.2-2.0μm esd.

More recent experiments by the Birmingham group (47) have investigated the influence on platelet orientation of a number of factors including (a) the nature of the exchange cation, (b) particle size fractionation, (c) ultrasonification prior to sample sedimentation and (d) film thickness, i.e. building up a slab sample from a large number of relatively thin films, down to approximately 10 microns in thickness. Of these the only factor which was found to significantly influence the orientation distribution function was the particle size. It was found that the finest particle fraction (<0.05 μm esd) of Na+ montmorillonites gave considerably enhanced preferential orientations characterized by rocking curves of approximately 15°-20° FWHM. For Ca^{2+} montmorillonites, the degree of orientation achieved was somewhat lower (25-30° FWHM) though considerably better than the larger particle size fractions or unfractionated clays. These results, though perhaps contrary to expectations, are entirely in agreement with earlier work of Roberson *et al.* (83) who studied the morphology of size-fractionated Na+ montmorillonites by electron microscopy and x-ray diffraction. They found that only the finest fraction (<0.05 μm esd) was composed entirely of separate, planar flakes. All larger fractions were composed of microaggregates even in dilute suspension, thus limiting the orientations which could be achieved. X-ray rocking curves of the (004) reflections in air-dried films confirmed the considerably enhanced orientation of the fine fractions. It is of interest to note that laponite (a synthetic hectorite, which is composed predominantly of extremely fine particles) also exhibits narrow rocking curves of FWHM 20° or less (62).

Hawkins (48) has apparently obtained a similar degree of orientation on an unfractionated montmorillonite. This sample had, however, been previously subjected to a technique aimed at deuterating the lattice OH groups. This technique, which involved heating to 350°C under vacuum, caused a color change probably due to oxidation of Fe^{2+} to Fe^{3+}. The effect of the deuteration technique on the nature of the clay surface or the microstructure of the sample is not well understood.

Consider now the general features of the ND patterns of partly-orientated montmorillonite-water systems, making comparisons with the more established XRD technique. The preliminary work done in this field (21, 42, 48) has led to the following general conclusions: (a) neutrons confirm x-ray measurements of the stepwise increase in basal spacing as a function of relative humidity in both monovalent (Li, Na, K, Cs) and divalent (Ca, Mg) exchanged montmorillonites. Moreover, good agreement is found between d-spacings measured by neutrons and x-rays on comparable samples; (b) for poorly orientated samples the number of orders of ($00l$) reflections observed with neutrons is somewhat fewer than for x-ray diffraction patterns of comparable samples. Moreover the breadth of the neutron peaks are larger than the corresponding x-ray peaks when plotted on the same momentum transfer scale (see Fig. 2-21); (c) the poorly oriented samples (mosaic FWHM 40°-50°) usually exhibit only the first 2-3 orders, of which only the (001) peak is of significant intensity (21, 42). Samples with more preferred orientation (47, 48) show up to six 00*l* reflections; and (d) replacement of interlamellar water by D_2O, though virtually eliminating the large incoherent background scattering beneath the Bragg peaks, also significantly reduces the intensity of the (001) reflection due to a loss of contrast variation. This arises due to the positive scattering length of the deuterium nucleus; as discussed in Section 2-2. Thus in montmorillonite-D_2O systems either no reflections are observed at all in poorly orientated samples (21) or else a series of fairly weak peaks is observed in more highly ordered samples (48).

The weakness of the higher order peaks may be attributed to (a) relatively low structure factors, (b) their breadth, and (c) the 'masking' effect of the large incoherent background scattering due to hydrogen. Since only the first six (00*l*) reflections are observable at the present time, it is not expected that any significant structural information can be deduced, since the resolution of the Fourier projection will be only 2.5-3Å at best. Hawkins (48) has calculated a one-dimensional Fourier projection for a D_2O-Na^+-montmorillonite system, but the interlamellar region shows only a few weak maxima which may not necessarily be due to preferred hydrogen positions (e.g. they could be termination-of-series effects). At present, beyond confirming the stepwise expansion of montmorillonite on hydration, no simple structure for the interlamellar water can be deduced by ND, even with the optimum degree of orientation currently possible. The situation is different with vermiculite single flakes (mosaic spreads 3° or less) and work is currently in progress in our laboratories on the analysis of the ND pattern of a flake of Kenya vermiculite, which has approximately 20 observable 00*l* orders. It may be possible to locate the hydrogen atom positions of the interlamellar water in this case, extending the structural information for the Mg^{2+} vermiculite system derived from X-ray single crystal studies (68, 94).

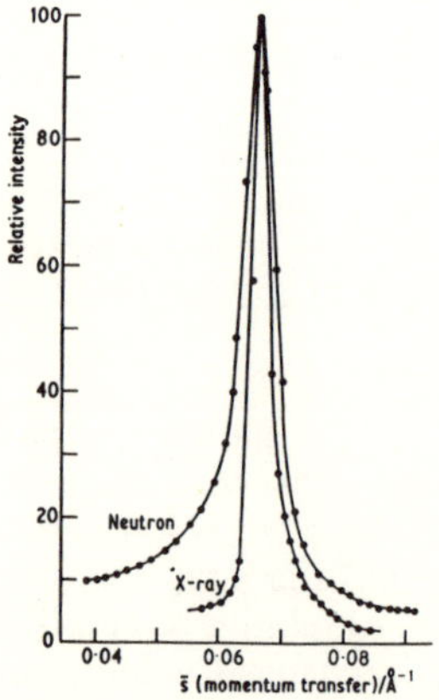

Figure 2-21. Comparison of x-ray and neutron diffraction patterns for Li* montmorillonite, showing the (001) reflection, plotted on identical momentum transfer scale.

Returning to the imperfectly orientated montmorillonite-water system, the chief differences between the results from ND and XRD are: (a) the ND reflections are somewhat broader and less well defined; and (b) the degree of orientation as measured with neutrons is sometimes apparently lower than that measured during X-rays. Thus neutrons (which diffract from the entire sample) appear to indicate a greater spread of d spacings and platelet orientations than X-rays (which penetrate the surface layers only). The surface layer of sedimented samples appears to be much more homogeneous and well-ordered than the bulk sample. It is possible that this is due to differential rates of sedimentation during sample preparation, the fine particle size fraction being concentrated near the top surface of the film.

Neutron Diffraction Studies of Intercalation Complexes of Clay Minerals with Organic Molecules. It is well known that smectites and vermiculite (and to a lesser extent clay minerals of the kaolinite group) form a wide range of intercalation complexes with organic molecules, showing well defined X-ray basal spacings. Nevertheless, very few accurate determinations of the structures of such intercalation complexes have been made. Molecular conformations are usually inferred from geometrical considerations alone, or perhaps aided by polarized infrared data. In cases where structural determinations have been made these have been almost exclusively by XRD. Consequently no information on hydrogen atom positions, considered essential for a complete characterization of the structure and bonding of such complexes, has been obtainable.

Among the reasons for the paucity of even one-dimensional structural determinations of clay intercalates are the problems discussed earlier in connection with clay-water systems, i.e. that the relatively small number of $(00l)$ reflections observable in disordered samples severely limits the resolution of the Fourier projections. This is a problem common to both XRD and ND measurements made on imperfectly orientated polycrystalline specimens.

The structures of only two clay-organic intercalates have been determined by ND: (a) the Na+ montmorillonite-pyridine system; and (b) the kaolinite-formamide intercalate. Na+ montmorillonite forms a stable intercalation complex with pyridine, having a d_{001} spacing of 14.8 Å. Earlier x-ray studies showed that this complex consists of a single layer of interlamellar pyridine molecules whose aromatic planes lie perpendicular to the silicate sheets (36, 37). Adams *et al.* (1) obtained 14 $(00l)$ reflections in orientated films of this complex using a two-circle neutron diffractometer having an incident wavelength of 1.18 Å. They obtained a one-dimensional Fourier projection perpendicular to the silicate sheets (Fig. 2-22). Calculations of the peak heights indicated that the C-C axis lay parallel to the silicate sheets. This conclusion is, however, at variance with earlier polarized infrared data which indicated that the long C-N axis lay parallel to the silicate sheets (93), so that these results may still be open to question.

More recent studies on the Na^+-montmorillonite-pyridine system have concentrated on the double-pyridine layer intercalate (d_{001} = 23.3 Å). Adams *et al.* (4) studied the rate of exchange of H_2O and D_2O in this complex by observing changes in the ND pattern as a function of time. This work was carried out on a multidetector diffractometer (D1B at ILL, Grenoble) which permits quite rapid measurements of diffraction patterns. From the observed 'half-life' of the exchange process (ca. 15 minutes) and the nature of the changes observed in the diffraction profiles they suggested that the mechanism was D for H rather than molecular replacement. Further kinetic diffraction studies of the exchange of pyridine and deuteropyridine in this system (6) have led to an estimate of a diffusion coefficient for pyridine molecules which is extremely slow ($\sim 10^{-16}$ m^2 s^{-1}).

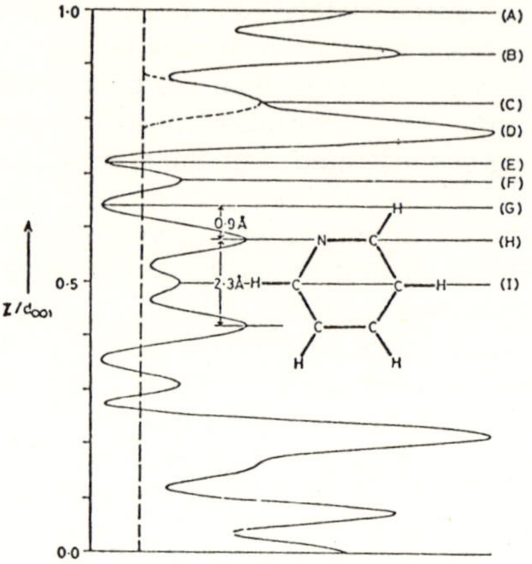

Figure 2-22. Fourier projection of nuclear density distribution along the c*-axis of the Na+-montmorillonite-pyridine intercalate.

Adams and Jefferson (2) performed a 3-dimension XRD crystal structure determination on the dickite-formamide intercalation complex. A later ND study of a preferentially orientated kaolinite-formamide specimen (3) enabled the conformations of the interlamellar formamide molecules with respect to the c*-axis of the complex to be deduced from a one-dimensional projection, assuming the z-coordinates of the atoms in the kaolinite unit cell to be identical to those determined for the dickite-formamide system.

2-4.3. Small Angle Neutron Scattering Studies of Colloidal Systems

The technique of small angle neutron scattering (SANS) is becoming increasingly important in the study of structural features of relatively large dimensions (50-5000 Å). These features include the study of pore size distribution in solids, the dimensions and conformation of polymers and macromolecules in the solid state or in solution, and the study of colloidal dispersions. Relatively little work based on this technique has been published on clay colloid chemistry. However Cebula *et al.* (20) have published some SANS data on montmorillonite sols (1% w/w suspensions) for a range of monovalent exchange cations. For Li+ montmorillonite the intensity of the small-angle scattering was approximately linear on a $\ln[Q^2 I(Q)]$ versus Q^2 scale, indicating that the data followed a relationship of the type, $I(Q) \propto \exp(-Q^2 H^2/12)/Q^2$, which is appropriate to thin disc-shaped particles of thickness H (see section 2-2). From the gradient Cebula *et al.* (20) obtained a value of H of approximately 10.3 Å, indicating that the system is well dispersed, single clay platelets. Marked changes in the form of the Guinier plots were observed with increasing cation size. Cs+-montmorillonite exhibited a Q^{-4} dependence (38) over most of the available momentum transfer range, the curve flattening somewhat at lower momentum transfers. From the value of Q at which this occurred, an upper limit of ∿ 40 Å for the mean platelet thickness could be calculated. More recent work (23) has compared the SANS from 0.8% w/w sols of Li+, K+, and Cs+ exchanged montmorillonite containing a range of relative H_2O/D_2O concentrations. The average structures were found to be (a) Li+-montmorillonite—completely dispersed platelets (H ∿ 10 Å); (b) K+-montmorillonite—two clay platelets interleaved with a double layer of water molecules (H ∿ 25 Å); and (c) Cs+-montmorillonite—three clay platelets interleaved with two double layers of water molecules (H ∿ 40 Å). One surprising result of these studies was the absence of significant exchange of H_2O and D_2O between the platelets over the time scale of the experiment (∿ 1 hour) whereas the effective diffusion coefficient measured recently by quasielastic neutron scattering (∿ 10^{-10} m^2 s^{-1}, see next section) would suggest an exchange time of a fraction of a second. The difference may be due to a barrier to diffusion occurring at the platelet edges due to localization of charge, while the intracrystalline diffusion is rapid but somewhat localized. This is in agreement with a detailed interpretation of the quasielastic neutron scattering data (see section 2-4.4).

There is a close analogy between the SANS technique and related techniques including the scattering of electromagnetic radiation such as γ-rays, X-rays (38) and light (Rayleigh scattering). Ramsay *et al.* (78) have utilized the last technique to investigate the structure of concentrated boehmite (AlOOH) sols as a function of solid and electrolyte concentration, in conjunction with quasielastic neutron scattering measurements.

Though the application of SANS to studies of the state of aggregation in clay colloids is clearly in its infancy, it appears to be of great potential usefulness for studies on these systems. For example, variation in mean particle size is clearly a factor which has an influence on the rheological properties of clay suspensions.

2-4.4. Quasielastic Neutron Scattering Studies of the Dynamics of Molecules Associated with Clay Surfaces

Clay-Water Systems. In this section, the experimental results which have been obtained to date on the dynamics of interlamellar water molecules in clays by quasielastic neutron scattering (QENS) will be discussed. The next section deals with some possible models which may explain these results, in the context of related information regarding the structure and mobility of the interlamellar water obtained using other techniques such as tracer diffusion studies, ESR, NMR and IR spectroscopy and diffraction studies.

The first studies of clay water systems by QENS were made by Olejnik, White and co-workers (55, 71, 72). They examined monovalent (Li+ and Na+) exchanged montmorillonite and vermiculite equilibrated to produce a wide range of water thicknesses. They derived 'effective' diffusion coefficients, D_{eff}, for these systems from the gradient of their broadening curves (plots of quasielastic energy broadening against Q^2) and showed that these values were quite strongly dependent on the water layer thickness. The values of D_{eff} showed an exponential dependence on inverse water layer thickness (see Fig. 2-23) the limiting value of D_{eff} for large spacings being in good agreement with the value for bulk water

($\sim$ 2.3 x 10^{-9} m^2 s^{-1}). These results were interpreted in terms of a macroscopic thermodynamic model assuming the interlamellar water to have a concave curved meniscus, which predicted an exponential dependence of D_{eff} upon 1/d, where d is the water layer thickness. The gradient of a plot of log(D_{eff}) against 1/d was in fair agreement (within 30%) of the theoretical value.

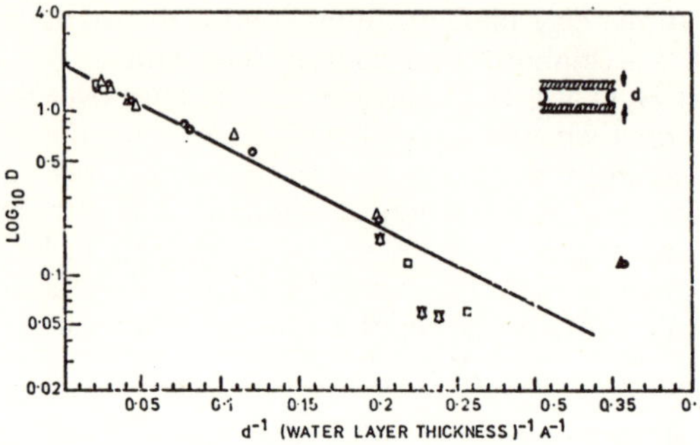

Figure 2-23. Dependence of apparent diffusion coefficient for water in clays on inverse interlayer expansion (logarithmic scale).

These results clearly show that (a) the interlamellar water is predominantly liquid-like rather than having an ice-like structure; (b) its mobility is somewhat restricted in comparison with bulk water, though increasingly less so as the interlayer thickness increases. Some care should be exercised in any interpretation of the quantitative conclusions of these workers for a number of reasons. First, the experimental energy resolution of the spectrometer employed was somewhat limited (ca. 250 μeV) and only rudimentary data analysis techniques were employed, so that the broadening measured may not be very accurate. Second, their interpretation was wholly in terms of translational macroscopic diffusion, and does not account for the possible influence of spatially restricted diffusion (41) or rotational motions (27) which may contribute both elastic and quasielastically broadened components to the observed data.

In addition, Low (67) has reviewed a number of measurements of the viscosity, η, of water in Na+-montmorillonite, derived either directly or indirectly by a number of techniques. Since the coefficient of viscosity for water in clays can be related to the water self-diffusion coefficient by a modified Stokes-Einstein equation, the D_{eff} values of Olejnik and White (72) could be used to calculate η. Although quite good agreement was found between the values of η derived from the QENS data and those derived by other techniques, the notion of a "curved meniscus" was dismissed as physically unreasonable since capillary action is not a principal mechanism of swelling.

P.G. Hall and co-workers (39) have also considered the thermodynamic model of Olejnik and White (72) in studies of QENS from thin water films adsorbed on silica. Though they found reasonable agreement with the model at high surface coverages (large p/p_o values) diffusion coefficients at lower coverages fell well below the theoretical predictions, suggesting that localized surface-adsorbate interactions must become important and that the bulk thermodynamic model is no longer applicable.

Recently, higher energy resolution measurements of QENS from Li+-montmorillonite (23) and Ca^{2+} and Mg^{2+} montmorillonite and vermiculite have been published (42, 44). The results for the divalent exchanged clays were obtained from orientated samples prepared by a sedimentation-suction technique (42). QENS measurements on the IN5 time-of-flight spectrometer (at energy resolutions down to 30 μeV) and the IN10 backscattering spectrometer (energy resolution ca. 1 μeV) were performed. The results indicated no significant anisotropy in quasi-elastic broadenings between transmission and reflection measurements (see Fig. 2-24). This indicates that the diffusive processes are insensitive to the direction of Q, i.e. to the direction of the molecular motions. Moreover, detailed analysis of the peak shape from the IN5 data revealed a number of facts: (a) The fraction of quasielastic to total (elastic plus quasielastic) intensity, X, was in all cases Q-dependent, following the variation expected for a predominantly localized diffusion process, such as molecular reorientations about fixed axes (14, 27) or diffusion within a restricted volume (41). The data for a two-layer Ca^{2+}-montmorillonite hydrate is illustrated in Fig. 2-25; (b) Though the average broadenings, and hence 'average' diffusion coefficients, D_{eff}, did increase with increasing water layer thickness, the best least squares fit to the data using the program QUELDA (45) indicated in all cases a purely elastic (unbroadened) component that was larger than could be predicted if all the water was undergoing a similar type of motion. The extra elastic fraction observed, over and above the elastic components due to a localized diffusion process and to the scattering from the clay lattice, must be attributed to a tightly bound fraction of water molecules immobile on the QENS timescale (having a residence time, or orientational correlation time, considerably longer than 10^{-11} seconds; (c) The limiting values at high Q of the relative intensity of the elastic and quasielastic fractions, after subtracting the elastic fraction due to the clay lattice itself, can be calculated from cross-sections (see Appendix 2). The values show good correlation with values for the relative populations of hydration and non-hydration shell water molecules, assuming divalent cations to exhibit 6-fold coordination when two or more molecular layers are present, (26, 94). Thus it appears that the relatively immobile fraction can be attributed to the inner hydration shell water molecules; (d) The IN10 data exhibited considerably lower quasi-elastic broadenings showing an approximately linear dependence of the broadenings ΔE on Q^2 up to about $Q^2 = 0.6$ Å^{-2}. From the points, one obtains a value of $D = 3.4 \times 10^{-10}$ m^2 s^{-1} for the two-layer hydrates of both Ca^{2+} and Mg^{2+} montmorillonite. This may well be attributable to truly macroscopic translational diffusion of the non-hydration shell water molecules, but certainly does not fit well to a two-dimensional model for diffusion parallel to the platelet surface only (101). Thus it appears that jumps both parallel and perpendicular to the silicate sheets occur. Moreover, the higher values of D_{eff} obtained from IN5 data, listed in Table 2-3, are almost certainly due predominantly to localized motions of non-hydration shell water having a correlation time of about 10^{-11} seconds; and (e) For the two layer Ca^{2+} vermiculite hydrate the average D_{eff} value is somewhat lower than for the two-layer Ca^{2+} montmorillonite (39).

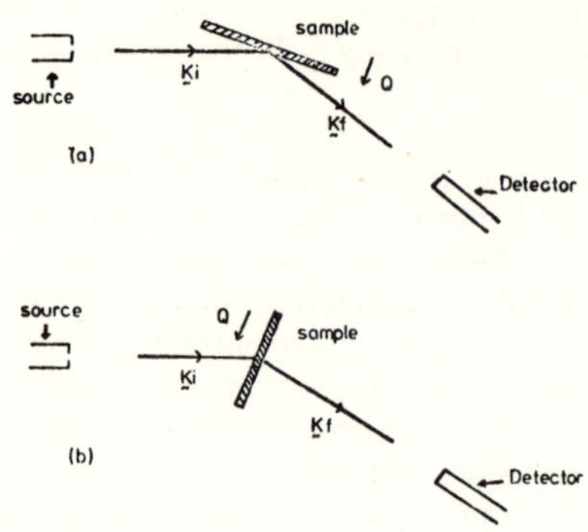

Figure 2-24. Scattering vector diagrams showing direction of momentum transfer for reflections and transmission geometries.

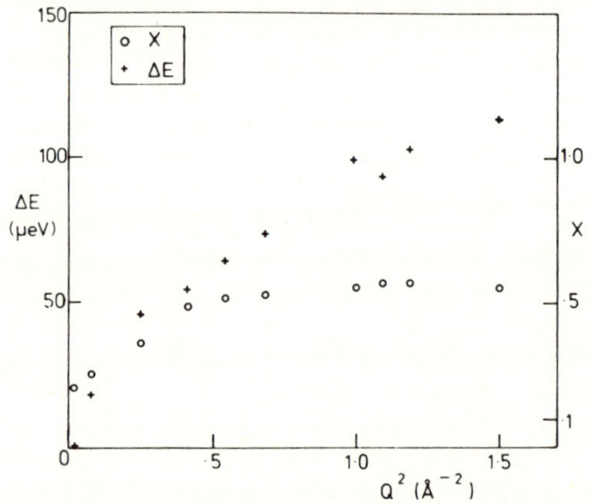

Figure 2-25. Variation of Quasi-elastic broadening, ΔE, and quasi-elastic fraction, X, with Q^2 for a Ca^{2+} montmorillonite two layer hydrate.

Measurements of QENS from the Li+ montmorillonite water system (23) gave similar results. These workers found values for the translational diffusion coefficient of 3.0, 4.0 and 5.2×10^{-10} $m^2\ s^{-1}$ in the one, two and three layer hydrates respectively, together with a similar rotational correlation time to that found for the divalent (Ca^{2+}, Mg^{2+}) montmorillonites. These workers did not observe the extra elastic component found by Hall *et al.* (42, 44) in the divalent cation-exchanged montmorillonite, which may perhaps be attributed to an "averaging" or exchange effect between the two different water environments due to the higher mobility of the small Li+ cations. Further experiments are required in order to clarify this point. It is of interest to note that neutron scattering studies of the water content of cements (46) have also been interpreted in terms of "free" and "bound" water components.

Adams, Breen and Riekel (6) have investigated the dynamics of water in the Na+-montmorillonite-pyridine-water system by high resolution QENS measurements. They derived a water self-diffusion coefficient of 6.1×10^{-11} $m^2\ s^{-1}$ at

Table 2-3. Apparent diffusion coefficients and asymptotic quasi-elastic fractions for clay-water samples

Sample	No. of water layers	Component	D_{eff} (m^2s^{-1} x 10^{-10})	Quasi-* elastic Fraction, X
Ca^{2+} montmorillonite	2	Broad	13.7	0.55
—	—	Narrow	3.4	—
Ca^{2+} montmorillonite	3	Broad	16.0	0.65
Mg^{2+} montmorillonite	2	Broad	15.0	0.60
—	—	Narrow	3.4	—
Mg^{2+} montmorillonite	3	Broad	16.0	0.68
Ca^{2+} vermiculite	2	Broad	10.3	0.40

*Asymptotic values measured at large Q.

room temperature, somewhat lower than for the pure cation-water-clay systems. The difference is probably attributable to the hindering effect of the large pyridine molecules. From measurements made at three different temperatures, they derived an activation energy for the diffusive process in this system of 18 ± 5 KJ mol^{-1}. It should be noted, however, that their data were analyzed on the assumption of 2 dimensional diffusion and without making allowance for the possibility of rotational motions. Thus their numerical values should be used with caution until verified.

Possible Models for Water Diffusion in Clays. This section outlines work currently in progress towards the construction of a quantitative model for the dynamics of the interlamellar water in montmorillonite and vermiculite. This is a difficult task since such a model must not only predict the observed QENS data at both high and medium energy resolutions, but also be consistent with the findings of a wide range of other techniques which have been applied to this problem, including other spectroscopic techniques, such as NMR and infrared absorption spectroscopy, and macroscopic methods such as radioactive tracer diffusion studies. This discussion presents some fairly qualitative ideas regarding the suggested model, and compares its predictions with the data provided by other techniques.

As discussed in section 2-2, the link between experiment and theory in the QENS experiment is the incoherent scattering function, $S^{inc}(Q,\omega)$ which describes the momentum and energy distribution of neutrons after being scattered by the sample. This function is related by a double Fourier transformation to the van Hove self-correlation function, $G_s(r,t)$, which quantitatively describes the space and time dependence of the diffusive process (105). The form of $S^{inc}(Q,\omega)$ is thus directly dependent on the nature of the diffusion mechanism. The goal of the experimentalist is to extract this information, the experimental scattering function being obtained from the data after supplying various correction factors and removing the finite instrumental resolution by deconvolution. In principle one could directly transform the experimentally derived $S^{inc}(Q,\omega)$ to obtain $G_s(\mathbf{r},t)$. However, as described more fully elsewhere (45), this is not easy in practice since experiments do not give a complete mapping of the scattering function at all values of energy and momentum transfer. One must therefore resort to constructing hypothetical models for $G_s(\mathbf{r},t)$ on the basis of various assumptions regarding the nature of the likely molecular motions, and hence obtain the theoretically predicted form of $S^{inc}(Q,\omega)$ for comparison with experiments. This is the more conventional data analysis procedure. For time-of-flight instruments some advantages may be gained by the alternative procedure of building all the necessary corrections into the theoretical model and simulating the observed time-of-flight spectrum completely, as described by Hall *et al.* (45). Here the parameters of the model are extracted by least squares fitting to the raw data.

In general the data are fitted to a model consisting of an elastic component plus a Lorentzian quasi-elastic component. The parameters obtained by this fitting process, particularly the quasi-elastic energy broadening, ΔE, and the fraction of quasi-elastic to total (elastic and quasi-elastic) scattering, X, can then be compared directly with values obtained by performing an identical fitting process on data simulated using any specific model for $G_s(\mathbf{r},t)$.

From the experimentally observed variation of ΔE and X with Q^2 for divalent cation-exchanged clays, the simple continuous diffusion model (Fick's Law) can be rejected, since this predicts a purely linear dependence of ΔE on Q^2 : $\Delta E = 2\hbar DQ^2$, where D is the diffusion coefficient and $\hbar = h/2\pi$ where h is Planck's constant. Moreover, this model predicts no variation of X with Q^2, in contrast to what is observed experimentally.

The first refinement to simple three-dimensional Fickian diffusion would be to assume that only two-dimensional diffusion (i.e. parallel to the interlamellar plane) is possible. Whether one assumes the diffusion mechanism to be continuous (27) or rapid jump diffusion between relatively longer periods at fixed sites (101) the scattering function for this type of motion is predicted to be of the form

$$S(Q,\omega) = \frac{1}{\tau} \frac{B \sin^2 \theta}{(B \sin^2 \theta)^2 + \omega^2} \qquad [2\text{-}123]$$

where θ is the angle between the direction of $\mathbf{Q}$ vector and the normal to the plane of the silicate sheets. For continuous diffusion $B = DQ^2$, where D is the two-dimensional diffusion coefficient. For jump diffusion $B = [1 - J_o(Qr)]/\tau$, where τ is the mean residence time between jumps of fixed length r. J_o is the Bessel function of the first kind of order zero.

For single crystals these models predict Lorentzian shaped quasielastic peaks whose widths are functions of the magnitude and direction of Q. For polycrystalline preferentially orientated samples, such as clay films, these models must be averaged over the observed platelet orientation distribution functions, $f(\theta)$, derived from ND rocking curves. Calculations on the basis of these models have indicated that even with $f(\theta)$ curves of FWHM 40°, the resultant scattering functions, $S^{inc}(Q,\omega)$ show considerably greater anisotropy than observed experimentally.

A more realistic calculation, allowing for restricted motions perpendicular to the silicate sheets, is to solve the diffusion equation within two rigid boundaries in one direction, representing the silicate surfaces. The scattering function $S^{inc}(Q,\omega)$ corresponding to this type of motion has recently been calculated (41), its form having been given in equations (2-85) and (2-86) above. Simulated data calculated on the basis of this model predicts not only a purely elastic component due to the restricted geometry in which the diffusion takes place, the intensity of which is Q dependent, but also that little or no anisotropy of ΔE is expected, in accordance with experiment. However, this model still predicts a fairly linear dependence of ΔE on Q^2. Though some improvement would clearly be obtained by modifying the equations of Hall and Ross (41) to incorporate a jump diffusion mechanism with allowed jumps both parallel and perpendicular to the silicate sheets, this model on its own would not correctly predict the difference in the magnitude of quasielastic broadenings observed between high and medium energy resolution experiments (42, 44), i.e. the observation of two distinct rate processes.

Clearly the model must incorporate two distinct types of motion, differing in both their time-scale and distance-scale, and assumed to be independent or uncorrelated. Both are attributed to the water molecules outside the cation hydration shells, while the water molecules directly coordinated to the cations are assumed to be immobile on the QENS time-scale ($10^{-11}-10^{-13}$ seconds). Neglecting the vibrational motions of water molecules at fixed sites, which influence only the intensity of the quasi-elastic peak via the Debye-Waller factor (see section 2-2), the scattering function for such a model will be a convolution in ω of the scattering functions for the two distinct phases of the motion. The details of the model will, of course, depend on the precise nature of the two types of motion. Two alternative interpretations of these motions will be discussed here.

The first interpretation assumes that the more rapid localised motions are rotations of water molecules about fixed oxygen positions, while the slower motions are translational (centre-of-mass) diffusion of the same molecules. The overall scattering function can then be written

$$S^{inc}(Q,\omega) = S_{trans}(Q,\omega) * S_{rot}(Q,\omega)$$

where the asterisk denotes the convolution in ω. The translational motions may be described by the lamellar diffusion model of Hall and Ross (41) or by some generalization of this model to jump diffusion processes. The rotational motions may be described by the Barnes (14) or Sears (89) models for diffusion on a circle or on a sphere respectively. Richardson (81) has shown that the polycrystalline average of the Barnes model (a jump reorientation among N equidistant sites on a circle) is indistinguishable from the continuous rotational diffusion model for sufficiently large N ($\geqslant 6$). Though space does not permit the full mathematical details of the combined model to be given here, essentially it contains three adjustable parameters: (a) a translational diffusion coefficient, D, (or alternatively a site residence time between jumps, τ_T, related to the diffusion coefficient by $D = \ell^2/6\tau_T$ where ℓ is the mean translational jump distance); (b) a rotational correlation time, τ_R; and (c) a rotational radius of gyration, r.

Figure 2-26 illustrates the comparison between the experimental quasi-elastic broadenings and the values obtained from simulated data calculated on the basis of a combined rotational-translational diffusion model consisting of a convolution of the Hall-Ross and Barnes models (14, 41) averaged over a particle orientation distribution of Gaussian shape and of FWHM 40°. The points are from IN5 ($\sim$ 30 μeV resolution) and IN10 ($\sim$ 1 μeV resolution) data for a two layer Ca^{2+} montmorillonite water system having d_{001} = 15.5 Å. The IN5 and IN10 data are shown in Figures 2-26a and 2-26b, respectively. The theoretical predictions (solid lines) were calculated for $D = 3.4 \times 10^{-10}$ m^2s^{-1} and $\tau_R = 1.0 \times 10^{-11}$ s. The model is in fairly good agreement with the observed broadenings, and accounts for the differences in the magnitudes of the broadenings measured at the two different energy resolutions. However the third adjustable parameter, r, while not significantly influencing the quasi-elastic peak widths, has a strong influence on the relative amplitude of the quasi-elastic peak, i.e. the Q-dependence of the parameter X. It is found that this variation can only be accurately described by the model by assuming a value of r which is significantly larger than the interprotonic distance in water, and therefore physically unreasonable.

The alternative model attributes the more rapid, localized motions to spatially restricted diffusion, i.e. to random motions of water molecules within bounded regions such as the spaces between the hydrated cations in the interlamellar region, or alternatively external micropores. On this model the slower motions are then attributed to more truly macroscopic translational diffusion, the time-dependence of which is determined by the frequency with which water molecules can escape between adjacent cages or micropores through relatively constricted channels. Calculations on the basis of such a model are currently in progress. However it has been established that diffusion within such restricted volumes can account correctly for the Q-dependence of the quasi-elastic peak intensity if the linear dimensions of the regions are taken to be approximately equal to the spaces between the hydrated cations.

For either model discussed above, the differences in the broadenings observed at the two different energy resolutions can be explained in the following way. The slower motions (i.e. narrower broadenings) are selectively observed at high energy resolution and small Q. The more rapid motions (larger broadenings) tend to give only a flat background on the narrow energy window of the IN10 spectrometer. In contrast the IN5 spectra at somewhat lower energy resolution tend to be dominated by the larger broadenings, while the narrower broadenings become indistinguishable from the instrumental resolution and merge with the elastic component.

Two points in favor of the second model discussed above are as follows. Firstly, it avoids the assumption that uncorrelated translational and rotational motions of water molecules can be separated. This assumption is not valid for bulk water on the basis of computer molecular dynamics simulations (76, 77), though whether the situation would be the same for water molecules in the more restricted environment of clays, influenced by the proximity of charged ions and the silicate surfaces, is not clear. Secondly, this model is equally applicable to water in interparticulate micropores, which may constitute a significant fraction of the water content of montmorillonite according to recent investigations.

The qualitative picture which emerges of the dynamics of water in clays may therefore be summarized as follows. (a) The inner hydration shell water molecules are essentially immobilized (by the strong local crystal fields) and give only elastic scattering on the QENS time-scale. (b) The non-coordinated water molecules exhibit relatively rapid localized motions and also somewhat slower, longer-range translational motions. (c) The slower motions must be pictured as being relatively tortuous and restricted both by the hydrated cations and by the silicate surfaces.

These concepts are in good agreement with data from other techniques, which are briefly summarized. Tracer diffusion measurements (61) indicate a value of approximately $10^{-12} m^2 s^{-1}$ for the diffusion coefficients of hydrated monovalent cations in montmorillonite and vermiculite. Similar values have been obtained from NMR measurements on Na^+ vermiculite (52), consistent with the assumption that the hydrated cations move on too slow a time-scale to give measureable broadening in our QENS experiments. Furthermore, the exchange rate which has been measured between the two water populations, 10^{-4} – 10^{-5} seconds, is likewise far too slow to have any influence on the QENS data.

Other studies (112) have led to the conclusion that water molecules exhibit preferential orientation on the NMR timescale (which is several orders of magnitude slower than the QENS timescale). This finding would still be consistent with our results if either (a) the observed preferential orientation was solely due to coordinated water molecules, or (b) the localized motions involved jumps between preferred orientations for hydrogen bonding. The latter conclusion would be consistent with the x-ray crystallographic data for Mg^{2+} vermiculite (94) and with infrared measurements (30).

Up to now we have considered only molecular motions. One cannot, however, rule out the possibility that free proton jumps contribute to the diffusion process, i.e. that a mechanism similar to that which occurs in ice takes place. In this process a proton jumps from an H_3O^+ molecule to a neighbouring H_2O molecule along a hydrogen bond, followed by a subsequent jump of any one of the three protons on the new H_3O^+ to a neighbouring H_2O, and so on. Now if the mean lifetime of an H_3O^+ ion, i.e. the mean proton residence time, were sufficiently short ($\sim 10^{-12}$ seconds) and the fraction of water molecules at the clay surface which are dissociated were sufficiently high ($\sim$ 1%) (32, 102) then virtually every proton in the system would have jumped once on average within 10^{-10} seconds. This process could then lead to measureable quasi-elastic broadening. Proton jump diffusion must therefore be considered as a possible factor contributing to the observed mobility which QENS measurements cannot easily distinguish from molecular motions. This distinction would require tracer diffusion studies using labelled oxygen.

Other evidence relating to the possible influence of free proton diffusion concerns determination of the macroscopic H_2O/D_2O exchange rate in clays. Here the amount of data available is somewhat limited. Adams *et al.* (5) found the exchange to take place quite rapidly in the Na^+-montmorillonite-pyridine-water system. However small angle neutron scattering measurements (23) and NMR studies (51) both apparently indicate a slower rate of exchange than those which would be expected on the basis of the QENS data. It is therefore not entirely clear

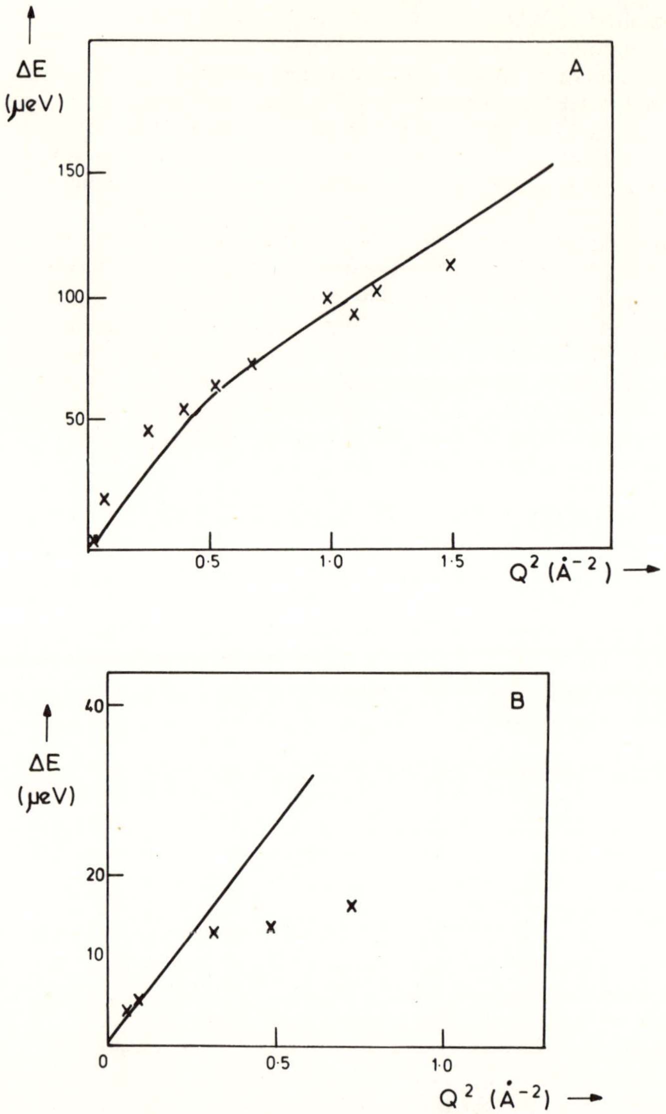

Figure 2-26. Fitted quasi-elastic broadenings for data calculated from a combined rotational-translational diffusion model. (a) IN5 simulation. (b) IN10 simulation.

whether these results support the hypothesis of proton jump diffusion, or whether they can be reconciled with the diffusion coefficients observed by neutron scattering by assuming high tortuosity factors for the long-distance scale diffusion, or perhaps a surface hindering effect, e.g. due to charges at particle edges. Clearly more experimental data is necessary to resolve these outstanding problems.

Dynamics of Hydrogenous Organic Molecules in Clays. The only published data in this field is the study of Adams *et al.* (5) on the Na+-montmorillonite-pyridine system. The diffusion of the pyridine molecules was found to be too slow to give measureable quasi-elastic broadening on the IN10 backscattering spectrometer, indicating that the value of D was $\leqslant 10^{-12}$ m^2 s^{-1}. This finding was in agreement with the relatively slow exchange rate between C_5H_5N and C_5D_5N observed by monitoring the time-dependent changes in the diffraction profile of the intercalate.

Measurements have been made (43) on the mobility of various alkyl-ammonium cations in montmorillonite. Both tetramethyl and tetraethyl cations exhibited relatively large quasielastic broadenings in samples equilibrated at 100% D_2O, but not in samples air-dried at 110°C, despite the fact that the basal spacings of the wet and dry systems are virtually identical. The Q-variation of the quasi-elastic fraction, X, was indicative of rotational motions and/or spatially restricted diffusion. Data for $(CH_3)_4N^+$-montmorillonite are illustrated in Fig. 2-27.

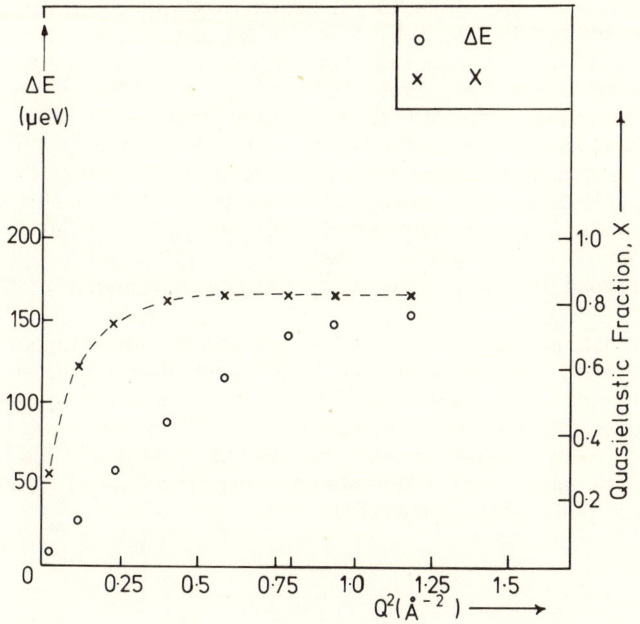

Figure 2-27. IN5 data for $(CH_3)_4N^+$-montmorillonite equilibrated at 100% D_2O.
O . . . quasi-elastic broadening. X . . . quasi-elastic fraction.

Clearly the hydration state of the clays has a strong, yet unexplained, influence on the mobility of the organocations. More experiments are needed but these results appear to be in good agreement with radioactive tracer diffusion measurements on similar systems by Gast and Mortland (33). These authors obtained translational macroscopic diffusion coefficients significantly below the range observable with neutron scattering, indicating that our results must be attributable to localized motions. Their translational coefficients did, however, depend significantly on the relative humidity at which the samples were equilibrated. The interactions between H_2O (or D_2O) and the organocations in these systems are complex and interesting, and more research is needed.

2-4.5. Inelastic Neutron Scattering Studies of Clay Minerals: Hydroxyl Vibrations

Inelastic incoherent neutron scattering can provide detailed information regarding low energy vibrational modes (below 1000 cm^{-1}), particularly weighted towards hydrogenous species, and unhampered by the usual selection rules which influence optical spectroscopy. Very little work on clay minerals has been published, however, and what little information is available was obtained some years ago on instruments having low energy resolutions.

Naumann *et al.* (70) obtained the neutron inelastic spectra for a range of clay minerals using a beryllium filter time-of-flight spectrometer. Detailed spectra were obtained, although in some cases with somewhat limited statistics. The chief point of interest is that considerable differences occur between the spectra of dioctahedral and trioctahedral structures, indicating that the OH vibrational modes are strongly influenced by the neighboring charge distribution and their orientation. Typical spectra for nuscovite and phlogopite micas are illustrated in Fig. 2-28. The reader is referred to the discussion of the neutron diffraction studies on micas reviewed in section 2-4.2. The same article (70) also gives some data on hydrated minerals, in which the characteristic modes of the lattice OH groups are superimposed on a broad frequency spectrum of vibrations characteristic of liquid water.

In the inelastic region of time-of-flight spectra of hydrated montmorillonites similar features are observed (71). It should be noted, however, that data from this type of spectrometer is of extremely low resolution at anything other than the lowest energy transfers, because the scale is non-linear in energy (see section 2-2). This area is virtually unexplored and much research is needed in the field of clay mineralogy. The results from future studies must be compared closely with the results of far infrared measurements (64).

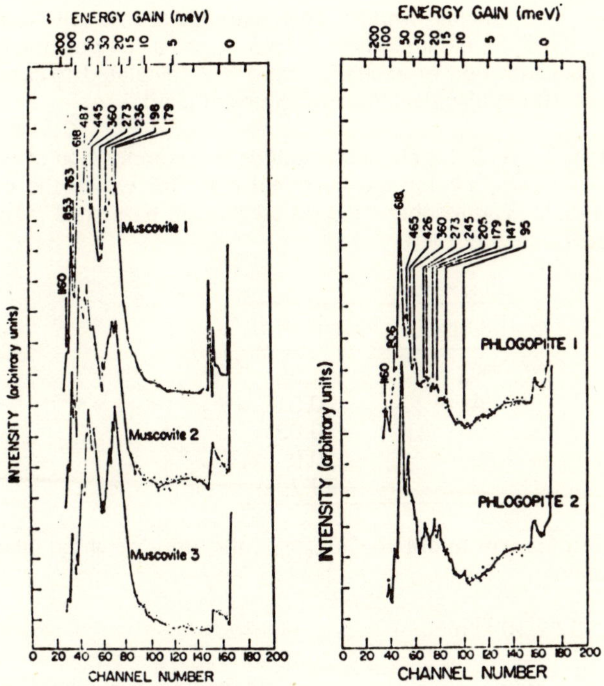

Figure 2-28. Inelastic scattering spectra of muscovite and phlogopite micas (70).

ACKNOWLEDGEMENTS

We should like to acknowledge the assistance of numerous people in the production of this chapter. In particular we have had fruitful discussions over a number of years with other members of the Birmingham and Oxford groups including Dr. M.H.B. Hayes, Mr. J.J. Tuck, Mr. R. Harrison, Dr. J.W. White, Dr. R.K. Thomas and Dr. D.J. Cebula.

We are also indebted to Dr. J.M. Adams of the University College of Wales, Aberystwyth, U.K. for useful discussions and making results available prior to publication, and to Dr. P.A. Egelstaff of the University of Guelph, Ontario, Canada for the communication of recent results.

Finally, we would like to express our thanks to Mrs. Sue Yeomans for clerical assistance.

APPENDIX 2-1. MACROSCOPIC CROSS SECTION FOR A MONTMORILLONITE-WATER SYSTEM

The macroscopic incoherent cross section, Σ_{inc}, for a two-layer Na+ montmorillonite with two interlamellar water sheets, assumes a unit cell formula of $Na^{+}_{0.36}Al_2\ Si_4\ O_{10}\ (OH)_2$, a basal spacing of 15.5 Å, and a composition of 0.2 g H_2O/1 g clay. (The concentration of exchange cations is calculated to correspond to a value of the cec of 100 milliequivalents/100 grams of clay).

It follows that 0.2 g H_2O/1 g clay corresponds to a composition of $Na^{+}_{0.36}\ Al_2\ Si_4\ O_{10}\ (OH)_2\ (H_2O)_{4.28}$ for the hydrated clay, for which the calculated formula weight is 445.7. The incoherent cross sections in barns ($10^{-28}\ m^2$) are listed below: (see also Table 2-2).

Element	σ_{inc} (barns)
Na	1.85
Al	0.01
Si	0.03
O	0.0
H	79.7

Summing over the formula unit to obtain $<\sigma_{inc}>_{av}$, the atom-weighted incoherent cross section, one obtains

$$<\sigma_{inc}>_{av} = 25.37 \times 10^{-28} m^2$$

Now

$$\Sigma_{inc} = N_A <\sigma_{inc}>_{av}$$

where N_A, the number of atoms/unit volume, will be given by

$$N_A = \frac{\rho_s \times 6.02 \times 10^{23}}{445.7} \times 33.2 \qquad [2\text{-}124]$$

$$= 4.484 \times 10^{22}\ \rho_s$$

where ρ_s is the sample density

↑	CLAY	9.6 Å	$\rho_C = 2.5 \times 10^6\ g/m^3$
d_{001} = 15.5 Å			
↓	WATER	5.9 Å	$\rho_W = 1 \times 10^6\ g/m^3$

Assuming the dimensions above, and a filled water layer, one can estimate

$$\rho_s = \frac{[(2.5 \times 9.6) + (1.0 \times 5.9)] \times 10^6 \text{ g/m}^3}{15.5} = 1.93 \times 10^6 \text{ g/m}^3. \qquad [2\text{-}125]$$

By substituting this value into equation [2-124] one obtains

$$\Sigma_{inc} \simeq 2.2 \times 10^2 \text{ metre}^{-1} = 0.22 \text{ mm}^{-1}.$$

APPENDIX 2-2. CALCULATION OF INCOHERENT SCATTERING INTENSITY RATIOS FOR A CLAY-WATER SYSTEM

The calculation of relative scattering intensities for a two-layer Ca^{2+}-montmorillonite water system is given. The following assumptions are made: (a) The value of Q does not overlap with a Bragg reflection, so that incoherent cross-sections may be used. (b) The cec of the montmorillonite is 100 meq/100 g and water content 0.19 g H_2O/g clay. (c) The cations are in 6-fold coordination. (d) Layer substitutions such as Mg, Fe, Li can be neglected. (These will not significantly influence the total incoherent scattering, which is dominated by the hydrogen atoms).

Using these assumptions a composition of approximately

$$Ca^{2+}_{0.18}\ Al_2\ Si_4\ O_{10}\ (OH)_2\ (H_2O)_{3.9}$$

may be calculated. From the incoherent cross sections given in Table 2-2 the following data may be calculated.

Component	σ^{tot}_{inc} (x 10^{28} m^2)	Fractional relative intensity
Clay lattice	159.5	0.20
Hydrated cations	181.7	0.23
Non-hydration water	444.7	0.57

REFERENCES

1. Adams, J.M., J.M. Thomas and M.J. Walters. 1975. Crystallographic, electron spectroscopic and kinetic studies of the sodium montmorillonite-pyridine system. *J. Chem. Soc.,* Dalton Trans. 1975: 1459-1463.
2. Adams, J.M. and D.A. Jefferson. 1976. The crystal structure of a dickite-formamide intercalate. *Acta. Cryst.* **B32**: 1180-1183.
3. Adams, J.M., P.I. Reid, J.M. Thomas and M.J. Walters. 1976. On the hydrogen atom positions in a kaolinite formamide intercalate. *Clays and Clay Minerals* **24**: 267-269.
4. Adams, J.M., C. Breen and C. Riekel. 1979a. Deuterium/Hydrogen exchange in interlamellar water in the 23.3-Å Na+-montmorillonite: pyridine/water intercalate. *J. Coll. Interface Sci.* **68**: 214-220.
5. Adams, J.M., C. Breen and C. Riekel. 1979b. Pyridine/deuteropyridine exchange in the 23.3Å Na+ montmorillonite/pyridine intercalate. *J. Coll. Interface Sci.* (in press).
6. Adams, J.M., C. Breen and C. Riekel. 1979c. The diffusion of interlamellar water in the 23.3Å Na+ montmorillonite: pyridine/H_2O intercalate by quasi-elastic neutron scattering. *Clays and Clay Minerals* **27**: 140-144.
7. Alefeld, B., M. Birr and A. Heidemann. 1969. Ein neues hochauflösendes neutronen kristallspektrometer und seine anwendung. *Naturwissenschaften* **56**: 410-412.
8. Anderson, I.S., C.J. Carlile and D.K. Ross. 1978a. The 50 K transition in β-phase palladium deuteride observed by neutron scattering. *J. Phys. C, Solid State Phys.* **11**: L381-L384.
9. Anderson, I.S., D.K. Ross and C.J. Carlile. 1978b. The structure of the γ phase of palladium deuteride. *Phys. Letters* **68A**: 249-251.
10. Bacon, G.E. 1972. Coherent neutron scattering amplitudes. *Acta. Cryst.* **A28**: 357-358.
11. Bacon, G.E. 1974. Coherent neutron scattering amplitudes. *Acta. Cryst.* **A30**: 847.
12. Bacon, G.E. 1975. Neutron Diffraction, 3rd edition. Clarendon Press, Oxford.
13. Barclay, L.M. and R.H. Ottewill. 1970. Measurement of forces between colloidal particles. *Spec. Disc. Faraday Soc.* **1**: 138-147.
14. Barnes, J.D. 1973. Inelastic neutron scattering study of the "rotor" phase transition in n-nonadecane. *J. Chem. Phys.* **58**: 5193-5201.
15. Baston, A. and D.H.C. Harris. 1978. Neutron beam instruments at Harwell. U.K. Atomic Energy Authority Report No. AERE-9278.
16. Born, M. and K. Huang. 1954. Dynamical theory of crystal lattices, Oxford University Press, England.
17. Carlile, C.J. 1974. Spectrum correction factors for sample holder and self-shielding effects for planar samples in thermal neutron scattering studies. Report No. RL-74-103, Rutherford Laboratory (U.K.)
18. Carlile, C.J. and D.K. Ross. 1975. Some aspects of the energy and momentum resolution of a cold-neutron rotating-crystal spectrometer. *J. Appl. Cryst.* **8**: 292-296.
19. Carpenter, J.M. 1977. Pulsed spallation sources for slow neutron scattering. *Nucl. Instr. Meth.* **145**: 91-113.

20. Cebula, D.J., R.K. Thomas, N.M. Harris, J. Tabony and J.W. White. 1978. Neutron scattering from colloids. *Faraday Disc. Chem. Soc.* **61**: 76-91.
21. Cebula, D.J., R.K. Thomas, S. Middleton, R.H. Ottewill and J.W. White. 1979a. Neutron diffraction from clay-water systems. *Clays and Clay Minerals* **27**: 39-52.
22. Cebula, D.J., R.K. Thomas, and J.W. White. 1979b. Small angle neutron scattering from dilute aqueous dispersions of clay. *J. Coll. Interface Sci.* (in press).
23. Cebula, D.J., R.K. Thomas and J.W. White. 1979c. The structure and dynamics of clay-water systems studied by neutron scattering. *Proc. Internat. Clay Conference* (Oxford, 1978), 111-120.
24. Chandrasehkar, S. 1943. Stochastic problems in physics and astronomy. *Rev. Mod. Phys.* **15**: 1-89.
25. Chudley, C.T. and R.J. Elliott. 1960. Neutron scattering from a liquid on a jump diffusion model. *Proc. Phys. Soc.* **77**: 353-361.
26. Clementz, D.M., T.J. Pinnavaia and M.M. Mortland. 1973. Stereochemistry of hydrated copper (II) ions on interlamellar surfaces of layer silicates. An electron spin resonance study, *J. Phys. Chem.* **77**: 196-200.
27. Dianoux, A.J., F. Volino and H. Hervet. 1975. Incoherent scattering law for neutron quasi-elastic scattering in liquid crystals. *Mol. Phys.* **30**: 1181-1194.
28. Dianoux, A.J. and F. Volino. 1977. Random motion of a uniaxial rotator in an N-fold cosine potential: correlation functions and neutron incoherent scattering law. *Mol. Phys.* **34**: 1263-1277.
29. Egelstaff, P.A. 1967. An Introduction to the Liquid State, Academic Press, London.
30. Farmer, V.C. and J.D. Russell. 1971. Interlayer complexes in layer silicates: the structure of water in lamellar ionic solutions. *Trans. Faraday Soc.* **67**: 2737-2749.
31. Favro, L.D. 1965. *In:* Fluctuations in Molecular Phenomena, ed. R.E. Burgess, Academic Press.
32. Fripiat, J.J., A. Jelli, G. Poncelet and J. André. 1965. Thermodynamic properties of adsorbed water molecules and electrical conduction in montmorillonites and silicas. *J. Phys. Chem.* **69**: 2185-2197.
33. Gast, R.G. and M.M. Mortland. 1971. Self-diffusion of alkylammonium ions in montmorillonite. *J. Coll. Interface Sci.* **37**: 80-92.
34. Giese, R.F. Jr. 1971. Hydroxyl orientation in muscovite as indicated by electrostatic energy calculations. *Science* **172**: 263-264.
35. Göltz, G., A. Heidemann, H. Mehrer, A. Seeger and D. Wolf. 1980. Self diffusion in sodium below the melting point. *Phil. Mag.*, to be published.
36. Greene-Kelly, R. 1955a. Sorption of aromatic organic compounds by montmorillonite. I. Orientation studies. *Trans. Faraday Soc.* **51**: 412-424.
37. Greene-Kelly, R. 1955b. Sorption of aromatic organic compounds by montmorillonite. II. Packing studies with pyridine. *Trans. Faraday Soc.* **51**: 425-430.
38. Guinier, A. and G. Fournet. 1955. Small angle scattering of X-rays. Wiley, New York.
39. Hall, P.G., A.J. Leadbetter, A. Pidduck and C.J. Wright. 1978. Quasi-elastic scattering from thin films of water absorbed at a silica surface. Neutron Inelastic Scattering, (*Proc. IAEA Symp.*, Vienna, 1977) **2**: 511-528.

40. Hall, P.G. and C.J. Wright. 1978. Neutron scattering from adsorbed molecules, surfaces and intercalates. *Chem Phys. Solids Surf.* **7**: 89-117.
41. Hall, P.L. and D.K. Ross. 1978. Incoherent neutron scattering function for molecular diffusion in lamellar systems. *Mol. Phys.* **36**: 1549-1554.
42. Hall, P.L., D.K. Ross, J.J. Tuck and M.H.B. Hayes. 1978a. Neutron Inelastic Scattering, (*Proc. IAEA Symp.*, Vienna 1977) **1**: 617-635.
43. Hall, P.L., M.H.B. Hayes, and D.K. Ross. 1978b. Diffusion of small organic molecules in clays. ILL Progress Report no. 09-07-012.
44. Hall, P.L., D.K. Ross, J.J. Tuck and M.H.B. Hayes. 1979a. Neutron scattering studies of the dynamics of interlamellar water in montmorillonite and vermiculite. *Proc. Internat. Clay Conf.* (Oxford, 1978), 121-130.
45. Hall, P.L., D.K. Ross and I.S. Anderson. 1979b. Direct model fitting of uncorrected time-of-flight data from quasi-elastic neutron scattering experiments. *Nucl. Instr. Meth.* **159**: 347-359.
46. Harris, D.H.C., C.G. Windsor and C.D. Lawrence. 1974. Free and bound water in cement pastes. *Mag. Concrete Res.* **26**: 65-72.
47. Harrison, R., P.L. Hall, M.H.B. Hayes and D.K. Ross. Neutron diffraction studies of preferential orientation in montmorillonite-water systems (in preparation).
48. Hawkins, R.H. 1978. A neutron diffraction study of clay-water structure. Ph.D. Thesis, University of Guelph, Ontario, Canada.
49. Hawkins, R.H. and P.A. Egelstaff. 1980. Interfacial water structure in montmorillonite from neutron diffraction experiments. *Clays and Clay Minerals* **28**:19-28.
50. Haywood, B.C.G. and D.L. Worcester. 1973. A simple neutron guide tube and diffractometer for small angle scattering of cold neutrons. *J. Phys. E* **6**: 658-671.
51. Hecht, A.M. and E. Geissler. 1973. Nuclear spin relaxation in a single and double layer system of adsorbed water. *J. Coll. Interface Sci.* **44**: 1-12.
52. Hougardy, J., W.E.E. Stone and J.J. Fripiat. 1976. NMR study of adsorbed water. I. Molecular orientation and protonic motions in the 2-layer hydrate of Na+ vermiculite. *J. Chem. Phys.* **64**: 3840-3851.
53. Howard, J., T.C. Waddington and C.J. Wright. 1976. Low frequency dynamics of hydrogen adsorbed upon a platinum surface. *J. Chem. Phys.* **64**: 3897-3898.
54. Hunt, D.G. and D.K. Ross. 1976. Optical vibrations of hydrogen in metals. *J. Less Comm. Metals* **49**: 169-191.
55. Hunter, R.W., G.C. Stirling and J.W. White. 1971. Water dynamics in clays by neutron spectroscopy. *Nature Phys. Sci.* **230**: 92-94.
56. Ibel, K. 1976. The neutron small-angle camera D11 at the high-flux reactor, Grenoble. *J. Appl. Cryst.* **9**: 296-309.
57. ILL (1977) Neutron beam facilities at the HFR available to users. (Available from the Scientific Secretary, ILL, Grenoble, France).
58. Jacrot, B. 1976. The study of biological structures by neutron scattering from solution. *Rep. Progr. Phys.* **39**: 911-953.
59. Koester, L. 1977. Neutron scattering lengths and fundamental neutron interactions. Springer Tracts in Modern Physics, (Springer-Verlag, Berlin) **80**: 1-55.
60. Kratky, O. and I. Pilz. 1978. A comparison of X-ray small-angle scattering results for crystal structure analysis and other physical techniques in the field of biological macromolecules. *Quart. Rev. Biophys.* **11**: 39-70.

61. Lai, T. and M.M. Mortland. 1968. Cationic diffusion in clay minerals. II. Orientation effects. *Clays and Clay Minerals* **16**: 129-136.
62. Langford, J.I., M.H.B. Hayes and W.R. Livingston. 1978. Diffraction studies of natural and synthetic swelling clay minerals, Conference on Applied Crystallography, Poland (preprint).
63. Livingston, R.E., J.M. Rowe and J.J. Rush. 1974. Neutron quasi-elastic scattering study of the ammonium ion reorientations in a single crystal of NH_4Br at 373 K. *J. Chem. Phys.* **60**: 4541-4546.
64. Loh, E. 1973. Optical vibrations in sheet silicates. *J. Phys. C.* **6**: 1091-1104.
65. Lorch, E. 1969. Neutron diffraction by germania, silica and radiation-damaged silica glasses. *J. Phys. C.,* **2**: 229-237.
66. Lovesey, S.W. and T. Springer. 1977. Dynamics of solids and liquids by neutron scattering, Springer-Verlag. Berlin.
67. Low, P.F. 1976. Viscosity of interlamellar water in montmorillonite. *Soil. Sci. Soc. Amer. J.* **40**: 500-505.
68. Matthieson, A.M. 1958. Mg vermiculite: a refinement and re-examination of the crystal structure of the 14.36 Å phase. *Amer. Min.* **43**: 216-227.
69. Muhlestein, L.D., E. Gurmen and R.M. Cunningham. 1972. Investigation of the phonon dispersion relations of paramagnetic and anti-ferromagnetic chromium, Neutron Inelastic Scattering (Proc. IAEA Symp., Grenoble) 53-59.
70. Naumann, A.W., G.J. Safford and F.A. Mumpton. 1966. Low frequency OH motions in layer silicate minerals. *Clays and Clay Minerals, Proc. Natl. Conf.* **14**: 367-383.
71. Olejnik, S., G.C. Stirling and J.W. White. 1970. Neutron scattering studies of hydrated layer silicates. *Spec. Discuss. Faraday Soc.* **1**: 194-201.
72. Olejnik, S., and J.W. White, 1972. Thin layer of water in vermiculites and montmorillonites – modification of water diffusion. *Nature Phys. Sci.* **236**: 15-16.
73. Placzek, G. 1952. The scattering of neutrons by systems of heavy nuclei. *Phys. Rev.* **86**: 377-388.
74. Podewils, P. and H.G. Preismeyer. 1978. Total neutron cross-section of protons bound in zirconium hydride at LN_2 temperature. Neutron Inelastic Scattering (*Proc. IAEA Symp.,* Vienna, 1977) **2**: 367-373.
75. Posner, A.M. and J.P. Quirk. 1964. The adsorption of water from concentrated electrolyte solutions by montmorillonite and illite. *Proc. Roy. Soc. London* **275A**: 35-56.
76. Rahman, A. and F.H. Stillinger. 1973. Hydrogen bond patterns in liquid water. *J. Amer. Chem. Soc.* **95**: 7943-7948.
77. Rahman, A. and F.H. Stillinger. 1975. Study of a central force model for liquid water by molecular dynamics. *J. Chem. Phys.* **63**: 5223-5230.
78. Ramsay, J.D.F., S.R. Daish and C.J. Wright. 1978. Structure and stability of concentrated boehmite sols. *Faraday Disc. Chem. Soc.* **61**: 65-75.
79. Rayner, J.H. 1974. The crystal structure of phlogopite by neutron diffraction. *Mineral. Mag.* **39**: 850-856.
80. Reynolds, R.C. 1976. The Lorentz factor for basal reflections from micaceous minerals in orientated powder aggregates. *Amer. Mineral.* **61**: 484-491.
81. Richardson, R.M. 1977. Neutron scattering in liquid crystals. Ph.D. Thesis, University of Bristol, England.

82. Rietveld, H.M. 1967. Line profiles of powder-diffraction peaks for structure refinement. *Acta Cryst.* **23**: 151-152.
83. Roberson, H.E., A.H. Weir and R.D. Woods. 1968. Morphology of particle-size fractionated montmorillonites. *Clays and Clay Minerals* **16**: 239-247.
84. Rodger, C.D., D.L. Worcester and A. Miller. 1977. Neutron diffraction of Lethocerus flight muscle. Insect Flight Muscle (Proc. Oxford Symp.), ed. R.T. Tregear, North-Holland, Amsterdam, 161-174.
85. Rothbauer, R. 1971. Untersuchung eines $2M_1$-Muscovit mit neutronenstrahlung. *Neues Jahrbuch Mineral. Monatshefte* **4**: 143-154.
86. Scherme, R. 1972. Fundamentals of neutron scattering by condensed matter. *Ann. Phys.* **7**: 349-370.
87. Schiff, L.I. 1955. Quantum mechanics, 2nd ed., McGraw-Hill, New York.
88. Schofield, P. 1960. Space-time correlation function formalism for slow neutron scattering. *Phys. Rev. Letters* **4**: 239-240.
89. Sears, V.F. 1966. Theory of cold neutron scattering by homonuclear diatomic liquids. II. Hindered rotation. *Can. J. Phys.* **44**: 1299-1311.
90. Sears, V.F. 1967. Cold neutron scattering by molecular liquids. III. Methane. *Can. J. Phys.* **45**: 237-254.
91. Sears, V.F. 1975. Slow neutron multiple scattering. *Advan. Phys.* **24**: 1-45.
92. Serratosa, J.M. and W.F. Bradley. 1958. Infra-red absorption of OH bonds in micas. *Nature* **181**: 111.
93. Serratosa, J.M. 1966. Infrared analysis of the orientation of pyridine molecules in clay complexes. *Clays and Clay Minerals, Proc. Natl. Conf.* **14**: 385-391.
94. Shirozu, H. and S.W. Bailey. 1966. Crystal structure of a two-layer Mg-vermiculite. *Amer. Mineral.* **51**: 1124-1143.
95. Singwi, K.S. and A. Sjolander. 1960. Diffusive motions in water and cold neutron scattering. *Phys. Rev.* **119**: 863-871.
96. Sköld, K. 1978. Quasi-elastic neutron scattering studies of metal hydrides. Topics in Applied Physics, (Springer-Verlag, Berlin), 28 (1): 267-287.
97. Springer, T. 1972. Quasi-elastic neutron scattering for the investigation of diffusive motions in solids and liquids, Springer-Verlag, Berlin.
98. Squires, G.L. 1978. Introduction to the Theory of Thermal Neutron Scattering, Cambridge University Press, England.
99. Stirling, G.C. 1973. Experimental techniques in "Chemical Applications of Thermal Neutron Scattering," ed. B.T.M. Willis. Oxford University Press, England, 31-48.
100. Stirling, G.C. 1978. The UK pulsed spallation neutron source. Neutron Inelastic Scattering (*Proc. IAEA Symp.,* Vienna, 1977) **1**: 25-37.
101. Stockmeyer, R. 1976. Zur bestimmung der beweglichkeit von kohlenwasserstuff molekülen auf katalysatoroberflächen mittels neutron enstreuung. *Ber. Bunsengen. Phys. Chem.* **80**: 625-629.
102. Touillaux, R., R. Salvador, C. van der Meersche and J.J. Fripiat. 1968. Study of water layers adsorbed on Na and Ca montmorillonite by the pulsed NMR technique. *Israel J. Chem.* **6**: 337-348.
103. Tsuboi, M. 1950. Positions of the hydrogen atoms in the structure of muscovite as revealed by infrared absorption study. *Bull. Chem. Soc. Japan.* **23**: 83-88.
104. Turchin, V.F. 1965. Slow Neutrons. Suron Press, Israel.

105. Van Hove, L. 1954. Correlations in space and time and Born approximation scattering in systems of interacting particles. *Phys. Rev.* **95**: 249-262.
106. Vineyard, G.H. 1953. Second-order scattering correction in neutron and X-ray diffraction. *Phys. Rev.* **91**: 239-240.
107. Vineyard, G.H. 1958. Scattering of slow neutrons by a liquid. *Phys. Rev.* **110**: 999-1010.
108. Wignall, G.D. 1967. Notes on the analysis of cold neutron scattering data. UK Atomic Energy Authority Report no. AERE-M-1928.
109. Willis, B.T.M. 1973. Chemical Applications of Thermal Neutron Scattering, Oxford University Press, England.
110. Windsor, C.G. 1973. Basic theory of thermal neutron scattering by condensed matter, in "Chemical Applications of Thermal Neutron Scattering," ed. B.T.M. Willis, pp 1-30.
111. Windsor, C.G. 1978. Pulsed neutron sources for epithermal neutrons. Neutron Inelastic Scattering (*Proc. IAEA Symp.,* Vienna, 1977) **1**: 3-23.
112. Woessner, D.E. and B.S. Snowden Jr. 1969. A study of the orientation of adsorbed water molecules on montmorillonite clays by pulsed NMR. *J. Coll. Interface Sci.* **30**: 54-68.

CHAPTER 3

INTRODUCTION TO X-RAY PHOTOELECTRON SPECTROSCOPY

C. DEFOSSE* AND P.G. ROUXHET
Groupe de Physico-Chimie Minerale et de Catalyse, Faculté des Sciences Agronomiques, Université Catholique de Louvain, Place Croix du Sud 1, B-1348 Louvain-la-Neuve, Belgium.
*Chargé de Recherches F.N.R.S. (Belgium)

3-1. INTRODUCTION

For solid samples, particularly those complex solids which are of interest in natural (rocks, soil aggregates) and industrial systems (adsorbents, pigments, catalysts, etc.), a satisfactory chemical characterization requires often more than an elemental or molecular analysis. It is frequently desirable to determine or estimate the nature and respective amounts of the phases present, the nature and extent of the surface developed by these phases, and the organization in space of the various phases with respect to each other.

While the identification of crystalline phases by x-ray diffraction is straightforward, identification of amorphous phases and evaluation of the respective amounts of different phases is still a delicate task which may require a skillful combination of various techniques.

Characterization of the overall surface area and pore size distribution of a sample can be performed on a routine basis. However, it is far more delicate to get deeper insight into the chemical nature, the properties and the respective contributions of the various types of surfaces present, and to establish a correspondence between the surfaces and the bulk phases. This is a field in which x-ray photoelectron spectroscopy appears as a particularly valuable tool. Its applications are numerous in many fields, from theoretical to analytical chemistry; however, here the main consideration will be the characterization of solid samples. Various examples will be given to illustrate the potential use of the method for the determination of surface chemical composition, for the study of adsorption properties, and for the investigation of the organization of various solid phases with respect to each other such as the coating of one phase by another.

3-1.1. Emission of Electrons

The emission of electrons by a substance (atom, molecule, or solid) can be provoked by the bombardment of electrons ($A + e_1^- \rightarrow A^+ + e_1^- + e_2^-$). One example of this principle is the amplification effect of dynodes in a photomultiplier. Emission of electrons can also arise as a result of the irradiation of a substance by

J. W. Stucki and W. L. Banwart (eds.), Advanced Chemical Methods for Soil and Clay Minerals Research, 169–203

photons of adequate frequency (photoionization : $A + h\upsilon \rightarrow A^{+} + e^{-}$). This process covers the well-known photoelectric effect, the study of which played an important role in the development of quantum theory, and is the basis of numerous devices of light detection, e.g. photography, photocells, photomultipliers, and ionization counters.

Electrons can also be emitted according to the Auger effect. This is a relaxation process involving an ion A^{+} in an excited state (A^{+*}), which is converted into an ion A^{++} in a lower energy state, with emission of an electron, the energy of which insures the overall energy balance ($A^{+*} \rightarrow A^{++} + e^{-}$).

3-1.2. The Principle of Photoelectron Spectroscopy

Photoelectron spectroscopy is based on the principle that, if a substance is irradiated by photons of sufficient energy, photoionization will occur. The kinetic energy of the photoelectron is measured in an electrostatic energy analyzer as illustrated in Fig. 3-1; the number of photoejected electrons is then plotted as a function of their kinetic energy, yielding a photoelectron energy spectrum.

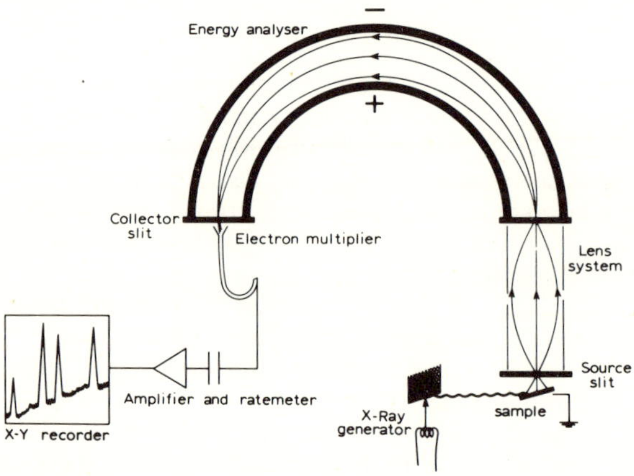

Figure 3-1. Schematic description of an x-ray photoelectron spectrometer.

Since energy conservation is required during photoionization, the kinetic energy, E_K, of the emitted electron is ideally related to its binding energy, E_b, in the molecular substance being analyzed by the relation

$$E_K = h\upsilon - E_b, \qquad [3\text{-}1]$$

where $h\upsilon$ is the energy of the photon. The binding energy represents the work required to bring the electron from its energy level in the atom or molecule to the level of zero attraction (zero binding energy). In practice, the measured kinetic energy differs from the ideal value given by equation [3-1] by terms which are related to experimental aspects. The actual relationship is

$$E_K = h\upsilon - E_b - \Phi_{sp} - E_C \qquad [3\text{-}2]$$

where Φ_{sp}, called the work function of the spectrometer, represents the work required to bring the electron from the zero binding energy level of the sample to the entrance of the kinetic energy analyzer of the spectrometer. E_C represents additional work expended due to the fact that, for insulating samples, the electron emission provokes sample charging.

3-1.3. Overview of Photoelectron Spectroscopy

From equations [3-1] and [3-2] it is evident that a kinetic energy spectrum (number of electrons emitted vs kinetic energy) provides information about the energy levels of the electrons in the sample. Examination of electrons removed from outer shells is very valuable and often used to improve understanding of the nature of chemical bonds and of the electronic properties of molecules and solids.

This Chapter and Chapter 4 are dedicated primarily to the photoelectron spectroscopy of inner shells or core levels. Since electron energy levels are quantized, each element in the periodic table will yield a unique XPS spectrum. Thus a sort of elemental analysis of the sample is possible; the position of the peaks identifies the elements present, the peak intensity is an indication of the relative abundances of the elements. Small shifts (0-5 eV) in peak position may reflect changes in the electronic environment surrounding the inner shells, such as valence shell population (oxidation state) or the nature of the ligand field (coordination environment). These shifts are of great interest to the theoretical chemist, but may also be used for analytical purposes. The analytical use motivated Siegbahn *et al.* (42), who introduced the technique as an analytical tool, to give to the technique the name of electron spectroscopy for chemical analysis (ESCA).

Removing electrons from the inner shells requires photons of high energy which are in the x-ray range; therefore, the method is also commonly referred to as x-ray photoelectron spectroscopy (XPS). This emphasizes the difference from ultraviolet photoelectron spectroscopy (UPS), where the ultraviolet source is more suitable for the study of electrons from outer shells.

A thorough presentation of the method and further details of general interest can be found in references 8, 9, 13, 12, and 42.

3-2. TREND OF XPS SPECTRA

3-2.1. Electronic Energy Levels

The central part of Fig. 3-2 presents a sketch of the lower electronic energy levels of a heavy atom, which is important to an understanding of the energy changes involving electrons. Such a diagram must be understood in the following way: the energy of atom A in the ground state (state of lower absolute energy) is considered as zero; the ordinate indicates, with respect to the chosen origin, the energy of the ion A^+ when one electron with the corresponding n,ℓ and j quantum numbers is missing. The difference between a given level and zero thus represents the energy required to remove one electron from that level of the atom, or the energy liberated when the ion A^+ with an electron vacancy at the given level recovers an electron at the same level.

By way of review, the meaning of quantum numbers as defined by the wave mechanical treatment of the one electron atom or hydrogen atom are summarized below. In parentheses in each case are given the terminologies according to the Bohr-Sommerfeld quantum theory.

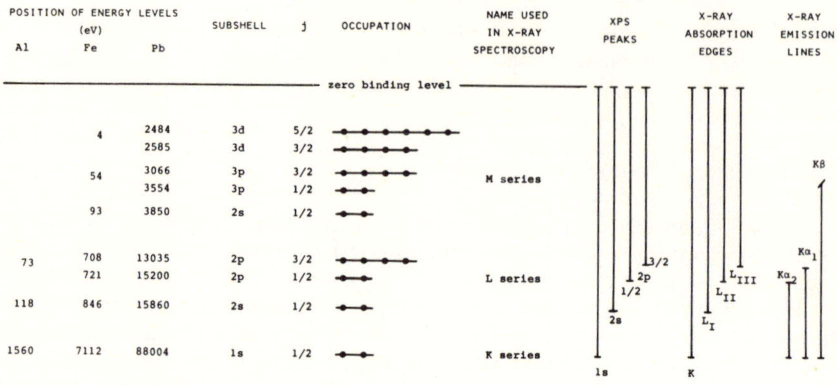

Figure 3-2. Sketch of the electronic energy levels. Relevant XPS lines, x-ray absorption edges and x-ray fluorescence lines are indicated. Numerical values of binding energies (in eV) are given for Al, Fe, and Pb.

Quantum number n. The value of n may be 1, 2, 3, 4, . . . which correspond to the shells K, L, M, N, ... n is related to the variation of the electron density as a function of distance from the nucleus (diameter of the orbit), and determines the eigenvalue, or energy, of the electronic state, or eigenfunction (orbit).

Quantum number ℓ. The value of ℓ may be any positive integer between 0 and n−1. It is related to the distribution of the electron density around the nucleus (orbit shape), determining the orbital angular momentum and the orbital magnetic moment.

Quantum number m_ℓ. The value of m_ℓ, the spatial quantum number, may be $-\ell, \ldots, 0, \ldots +\ell$. It is related to the orientation of the orbital magnetic moment with respect to an external magnetic field.

Quantum number m_s. The spin quantum number, m_s, may have values of only −1/2 or +1/2. It is related to the orientation of the spin magnetic moment with respect to an external magnetic field.

For multi-electron atoms, the atomic states of different ℓ values for the same shell (same n) have no longer the same energy, i.e. are non-degenerate, because of the influence of the coulombic interaction of any electron with the other electrons of the atom. In addition, there is a coupling between the orbital magnetic moment (if $\ell \neq 0$) and the spin magnetic moment. This coupling makes the orientation of each vector dependent on the orientation of the other. The result is a total angular momentum, the value of which is characterized by the quantum number $j = \ell + ½$ or $\ell - ½$, depending on whether the orbital and spin magnetic moments are parallel or antiparallel to each other. Spin-orbit coupling and relativistic effects are responsible for the splitting of any subshell ℓ into the two levels, $\ell + ½$ and $\ell - ½$, the energy of the former being higher than the energy of the latter.

The orientation of the total angular momentum with respect to an external field is characterized by quantum number $m_j = -j, -j+1, \ldots, 0, \ldots, j-1, j$. The multiplicity of the levels, i.e. the number of electrons which can be associated with the levels as schematized in Fig. 3-2, is given by the number of different values which m_j can have, i.e. 2j+1.

The left side of Fig. 3-2 indicates the position of the energy levels for three typical elements of different atomic weight: Al, Fe, Pb (3). The right side illustrates the meaning of the binding energy measured by XPS. It shows that on an energy scale the position of XPS peaks corresponds to the position of the edges observed on an x-ray absorption spectrum. In fact, the edge occurs when the energy of the photon is just sufficient to provoke removal of an electron from a given level. Note that the photon wavelength, λ^* (in Å units), may be converted to energy, E (in eV units), by the formula $E = 12400/\lambda^*$. On the right side of Fig. 3-2, the transitions involved in the production of x-rays (fluorescence or electron excitation) are recalled.

It should be noted that the relationship between the photoelectron spectrum and the diagram of energy levels is based on the hypothesis that during the photoemission process there is no rearrangement of other electrons over the energy levels. The approximation is satisfactory except in certain cases that will be mentioned below. This means that Koopman's theorem used in quantum theory is applicable, i.e. that the atomic orbitals are frozen during the photoemission process.

3-2.2. Observed Spectra

Fig. 3-3 presents the spectrum of alumina on which molybdenum has been deposited. Note that there are two peaks for the Mo 3p level, corresponding respectively to j quantum numbers 3/2 and 1/2. The doublet for the Mo 3d level is unresolved in the general spectrum, but is resolved into its 5/2 and 3/2 components in the more detailed record shown. The relative peak intensities of a detailed spectrum correspond roughly to the occupation of the levels, i.e. 2:1 for the 2 peaks of a p doublet and 3:2 for those of a d doublet. Smaller peaks will be discussed later.

3-2.3. Baseline

The main peaks observed in an XPS spectrum are due to electrons which have suffered no inelastic collisions from the time of their ejection from the atom until their detection (no-loss peaks). Only a small proportion of the total number of photoelectrons produced actually escape with no energy change.

The interaction of charged particles with matter is much stronger than that of photons. In the case of a solid, the mean free path, i.e. the average distance travelled between two inelastic collisions, is much shorter for electrons than for x-rays. In aluminum for instance, the mean free path of x-rays is about two orders of magnitude greater than for electrons in the energy range of 10^2-10^3 eV. Consequently, if an electron is photoejected from below a certain depth, where the x-rays are still slightly attenuated, the probability of the electron escaping the solid without suffering some energy loss is virtually zero. During its travel toward the surface, the electron will undergo several inelastic processes and finally will contribute to the general background rather than to the no-loss peak. This explains why the overall intensity of the background is so much higher than the cumulated intensity of the peaks. This also explains why the baseline is enhanced on the high binding energy side, i.e. the low kinetic energy side, of each peak. The baseline difference between the high and low sides of a peak (O_{1s} for instance) is due to the electrons originating from the level giving rise to the peak (O_{1s} in this case) that have suffered inelastic collisions.

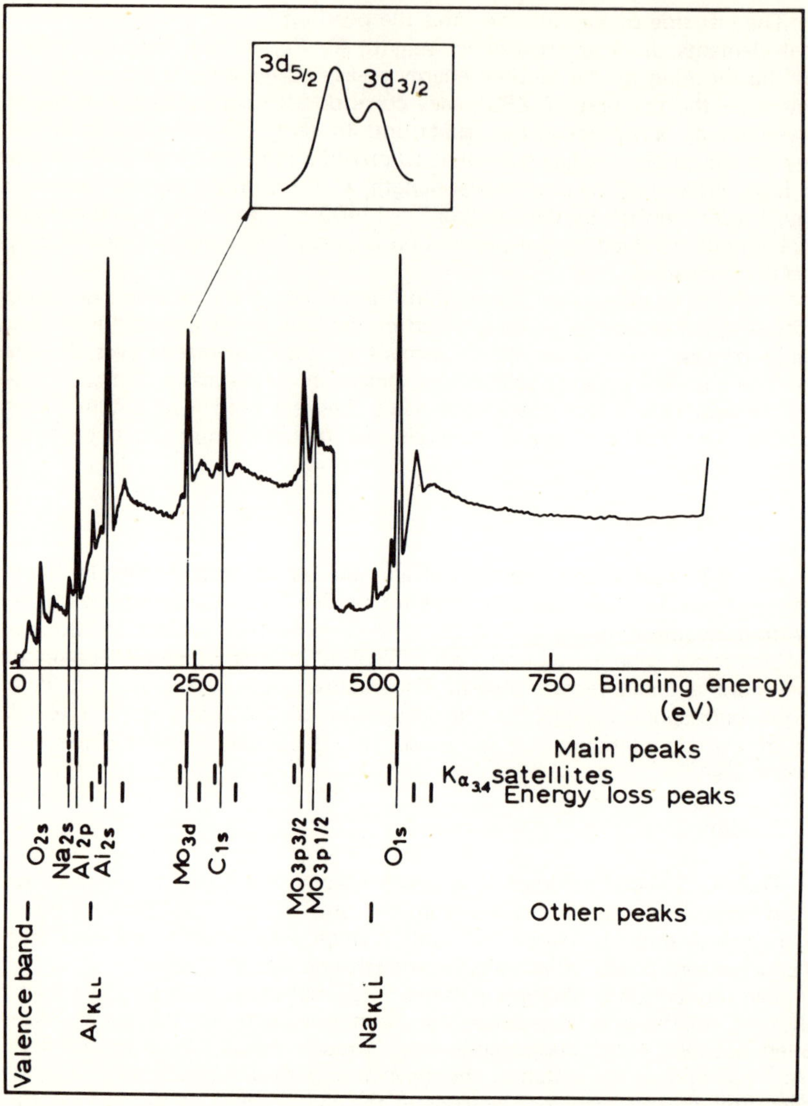

Figure 3-3. Typical XPS spectrum. The analyzed solid was an alumina on which molybdenum oxide had been deposited. The insert shows the partially resolved Mo_{3d} doublet; the intensity scale was changed by a factor of 3.5 at E_b = 450 eV.

3-3. INSTRUMENTATION

3-3.1. Vacuum Requirement

The analyzing chamber of a photoelectron spectrometer must be under a high vacuum. A residual pressure above about 10^{-6} torr would provoke excessive inelastic scattering of the electrons by the gaseous molecules surrounding the sample, such that they would either fail to reach the detector or, at least, would contribute to the background rather than to the no-loss peak. In fact, the vacuum must be better than 10^{-6} torr in order to limit surface contamination, which is extremely important as will be discussed later. Therefore, a vacuum of 10^{-8} to 10^{-10} torr is more desirable. The design of the pumping system is also very important for minimizing surface contamination.

3-3.2. X-ray Photon Source

For XPS, the photon source is an x-ray tube separated from the sample by a thin window that is transparent to x-rays but opaque to electrons. Fig. 3-4 recalls the shape of the emission spectrum of an x-ray tube. If the entire x-ray spectrum is used as the source to excite photoelectrons in a sample, the main $K\alpha$ line will be responsible for the main XPS peaks, but other x-ray lines could also induce excitations resulting in the appearance of satellite peaks at certain distances from each parent peak. The continuous emission or Bremsstrahlung, also called white radiation, will contribute to the production of photoelectrons over a broad kinetic energy range and enhance the level of the baseline.

The material commonly chosen for the anode is Mg or Al. These sources each give a sharp $K\alpha_{1,2}$ line (width of 0.8-0.9 eV), the intensity of which is about half the total emitted intensity. The position of the main x-ray lines of these elements are the following:

Mg : $K\alpha_1$ = 1253.7 eV; $K\alpha_2$ = 1253.4 eV

Al : $K\alpha_1$ = 1486.7 eV; $K\alpha_2$ = 1486.3 eV

The relative intensities of the two components are in the ratio of 2:1 for $K\alpha_1$ over $K\alpha_2$. The $K\beta$ line has an intensity that is only 2% of the $K\alpha_{1,2}$ line, so that peaks due to $K\beta$ excitation are indistinguishable from the general background. There are also other emission lines of low intensity that are due to complex relaxation processes. Among them is the $K\alpha_{3,4}$ doublet which arises from the relaxation of a doubly ionized atom at an energy about 10 eV greater than, and an intensity about 10% of, the $K\alpha_{1,2}$ line. The $K\alpha_{3,4}$ line is responsible for the satellite peaks shown in Fig. 3-3. The $K\alpha$ line of either of these anodes is unable to eject photoelectrons from the 1s level of the heavier elements.

A typical source for UPS is a helium discharge lamp which emits an intense line at 584.3 Å or 21.22 eV. This has a spectral width of a few meV and thus allows the low binding energy zone to be scanned with satisfactory resolution.

3-3.3. The Electron Analyzer

The role of the electron analyzer is to separate in space electrons of different kinetic energy so that the detector can count them separately and thereby provide a record of the number of electrons as a function of their kinetic energy. The most commonly used analyzer is the electrostatic type, with retarding potential. The

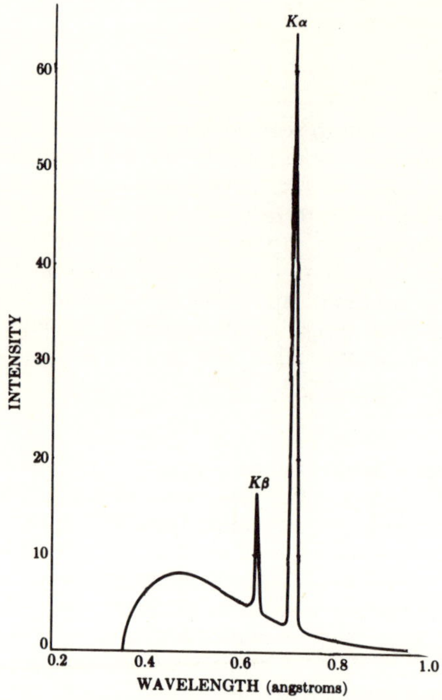

Figure 3-4. Emission spectrum for an x-ray tube with a Mo target.

scheme of the electrostatic analyzer is presented in Fig. 3-1 and consists of two concentric hemispheres of radius R_1 and R_2 between which is applied a potential difference, V′. The electrons entering the analyzer through the entrance slit are focused on the exit slit only if their kinetic energy obeys the relationship

$$E'_K = V'e\frac{R_1 R_2}{R_1^2 - R_2^2}\,, \qquad [3\text{-}3]$$

where e and E'_K are the charge and kinetic energy of the photoelectron, respectively. Electrons with a lower or higher kinetic energy will be deflected by the applied potential either too much or too little to reach the exit slit. The hemispherical analyzer is often combined with a lens allowing application of a controlled retarding potential, E_{ret}, as illustrated by Fig. 3-1. In that case the energy of the electron, E_{an}, at the entrance of the hemispherical analyzer is given by

$$E_{an} = h\upsilon - E_b - \Phi_{sp} - E_C - E_{ret}\,. \qquad [3\text{-}4]$$

If equation [3-2] is substituted into [3-4] the result is

$$E_{an} = E_K - E_{ret}\,. \qquad [3\text{-}5]$$

A spectral scan may be performed by varying either the potential of the hemispherical plates or the retarding potential.

3-3.4. Detector

The detector used is an electron multiplier. A common type is a channeltron which is essentially a glass tube of horn shape, the inside wall of which is coated with an adequate substance; a difference of potential is applied between the entrance and the bottom of the horn. When an electron strikes the internal wall of the tube it provokes the emission of other electrons, which are accelerated until they in turn also strike the wall thus provoking the emission of additional electrons, etc. Any electron entering the detector thus provokes a cascade of electrons which is observed as a current pulse.

3-3.5. Data collection and handling

The XPS spectrum is obtained on an X-Y recorder that records the number of pulses reaching the detector as a function of E_K. In order to reach a sufficient sensitivity and to improve the signal/noise ratio, it is often suitable to extend collection of data over a long period of time. This is done by repetitively scanning the spectrum and storing the data; the energy range is then covered by a number of channels each accumulating a number of pulses. Such spectra accumulators are essential for decent performance of the spectrometer.

More sophisticated data handling systems allow subjection of the data to various operations such as automatic peak area measurement, line separation, etc. As in other forms of spectroscopy, separation of partially overlapped bands should be used with much caution due to the high number of variables often involved, e.g., position, width, height, and shape of the peaks. Ancillary information from other sources is helpful, and good sense is required in evaluating the variability in some of these parameters. An important point to consider is that computer optimization of fitting an experimental spectrum to a spectrum reconstituted from individual components does not necessarily mean that the solution is unique and/or that it has physical meaning.

3-3.6. Sample

This and the succeeding chapter are dedicated to the application of XPS to solids. Coherent solids are easily handled; powders may be pressed in a pellet or distributed on an adhesive tape or a mesh. Deposition on an adhesive tape may be useful to avoid displacement of the particles during evacuation. Particular aspects related to sample preparation will be discussed in Chapter 4.

Note that XPS spectra can also be obtained for a gas, in which case efficient differential pumping is required in order to obtain a sufficient pressure in the sample compartment ($\sim 10^{-2}$ torr) while keeping a satisfactory vacuum (10^{-5}-10^{-6} torr) in the analysis compartment. Liquid samples can also be studied, provided their vapor pressure is sufficiently low.

3.4. PEAK POSITION

3-4.1. Information Provided by the No-Loss Peaks

It has been shown above and illustrated by Fig. 3-3 that the position of a no-loss peak is characteristic of a given element. Fig. 3-5 illustrates that the

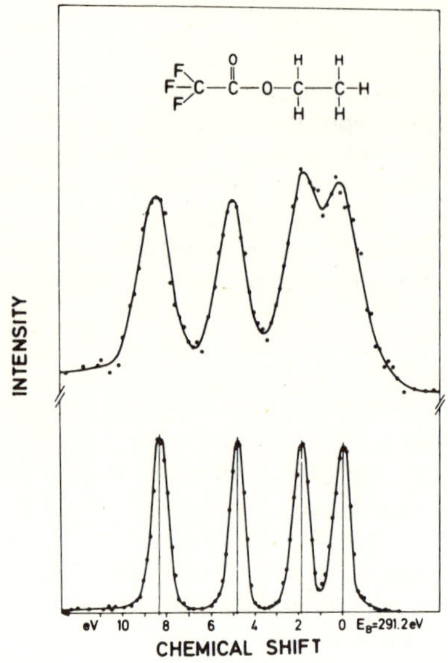

Figure 3-5. X-ray photoelectron spectrum, at two different resolutions, of the C_{1s} level in ethyltrifluoroacetate. (from Siegbahn, 1974).

position of a peak may reflect the state of the atom, in particular the electron density on that atom. This can be partially accounted for in terms of formal oxidation state: + 3 for C in $-CF_3$ and 0=C–0; –1 for C in $-0-CH_2$; –3 for C in $-CH_3$. However, the difference between the positions of the two former C atoms shows the importance of the electronegativity of the adjacent atoms. In general, the higher the electron density on the atom, the lower the binding energy. The shift of the peak due to the state of the atom and its environment is often called the chemical shift.

Further illustration of the effect of oxidation state is provided by Fig. 3-6, which shows the progress of oxidation as germanium is exposed to oxygen. Fig. 3-7 gives spectra obtained for samples of Al_2O_3 treated by increasing the amounts of NH_4F and calcined, and of pure AlF_3. The binding energy of Al_{2p} electrons in samples with a high fluoride content is clearly higher than in alumina, which reflects the effect of the higher electronegativity of fluorine as compared to oxygen. In other words, the Al-F bond has more ionic character than the Al-O bond.

The sensitivity of the chemical shift to the oxidation state or to the atomic environment varies according to the nature of the element.

3-4.2. Peak Width

The width of a recorded peak is usually measured by the width at an intensity which is half the maximum (full width at half maximum, FWHM). The mathematical function describing the peak envelope is the result of the convolution of various functions that describe the effects due to different factors. A fundamental factor is the width of the energy level from which the photoelectron is ejected. A general illustration of the role of the width of the energy levels is provided by Fig.

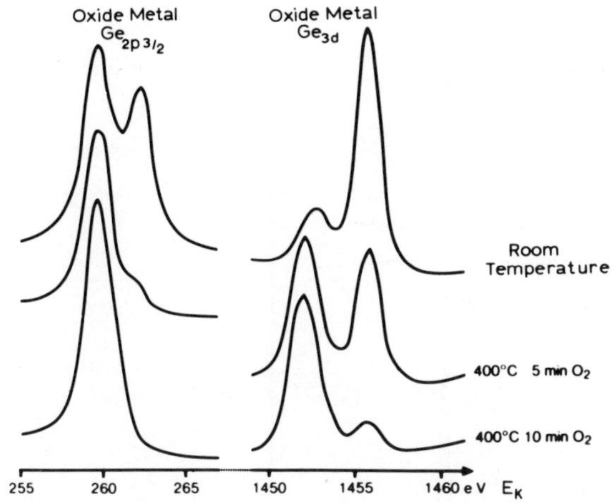

Figure 3-6. Germanium $2p_{3/2}$ and 3d peaks for increasing oxidation of metallic Ge. (from Holm, 1978).

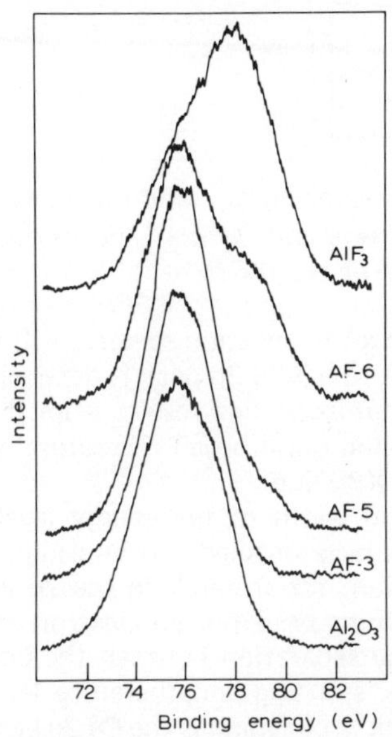

Figure 3-7. Al_{2p} peaks of aluminum fluoride and alumina treated by increasing amounts of fluoride; fluorine content expressed as weight % AlF_3 and equal to 0(Al_2O_3), 3.7(AF-3), 17.7 (AF-5), and 36.8 (AF-6). (from Scokart *et al.,* 1979).

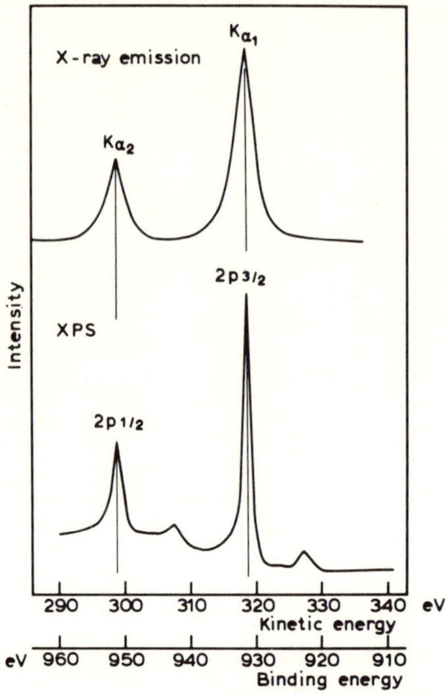

Figure 3-8. Upper part: x-ray emission (fluorescence) $K\alpha_1$ and $K\alpha_2$ lines for copper. Lower part: photoelectron Cu $2p_{3/2}$ and Cu $2p_{3/2}$ peaks obtained with Mg $K\alpha$ radiation. (from Siegbahn *et al.,* 1967).

3-8, which shows a comparison between an x-ray emission spectrum (fluorescence) and an XPS spectrum of copper. The width of a XPS 2p peak is determined by the width of the 2p level; the width of a $K\alpha$ emission line results from the width of both the 2p and the 1s levels, the latter being quite broad. Therefore, the fluorescence line is much broader than the related XPS line.

A phenomenon that can broaden the width of the energy level is the so-called multiplet splitting. Transition metals have unfilled d orbitals and if an additional vacancy is created by photoionization, for instance in the 2p level, a spin coupling will take place between the remaining unpaired 2p electron and those in the incompletely filled 3d shell. The energy separation between the possible final states is usually insufficient to cause peak splitting and therefore leads to peak broadening. An example of this type of peak broadening is the Cu 2p level, which is noticeably wider for Cu^{2+} (one unpaired electron in the d shell) than for Cu^+ (completely filled d orbitals).

Another determining factor is the energy profile of the $K\alpha$ line of the source. The width of an XPS peak cannot be smaller than the spectral width of the source which is about 0.8 eV. Resolution can be improved by using an x-ray monochromator, however monochromatisation cuts down the intensity and may be unfavorable if the sensitivity is critical. For the high resolution spectrum presented in Fig. 3-5, the use of a monochromator allows the peak width to be reduced to that imposed by the width of the energy level. Note that the monochromator also removes the $K\alpha_{3,4}$ satellites; it reduces the background, and consequently improves the signal to noise ratio, by removing the white radiation.

The dispersion of the electron analyzer is a characteristic of the instrument that also contributes to line broadening. The resolution of the analyzer increases, and subsequently the width of the peak decreases, as the kinetic energy at the exit slit, E_{an}, decreases.

For solids, sample charging often gives rise to peak broadening due to its variability throughout the sample.

3-4.3. Measurements of Peak Position

In an ideal situation such as metallic samples, measurement of peak position is a matter of calibrating the work function, Φ_{sp}, which is constant for a given spectrometer. This can be done by using a reference element, such as gold, and recording its spectrum.

For insulators however, calibration requires measurement of both Φ_{sp} and E_c. It must be realized that the charging term, E_c, varies from one sample to another and may vary with time. In particular, it may be different before and after gold evaporation or change during the analysis if the state of the surface (e.g. the contamination) varies with time. In order to suppress or at least limit sample charging, a slow electron gun (the so-called flood gun) is sometimes used. However, this use is delicate and even questionable as excessive electron flux will charge the sample negatively.

Measurement of the "absolute value" of an XPS peak position is performed with respect to the Fermi level of the reference element and assumes that the Fermi levels of the sample, of the reference and of the spectrometer coincide. Obviously, determination of the absolute position is therefore difficult and any such measurement must be considered very critically. This will be further illustrated in Chapter 4.

For many purposes, the relative position of the peak is sufficient (see Figs. 3-5, 3-6 and 3-7). This is commonly done with respect to, or referenced to, one of the following peaks: (i) a peak of gold, in which case the spectrum is scanned before and after evaporation, and it is optimistically assumed that a good electrical contact between the sample and the reference is achieved; (ii) a carbon peak from contaminating substances; or, (iii) a peak of an element present in the sample. The choice of the reference will depend on the intended use of the XPS information. The reference used should always be stated in reports and papers, and the results should always be considered critically by the author and by the reader.

3-4.4. Satellite Peaks

We have already mentioned satellites due to excitation by the $K\alpha_{3,4}$ line. Satellites also may be due to a discrete energy loss, as shown in Fig. 3-3. The presence of a satellite for every peak is typical for alumina, and indicates that the energy loss is not limited to an atom but is due to collective excitation. Such collective excitation is common for metals (plasmons) but has not been investigated in detail for insulators.

Energy loss satellites restricted to a given atom may also be observed due to more complex photoionization processes called shake-up. In this case, photoejection is accompanied by excitation of another electron and the kinetic energy of the ejected electron is decreased by this excitation energy; consequently, the corresponding peak appears at an apparent higher binding energy.

Note that, in a process called shake-off, two electrons are ejected by the photon; in this case the distribution of the energy over the two electrons is not fixed, therefore they are not responsible for the appearance of a peak but contribute to the background.

3-4.5. Auger Peaks

The excited ions produced by photoionization are subject to deexcitation, either through photon emission (fluorescence) or through the Auger effect (see Sections 3-1 and 3-7 for details). The latter gives rise to the appearance of Auger peaks in the XPS spectrum, as illustrated in Fig. 3-3. The kinetic energy associated with an Auger electron is, of course, independent of the nature of the x-ray source.

3-5. EXPLORED DEPTH

3-5.1. Theoretical aspects

The main characteristic of XPS that must be especially stressed is its surface sensitivity, i.e. the fact that it probes the surface or at least the first few layers below the solid-vacuum interface. This feature is particularly important to recognize if one wishes to assess correctly the information yielded. The above explanations concerning the spectral baseline (Section 3-2.3) indicate that only the electrons photoejected at a small depth have a non-negligible probability of escaping the solid without energy loss, i.e. of contributing to the no-loss peak. The implications of this point will now be discussed in some detail.

Consider a solid with a perfectly flat surface. The probability, Q, of an electron escaping without inelastic scattering is given by

$$Q = \exp\left[-x/(\lambda \sin\theta)\right] \qquad [3\text{-}6]$$

where λ is the electron mean free path in the solid considered, x is the depth at which photoejection occurred and θ is the angle between the direction of electron collection and the surface. Numerous compilations of λ in different solids have been published (8, 25, 35, 36). Typical data are presented in Fig. 3-9. Although λ is influenced by the matrix (47), it depends mainly on the kinetic energy of the electron. It is approximately equal to 9 Å for 200 eV and 20 Å for 1500 eV. In this range a proportionality exists between λ and $(E_K)^{1/2}$, which holds especially well inside a given class of solids (oxides or metals for instance).

The intensity, $I_{X\,y}$, for the y peak of element X can be obtained from equation [3-6] by integrating over the thickness of the sample:

$$I_{X\,y} = P_{X\,y}\,\frac{1}{\sin\theta}\int C_X \sigma_{X\,y} \exp\left[-x/(\lambda_{X\,y}\sin\theta)\right]\,dx \qquad [3\text{-}7]$$

where C_X is the volume concentration of element X in the solid (moles/cm^3); $\sigma_{X\,y}$ is the photoelectric cross section for level y of element X; P is a constant to be discussed below, which involves the examined sample area, the x-ray intensity and the sensitivity. Due to the small value of the x-ray wavelength, λ^*, the attenuation of the x-ray intensity as a function of depth may be neglected.

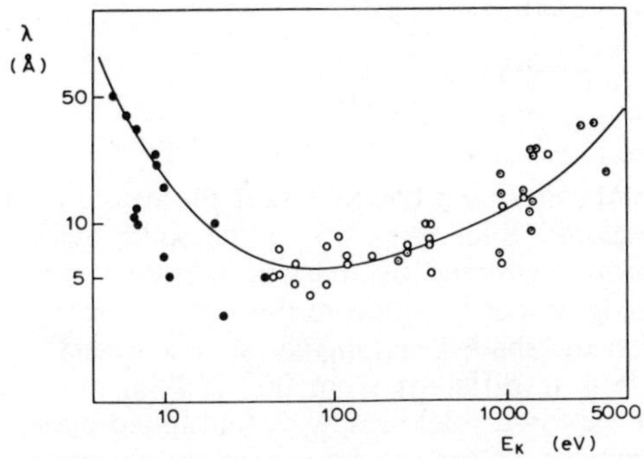

Figure 3-9. Compilation showing the variation of λ with kinetic energy obtained for various solids (from Brundle, 1974).

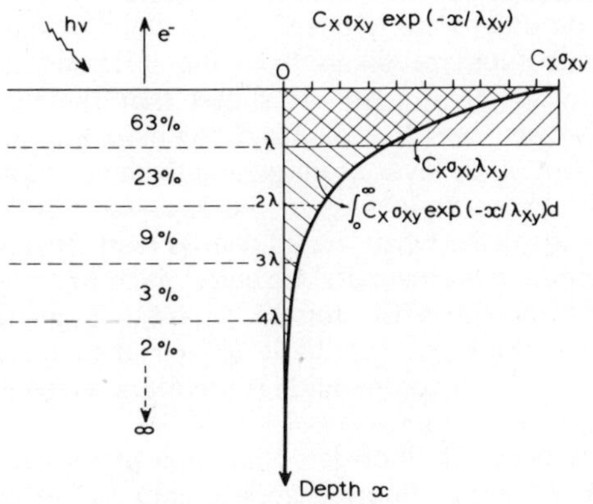

Figure 3-10. Left side: contribution of successively deeper layers of thickness λ to the total XPS signal intensity; θ = 90° (from Friedman, 1973). Right side: attenuation of the photoelectron intensity with escape depth x for θ = 90°; illustration of the concept of apparent explored thickness (see text).

The contribution to I_{XY} of successive layers at increasing depth can be obtained by performing the integration between various intervals. This is illustrated by Fig. 3-10 (left side) for the case where the photoelectrons are collected in the direction normal to the surface (θ = 90°):63% of the intensity arises from the first layer of thickness λ, 23% from the next layer, 9% from the next layer, and 5% from material deeper than 3λ.

Due to the small value of λ, integration of equation [3-7] over the actual sample thickness is close to integration from 0 to ∞. It is interesting to point out that, if the solid is homogeneous and $\theta = 90°$, equation [3-7] then becomes

$$I_{X\,Y} = P_{X\,Y} \int_0^{\infty} C_X \sigma_{X\,Y} \exp(-x/\lambda_{X\,Y}) dx$$

or [3-8]

$$I_{X\,Y} = P_{X\,Y} C_X \sigma_{X\,Y} \lambda_{X\,Y} \equiv P \int_0^{\lambda_{X\,Y}} C_X \sigma_{X\,Y} dx.$$

Thus, for a homogeneous solid everything happens as if the whole intensity originated from an apparent explored thickness $\lambda_{X\,Y}$ of the solid, inside which the photoelectron intensity is not attenuated by inelastic scattering, and the layers situated deeper than $\lambda_{X\,Y}$ made no contribution to the signal. This is illustrated by Fig. 3-10 (right side) in which the shaded rectangular area is indeed equivalent to the integral of the curve. If θ is different from 90° (45° in most commercial spectrometers) the apparent explored thickness, λ, is multiplied by $\sin\theta$ and the selectivity of the intensity towards the surface zone increases, the more grazing the emission angle and the more marked the surface character of the XPS information (24).

3-5.2. Illustration of Surface Analysis Capability of XPS

The two following examples illustrate the surface character of the information provided by XPS and its use to study chemical reactions of solids. Fig. 3-6 gives peaks recorded for germanium as taken from the shelf and subsequently submitted to successive oxidizing treatments. It shows that the surface of the starting material had already been partially oxidized. As oxidation proceeds further, the metal is covered by an oxide layer of increasing thickness. Comparison of the relative intensities of the two components in the $2p_{3/2}$ and 3d peaks shows that the oxide to metal ratio is smaller when the 3d peak is used. This is due to the fact that the oxide forms a coating on the metal through which the more energetic electrons escape more efficiently. Since E_K for 3d electrons is greater than for $2p_{3/2}$ electrons, the sampling thickness for 3d is greater than for $2p_{3/2}$. The contribution of the metal with respect to the oxide is therefore relatively larger for 3d than for $2p_{3/2}$ (27).

The evolution of the spectra of fluoridated alumina presented in Fig. 3-7 shows that peaks typical of aluminum fluoride appear only for an overall AlF_3 content above 20%. This indicates that the aluminum fluoride phase is not coating the alumina phase but appears as separate, presumably bulky, particles. Note on the other hand that the peak recorded for aluminum fluoride contains a contribution at the same binding energy as alumina, which indicates that the surface has been hydrolyzed.

3-5.3. Surface Contamination and Modification

The surface sensitivity of XPS raises the problem of surface contamination. Any surface, as seen by photoelectron spectroscopy, is covered by a carbonaceous overlayer which is responsible for the presence of carbon peaks, such as the C 1_s peak observed in Fig. 3-3. Consequently, the intensity of any line X originating from the sample itself is changed by a factor $\exp[-t/(\lambda'_X \sin\theta)$, where t is the contaminating overlayer thickness and λ'_X is the mean free path in the contaminant layer. If the contamination builds up with time, the line intensity from the underlying sample will decrease accordingly, making the signal to noise ratio worse. Comparison of data for Na_{1s} and F_{1s} or Na_{2s} in Fig. 3-11 illustrates the influence of E_K on the intensity attenuation resulting from contamination, i.e. on λ'_x.

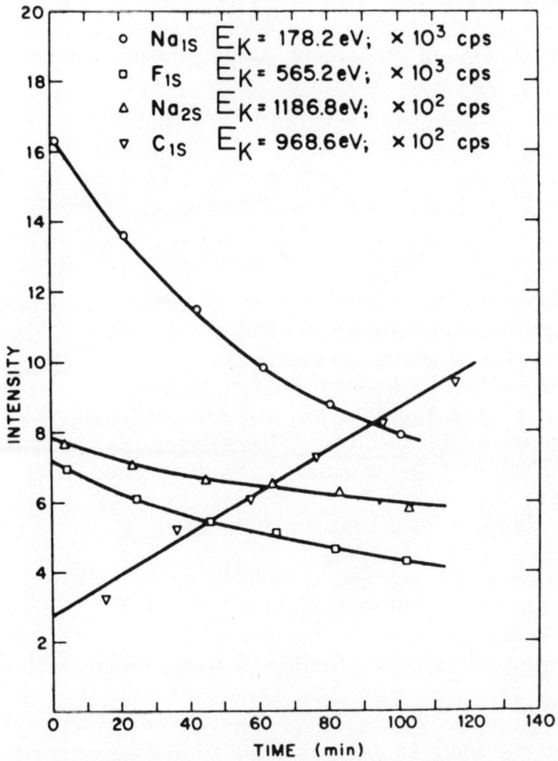

Figure 3-11. The effect of surface contamination, increasing with time, on the photoelectron intensities of NaF (from Swingle, 1975).

Due to the surface-sensitivity of the information provided by an XPS spectrum, the results may be strongly influenced by modifications of the sample such as drying, air oxidation, other types of contamination, and removal or deposition of colloidal particles. The analysis and interpretation of the data should be performed with careful consideration of sample history; this is particularly critical for wet samples, which are necessarily dried for the XPS analysis.

3-6. PEAK INTENSITY

The quantitative use of photoelectron spectroscopy has emerged only recently because this aspect is not as straight-forward as most of the chemical shift interpretations and, furthermore, several problems still limit both the accuracy and the interpretation of the quantitative XPS analysis. After pointing out the practical and fundamental problems related to intensity measurements, a few illustrations will be given that demonstrate the value of this tool, despite these limitations, for obtaining information on complex solids.

3-6.1. Theoretical Aspects

Consider first a sample with a perfectly flat surface. Neglecting the x-ray beam attenuation over the depth probed by XPS, the contribution to the intensity of peak y of element X, I_{Xy}, of the layer dx located at depth x is given by (7, 24):

$$dI_{Xy} = \frac{F\delta\sigma_{Xy}\Omega}{\delta\Omega} \left(C_X \frac{A}{\sin\theta} dx\right) \exp[-x/(\lambda_{Xy} \sin\theta)] \, L_{Xy}$$

$$\exp[(-t/\lambda'_{Xy} \sin\theta)] \qquad [3\text{-}9]$$

where

F = the photon flux, which is assumed to be uniform at the sample surface;
σ_{Xy} = the photoelectron cross-section of level y of element X for the photon energy of the particular anode material used;
Ω = the solid angle for photoelectron collection;
C_X = the volume concentration of element X;
A = the area of the sample involved in the analysis;
θ = the angle between the direction of photoelectron collection and the surface of the solid;
$\exp[-x/(\lambda_{Xy} \sin\theta)]$ = the attenuation produced by thickness x of the solid; and
$\exp[-t/(\lambda'_{Xy} \sin\theta)]$ = the attenuation produced by the contamination overlayer.

$\frac{\delta\sigma_{Xy}}{\delta\Omega}$ takes into account the fact that photoejection may be anisotropic.

The parameter L is the analyzer luminosity and, for spectrometers with a pre-retardation voltage, it is approximately accounted for by E_{an}/E_K (26), where E_{an} is the fixed kinetic energy of the photoelectrons as they travel inside the analyzer, the so-called analyzing energy, and E_K is the kinetic energy of these electrons when they leave the sample.

Even if most terms in the equation [3-9] are known or could be evaluated, the photon flux and the contaminant thickness are almost impossible to estimate. Therefore one has always to deal with the ratio R_X of the intensities of two peaks defined as

$$R_X = \frac{I_{Xy}}{I_{Rz}} \qquad [3\text{-}10]$$

where I_{Rz} is the intensity of peak z of a reference element R.

The anisotropy of σ may be taken into account by substituting

$$\frac{\delta\sigma}{\delta\Omega} = \frac{\sigma}{4\pi} \left[1 + \frac{1}{2}\beta \left(\frac{3}{2}\sin^2\alpha - 1\right)\right] \qquad [3\text{-}11]$$

into [3-9], where σ, β and α are the photoelectric cross-section, the so-called asymmetry parameter, and the angle defined by the x-ray beam and electron emission direction, respectively.

Fundamental explanations and values of β for different levels and exciting radiations are given in the literature (30, 38). The ratio of the expressions between square brackets in equation [3-11] can be factored out of the right hand side of equation [3-10] and denoted R'_A. This leads to the result

$$R_X = R'_A \; R'_L \; \frac{\sigma_{X\,y} \int_0^\infty \; C_X \; \exp[-x/(\lambda_{X\,y} \sin\theta)]\, dx}{\sigma_{R\,z} \int_0^\infty \; C_R \; \exp[-x/(\lambda_{R\,z} \sin\theta)]\, dx} \; R'_c \qquad [3\text{-}12]$$

where

$$R'_L = \frac{L_{X\,y}}{L_{R\,z}} \qquad [3\text{-}13]$$

and

$$R'_C = \exp\left[-t\left(\frac{1}{\lambda'_{X\,y}} - \frac{1}{\lambda'_{R\,z}}\right)/\sin\theta\right]. \qquad [3\text{-}14]$$

If the contaminant thickness is appreciable, R'_C can significantly differ from unity; the more so, the more $\lambda'_{X\,y}$ differs from $\lambda'_{R\,z}$. From the relationship between λ and the kinetic energy, it appears (Fig. 3-9) that the two peaks should have kinetic energies as close as possible to each other, in order to make R'_C close to one; this is also illustrated by Fig. 3-11. Increasing hydrocarbon cracking on nickel-exchanged zeolites, for instance, has been shown to have a significant and troublesome influence on the intensity of the Ni_{2p} peak, measured with respect to the Al_{2p} peak as a reference (18).

These relationships apply to the ideal case of a flat surface, the flatness being evaluated at the scale of λ. Microroughness induces deviations which can still be adequately described (23). Macroroughness, as obtained with a powder either pressed or sprinkled on adhesive tape, cancels or attenuates the dependence of intensity with respect to the orientation of the sample. For treating the following examples which deal with powders, we shall drop sin θ assuming that λ represents an effective mean free path.

The above relationships also suppose that there is no preferential path for electrons (channeling) in the sample. This is realistic for powders as statistical averaging over the particles will take place anyway.

3-6.2. Peak Intensity Measurement

The intensity of a line is the surface under the peak envelope, provided the background has been subtracted. The area measured must be normalized taking into account the different scale sensitivities, the sweep width and, when a signal averager is used, the number of sweeps and the time per channel. The peak height cannot be considered as a valid intensity measurement since for the same intensity value, different peak heights can be obtained according to the value of the full width at half maximum (FWHM). This latter can be influenced by several factors discussed above and may vary.

Before proceeding further, it is important to stress several factors which may limit the precision, depending on how they are handled. One of these factors is the background subtraction. As the exact shape of the background under the photoelectron peak is ignored, some rather arbitrary assumptions must be made. Usually, a linear interpolation is performed, using averaged values of the background on both sides of the peak but more sophisticated methods have been used (2).

A lack of precision can also arise from incorrect estimation of the extent to which electron shake-off occurs and how it varies from one sample to another (44). Similarly, if a given line is accompanied by shake-up or energy loss peaks, the correct intensity for the level considered is the sum of the main line plus the satellite peaks. In practice, energy loss satellites due to collective processes are almost never considered because their very broad shape makes it particularly difficult to measure their intensity accurately. Neglecting their contribution to R is equivalent to assuming that they represent a similar fraction of the intensity in both the X and R lines. This assumption is probably correct as a first approximation, but certainly brings about some lack of precision as well.

3-6.3. Approach from the Model of a Homogeneous Bulky Solid

In the case of a powdered sample which is or can be tentatively considered as a homogeneous bulky solid, the intensity ratio of a peak Xy with respect to a reference peak Rz may be written as

$$R_X = R'_C \left(R'_A\, R'_L \frac{\sigma_{Xy}\, \lambda_{Xy}}{\sigma_{Rz}\, \lambda_{Rz}}\right) \frac{C_X}{C_R} = \frac{i_{Xy}}{i_{Rz}} \frac{C_X}{C_R} \quad . \qquad [3\text{-}15]$$

Usually R'_C is considered close to unity under the conditions discussed above. The i_{Xy}/i_{Rz} ratio can be determined experimentally by running standards of known composition. The evident pitfall of this approach is the quality of the standards used, as it is implicitly assumed that their surface composition is identical to that of their bulk. In this respect, the choice of standards may appear to be a lottery, and there is often no argument to justify discarding one standard versus another in case of discrepancies. Measurements of i_{Xy}/i_{Rz} have been carried out, however, for the more intense lines of most elements (4, 11, 14, 31, 32, 33, 46). The use of these values should be considered critically with consideration of the possible influence of the spectrometer and of experimental factors.

Another approach for determining i_{Xy}/i_{Rz} consists of computing R'_A, R'_L and $\lambda_{Xy}/\lambda_{Rz}$, then using the cross sections σ_{Xy} and σ_{Rz} which are found in the tabulation by Scofield (39). This method also suffers some limitations in precision because the values of R'_A and R'_L are approximate.

Even with all these limitations, XPS has been very successfully applied in place of standard analytical procedures, such as atomic absorption, for the quantitative analysis of elements such as lead (47) and arsenic (15). Agreement within 10% between XPS and atomic absorption is achieved in the ppb range and ultimate detection limit is of the order of 300 ppt for As. The analysis of Pb involved preparation of a mechanical mixture of $PbSO_4$ with another powder. For this particular case of mechanical mixture, the theoretical estimation of i_{Xy}/i_{Rz} has been shown to fail (47). This is readily understood if one recalls that photoelectrons originate from different phases. On the other hand, experimental determination of i_{Xy}/i_{Rz} by means of mixtures of the same phases as exist in the sample gives excellent results. The analysis of a mixture of solids will be further discussed in the following paragraphs.

Analysis of Na and Ca in glasses has been reported (21). This example underscores the fact that, in direct analyses of solids, the phase present is not always known or well defined, and finding adequate standards may be difficult. On the other hand, theoretical estimation of $\lambda_{X_y}/\lambda_{R_z}$ is satisfactory when the photoelectrons corresponding to both peaks travel in the same phase.

In many applications the atomic ratio C_X/C_R is calculated on the basis of a homogeneous bulky solid. This atomic ratio as measured by XPS is then considered as characteristic of a surface zone and compared with the overall composition. Deviation of the C_X/C_R ratio determined by XPS with respect to the overall ratio obtained for the whole sample by conventional analysis can be considered to be meaningful if it exceeds at least 20-30%. However, one can be more confident in smaller deviations that occur if a trend in a series of samples is established. In many problems, particularly those related to natural samples or to experimental studies of solids of mineralogical or industrial interest, indeed emphasis is not on the precision of isolated C_X/C_R determinations but rather on the trend followed by the apparent value of C_X/C_R within a series of samples.

TABLE 3-1. Examination of pigment (Chrome Yellow) coating by electron microscopy and XPS.

		XPS – Atomic Ratios				
Sample	Electron microscopy	$\frac{Pb}{O}$	$\frac{Cr}{O}$	$\frac{Si}{O}$	$\frac{Al}{O}$	$\frac{Pb}{Cr}$
1 untreated	——	0.18	0.12	–	–	1.5
2	no coating	0.11	0.07	0.07	0.05	1.6
3	observed	0.14	0.09	0.06	0.04	1.5
4	irregular coating + colloidal particles	0.08	0.03	0.05	0.03	2.3
6	good coating with colloidal	0.01	0.007	0.23	0.06	1.4
8	particles	0.004	n.m.*	0.17	0.095	n.m.*
9	thin coating, hardly visible	0.025	0.014	0.13	0.024	1.7
10	very regular coating	0.011	0.011	0.21	0.092	1

*n.m.: not measured

The example illustrated in Table 3-1 shows that XPS is an interesting tool for studying the coating of a solid by another phase. This case is concerned with industrial pigments of the Chrome-Yellow type $[Pb(CrO_4)_{0.5\text{-}1}\ (SO_4)_{1\text{-}0.5}]$ which are submitted to chemical treatments designed to increase the stability of the pigment to light and atmospheric exposure. In these treatments, the pigment crystals are coated by a layer of amorphous oxides such as silica-alumina. The atomic ratios of the treated materials were deduced from i_X/i_R values reported in the literature (46). Samples were also examined by electron microscopy and the results are summarized in Table 3-1. Clearly, the attenuation of the Pb and Cr signals reflects the quality of particle coating. In the case of sample 9, the coating layer is hardly visible by electron microscopy but its presence is clearly evident in the XPS spectrum.

Consider next the alteration of obsidian by water (45). The evolution of the Al_{2p}/Si_{2p} intensity ratio as a function of time follows a peculiar trend shown in Fig. 3-12, with a slow decrease followed by a sharp increase. Such an evolution of the Al/Si ratio could arise if the surface composition of the starting material is different from the bulk composition. However, this is ruled out by the fact that the sample investigated is a single particle obtained by fracture of a bigger piece. Therefore it may be concluded that, during the first stage of the alteration, aluminum dissolves preferentially while after a certain time the surface becomes aluminum enriched. A possible explanation would be that the rate of Al versus Si dissolution changes as a function of time due to a pH change; however, this is incompatible with the steep rise of the Al/Si ratio after 1000 min. Another explanation, which is supported by observations performed on the alteration of feldspars (10), is that aluminum dissolves quickly at the beginning and reprecipitates to produce a surface layer of alumina-silica gel.

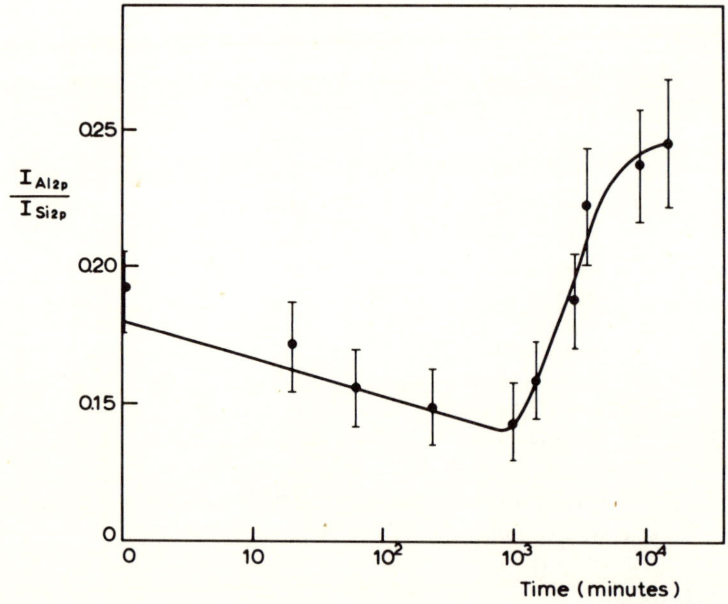

Figure 3-12. Variation of the Al_{2p}/Si_{2p} intensity ratio of an obsidian during its alteration in water (from Thomassin *et al.*, 1976).

3-6.4. Mixture of Solid Phases

The illustrations given in Section 3-6.3 have shown that the interpretation of XPS data, starting from the model of a homogeneous bulky solid, provides qualitative information on the organization of a heterogeneous solid sample. This potential ability of XPS to characterize the organization of complex solids as encountered in rocks and soils, warrants further discussion on the manner in which intensity ratios are formulated for random mixtures of powders.

Let X and R be two elements giving XPS peaks and let I and II be two different solid phases which are not porous. Consider first the case where X is only present in phase I and R only in phase II. Then,

$$\frac{I_X}{I_R} = R'_C R'_A R'_L \frac{\sigma_X}{\sigma_R} \frac{\lambda_{X\,I} C_{X\,I}\, S_I\, C'_I}{\lambda_{R\,II}\, C_{R\,II}\, S_{II}\, C'_{II}} \qquad [3\text{-}16]$$

where $C_{X\,I}$ is the volume concentration of X in phase I; C'_I, the weight concentration of phase I in the mixture; and S_I, the surface area developed by one gram of particles in phase I. For given values of C'_I and C'_{II}, the ratio I_X/I_R increases if S_I increases with respect to S_{II}, that is if the particle size of I decreases as compared to the particle size of II. This is illustrated by Fig. 3-13 for a catalyst where X and I are rhodium and where R and II are the carbon support (6).

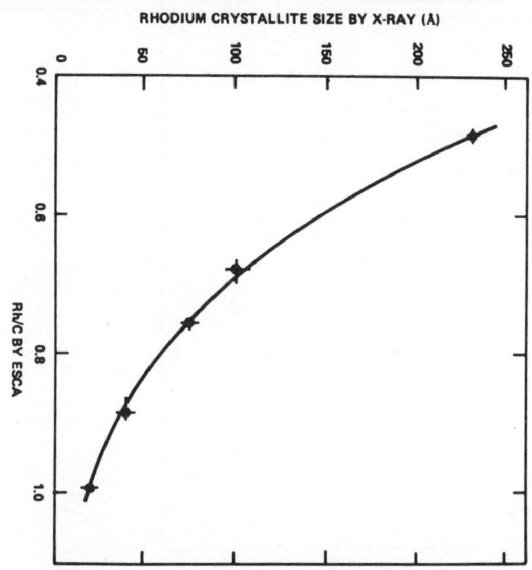

Figure 3-13. Correlation between Rh crystallite size determined from x-ray diffraction and the Rh/C XPS intensity ratio in a charcoal supported rhodium catalyst (from Brinen *et al.*, 1975).

Consider now the case where X is present only in phase I and R is present in both phases, giving

$$\frac{I_X}{I_R} = R'_C R'_A R'_L \frac{\sigma_X}{\sigma_R} \frac{(\lambda_{X\,I} C_{X\,I} S_I\, C'_I)}{(\lambda_{R\,I}\, C_{R\,I}\, S_I\, C'_I) + (\lambda_{R\,II}\, C_{R\,II}\, S_{II}\, C'_{II})} \qquad [3\text{-}17]$$

This relationship shows that, if $S_I < S_{II}$ or $S_I \simeq S_{II}$, the variation of I_X/I_R versus the overall chemical composition will be monotonous. On the other hand, if $S_I > S_{II}$, I_X/I_R will increase sharply as the phase I content increases with respect to phase II, and will reach a plateau.

With these intensity ratios formulated, it is interesting to discuss further the case of fluorinated alumina (40) from Sections 3-4.1 and 3-5.2 (Fig. 3-7). The apparent atomic ratio of F/Al in the zone explored by XPS has not been calculated, but the intensity ratio is simply plotted as a function of the overall atomic ratio (Fig. 3-14). The monotonous character of the plots above Al/F = 0.2 is in agreement with a mixture of distinct particles of alumina and aluminum fluoride indicated by x-ray diffraction and electron microscopy. The sharp rise observed below F/Al = 0.2 suggests that, up to 5% AlF_3, fluorine is better dispersed either because aluminum fluoride formed under these conditions has a very small particle size (see comments on equation [3-17]) or because it is coating alumina (see section 3-6.3 and Table 3-1). The existence of a thick coating of AlF_3 on the alumina surface was ruled out in Section 3-5.2; however, according to an infrared study of the surface acidity, it turns out that fluoride ions added in small amounts are dispersed on the alumina surface, forming less than a monolayer. This explains then why the Al_{2p} peak position is still characteristic of alumina and not of aluminum fluoride.

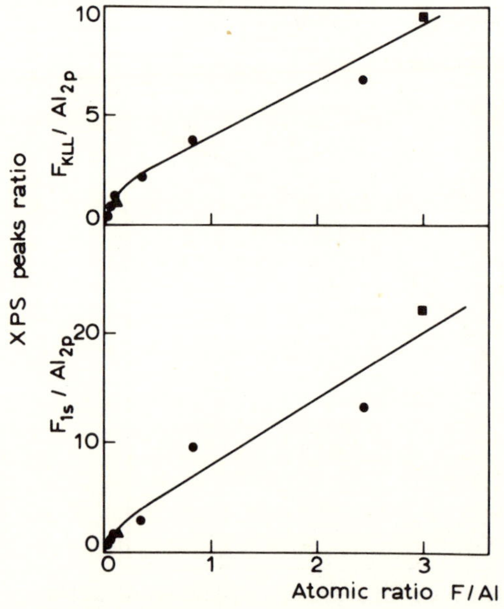

Figure 3-14. Variation of F_{1s}/Al_{2p} and Auger F_{KLL}/Al_{2p} intensity ratios in a series of aluminas treated with increasing amounts of fluoride (from Scokart *et al.*, 1979).

3-6.5. Detailed Study of High Surface Area Solids

Quantitative application of XPS to high surface area solids has been developed mainly in the field of heterogeneous catalysis, in particular to estimate the degree of dispersion of a solid phase supported by another phase. The example presented below illustrates, for a simple case, the approach required by such quantitative applications. This requires the establishment of a model for the high surface area solid and then working out the expressions for the peak intensity, keeping in mind the limitations and implications inherent to the model.

For illustration, consider a study by XPS of the adsorption of pyridine (a molecule chemisorbing on strongly acidic sites) on coprecipitated silica-alumina preheated at high temperature (550° C) (17). In this case

$$\frac{I_{N1s}}{I_{Al2p}} = \frac{i_{N1s}}{i_{Al2p}} \frac{C_N}{C_{Al}} \frac{1}{r} \qquad [3\text{-}18]$$

where r accounts for the fact that the solid is porous and that nitrogen is necessarily on the surface of the solid. The ratio C_N/C_{Al} denotes the concentration of chemisorption sites per Al atom. This ratio can be converted into units of sites per nm^2 by multiplying by $C'_{Al}N/S \cdot 10^{20}$, where C'_{Al} is the Al concentration in moles/g, N is Avogadro's number and S is the surface area in m^2/g. The value of r has been computed using geometrical models which simulate the actual porous solid as an aggregate of elementary particles. The size of these particles has no relationship to the size of the grains in the actual sample but is chosen so that the surface to mass ratio of the elementary particles of the model is equal to the surface area, S, of the actual samples.

Three models, illustrated in Fig. 3-15, were considered: (i) sheets of thickness $d = 2/\rho S$, ρ being the true density of the solid; (ii) cubes of edge $a = 6/\rho S$; and (iii) spheres of radius $q = 3/\rho S$. Computed values for r allow the surface density of acidic sites to be estimated and plotted as a function of the SiO_2 percentage in silica-alumina. The results (Fig. 3-17) show that the XPS data are in good agreement with independent measurements obtained by infrared spectroscopy.

Fig. 3-16 gives the variation of r as a function of the surface area. It shows that, even if one of the species analyzed is a molecule adsorbed on the surface, and the other species belongs to the solid, XPS analysis is actually similar to an overall analysis if the surface area exceeds 500 m^2/g for a density of 2.5. In fact, in this case the concepts of bulk and surface no longer have any meaning.

3-7. OVERVIEW OF METHODS OF CHARACTERIZATION OF SOLIDS BASED ON X-RAY, ELECTRON AND ION BEAMS

Table 3-2 presents a classification of these methods according to the type of primary beam used, the kind of secondary beam detected and the nature of the interaction with the sample. These methods will be briefly presented and their potential for characterization of surface properties and of the organization of complex solids will be pointed out.

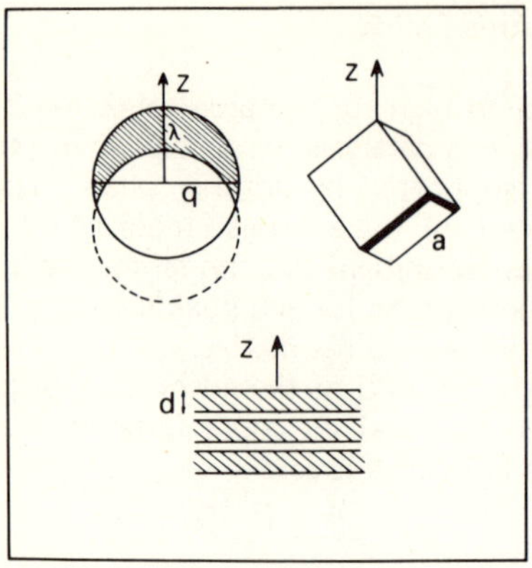

Figure 3-15. Geometric models used to simulate porous solids; Z represents the direction of photoelectron collection (from Defossé *et al.*, 1978).

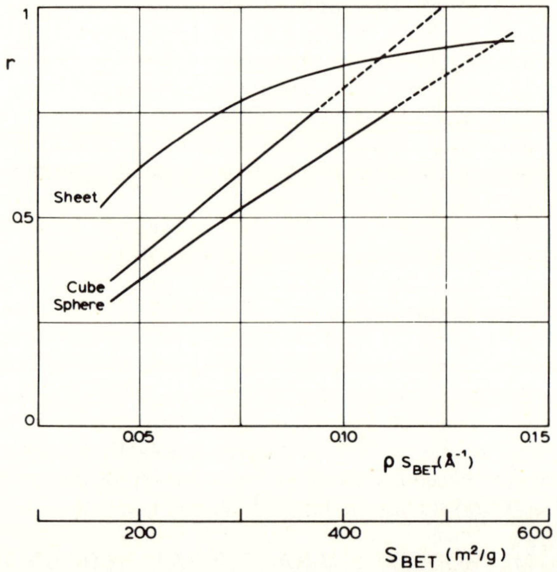

Figure 3-16. Dependence of the correction factor r on the universal parameter ρS, for the different geometric models considered. The lower abscissa is given in terms of m^2/g for a fixed density of 2.5 g/cm^3 (from Defossé *et al.*, 1978).

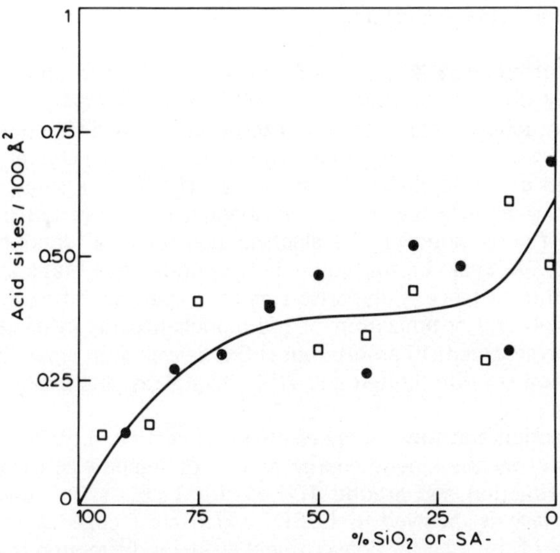

Figure 3-17. Strong acid sites density as determined by XPS (•) and IR (□) in a series of silica-aluminas (from Defossé *et al.*, 1978).

TABLE 3-2. Comparative view of methods of characterization of solids based on x-ray, electron and ion beams.

Characteristics	Incident beam: hυ	Incident beam: Electrons	Incident beam: Ions
Process without energy exchange	X-ray diffraction Radial electron distribution	Electron diffraction LEED	
Process with energy exchange			
Collection of the incident particles	X-ray absorption EXAFS	Transmission electron microscopy	ISS
Collection of the particles resulting from the interaction			
hυ	X-ray fluorescence	Electron microprobe analysis	
e	XPS	Scanning electron microscopy Auger spectroscopy	
Ions			SIMS

3-7.1. Processes Without Energy Exchange

X-rays and electrons may be scattered elastically by a solid. In this category we can include x-ray diffraction, which is a well-known method for the identification of crystalline solids. Recall that the interpretation of diffraction patterns leads to the determination of the arrangement of the atoms in the crystal.

In the method of radial electron distribution (RED), the x-ray scattering diagram (intensity vs θ as in x-ray diffraction diagram) is converted by Fourier transformation into a curve showing the electron density as a function of interatomic distance (37). A peak in the curve corresponds to a statistically well-represented interatomic distance. Comparison of the experimental data with data calculated from models and optimization of the models provide information concerning the atomic arrangement in amorphous solids as well as in crystalline solids. The use of this method is quite limited and RED, like x-ray diffraction, is a bulk technique.

Electron diffraction and low energy electron diffraction (LEED) differ from one another essentially by the kinetic energy of the impinging electrons – several KeV for electron diffraction and around 100 eV for LEED (20). Consequently, only the sample surface is involved in LEED, which gives clues as to the two-dimensional structure of the surface. Conventional electron diffraction is applicable only to thin crystallites. LEED requires single crystals and ultra-high vacuum (10^{-10} torr).

3-7.2. Process With Energy Exchange – Collection of Incident Particles

X-rays. In several methods, the only particles collected are those which have suffered no energy exchange during the process of interaction with matter. In x-ray absorption the decrease of intensity of the primary x-ray beam is measured, whereas extended x-ray absorption fine structure (EXAFS) is concerned with the measurement and interpretation of the fine structure on the high energy side of the x-ray absorption edge (43). Fourier transformation of that fine structure gives an electron density distribution curve versus the interatomic distance which is very similar to the plot obtained from RED, but is more specific because the atom used as the origin, rather than being undefined, is the one responsible for the absorption edge. The EXAFS method, which requires an intense source of x-rays, has developed only in the last few years, and serious difficulties still exist in reducing the experimental data to useful information.

Electrons. In transmission electron microscopy (TEM), an image of the sample is formed thanks to the contrast generated by differences in transmitted intensity of the primary electron beam, which is due to inelastic collisions and diffraction within the sample (29).

Combination of electron transmission and electron diffraction (see Section 3-7.1) allows imaging as well as structural information. In high resolution electron microscopy, the image is made by allowing the transmitted and diffracted beams to recombine to form the image. This provides a visualization of atomic planes or atoms with a higher resolution than conventional electron microscopy. In dark field electron microscopy, the image is formed by a diffracted beam only. Even if the material is poorly crystallized, microdomains giving rise to coherent scattering, when oriented adequately, appear as bright spots on a black background. The technique has been widely used for metals and, more recently, for crystalline oxides. Although it has not been much used for that purpose, it could be an interesting method for characterizing the structural organization of poorly crystallized oxides such as those that give only a broad x-ray diffraction band. Works performed on carbonaceous solids are indeed quite encouraging (5, 34).

3-7.3. Processes with Energy Exchange – Collection of Particles Resulting From Interaction Processes

Electrons. In scanning electron microscopy, a beam of high energy electrons is directed onto the sample where it produces a cascade of low energy electrons (29). These are collected at a somewhat grazing angle by a detector (Fig. 3-19). By scanning the sample spatially an image can be formed, the contrast of which gives a three-dimensional visual impression.

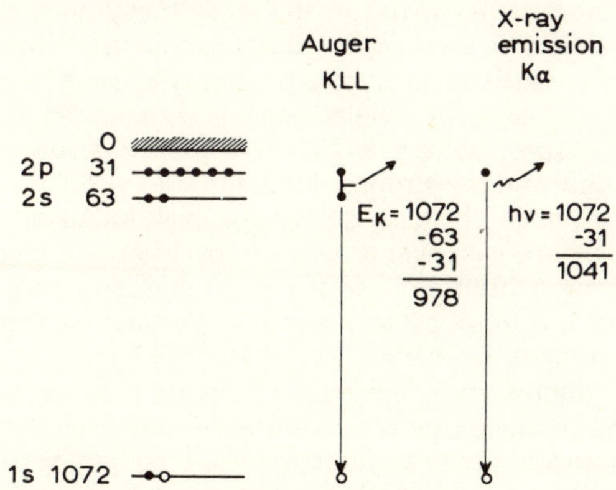

Figure 3-18. Schematic representation of Auger and x-ray emission deexcitation processes for sodium.

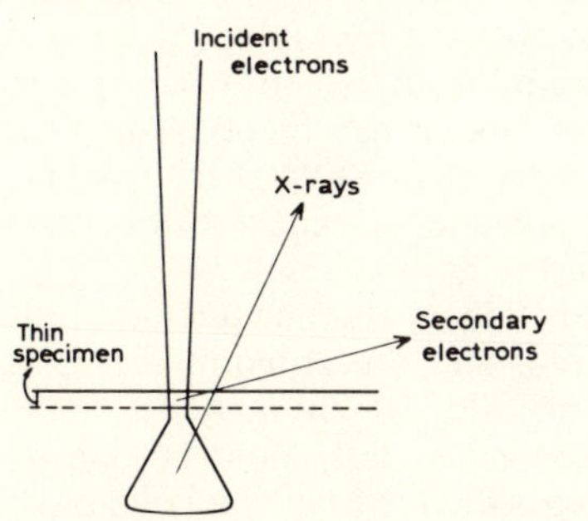

Figure 3-19. Illustration of the zone of x-ray emission (microprobe analysis) and of the emission of secondary electrons (scanning electron microscopy) under electron bombardment.

The Auger effect, as mentioned in the introduction of this chapter, is a deexcitation process of an ionized atom giving rise to removal of a second electron. An Auger electron is characterized by 3 letters, the first of which designates the non-occupied energy level (hole) in the singly ionized atom, while the two following letters designate the levels from which electrons are removed as a result of the deexcitation process. Fig. 3-18 illustrates the meaning of the kinetic energy of the Auger electron and allows a comparison with the energy of the photon (x-ray) emitted in the radiative deexcitation process. The two deexcitation processes are competitive; as the atomic number increases the importance of radiative deexcitation increases while the importance of Auger deexcitation decreases.

In Auger spectroscopy, electrons are used for excitation of the sample and the energy of the secondary electrons is analyzed (12). There are similarities between Auger and x-ray photoelectron spectroscopy wherein they both have about the same surface sensitivity; but they differ in the sense that Auger uses electrons as the excitation source and XPS uses x-rays. Both can be used for quantitative surface analysis but Auger is sensitive to trace amounts whereas XPS is not; this is due to the much higher flux of the primary electron beam compared to x-rays. One of the drawbacks of Auger spectroscopy is that it is more destructive than XPS because of both the high flux and the stronger interaction of electrons with matter as compared to x-rays. Although chemical shifts also exist in Auger, their explanation is much less straightforward because two energy levels are involved in the Auger process. Intrinsic peak widths are also wider for the same reason and thus resolution is poorer. Auger spectroscopy and scanning electron microscopy can be combined in the same instrument.

X-rays. Like x-ray fluorescence, electron microprobe analysis (EMPA) is based on analysis of x-rays produced by deexcitation of ionized atoms (29). However, EMPA has two advantages over x-ray fluorescence. First, as electrons are used as the primary beam instead of x-rays, the beam flux can be much higher making the absolute detection limit lower. However the same drawback appears as in Auger, namely the destructive effect of the electron beam. Second, EMPA has a very high spatial resolution as compared to x-ray fluorescence because the electron beam can be focused down to a very small diameter. Thus quantitative and qualitative analysis of grains can be carried out; and by tuning the x-ray analysis device to a particular transition and then scanning the electron beam across the grain, concentration profiles or concentration images can be obtained. Since the energy of the incoming beam is larger in EMPA (100 KeV) than in x-ray fluorescence, deeper electronic levels can be excited and, except for the heaviest elements ($Z > 80$), a K transition can be observed.

In a study of the distribution of Ni in an alumina grain (19), the joint use of EMPA and XPS was particularly helpful and prevented misinterpretation of previously obtained XPS data. Indeed, the XPS Ni $2p_{3/2}$/Al 2p intensity ratio was substantially higher than the expected value; this could be due to a dispersion higher than the expected one or to segregation of NiO particles outside the porosity of the grains. EMPA showed that it was caused instead by a Ni enrichment in the outer regions of the grains, Ni being still distributed in the pores.

EMPA is very often coupled with electron microscopy, this combination is referred to as Analytical Electron Microscopy. It is important to point out that the spatial resolution of EMPA depends on the thickness of the analyzed sample. This is illustrated by Fig. 3-19. Due to inelastic collisions of the electrons, the zone from which x-rays are emitted has a pear-like shape, the large section of the pear having an area of about 1 μm^2 and the length being of the order of 1 μm. For samples of a thickness of the order of 1 μm and above, the spatial resolution is thus of the order of 1 μm. On the other hand, for a very thin sample better spatial resolution may be obtained; values of the order of 200 Å can be reached for the thinnest samples, with an electron beam of the order of 50 Å.

3-7.3. Processes Involving Ions

Argon Etching. Ion bombardment of a surface is often used in conjunction with XPS or Auger as a way to etch the sample. A beam of Ar is produced in the system by a microleak so that the dynamic pressure rises to the 10^{-6}-10^{-4} torr range. Before entering the vessel, the beam is ionized, accelerated and finally focussed onto the sample. In that way the sample surface is eroded by simple ion impact (28).

It can be used for two different purposes: one is cleaning the surface to get rid of carbon contamination or oxide overlayer so that a fresh surface is exposed; the other consists of performing surface analysis of the sample, e.g. by XPS or Auger, between successive etchings – in that way a concentration profile can be obtained. This application requires a rather precise knowledge of the etching rate which is not always easy to estimate. For both types of uses, extreme caution should be exercised as the argon etching very often perturbs the organization of the solid. Chemical reduction, especially of transition metals, induced by ion bombardment has been reported in numerous papers, and a modification in concentration profile resulting from preferential ion sputtering has also been reported.

Secondary ion mass spectrometry (SIMS). SIMS consists simply of performing a mass spectral analysis of the secondary ions produced by argon etching, while the surface is being eroded (16). A typical spectrum is presented in Fig. 3-20. The method gives a direct, qualitative elemental analysis of the surface and is much more surface selective than XPS or Auger; a few percent of a monolayer can be detected.

Thanks to the sensitivity of the currently available detectors, the etching rate can be lowered to a point where removing the equivalent of one monolayer takes several hours, so that, contrary to what might be expected, SIMS can be virtually a non-destructive method (so-called static SIMS). Another advantage is that SIMS does detect hydrogen, whereas Auger and XPS do not.

Quantitative analysis by SIMS is feasible but not easy. Gaining information on the chemical state of the surface is, in principle, possible by a careful examination of the clusters extracted from the surface. However, recombination of secondary ions when leaving the surface makes the data ambiguous and the interpretation difficult.

Ion microprobe analysis is a later development of SIMS with an ion beam diameter of about 1 μm. This allows a high spatial resolution that is not accessible by conventional SIMS. Consequently the beam can be scanned on the sample as is done in EMPA.

Ion scattering spectroscopy (ISS). ISS is similar to static SIMS as far as basic principles are concerned, but where SIMS measures the mass spectral distribution of the secondary ions, ISS determines the kinetic energy distribution of the primary scattered ion beam (1). When colliding with a surface atom, part of the kinetic energy of the incoming ion can be released to the surface atom, just as is described by the classical treatment of ball collisions in mechanics. The energy loss of the incoming ion depends thus on the mass of the surface ion encountered, and the energy distribution of the scattered beam provides a qualitative elemental analysis of the first monolayer.

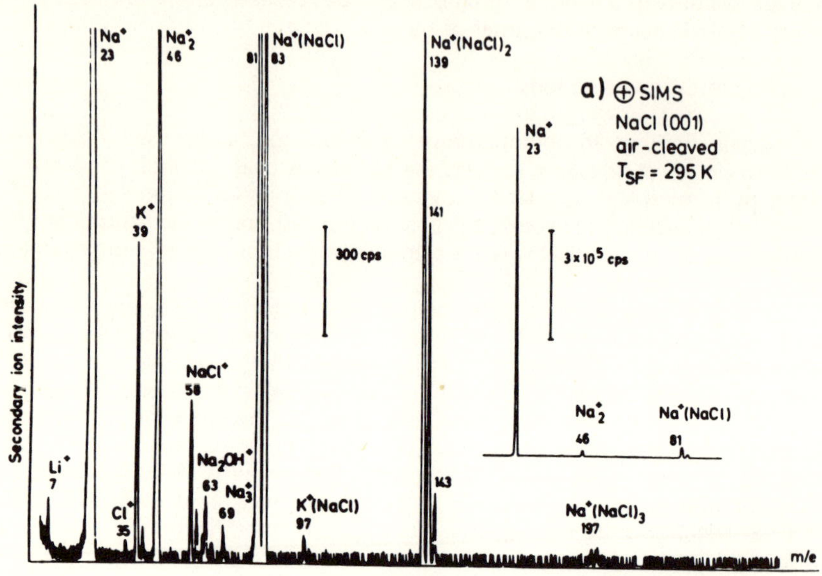

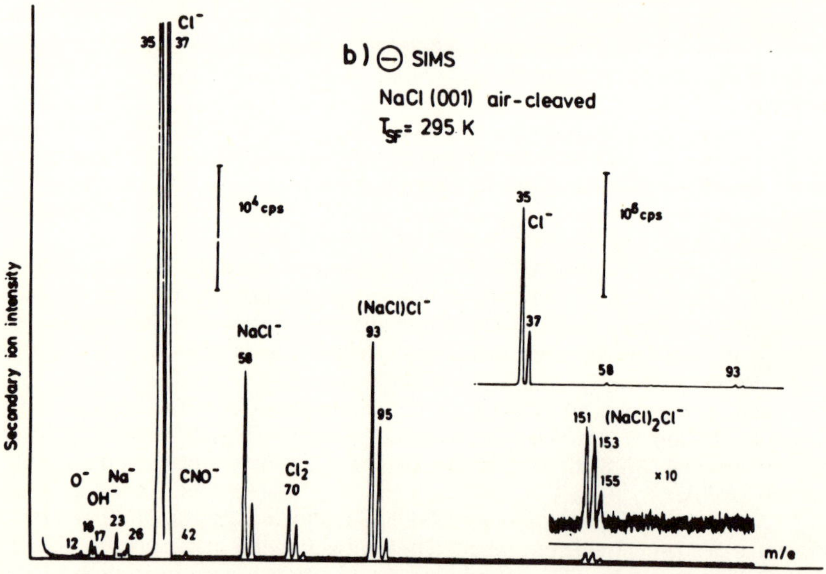

Figure 3-20. Secondary ion mass spectra of air cleaved NaCl(001) surface; the positive and negative ion spectra are displayed (from Estel *et al.,* 1976).

REFERENCES

1. Ball, D.J., T.M. Buck, D. Macnair, and G.H. Wheatley. 1972. Investigation of low energy ion scattering as a surface analytical technique. *Surf. Sci.* **30**: 69-90.
2. Barrie, A. and F.J. Street. 1975. An Auger and x-ray photoelectron spectroscopic study of sodium metal and sodium oxide. *J. Electron Spectrosc. Rel. Phenom.* **7**: 1-31.
3. Bearden, J.A. 1967. X-ray wavelength and x-ray atomic energy levels. Nat. Standard Ref., Data Series – NBS 14; *Rev. Mod. Phys.* **31**(1).
4. Berthou, H. and C.K. Jorgensen. 1975. Relative photoelectron signal intensities obtained with a magnesium x-ray source. *Anal. Chem.* **47**: 482-488.
5. Boulmier, J.L., A. Oberlin and B. Durand. 1977. Etude structurale de quelques séries de kérogènes par microscopie électronique: relations avec la carbonisation. *In* R. Campos and J. Goni, Eds. Advances in Organic Geochemistry 1975. Enadimsa, Madrid. pp. 781-796.
6. Brinen, J.S., J.L. Schmitt, W.R. Doughman, P.J. Achorn, L.A. Siegel and W.N. Delgass. 1975. X-ray photoelectron spectroscopy study of the rhodium on charcoal catalyst. II. Dispersion as a function of reduction. *J. Catal.* **40**: 295-300.
7. Brion, D. and J. Escard. 1976. Application de la spectroscopie des photoélectrons à l'analyse quantitative des surfaces. *J. Microsc. Spectrosc. Electron* **1**: 227-246.
8. Brundle, C.R. 1974. The application of photoelectron spectroscopy to surface studies. *J. Vac. Sci. Technol.* **11**: 212-224.
9. Brundle, C.R. and A.D. Baker, Eds. 1979. Electron spectroscopy: theory, techniques and applications. Vol. 1, 2 and 3. Academic Press Inc.
10. Busenberg, E. and C.V. Clemency. 1976. The dissolution kinetics of feldspars at 25°C and 1 atm CO_2 partial pressure. *Geochim. Cosmochim. Acta* **40**: 41-49.
11. Calabrese, A. and R.G. Hayes. 1974. Photoelectric cross sections of some atom-like valence levels for MgKα radiation and comparison with OPW estimates. *Chem. Phys. Lett.* **27**: 376-379.
12. Carlson, T.A. 1975. Photoelectron and Auger spectroscopy. Plenum Press, New York.
13. Carlson, T.A., Ed. 1978. X-ray photoelectron spectroscopy. Academic Press Inc.
14. Carter, W.J., G.K. Schweitzer and T.A. Carlson. 1974. Experimental evaluation of a simple model for quantitative analysis in x-ray photoelectron spectroscopy. *J. Electron Spectrosc. Rel. Phenom.* **5**: 827-835.
15. Carvalho, M.B. and D.M. Hercules. 1976. Trace arsenic determination by volatilization and x-ray photoelectron spectroscopy. *Anal. Chem.* **50**: 2030-2034.
16. Czanderna, A.W., Ed. 1975. Methods of surface analysis. Elsevier.
17. Defossé, C., P. Canesson, P.G. Rouxhet and B. Delmon. 1978. Surface characterization of silica-aluminas by photoelectron spectroscopy. *J. Catal.* **51**: 269-277.
18. Defossé, C., R.M. Friedman and J.J. Fripiat. 1975. Etude préliminaire des conditions de prétraitement, de réduction et d'adsorption des zéolites Y échangées au nickel par spectroscopie des photoélectrons. *Bull. Soc. Chim. France* **7-8**: 1513-1518.

19. Delannay, F., M. Houalla, D. Pirotte and B. Delmon. Critical assessment by analytical electron microscopy of the significance of XPS measurements of the dispersion of supported catalysts. *Surf. Interf. Analysis* **1**: 172-174.
20. Ertl, G. and J. Küppers. 1974. Low energy electrons and surface chemistry. Monograph in Modern Chemistry, 4, Verlag, Weinheim.
21. Escard, J. and D. Brion. 1973. Possibilités d'analyse quantitative des intensités en spectroscopie électronique: application aux verres. C.R. Acad. Sci., Paris, *Ser. B* **276**: 945-947.
22. Estel, J., H. Hankes, H. Kaarmann, H. Nahr and H. Wilsch. 1976. On the problem of water adsorption on alkali halide cleavage planes, investigated by secondary ion mass spectroscopy. *Surf. Sci.* **54**: 393-418.
23. Fadley, C.S., R.J. Baird, W. Siekhaus, T. Novakov. S.Å.L. Bergström. 1974. Surface analysis and angular distributions in x-ray photoelectron spectroscopy. *J. Electron Spectrosc. Rel. Phenom.* **4**: 93-137.
24. Fraser, W.A., J.V. Florio, W.N. Delgass and W.D. Robertson. 1973. Surface sensitivity and angular dependence of x-ray photoelectron spectra. *Surf. Sci.* **36**: 661-674.
25. Friedman, R.M. 1973. The application of x-ray photoelectron spectroscopy to the study of surface chemistry. *Silicates Industriels* **39**: 247-253.
26. Helmer, J.C. and N.H. Weichert. 1968. Enhancement of sensitivity in ESCA spectrometers. *Appl. Phys. Lett.* **13**: 266-268.
27. Holm, R. 1978. Analyse de surface dans le domaine des couches monomoléculaires: un nouveau champ d'application des méthodes de travail en microchimie. *Actualité Chimique* **Jan. 1978**: 13-23.
28. Holm, R. and S. Storp. 1977. ESCA studies on changes in surface composition under ion bombardment. *Appl. Phys.* **12**: 101-112.
29. Hren, J.J., J.I. Goldstein and D.C. Joy, Eds. 1979. Introduction to analytical electron microscopy. Plenum, New York.
30. Kennedy, D.J. and S.T. Manson. 1972. Photoionization of the noble gases: cross sections and angular distributions. *Phys. Rev. A* **5**: 227-247.
31. Nefedov, V.I., N.P. Sergushin, I.M. Band and M.B. Trzhaskovskaya. 1973. Relative intensities in x-ray photoelectron spectra. *J. Electron Spectrosc. Rel. Phenom.* **2**: 383-403.
32. Nefedov, V.I., N.P. Sergushin, Y.V. Salyn, I.M. Band and M.B. Trzhaskovskaya. 1975. Relative intensities in x-ray photoelectron spectra. Part II. *J. Electron Spectrosc. Rel. Phenom.* **7**: 175-185.
33. Nefedov, V.I. and V.G. Yarzhemsky. 1977. Relative intensities in x-ray photoelectron spectra. Part III. *J. Electron Spectrosc. Rel. Phenom.* **11**: 1-11.
34. Oberlin, A., J.L. Boulmier and M. Villey. 1980. Electron microscopic study of kerogen microstructure; selected criteria for determining the evolution path and evolution stage of kerogen. *In* B. Durand, Ed., Kerogen. Technip, Paris. Chapter 7. pp. 191-241.
35. Penn, D.P. 1976. Quantitative chemical analysis by ESCA. *J. Electron Spectrosc. Rel. Phenom.* **9**: 29-40.
36. Powell, C.J. 1974. Attenuation lengths of low energy electrons in solids. *Surf. Sci.* **44**: 29-46.
37. Ratnasamy, P. and A.J. Léonard. 1972. X-ray scattering techniques in the study of amorphous catalysts. *Catal. Rev.* **6**: 293-322.
38. Reilman, R.F., A. Msezane and S.T. Manson. 1976. Relative intensities in photoelectron spectroscopy of atoms and molecules. *J. Electron Spectrosc. Rel. Phenom.* **8**: 389-394.

39. Scofield, J.H. 1976. Hartree-Slater subshell photoionization cross sections at 1254 and 1487 eV. *J. Electron Spectrosc. Rel. Phenom.* **8**: 129-137.
40. Scokart, P.O., S.A. Selim, J.P. Damon and P.G. Rouxhet. 1979. The chemistry and surface chemistry of fluorinated alumina. *J. Colloid Interface Sci.* **70**: 209-222.
41. Siegbahn, K. 1974. Electron spectroscopy – an outlook. *J. Electron Spectrosc. Rel. Phenom.* **5**: 3-97.
42. Siegbahn, K., C. Nordling, A. Fahlman, R. Nordberg, K. Hamrin, J. Hedman, G. Johansson, T. Bergmark, S.E. Karlsson, I. Lindgren and B. Lindberg. 1967. ESCA: atomic, molecular and solid state structure studied by means of electron spectroscopy. Nova Acta Regiae Soc. Sci. Uppsaliensis, Ser. IV, Vol. 20.
43. Stern, E.A. 1974. Theory of the extended x-ray absorption fine structure. *Phys. Rev. B* **10**: 3027-3037.
44. Swingle, R.S. 1975. Quantitative surface analysis by x-ray photoelectron spectroscopy (ESCA). *Anal. Chem.* **47**: 21-24.
45. Thomassin, J.H., J.C. Touray and J. Trichet. 1976. Etude par spectrométrie ESCA des premiers stades d'altération d'une obsidienne: le comportement relatif de l'aluminium et du silicium. C.R. Acad. Sci. Paris, Ser. C, **282**: 1229-1232.
46. Wagner, C.D. 1972. Sensitivity of detection of the elements by photoelectron spectrometry. *Anal. Chem.* **44**: 1050-1053.
47. Wyatt, D.M., J.C. Carver and D.M. Hercules. 1975. Some factors affecting the application of electron spectroscopy (ESCA) to quantitative analysis of solids. *Anal. Chem.* **47**: 1297-1301.

CHAPTER 4

APPLICATION OF X-RAY PHOTOELECTRON SPECTROSCOPY TO THE STUDY OF MINERAL SURFACE CHEMISTRY

Mitchell H. Koppelman
Georgia Kaolin Research
25 Route 22 East
Springfield, NJ 07081

4-1. UNIQUENESS OF XPS FOR THE INVESTIGATION OF MINERAL SURFACE PHENOMENA – PROBING DEPTH

A majority of the chemical reactions generally associated with clay minerals, metal oxides, and other soil-related minerals occur at the interfaces between the minerals and their surroundings. Two of these interfaces occur at the gas/solid, and liquid/solid reaction sites with the common component being the solid mineral surface. Few reactions related to soil environments involve chemical reactions generated at atom sites in a mineral's bulk. Therefore, much useful chemical insight can be achieved by examining the chemistry of atoms and reaction products associated with the surface region of these minerals.

Techniques such as x-ray powder diffraction, electron spin resonance spectroscopy, infrared, ultra-violet and visible spectroscopy (transmission), and Mössbauer spectroscopy, are highly sensitive and informative techniques, but all reveal properties and information related to the entire or bulk mineral phase. Since a mineral surface represents only a relatively small portion of a mineral's bulk, only a fraction of the information obtained through these techniques may be due to surface contributions.

The use of x-ray photoelectron spectroscopy (XPS) in the study of mineral surfaces has afforded a method of direct examination of the chemistry at mineral interfaces. Applications may be classified into those with analytical intentions and those affording an insight into the chemical bonding state of the elements present.

The uniqueness of XPS for studies of this type arises from its effective sampling depth. X-rays generated by targets such as aluminum (K_{α_1,α_2}, $h\upsilon$ = 1486.6 eV) or magnesium (K_{α_1,α_2}, $h\upsilon$ = 1253.6 eV) have sufficient energy to penetrate deep into the bulk of a mineral sample. However, the surface analysis capabilities of XPS arises from the limited escape depth of the photoejected electrons. Inelastic collisions with atoms surrounding the electron emitter result in an effective photoelectron sampling depth of generally less than 50 Å (Fig. 4-1). This photoelectron escape depth will vary from sample to sample and is dependent upon the energy of the incident radiation, the kinetic energy of the ejected electron, the crystallinity, and density of the sample material.

J. W. Stucki and W. L. Banwart (eds.), Advanced Chemical Methods for Soil and Clay Minerals Research, 205–243.

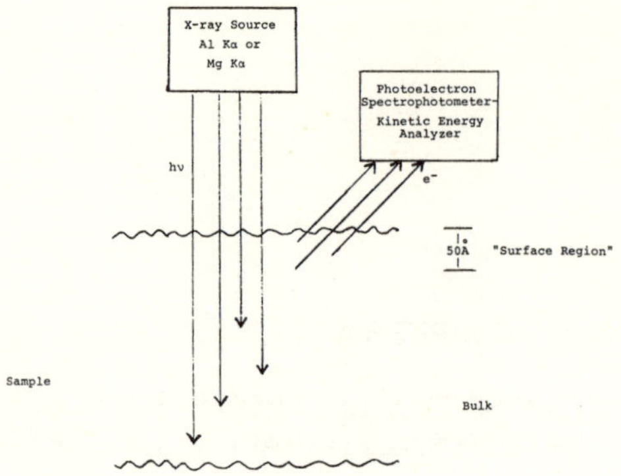

Figure 4-1. Effective photoelectron sampling depth.

4-2. SAMPLE HANDLING TECHNIQUES

When x-ray radiation ($h\nu$) of sufficient energy bombards a sample, inner shell, non-valence electrons are photoejected with kinetic energies related to the initial binding energies of these electrons. XPS sampling for most soil minerals including clays results in electrically insulating specimens. During the XPS measurements on insulating samples, electrostatic charge can build up at the sample surface due to the electron ejection process and the poor electrical conductivity between the sample and the spectrometer. This situation is commonly referred to as sample charging and, experimentally, is the most common sampling difficulty likely to be encountered in electron binding energy determination in mineral examination (Fig. 4-2).

Generally speaking, the electron binding energy for a core electron is related to the photon energy, measured electron kinetic energy and spectrometer work function by the following equation:

$$E_{binding} = E_{h\nu} - E_{kinetic} - \Phi_{sp} \qquad [4\text{-}1]$$

For non-conducting (mineral) samples, the sample surface charging or electron retarding energy, E_c, must be included in this expression resulting in equation [4-2]:

$$E_{binding} = E_{h\nu} - E_{kinetic} - \Phi_{sp} - E_c \qquad [4\text{-}2]$$

Many sample handling techniques have been used to both minimize and quantify E_c (see Table 4-1). Any sampling mode which improves the surface conductivity and charge equilibration between the sample and the spectrometer reduces the absolute value of E_c.

One of the more common methods used to calibrate the spectra of insulating samples has been the use of the carbon 1s signal from the spectrometer background contamination (often pump oil) (16, 21). This technique yields an obvious disadvantage. If the sample itself contains carbon, the signal from the surface deposited carbon may be masked, making calibration extremely difficult. A second disadvantage of this method of calibration is the uncertainty of the actual identity

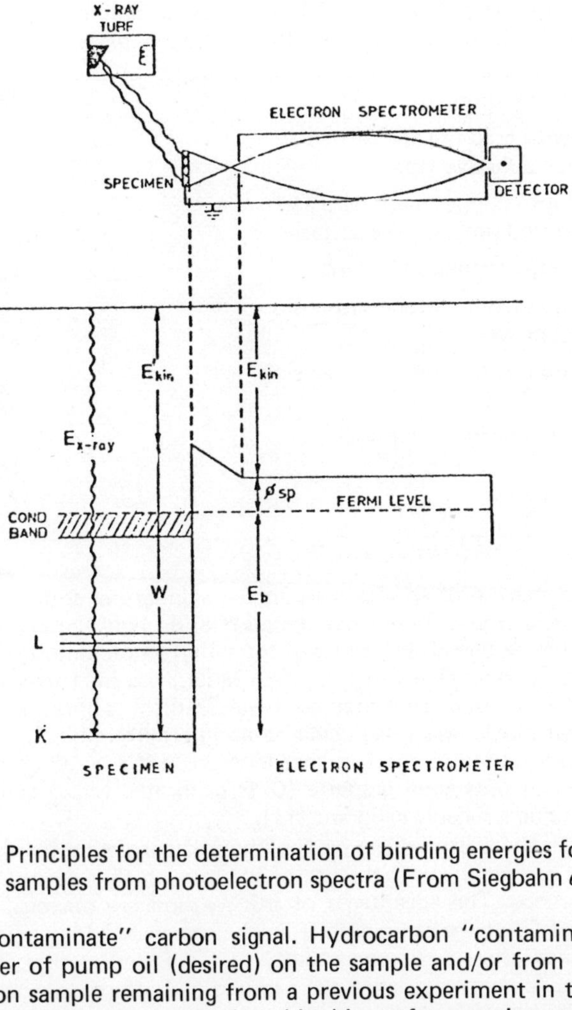

Figure 4-2. Principles for the determination of binding energies for non-conductive samples from photoelectron spectra (From Siegbahn *et al.*, 1967).

of this "contaminate" carbon signal. Hydrocarbon "contamination" may result from a layer of pump oil (desired) on the sample and/or from a small amount of hydrocarbon sample remaining from a previous experiment in the instrument (undesired). It is also conceivable that this thin surface coating could be affected by reactions with the sample surface onto which it has been deposited.

Another commonly employed method for charge correction is the use of gold, deposited onto the sample surface from the vapor phase. The gold actually forms "islands" on the surface, and is in electrical equilibrium with the sample surface and the spectrometer. The gold 4f level can correct for charging. Recently, however, various workers have questioned the universal applicability of this gold standard because of its chemical reactivity with inorganic solids, surface coverage problems with polymers, interactions with phosphorus compounds, and possible variations in the amount of gold deposited (10).

The placing of a thin film of sample, either by evaporation from a solvent or by sublimation, onto a gold surface is another method which has been used to calibrate XPS spectra. Presumably, if the film is thin enough, both the sample and gold are in electrical equilibrium (37). The gold 4f levels are then used to calibrate the sample.

Table 4-1. Sample Handling Techniques for XPS Studies of Mineral Samples (From Jaegle *et al.* (1978), and Koppelman, 1976).

Method	Calibrant
Sample mixed with graphite and dusted onto double-stick adhesive tape	C 1s
Sample pressed into copper mesh wire and gold vapor deposited onto sample surface	Au 4f
Sample pressed into copper wire mesh	C 1s
Sample evaporated from acetone suspension onto gold plated probe	Au 4f
Sample evaporated from acetone suspension onto probe	Si 2p
Pure sample dusted onto double stick adhesive tape	Si 2p
Electron flood gun	C 1s

In a study of the use of an internally mixed standard as compared to vapor deposition procedures, it was found that the correlation between Pauling electronegativities of the halide ligands in a series of tetraethyl-ammonium tin halides and the Sn $3d_{5/2}$ binding energies was better when MoO_3 was internally mixed with the sample than when vacuum deposited gold was used for calibration (41). It has also been found that MoO_3 was not reliable as an internally mixed calibrant, and this observation is attributed to the fact that MoO_3 is not a conductor (16). More recently the mixing of powdered graphite (C 1s calibrated versus gold) with the sample has proved to be a reliable calibrant (11).

It would seem that perhaps the best method of correcting for the charging effects of a sample surface involves the use of an atom which is part of the sample as an internal reference. The advantages of this method are obvious, the Fermi levels of the reference and the sample must be the same since they are part of the same molecule; hence, there is no uncertainty in the E_b correction. There is no problem accounting for charging since the samples and reference *must* charge to the same extent. This technique also provides the additional advantage of higher count rates, since the sample does not have to be diluted with an internally mixed standard.

Charging can usually be detected readily by an examination of the value of E_c for a given set of data. A higher E_c is indicative of a larger degree of charging by the sample. E_c values indicative of only small degrees of charging are generally between 3 and 4 eV. Abnormally broad peaks which become narrower when investigated by another, perhaps more reliable sample handling technique, are another sign of sample charging. Double peaks may occur if surface particles or insulating regions charge to different potentials (heterogeneous). An important criterion for the validity of a given method of charge correction is that the data obtained must be reproducible. The reproducibility of data can be determined by performing several measurements and calculating the standard deviation associated with the distribution of values.

During some preliminary investigations of the XPS spectra of clay minerals, it was found advantageous (increase count rate, retain "purity" of the sample, etc.) to calibrate the binding energy of an element common to all clays studied (kaolinite, chlorite, and illite) and to use that element as an internal standard to determine the binding energies of the other various lattice and adsorbed elements (22). Silicon was chosen as that internal standard for three reasons: (i) high concentration in all three clay lattices, thus allowing scanning of the silicon binding energy region for shorter periods of time and thereby permitting longer counting intervals on elements of lower concentrations, (ii) silicon substitutes in only the tetrahedral lattice position; hence, any additional peaks due to silicon in a different silicon environment or broadening of the single photopeak was not anticipated, and (iii) solution treatment of the clays was found not to alter the nature of the silicon lattice site and thus the silicon photopeak (B.E.). Specifically, the Si 2p level was selected for use as the internal standard energy level.

Initial attempts at determining the absolute binding energy for silicon 2p electrons in the three clays were conducted by mixing the clay sample and graphite in a ratio (by weight) of 2:1, clay:graphite. The binding energies determined for the Si 2p electrons in chlorite, illite and kaolinite using the previously mentioned sample handling techniques are reported in Table 4-2.

Table 4-2. Calibration of internal Standard Binding Energies for the Si $2p_{1/2,3/2}$ Level Using Various Sample Handling Techniques (from Koppelman, 1976).

Method	Chlorite		Kaolinite		Illite	
I. Graphite Admixture						
Sample:Graphite Ratio-	Si $2p_{1/2,3/2}$	C 1s FWHM	Si $2p_{1/2,3/2}$	C 1s FWHM	Si $2p_{1/2,3/2}$	C 1s FWHM
1:1	102.4 ± 0.1	1.3	105.7 ± .20	2.1	105.2 ± 0.1	2.0
2:1	102.5 ± 0.1	1.3	105.8 ± .25	2.2	105.5 ± 0.1	2.2
9:1	102.5 ± 0.1	1.6	105.7 ± .20	2.1	105.6 ± 0.1	2.2
II. Evaporation from Acetone Suspension	102.5 ± 0.1		103.8 ± .15		104.1 ± 0.1	
III. Vapor Deposition of Gold	102.1 ± 0.1		102.7 ± 0.1		102.5 ± 0.1	
IV. Background Carbon "Contamination"	102.1 ± 0.1		102.7 ± 0.1		102.5 ± 0.1	

Using the graphite admixture technique (virtually independent of the sample:graphite ratio) it was observed that the binding energies for the silicon 2p electrons in chlorite, compared to kaolinite and illite, were very different (21). This was not expected since the geometry and oxygen coordination to silicon in all three clays is virtually the same, with the only difference being the stacking array of layers in the sheet structure.

To test the validity of the graphite admixture technique as applied to XPS of clays, a sample was prepared which contained 25% chlorite, 25% kaolinite and 50% graphite by weight. If the graphite calibration of silicon were accurate, one would expect the spectrum to show two different silicon peaks (environments) separated by approximately 3.3 eV. Only one peak in the binding energy region of silicon was observed. The binding energy, calibrated with the graphite, was 105.3 ± 0.1 eV. Probably more significant was the fact that the full width at half maximum (FWHM) was 2.6 eV. The FWHM of the Si $2p_{1/2,3/2}$ photopeak in pure chlorite was 2.4 eV, and in pure kaolinite was 2.2 eV. This increase in width of the photopeak for the chlorite-kaolinite mixture was attributed to a small difference in binding energy for silicon 2p electrons in chlorite and kaolinite. The difference in electronic environments for silicon in chlorite and kaolinite is probably small since only one peak, instead of the anticipated two was observed.

This evidence, combined with the difficulty in obtaining reproducible results for silicon in kaolinite, and the variability of the FWHM of the graphite C 1s peak through the series of clay samples, indicated that the graphite-admixture technique was not minimizing charging, nor was it useful in correcting for charging through calibration. Failure of this technique was attributed to differences in particle size between the clay particles and the graphite powder (21). In order for the graphite admixture technique to work, it is assumed that all particles (graphite and sample) are in intimate contact, so that charging will not be localized on specific particles. If, however, the particle sizes are not the same, this intimacy may not be achieved.

To determine the absolute binding energy for the silicon 2p level, an acetone suspension of each clay mineral was evaporated onto a clean gold plated probe to obtain a very thin sample film. It was anticipated that if the film was thin enough, permitting observation of the underlying gold 4f level, a positive charge would not build up on the sample surface (21). The spectrum of the gold 4f electrons was then used to calibrate the binding energy of the Si 2p electrons in the sample film (Table 4-2).

While there was a reduction in binding energy for the Si 2p level in all three clay samples, there was still a binding energy difference of 1.5 eV between chlorite silicon 2p electrons and kaolinite silicon 2p electrons. Since the mixed (chlorite-kaolinite) sample indicated there was not a significant difference between the two silicon environments, it was obvious that the samples were still charging.

A sample handling technique in which a very thin film of gold was vapor deposited on the sample surface was then tried (21). It was anticipated that, while the gold may not reduce sample charging, it should charge to the same extent as the sample surface. The Si 2p level in the sample could then be calibrated using the gold 4f level from the gold on the surface. Similarly, background carbon (probably diffusion pump oil, a polyphenyl ether) was allowed to build up on the sample surface. The C 1s photopeak from this carbon buildup was then used to calibrate the Si 2p level in the sample. The results from these two surface deposition techniques are summarized in Table 4-2.

The results indicated that charging had been reduced, or at least a better correction was made for the effects on the Si 2p binding energy. The reproducibility of both of these techniques, as well as the agreement in binding energy between the two techniques, were strong indicators that corrections had been made for surface charging.

4-3. ANALYTICAL APPLICATIONS

4-3.1. General

Siegbahn *et al.* (38) summarized the following virtues of XPS in light of its analytical implications:

1) Heavy and light elements alike may be studied (with the exception of hydrogen).
2) The absolute sensitivity is high, i.e. the amount of material required for obtaining an XPS spectrum is small ($< 10^{-8}$ grams).
3) Amorphous as well as crystalline samples may be investigated.
4) XPS, in general, is a non-destructive method.

5) The spectral (binding energy) position of an XPS line may depend upon the valence state of the corresponding atom resulting in quantitative oxidation state analysis.

The intensity (in counts per second, or peak area for composite photopeaks) of an XPS photosignal is given by equation [4-3], and is related to elemental concentration by factors including the photobeam intensity, photoelectric cross section, effective sampling depth and the mean free path, *viz.*

$$I = I_o \, \sigma_A \, C_A \, \lambda_A \qquad [4\text{-}3]$$

where I is the signal intensity (in counts/sec) of a specific peak for a given element (A); I_o, the incident photon beam intensity at the sample surface, which is a constant for the whole spectrum; σ_A, the photoelectric cross sections for a specific level; C_A, the concentration of element A; and λ_A, the mean free path for electrons of specific kinetic energy.

Most analytic determinations of elemental concentrations as related to mineral systems are done on a relative basis. Intensity ratios, I_A/I_B, for elements A and B in a sample are determined by substitution into equation [4-3] for species A and B. The resulting equation for the intensity ratio for elements A and B in homogeneous polycrystalline solid material $A_n B_m$ is:

$$\frac{I_A}{I_B} = \frac{n}{m} \cdot \frac{\sigma_A}{\sigma_B} \cdot \left(\frac{E_A}{E_B}\right)^{0.5} \qquad [4\text{-}4]$$

where E_A, E_B are the respective kinetic energies of the ejected photoelectrons. Within a limited kinetic energy range, E_A and E_B are found to be proportional to λ_A and λ_B respectively (Chapter 3).

In actuality, the relationship expressed in equation [4-4] should be multiplied by a factor R_A, taking into account the anisotropy of the photoejection process as discussed in Chapter 3.

4-3.2. Comparison of Bulk and Surface Chemical Compositions

The bulk sensitivity of XPS is limited to concentrations of approximately 0.5% based on bulk percentage. If, however, the element under investigation is found primarily in the surface (< 20 Å) region, it may be detected by XPS in amounts as small as approximately 0.01 of an adsorbed monolayer (10^{-9} g/cm^2) (18).

Koppelman (21) attempted to correlate relative abundances of the elements Al, K, Mg and Fe in a Fithian illite sample with those obtained from bulk composition data. Elemental composition analyses are reported relative to the silicon abundance in the sample. Experimental values for the Si/Al, Si/Mg, Si/Fe and Si/K ratios, and bulk composition values for these ratios are listed in Table 4-3. Fig. 4-3 shows a 36 eV section of the XPS spectrum of illite. The peak at higher binding energy is Si 2p and that at lower binding energy is Al 2p. This spectrum was used to determine the Si/Al ratio in illite. The experimental values for these ratios were calculated using the following formula:

$$\text{Si/Element} = \frac{\text{relative intensity of the silicon photopeak}}{\text{relative intensity of the photopeak of the element in question}} \times \frac{\lambda\ \text{element}}{\lambda \text{Si}} \qquad [4\text{-}5]$$

$$\text{relative intensity} = \text{(net XPS counts for photopeak x height of photopeak x FWHM)/(time per channel x photoionization cross section for the particular element)} \qquad [4\text{-}6]$$

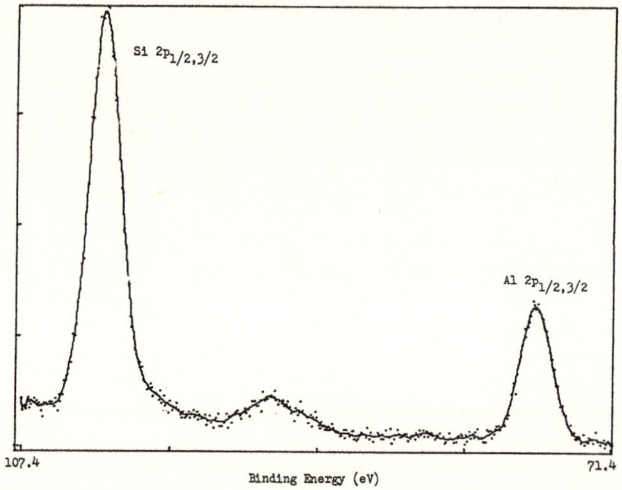

Figure 4.3. Silicon and aluminum photopeaks in illite (From Koppelman, 1976).

Table 4-3. Elemental Composition Analysis of Illite (From Koppelman, 1976).

Ratio	Data Source	
	Bulk Analysis	XPS Relative Intensities
Si/Al	1.9	1.8
Si/Fe	7.5	7.8
Si/Mg	20.1	18.8
Si/K	5.3	8.4

It was shown that for elements (Mg, Al, Si and Fe) which are constituents of the tetrahedral or octahedral layers, the Si/element ratios obtained by XPS were within 6% of the bulk chemical composition. The Si/K ratio obtained by XPS, however, differed greatly (approx. 50%) from that determined by bulk chemical analysis. It was suggested that this deviation arose because no correction for elemental depth in the sample was made. Since potassium is in the interlayer in illite, it is at a depth considerably deeper than any metal in an exposed octahedral or tetrahedral layer. Therefore, the relative intensity of the potassium photopeak would be reduced producing a high Si/K ratio as observed by XPS. Koppelman (21) also observed excellent (< 5% differential) agreement for XPS measurement versus bulk chemical composition for the Si/Al ratio in kaolinite and chlorite.

Adams *et al.* (2) evaluated XPS as a quantitative technique for surface analysis of aluminosilicate minerals. They determined experimentally the relative cross-sections for the 1s (Li-F), 2s (Na-K) and 2p (Na-K) subshells and used these measurements to evaluate Si/metal atom ratios in ground polycrystalline samples of kaolinite, montmorillonite and other minerals as well as freshly cleaved single crystals of lepidolite, muscovite and phlogopite. They concluded that XPS is capable of providing bulk quantitative analyses of air-stable homogeneous solids (specifically aluminosilicates), accurate to 5% on the average for main group elements (Table 4-4).

Table 4-4. Comparison of XPS and Wet-Chemical Analyses for Selected Polycrystalline Minerals (from Adams *et al.*, 1977).

(1) Mineral	(2) Level	(3) XPS Peak Area Ratio	(4) Derived Atom Ratio	(5) Mean XPS Atom Ratio	(6) Atom Ratio From Wet Analysis	(7) % Diff., Cols. 5 and 6
Kaolinite	Si 2p	1.00	1.00 }	1.00	1.00	
(St. Austell)	Si 2p	0.81	1.02			
	Al 2s	0.69	0.95 }	0.92	0.96	4
	Al 2p	0.63	0.90			
Kaolinite	Si 2p	1.00	1.00			
No. 7 (Bath, S. Car.)	Si 2s	0.84	1.06 }	1.00	(1.00)	
	Al 2s	0.68	0.94			
	Al 2p	0.65	0.93 }	0.91	(0.89)	2
Talc	Si 2p	1.00	1.00 }	1.00	1.00	
	Si 2s	0.81	1.03			
	Mg 2s	0.35	0.51 }			
	Mg 2p	0.25	0.54	0.52	0.52	0
Montmorillonite	Si 2p	1.00	1.00	1.00	1.00	
No. 23 (Chambers, Ariz.)	Si 2s	0.77	0.97 }			
	Al 2s	0.26	0.35 }			
	Al 2p	0.25	0.36	0.36	0.36	0
Montmorillonite	Si 2p	1.00	1.00 }	1.00	1.00	
(No. 22a (Amory, Miss.)	Si 2s	0.82	1.04			
	Al 2s	0.27	0.37 }			
	Al 2p	0.27	0.39	0.37	0.38	3
Montmorillonite	Si 2p	1.00	1.00 }	1.00	1.00	
(Belle Fourche, S. Dak.)	Si 2s	0.77	0.97			
	Al 2s	0.25	0.34 }			
	Al 2p	0.24	0.38	0.37	0.39	5
Lepidolite	Si 2p	1.00	1.00 }	1.00	1.00	
(Brazil)	Si 2s	0.82	1.04			
	Al 2s	0.35	0.48 }			
	Al 2p	0.33	0.49	0.48	0.46	4
	K 2s	0.22	0.21 }			
	K 2p	0.97	0.24	0.22		
	F 1s	0.64	0.42	0.42		

Defossé *et al.* (15) utilized XPS for surface characterization and analytical bulk comparison of silica-aluminas. It was concluded from the examination of Si 2p and Al 2p peaks that there is no relative enrichment of SiO_2 or Al_2O_3 near the surface of the grains of the powdered silica-alumina. No superficial segregation of either SiO_2 or Al_2O_3 was observed, and the existence of a distinct alumina phase below 30% SiO_2 was supported by the appearance of an energy loss peak.

4-3.3. Quantitative Measurements of Surface Adsorbed Species

The extreme high surface sensitivity of XPS has been demonstrated both for the adsorption studies of gases on metal surfaces and for the detection of ions in solution in the ppb range by adsorption onto a solid surface. In a study by Ban-

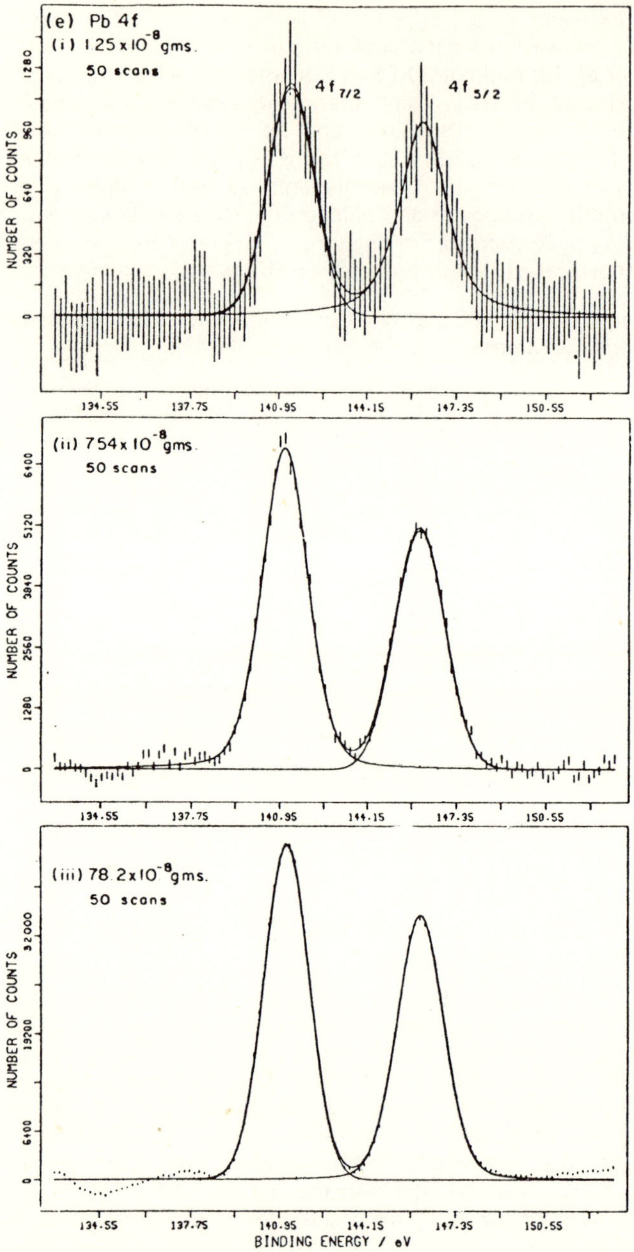

Figure 4-4. ESCA spectra for the Pb 4f level of lead adsorbed on calcite at different $[Pb^{2+}]_{initial}$ (From Bancroft *et al.*, 1979).

croft *et al.* (7), lead and barium nitrate solutions were microsyringed onto cleaved calcite surfaces (Fig. 4-4). Utilizing calibration plots to convert relative peak areas to weights such as in Fig. 4-5, the possibility for quantitative analyses of species adsorbed at mineral surfaces was demonstrated (8). Similarly, XPS has been utilized to quantify the kinetics of the adsorption of Ba^{2+} onto calcite (Fig. 4-6) (9).

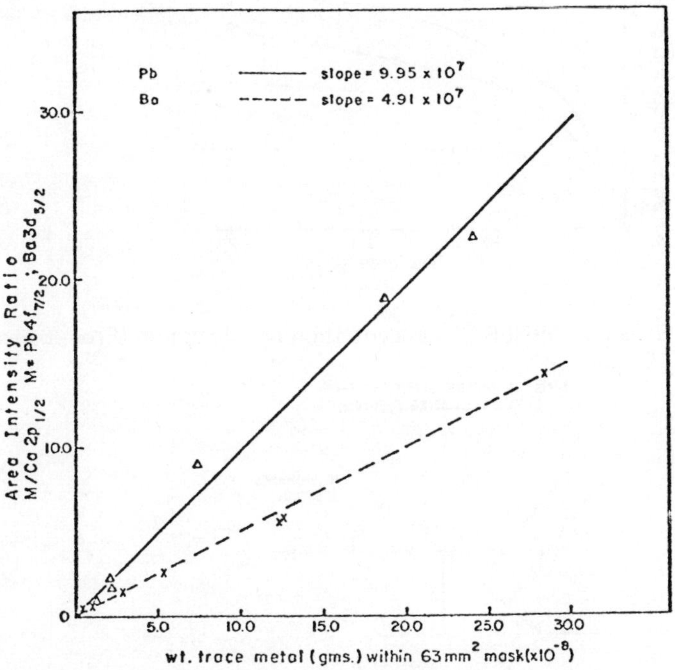

Figure 4-5. Calibration plots for Pb^{2+} and Ba^{2+} adsorbed on calcite (From Bancroft *et al.*, 1977b).

4-4. ELECTRON TAKE-OFF (GRAZING) ANGLE ANALYSIS APPLICATIONS

The enhancement of surface sensitivity in XPS may be achieved by utilizing variations in grazing angle of electron escape from the surface of solid samples. The presence of a significant amount of surface contour irregularity or roughness renders the interpretation of XPS data more difficult, because the true photoelectron escape angles are not directly measurable and shading of certain surface regions may occur for both incident photon (x-ray) and electron exit. Characteristic sample surface roughness dimensions need only be somewhat greater than electron attenuation lengths (10-50 Å) in order to influence angular distributions. It is clear, however, that surface profile variations can dramatically alter the form of surface sensitivity variations with electron emersion angle (Fig. 4-7).

By substituting the values for θ listed in Fig. 4-7 into equation [4-7], the variation in effective sample depths due to take-off angle can be ascertained. As this angle is reduced, significant increases in surface sensitivity are observed. By comparing the relative photopeak intensity of a bulk element with that of a

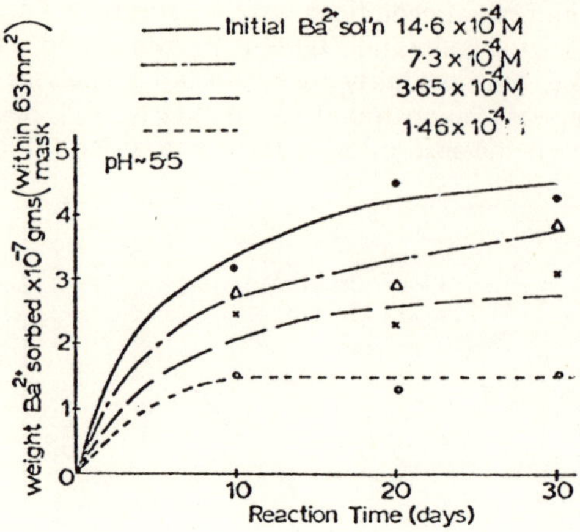

Figure 4-6. Effect of initial Ba^{2+} concentration on adsorption (From Bancroft *et al.*, 1977a).

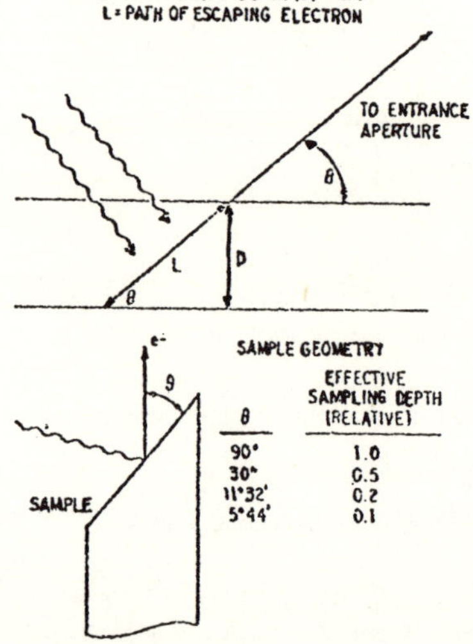

θ	EFFECTIVE SAMPLING DEPTH (RELATIVE)
90°	1.0
30°	0.5
11°32'	0.2
5°44'	0.1

Figure 4-7. Definition of electron take-off angle θ (From Koppelman and Dillard, 1977a).

suspected surface element at two different values of θ, a relative peak enhancement ratio is established. Substitution of the photopeak intensities for elements N and M at two different values for the electron take-off angle (θ_1, θ_2) into equation [4-4] results in the following relative peak enhancement (RPE) ratio:

$$\mathrm{RPE} = \frac{(I_N/I_M)_{\theta_1}}{(I_N/I_M)_{\theta_2}} \qquad [4\text{-}7]$$

A value of 1.0 for this ratio would indicate that elements N and M are uniformly distributed throughout the sampling region; a value <1.0 would indicate that element N is concentrated deeper in the sampling region relative to element M, while a value >1.0 would indicate outermost surface concentrating of element N relative to element M.

If more than one element is adsorbed onto the substrate, theory predicts that, when bombarded by x-rays, the photoelectrons ejected from the elements at the surface will escape with less probability of inelastic loss than those farther from the surface. It has been demonstrated that, by utilizing low angles of electron escape from a solid surface (that is, electron velocity vector nearly parallel with the surface), the relative XPS intensities from surface-layer atoms could be augmented by roughly an order of magnitude.

Baird *et al.* (6) utilized RPE ratios to study the reactions between a crystalline Al_2O_3 powder, and solutions of $Si(OH)_4$ and $CaCl_2$. Using take-off angles (θ) of 5° and 38.5°, they observed spectra as in Fig. 4-8 and were able to compile the data listed in Table 4-5.

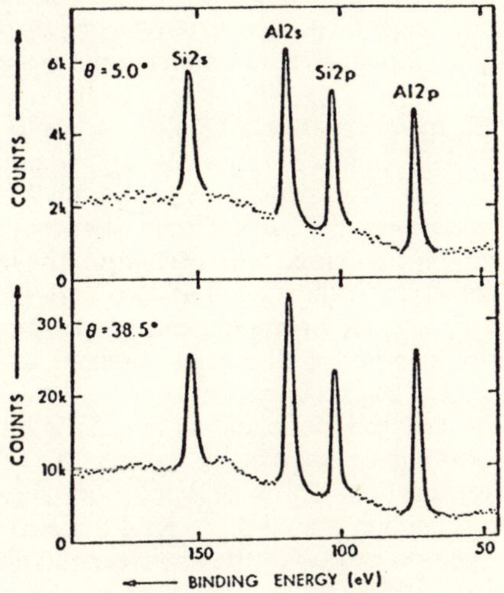

Figure 4-8. XPS spectra for the angular dependence on an alumina specimen with adsorbed silicon and calcium (From Baird *et al.*, 1976).

Table 4-5. Core-Level Intensities Relative to Al 2s at θ = 38.5°C and Relative Intensity Enhancement Ratios Between θ = 5° and 38.5° (From Adams and Evans, 1979).

	Untreated Al_2O_3		Al_2O_3 + Si + Ca	
k	$(k/Al\ 2s)_{38.5°}$	$(k/Al\ 2s)_{5°}$ / $(k/Al\ 2s)_{38.5°}$	$(k/Al\ 2s)_{38.5°}$	$(k/Al\ 2s)_{5°}$ / $(k/Al\ 2s)_{38.5°}$
C 1s	0.13	3.0 ± 0.4	0.29	2.6 ± 0.2
Ca 2p	—	—	0.03	1.8 ± 0.5
Si 2s	—	—	0.56	1.2 ± 0.1
Si 2p	—	—	0.62	1.2 ± 0.1
Al 2s	1	1	1	1
Al 2p	0.74	1.1 ± 0.1	0.82	0.9 ± 0.1
O 1s	12.16	0.9 ± 0.1	15.32	0.9 ± 0.1

For chemically treated Al_2O_3 samples, the Si peaks are quite intense, indicating that a significant amount has remained on the powder even after washing. Also, there is a small, but unambiguous, enhancement of the relative intensities of the Si peaks when the emission angle is lowered from 38.5° to 5°. This enhancement indicates that the Si is on the average nearer the surface than the Al derived from the substrate, as would be expected for a surface-adsorbed species.

The Al 2p enhancement ratios are within experimental error of unity, since Al 2p originates in the same atom as the reference Al 2s and also has nearly the same kinetic energy. The O 1s enhancement ratios for all specimens are also very near unity, indicating that essentially all of the oxygen is associated with the substrate Al_2O_3. For C 1s, on the other hand, the enhancement ratios are much larger with values of 3.0 and 2.6. This finding suggests that C is present in an outermost contaminant layer, as expected from the mode of specimen preparation and analysis. Furthermore, no fine structure due to chemical shifts was distinguishable in the O 1s peak. This result is consistent with a relatively well-defined chemical state. The O 1s/Al 2p ratio was also essentially identical for both treated and untreated specimens, indicating relatively little modification of the near-surface stoichiometry by the adsorption. Thus, distinctly different surface species containing oxygen do not appear to play a major role in altering the O 1s enhancement ratio.

The Ca 2p, Si 2s, and Si 2p ratios associated with atoms in the treatment solution are above unity (1.8, 1.2, and 1.2, respectively), indicating that these species are primarily surface-adsorbed. The larger value for the Ca 2p ratio may also suggest that Ca is on the average nearer the surface than Si, although the quoted error limits do not permit this to be concluded with certainty. The largest ratio for the treated specimens is 2.6 for C 1s, indicating that C occurs in an outermost contaminant layer. Thus, these results for powdered specimens at two angles also permit a qualitative concentration profile for all species observed, and indicate that Ca and Si are tenaciously adsorbed at the surface.

For a group of peaks with a large kinetic energy range, the interpretation of subtle changes in enhancement ratio may be difficult, however. For example, the 0 1s ratio is consistently slightly less than unity, and the normalized 0 1s/Al 2p (oxide) ratio for the aluminum specimen scanned in Fig. 4-8 is also somewhat less than unity at 10°. These effects could be due to the lower attenuation length for 0 1s photoelectrons (kinetic energy ≃ 955 eV) as compared to the reference Al 2s photoelectrons (kinetic energy ≃ 1365 eV), which would cause more attenuation of 0 1s in any overlayers present on the oxide. Such differences in attenuation would be amplified at lower angles. Alternatively, it was suggested that these enhancement ratios of slightly less than unity could be due to a slight reduction in the 0/Al concentration ratios near the surface. Validity of this effect may be achieved by measuring the low-energy band of 0 2s states, observed in the valence region of Al_2O_3, which possess kinetic energies greater than that of Al 2s. This band, however, is of much lower intensity than 0 1s and thus is difficult to measure with sufficient precision.

Take-off angle variation has also been used to study the reactions between dissolved Co(II) species and chlorite and illite (26). The enhancement of the Co 2p signal, relative to substrate Al 2s indicates that adsorbed cobalt is predominantly a surface species (Table 4-6). The measured binding energies for cobalt determined at 11° are equivalent to those measured at 90°. Additionally, the Co $2p_{1/2}$, Co $2p_{3/2}$ energy separation (16 eV) is unchanged at 11° compared to 90°. These results are consistent with the notion that cobalt is adsorbed as a $Co(H_2O)_6{}^{2+}$ ion and not as $Co(OH)_2$ as Tewari and co-workers (42, 43) observed for the Co(II)/alumina systems. Furthermore, there is no evidence that there has been surface initiated cobalt oxidation at adsorption sites near the clay surfaces as observed with the Co(II)/ MnO_2 system (33). The binding energy results for aluminum indicate that the chemical nature of surface aluminum is similar to that deeper in the sample. It is noted that significant surface enhancement for aluminum is not realized. This fact indicates that aluminum is homogeneously distributed in the surface region and may resemble bulk aluminum.

Table 4-6. Grazing Angle Measurements for Chlorite and Illite Clays: Relative Peak Enhancement (RPE) and Binding Energy (B.E.) Results (From Koppelman and Dillard, 1979).

	Cobalt		Aluminum	
	RPE (11°)	B.E.	RPE (11°)	B.E.
Chlorite	1.22	782.1	0.98	74.1
Illite	1.31	782.2	1.01	74.2

4-5. QUALITATIVE BONDING INVESTIGATIONS

The information XPS provides is not only analytical, but also can give insight into the bonding nature of the element in question. The binding energy of a photoejected electron is dependent upon the chemical environment of the orbital from which the electron was removed. Oxidation state, type of bonding, (i.e. ionic versus covalent) spin state, and nearest neighbor atoms are some chemical factors which can influence the binding energy of an electron.

4-5.1. Chemical Nature of Silicate Lattice Elements

XPS has been used to study the nature of oxygen atoms in olivines and pyroxenes (48). The oxygen 1s spectrum for olivines exhibited only one narrow oxygen 1s photopeak whereas pyroxenes contained two distinguishably different oxygen photopeaks (Fig. 4-9). The intensity ratio of the two components in pyroxene was 2:1 with an energy separation of about 1 eV. It was suggested that the two oxygen components were the result of a difference in binding energy between bridging and non-bridging oxygen atoms within a silicate chain in the pyroxene structure.

Adams *et al.* (1) measured the core electron binding energies for Fe, Mg, Al, Si, and O in a number of well-characterized silicate minerals. Adams was unable to correlate Fe 2p binding energies with iron oxidation state in the minerals examined (Table 4-7, Fig. 4-10). It was also observed that O 1s peak widths for minerals with only one type of oxygen were generally narrower than those containing oxygen in more than one type of chemical environment (Tables 4-8, 4-9). Small differences in Al 2p binding energy for aluminum in four coordination and aluminum in six coordination were reported.

Nicholls *et al.* (34) studied a series of magnesium and aluminum compounds with XPS and x-ray emission spectroscopy. They concluded that increasing the coordination number from four to six increased the binding energy of both magnesium and aluminum electrons. It was also noted that increasing the electronegativity of the ligand from oxygen to fluorine further increased the magnesium and aluminum binding energy.

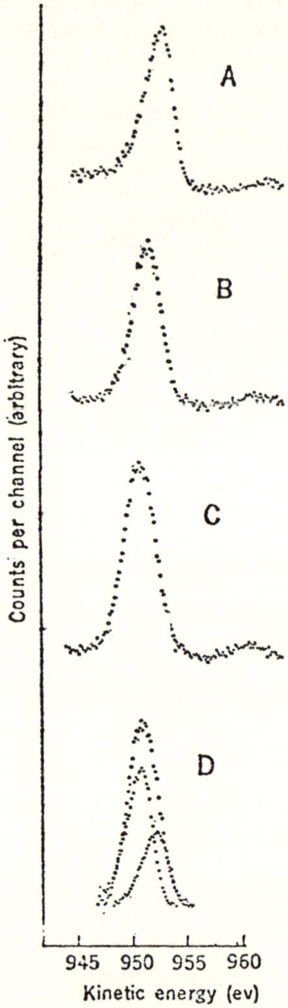

Figure 4-9. O 1s spectra in olivine and pyroxene (From Yin *et al.,* Copyright 1971 by the American Association for the Advancement of Science).

Anderson and Swartz (5), upon examining the minerals kyanite, sillimanite, and mullite with XPS, found that the Al 2p binding energy for sillimanite, with aluminum in both fourfold and sixfold coordinations, was experimentally identical with that in kyanite, where aluminum is in only sixfold coordination (Table 4-10). It was concluded that XPS could not be used to differentiate between aluminum atoms in different coordinations. This conclusion was strengthened by the XPS data for mullite where the Al 2p binding energy and peak shape were identical to those of kyanite and sillimanite.

Table 4-7. Fe and Si Binding Energies (B.E.) and Full Widths at Half Height (FWHM) in eV (from Adams *et al.*, 1972).

Mineral	Fe* Composition	Fe $2p_{1/2}$ B.E.	Fe $2p_{1/2}$ FWHM	Si 2p B.E.	Si 2p FWHM	C 1s B.E.	C 1s FWHM	O 1s FWHM
Hedenbergite	54% Fe^{2+}	723.1 ±0.6	8.0	102.1 ±0.6	2.6	284.4 ±0.4	2.2	3.2
Hedenbergite	77% Fe^{2+}	723.3 ±1.5	6.7	102.1 ±0.3	2.6	Not Recorded		
Hedenbergite	85% Fe^{2+}	722.7 ±0.6	6.8	102.0 ±0.5	2.5	284.6 ±0.4	2.4	3.2
Crocidolite	55% Fe^{2+} 40% Fe^{3+}	723.4 ±0.8	7.0	102.4 ±0.5	2.4	284.8 ±0.4	2.3	2.9
Epidote	86% Fe^{3+}	723.1 ±0.7	5.7	101.8 ±0.5	3.1	284.5 ±0.4	2.2	3.0

*Percentage of available cation sites filled by iron.

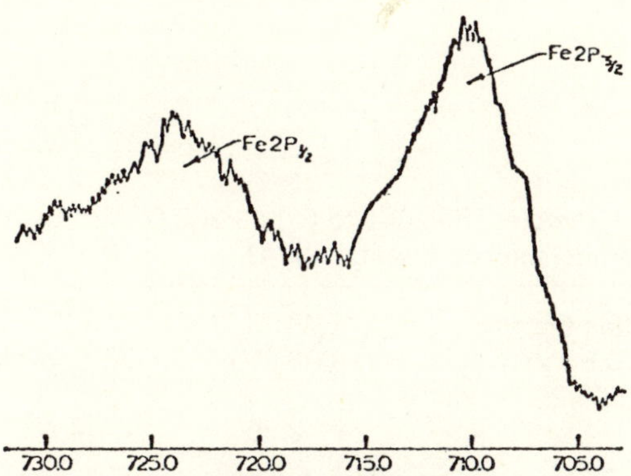

Figure 4-10. XPS spectrum of Fe 2p levels in hedenbergite (From Adams *et al.*, 1972).

Table 4-8. Si 2p, Mg 2s and O 1s Binding Energies (B.E.) and Widths (FWHM) (from Adams *et al.*, 1972).

Mineral	Mg Composition	Si 2p B.E.	Si 2p FWHM	Mg 2s B.E.	Mg 2s FWHM	O 1s B.E.	O 1s FWHM
Enstatite	86%	102.0	3.3	88.3	3.2	531.4	3.0
Hedenbergite	46%	102.3	2.6	88.5	3.0	531.0	3.2
Anthophyllite	77%	101.9	2.6	88.4	3.1	531.1	2.8

All binding energies are referenced to the C 1s contaminant line as 284.6 ± 0.5 eV.

Lindsay (31) reasoned that the significant difference in Al 2p binding energy between microcline (aluminum in fourfold coordination) and Al_2O_3 (aluminum in sixfold coordination) observed by Nicholls *et al.* (34) could be explained by using ionic model concepts. He indicated that the presence of additional potassium cations in the crystal lattice of microcline had the effect of reducing the electron-

Table 4-9. Al 2p, and O 1s Binding Energies (B.E.) and Widths (FWHM) Referenced to Si 2p as 102.0 eV (From Adams *et al.*, 1972).

	Al 2p		O 1s		C 1s	
Mineral	B.E.	FWHM	B.E.	FWHM	B.E.	FWHM
Albite $NaAlSi_3O_8$	73.8	2.4	531.0	2.4	284.3	2.2
Garnet $(MgFe)_3Al_2(SiO_4)_3$	74.5	**	531.1	2.7	284.7	2.1

Table 4-10. Al 2p Binding Energies (Relative to C 1s = 285.0 eV) For Aluminosilicates (From Anderson and Swartz, 1974)

Mineral	Binding Energy (±0.5 eV)	FWHM (±0.05 eV)	Al Coordination
Kyanite	74.9	2.25	6-fold
Sillimanite	74.9	2.16	50% 6-fold 50% 4-fold
Mullite	75.0	2.28	41-56% 6-fold 59-44% 4-fold

attracting ability of the oxygen atoms. This would result in a decrease in Al 2p binding energy as the number of positively charged ions increased, and therefore, could account for the Al 2p binding energy in microcline being 1.4 eV lower than in alumina.

Urch and Murphy (47) determined the Al 2p and 2s binding energies for a series of aluminosilicate minerals which included microcline and alumina. He observed a 0.5 eV increase in Al 2p and 2s binding energy in going from microcline (Al-O bond length of 1.75 Å) to α-Al_2O_3 (Al-O bond length of 1.92 Å). It was concluded that there was a correlation between bond length and orbital ionization (binding) energies.

Schultz *et al.* (36), in an effort to identify silicate minerals in respirable coal dust, used XPS to measure the Si 2p binding energy in a series of aluminosilicate minerals. He observed five different silicon chemical environments in coal, and three different silicon environments in respirable coal dust. It should be noted that Schultz *et al.* (36) observed a 6.0 eV range in Si 2p binding energies for the various minerals he examined (Table 4-11). This large, unanticipated range in Si 2p binding energy may have been a result of inadequate compensation (calibration) for sample charging as previously discussed.

Table 4-11. Electron-Binding Energies for Silicon Minerals (From Schultz *et al.*, 1974).

Sample	Electron-Binding Energies (eV) Si 2s	Si 2p	O 1s	Formula	Crystalline System
Silicon Metal (Standard)		99.0			
Coesite	155.1	104.3		SiO_2 (High Pressure Phase)	Monoclinic
Cristobalite		104.4	535.6	SiO_2	Cubic
Keatite	155.3	104.5		SiO_2 (High Pressure Synthetic Phase)	Tetragonal
α-Quartz	160.2	105.2	534.5	SiO_2	Trigonal
Kaolinite	155.6	107.4		$Al_4(Si_4O_{10})(OH_8)$	Triclinic
Muscovite	159.5	108.3		$KAl_4(Si_6Al_2O_{20})(OH, F)_4$	Monoclinic
Illite		109.3		$KAl_4(Si_7AlO_{20})(OH)_4$	Monoclinic
Montmorillonite		110.0		$NaAl_4Si_8O_{20}(OH)_4 \cdot nH_2O$	Monoclinic
Coal		107.4		–	–
		105.5		–	–
		104.1		–	–
		102.9		–	–
		99.5		–	–
Respirable Coal Dust		112.2		–	–
		107.5		–	–
		105.3		–	–

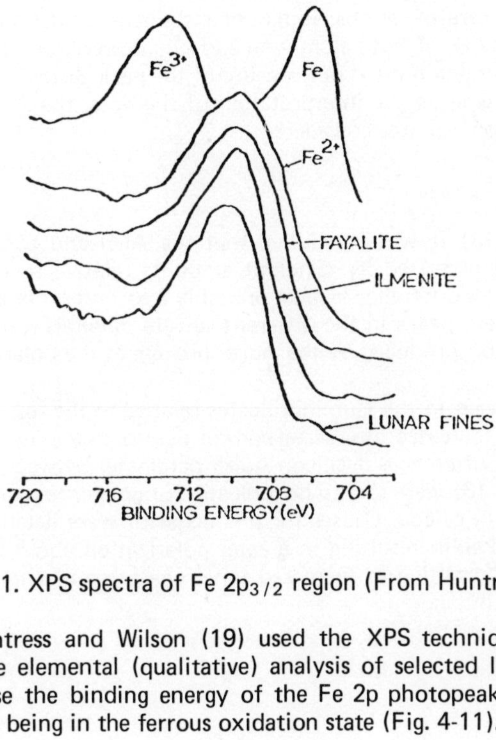

Figure 4-11. XPS spectra of Fe $2p_{3/2}$ region (From Huntress and Wilson, 1972).

Huntress and Wilson (19) used the XPS technique to obtain rapid, non-destructive elemental (qualitative) analysis of selected lunar samples. They were able to use the binding energy of the Fe 2p photopeak to identify iron in lunar samples as being in the ferrous oxidation state (Fig. 4-11).

Koppelman and Dillard (22) observed that the binding energies for Si, Al and O, three major lattice constituents of kaolinite, chlorite, and illite, varied little from mineral to mineral. The binding energy for the Si 2p electrons (average of 102.5 eV) was in good agreement with values published previously by Huntress and Wilson (19) and Adams *et al.* (1) (Table 4-12). This value has been confirmed by Carriere and Deville (12).

Table 4-12. Core Electron Binding Energies (B.E.) and FWHM for Mineral Lattice Elements (in eV ± 0.1) (From Koppelman and Dillard, 1975).

	Chlorite		Illite		Kaolinite	
	B.E.	FWHM	B.E.	FWHM	B.E.	FWHM
Si $2p_{1/2, 3/2}$	102.1	2.4	102.5	2.5	102.7	2.2
Al $2p_{1/2, 3/2}$	74.2	2.5	74.3	2.5	74.4	2.1
O $1s_{1/2}$	531.4	2.8	531.7	2.9	531.9	2.4
K $2p_{3/2}$	292.8	2.2	293.2	2.2	–	–
Ca $2p_{1/2, 3/2}$	350.9	3.5	350.9	3.6	–	–
Mg $2p_{1/2, 3/2}$	50.0	2.1	49.7	3.4	–	–

Recently, Zr Lα ($E_{h\upsilon}$ = 2042.4 eV) radiation was used to excite the 1s peaks and KLL Auger spectra for aluminosilicates (13). The Auger parameters were found to be of value because they are independent of sample electrostatic charging. A comparison of the KLL Auger spectrum and the Si 1s peaks for silicon in silicon metal and silicon dioxide appears in Fig. 4-12. Table 4-13 lists the positions of the principal peaks in the spectra of various aluminum and silicon oxides and aluminosilicate minerals. The 1s peak of both aluminum and silicon were found to vary up to 4 eV. Electrostatic charging is cited as the cause of the peak shifts. To eliminate the effect of sample charging on the interpretation of the data, the Auger parameter, α, defined by equation 4-8, was considered.

$$\alpha = E_{1s} + E_{KLL} - E_{photon} \qquad [4\text{-}8]$$

From this work (13) it was concluded that the Al-O and Si-O bonds in different compounds are polarized by different amounts relative to each other. Furthermore, the Al-Si peak difference is most probably a second order effect. This ability to differentiate these peaks in the different silicate minerals was attributed to the sensitivity of the ion produced by the Auger process to the polarizability of the surrounding bonds.

The decrease in α seen in the aluminosilicates relative to the respective pure metals was attributed to increased atomic relaxation due to charge neutralization around the cations (13). Differences in silicon Auger parameter between muscovite mica and kaolin (Table 4-13) were said to be indicative of greater relaxation of the atomic levels in the kaolin silicon. Causes for this behavior were listed as shorter interatomic distances in kaolin resulting in greater polarization about Si^{4+} or increased lattice hydration resulting in increased ionic character of the bonds surrounding the silicon ion.

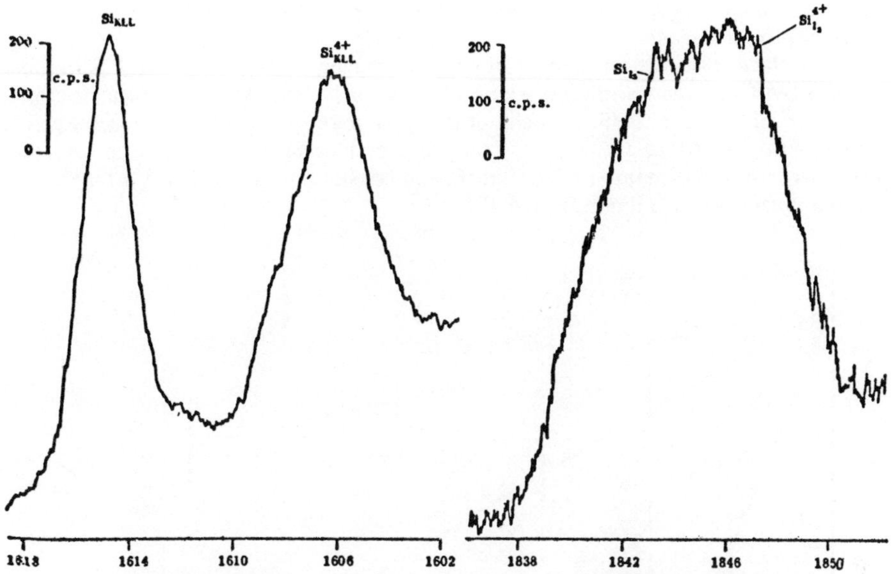

Figure 4-12. Si 1s and KLL Auger peaks in silicon metal and silicon dioxide (From Castle *et al.,* 1979).

Table 4-13. Positions of Principal Peaks in the Photoelectron Spectra With Values of the Auger parameter, α, for Some Al and Si Compounds (From Castle *et al.,* 1979).

Sample	Si (1s) BE (eV)	Si (KLL) KE (eV)	α (Si) (eV)	Al (1s) BE (eV)	Al (KLL) KE (eV)	α (Al) (eV)	Al (2s) BE (eV)	Al (2p) BE (eV)	O (1s) BE (eV)	C (1s) BE (eV)	Si (KLL)-Al(KLL) (eV)
Al				1558.2	1395.0	910.8					
Al_2O_3				1561.2	1388.5	907.3					
Si	1839.8	1616.5	1413.9								
SiO_2	1845.2	1607.2	1410.0						531.2		
Muscovite Mica (vacuum cleaved)	1845.4	1607.1	1410.1	1564.3	1384.5	906.4	103.7	74.5	533.2	284.8	222.6
Mica (hydrated)	1844.2 (±0.3)	1607.5	1409.3 (±0.3)	1563.8	1384.6	906.0					222.9
Kaolin 1	1843.3	1608.2	1409.1	1563.4	1385.5	906.5	103.1	73.4	533.3	285.2	222.5
Kaolin 2	1843.0	1608.8	1409.4	1562.0	1387.0	906.6			532.4		221.8
Kaolin 3	1842.6	1608.6	1408.8	1562.2	1386.0	905.8			532.7		222.6
Molecular Sieve	1843.5	1608.7	1409.8	1561.6	1385.8	905.0			532.3	285.3	222.9

4-5.2. Application of XPS to Study Mineral Reactivity

XPS has also been employed to study the dissolution mechanism of feldspars (35). Examination of the K, Al and Si content of the surface of feldspar grains with XPS both prior to and after dissolution revealed no evidence for silica or potassium depletion relative to aluminum within the outermost 10-20 Å. It was shown that the surface of the reacted feldspar had the same composition, within experimental error, as unreacted feldspar. This evidence led to the conclusion that the kinetics of feldspar dissolution (on a laboratory time scale) are not controlled by diffusion through a tightly adhering protective layer of hydrous aluminum oxide, kaolinite, or decationated feldspar, but rather through processes occurring at the fresh feldspar/solution interface.

Surface dissolution and diffusion in Mg and Al-silicates have also been studied by Thomassin and co-workers (44, 45, 46). To monitor surface reactions, Al 2p/Si 2p and Mg 2p/Si 2p peak area ratios were measured after leaching. The results gave insight as to whether congruent or incongruent dissolution processes were occurring. Estimates of Mg^{2+} diffusion coefficients and rates as a function of temperature were also made (Fig. 4-13).

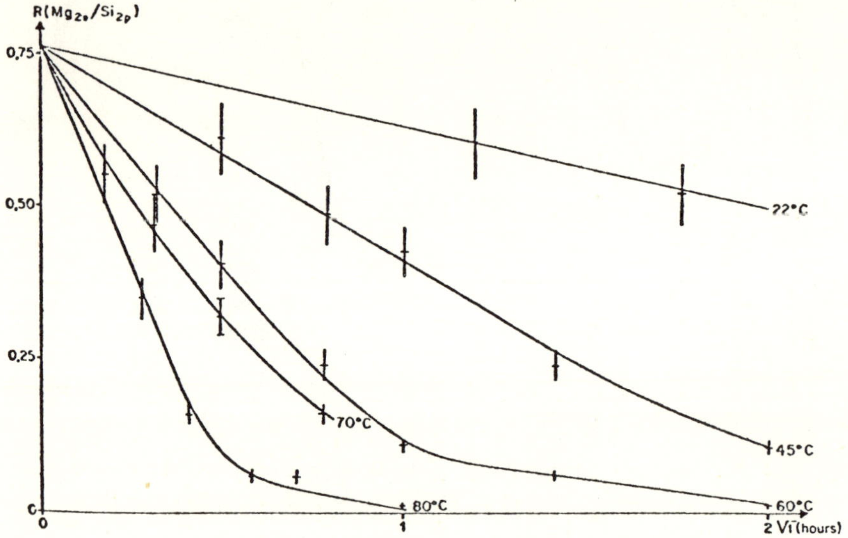

Figure 4-13. Kinetic curves of leaching chrysotile with oxalic acid (0.1 N) (From Thomassin *et al.*, 1977).

XPS has also been employed to estimate the cation exchange capacities of cation-exchanged beidellites (3). In addition, this study also examined the extent of surface (as opposed to interlamellar) uptake of Na^+, K^+, Ca^{2+}, Pb^{2+} and Ba^{2+} by the $< 2.0\mu$ fraction of a well-characterized beidellite (0.43 charges per $(Si, Al)_4$ unit). Utilizing equation [4-4], atom ratios of both lattice and adsorbed species were compiled and compared to those calculated from analytical data (Table 4-14, 4-15). Consideration of the XPS data in Tables 4-14 and 4-15 suggests that Na^+ and Ca^{2+} exchange in beidellite is consistent with the independently determined layer charge for the mineral (30). Exchange or uptake with K^+, Pb^{2+} and Ba^{2+}, however, occurs to a substantially greater extent. Residual salts cannot explain this anomalous uptake since respective salt anions were not detected.

Preferential external surface adsorption (as opposed to interlamellar) of these ions was suggested to explain the above data (3). The uptake of Ca^{2+} equals that of a purely Ca^{2+}-exchanged sample, while the Ba^{2+} uptake, assumed to be exchanged on external surfaces, is equal to the excess uptake found when Ba^{2+} salts alone are used in the exchange process.

Table 4-14. Chemical Analyses of Beidellite (From Adams and Evans, 1979).

	Natural Beidellite				Ca-Clay	Ba-Clay	Mixed Ca/Ba Clay	
	Si	Al	Mg	Fe	Ca	Ba	Ca	Ba
% Oxide*	53.3	27.3	1.4	0.03**	3.02	8.44	†	1.49
Atom Ratio (A.R.)	1.0	0.604	0.039	0.0004	0.060(.004)	0.060(.004)	†	0.011(.002)
Mean A.R. by XPS (estimated standard deviations in parentheses)	1.00 (2p)	0.59(.03) (2s,2p)	–	–	0.050(.004) (2s,2p)	0.079(.008) (3d,4d)	0.053(.004) (2p)	0.030(.003) (3d,4d)

*Air-dried clay containing 8% loosely bound water removable at 70-100° C (estimated by thermogravimetric analysis).
**Fe_2O_3.
†There was insufficient material for the adequate determination of both Ca and Ba in this sample.

Table 4-15. Atom Ratios for Cation-Exchanged Beidellite by X-ray Photoelectron Spectroscopy (From Adams and Evans, 1979).

	Relative Atomic Abundance				
	Layer				Apparent
Exchangeable Cation	Si 2p	Al 2s	Al 2p	Mean Cation	Layer Charge per Si
Na	1.00	0.64	0.55	0.10(.02)	0.10(.02)
Ca	1.00	0.60	0.58	0.050(.004)	0.10(.01)
K	1.00	0.61	0.54	0.15(.02)	0.15(.02)
Pb	1.00	0.62	0.57	0.073(.006)	0.15(.01)
Ba	1.00	0.58	0.57	0.079(6)(.006)	0.16(.01)
Ca, Ba }	1.00	0.60	0.56	0.05 (Ca), 0.03 (Ba) }	0.16(.02
Mean	1.00	0.59			

Estimated standard deviations in parentheses.

It was suggested that since the extent of excess uptake of K^+, Pb^{2+} and Ba^{2+} was the same, perhaps the same external surface sites may be active in all three cases. The authors (3) also noted that although the adsorption of hydrolyzed species such as $BaOH^+$ and $PbOH^+$ could explain excess uptake data, Pb or Ba values obtained by bulk chemistry should be of the same magnitude as XPS values, which was not the case.

4-5.3. Redox Chemistry of Iron Bearing Minerals

The desire to qualify the oxidation state of iron and to quantify it in both the bulk and surface regions of a mineral has long been a goal of geochemists. In this regard, the XPS examination of the binding energy of the Fe 2p level has proven to be of considerable interest.

Although attempts by Adams *et al.* (1) to use Fe 2p binding energies to differentiate iron oxidation states were, to some degree, futile, research efforts in this direction did not terminate. In their study only composite Fe $2p_{3/2}$ photopeak positions (binding energies), having extremely high ($\simeq$ 7.0 eV) peak widths, were tabulated. No attempts at photopeak deconvolution were made. Furthermore, tabulated Fe(II) and Fe(III) compositions were based on bulk analysis, which may not be representative of surface region oxidation state compositions.

Koppelman and Dillard (21, 22) compared the Fe 2p binding energy for nontronite (determined by Mössbauer spectroscopy to contain only Fe^{3+}) to that of chlorite (Fe^{2+} only), and observed a difference of 1.9 eV (Figs. 4-14, 4-15, Table 4-16). Using the binding energies for the Fe 2p photopeaks of chlorite and nontronite, the rather broad (6.4 eV) Fe 2p photopeak of illite was deconvoluted into its ferric and ferrous components, including satellite structure (Figs. 4-16, 4-17). Comparison of the Fe^{2+}/Fe^{3+} ratios obtained by wet chemical analysis and Mössbauer spectroscopy (0.20) with that obtained by XPS measurements (0.35) indicated only fair agreement (21). It was suggested that this may be an indication of a difference in iron oxidation in the surface region as compared to the bulk mineral phase. Koppelman (22) was unable to detect an Fe 2p photopeak for kaolinite, although bulk chemical analysis revealed approximately 0.5% Fe_2O_3. This lack of sensitivity was attributed to iron in kaolinite being located well within the bulk of the mineral.

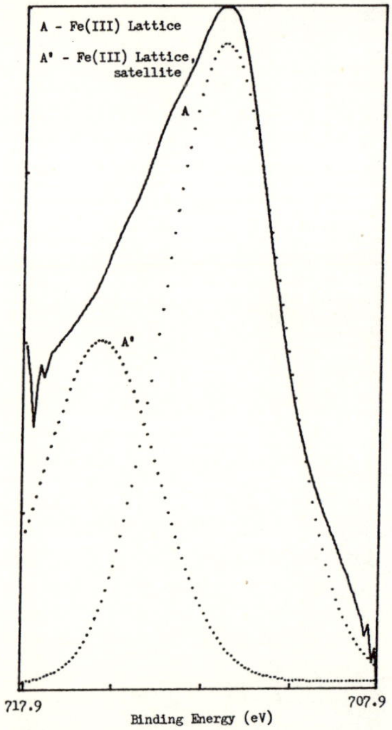

Figure 4-14. Fe $2p_{3/2}$ photopeak (deconvoluted) for nontronite (From Koppelman, 1976).

Stucki *et al.* (40) used XPS to examine the redox reactions of nontronite and biotite. For the unaltered minerals, a difference of 1.8 eV between the Fe 2p photopeak of nontronite (Fe^{3+}) and that of biotite (Fe^{2+}) was noted (Table 4-17). Upon reduction of the nontronite sample with either hydrazine or dithionite, peak broadening of the Fe 2p photopeak was observed with a shoulder at lower binding energy. Ferric-ferrous iron ratios were determined colorimitrically and agreed favorably with photopeak intensities (Table 4-18).

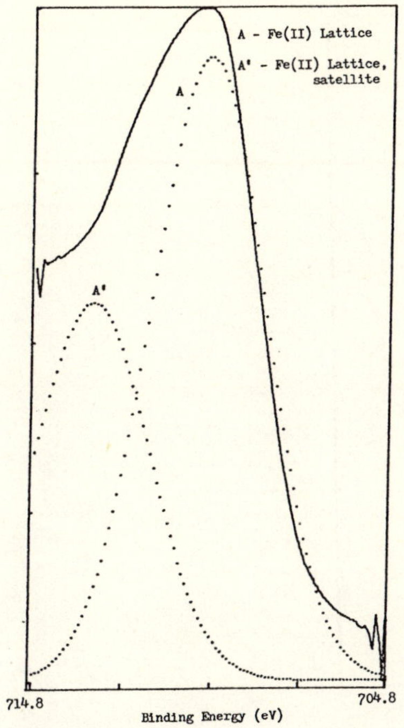

Figure 4-15. Fe $2p_{3/2}$ photopeak (deconvoluted) for chlorite (From Koppelman, 1976).

Table 4-16. Binding Energies for Fe $2p_{3/2}$ Electrons (From Koppelman and Dillard, 1975).

	Binding Energy (±0.1 eV)	FWHM (eV)
Illite (Fe^{3+}, Fe^{2+})	712.6	6.4
Chlorite (Fe^{2+})	710.6	5.2
Amorphous Ferric Hydroxide	711.9	4.3
Fe_2O_3	711.1	4.9
Nontronite (Fe^{3+})	712.5	4.9

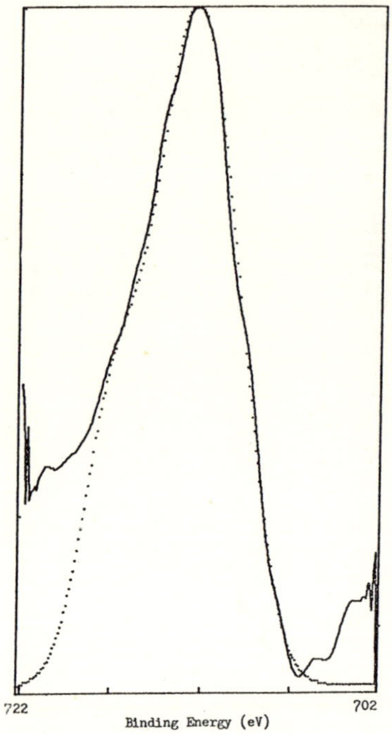

Figure 4-16. Deconvolution of Fe $2p_{3/2}$ peak for illite. Comparison between experimental (—) and calculated (. . .) peaks (From Koppelman, 1976).

Oxidation of biotite in heated bromine water caused the Fe 2p photopeak to broaden and shift to higher binding energy. In a later study, Stucki and Roth (39) were able to use XPS in conjunction with infrared and Mössbauer spectroscopy to postulate a mechanism of iron reduction in nontronite. The spectroscopic results were supportive of a two-step mechanism that involves an initial reduction of Fe^{3+} to Fe^{2+} with an accompanying increase in layer charge and no structural changes. In a second step, a further reduction of Fe^{3+} was postulated with layer charge remaining constant through elimination of structural OH and alteration of iron coordination number.

4-4.4. Bonding Nature of Adsorbed Species

Due to the relatively small percentage of the total mineral composition that a species adsorbed on a mineral surface represents, it is difficult to "directly examine them" by conventional (bulk) spectroscopic techniques. The ability of XPS to examine only the surface region of a mineral sample enables the investigator to "look at" adsorbed layers and compare their chemical nature with that in the first few crystal or atomic layers of the mineral substrate.

XPS has been used to examine the chemical nature of lead adsorbed on montmorillonite (14). Comparison of the Pb 4f photopeaks for the Pb-montmorillonite sample with those of elemental lead, lead oxide (PbO) and lead dioxide (PbO_2) indicated that lead adsorbed on montmorillonite was in a similar bonding state as lead in lead oxide.

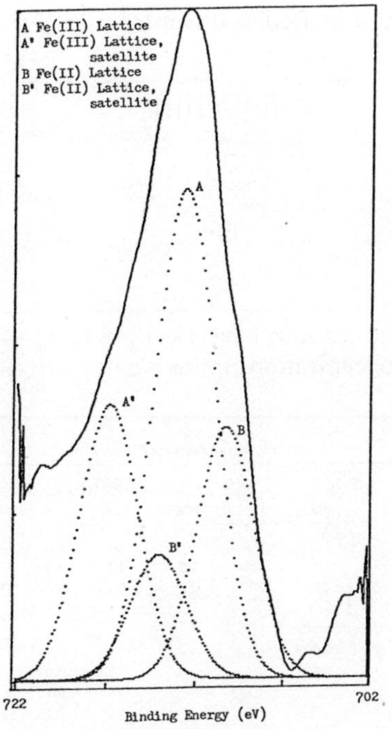

Figure 4-17. Deconvolution of Fe $2p_{3/2}$ photopeak for illite (From Koppelman, 1976).

Table 4-17. Iron Redox Chemistry (From Stucki *et al.*, 1976).

Sample	Fe $2p_{3/2}$ Binding Energies (eV)	
	Fe(III)	Fe(II)
Nontronite – unaltered	711.8	–
– hydrazine-reduced	711.8	708.6
– dithionite-reduced	711.0	708.6
Biotite – unaltered	–	710.0
– oxidized	711.0	709.0

The interaction of gibbsite with $Ca(H_2PO_4)_2$, $Si(OH)_4$, $CaSiO_3$ and $Ca(NO_3)_2$ has been examined using the XPS technique (4). Gibbsite samples treated with $Si(OH)_4$ and $Ca(NO_3)_2$ revealed little or no detectable Si, Ca, or N photopeaks. However, upon treatment with $CaSiO_3$, significant calcium and silicon photopeaks were noted. Similarly, after treatment with $Ca(H_2PO_4)_2$, Ca and P photopeaks were also observed. From the absence of detectable calcium or silicon signals after respective $Ca(NO_3)_2$ and $Si(OH)_4$ treatments, and the detection of Ca and Si signals after $CaSiO_3$ treatments, it was suggested that a silicon adsorption

Table 4-18. Ferric-Ferrous Iron Ratios in Reduced Nontronites (From Stucki *et al.*, 1976).

Mineral Treatment	Fe(III)/Fe(II)	I_{3+}/I_{2+}
Nontronite Unaltered	13.1	No Fe(II) Photopeak Observed
Nontronite Hydrazine-Reduced	6.4	6
Nontronite Dithionite-Reduced	2.2	2.5

Table 4-19. Relative Peak Intensities for Various Levels k; $I'_k = I_k/I_{A\,1\,2s} \times 100$ And Estimated Atomic Concentration Ratios $p'_A = p_A/p_{A\,l}$ (From Alvarez *et al.*, 1976).

Atom (A)	Level (k)	Binding Energy E_b (eV) Observed Values	Untreated $Al(OH)_3$		$Al(OH)_3$ Plus Adsorbate Treatments: $CaSiO_3$		$Si(OH)_4$		$Ca(PO_4H_2)_2$		$Ca(NO_3)_2$	
			I'_k	p'_A	I'_k	p'_A	I'_k	p'_A	I'_k	p'_A	I'_k	p'_A
Al	2S1/2	118*	100	1	100	1	100	1	100	1	100	1
Al	2S1/2, 3/2	73*	72	1	79	1.1	85	1.2	83	1.2	77	1.1
O	1S1/2	533*	1230	2.7	1170	2.5	1200	2.6	1200	2.6	1200	2.6
Si	2S1/2	151			10.4	0.06						
Si	2P1/2,3/2	104			8.3	0.05						
Ca	2P1/2	351†			1.8	0.007						
Ca	2P3/2	349†			3.6	0.007			3.4	0.007		
P	2S1/2	193							19	0.15		
P	2P1/2,3/2	133							25	0.15		
C	1S1/2	284*	38	0.26	55	0.39	50	0.35	52	0.36	24.5	0.17

*For Untreated $Al(OH)_3$

†For Specimen Treated with $CaSiO_3$

mechanism onto gibbsite is dependent upon the availability of calcium ions (Table 4-19).

The interaction of Ba^{2+} with ripidolite as a function of pH has been examined with XPS (17). XPS spectra of these barium-treated clay species were virtually unchanged throughout the pH range examined. Furthermore, from the Si/Ba XPS ratios obtained, the amount of barium adsorbed on the mineral surface was dependent upon pH, with the degrees of adsorption increasing with increasing pH.

The chemical nature of adsorbed iron species was probed by XPS in the work of Koppelman (21). Untreated kaolinite, which contained no XPS detectable surface iron species was reacted with $Fe(NO_3)_3$ solutions at pH values low enough to prevent hydroxide precipitation. XPS examination of this iron-treated kaolinite sample revealed a distinct Fe 2p photopeak at a binding energy 1.1 eV lower than lattice Fe^{3+} in chlorite (Fig. 4-18). Deconvolution of the rather broad Fe 2p photopeaks of chlorite and illite which had been subjected to similar Fe^{3+} treatment revealed adsorbed iron at the same binding energy as iron adsorbed on kaolinite (Table 4-20). The lowering of binding energy for adsorbed Fe^{3+} relative to lattice Fe^{3+} was interpreted to indicate that electron density in the Stern layer where ions are adsorbed is shifted to the metal ion, thus lowering its binding energy.

A similar reduction in binding energy for chromium adsorbed on kaolinite, chlorite and illite was also observed (29). These three minerals were reacted with

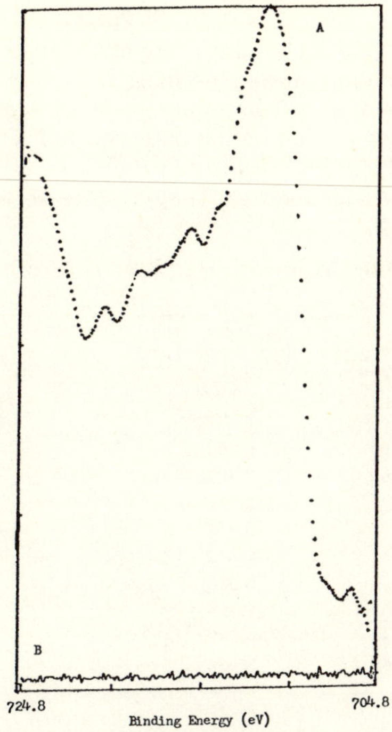

Figure 4-18. Fe $2p_{3/2}$ photopeaks; (A) Kaolinite with adsorbed Fe(III); (B) Pure Kaolinite (From Koppelman, 1976).

Table 4-20. Binding Energies for Fe $2p_{3/2}$ Electrons (From Koppelman, 1976).

		Assignment	Binding Energy (eV)	FWHM (eV)
Chlorite	(Fe^{2+})	Native:		
		composite experimental peak	710.6	5.2
		lattice Fe^{2+}	710.3	3.5
		With Adsorbed Fe^{3+}:		
		composite experimental peak	711.3	6.3
		lattice Fe^{2+}	710.3	3.7
		adsorbed Fe^{3+}	711.4	3.9
Nontronite	(Fe^{3+})	Native:		
		lattice Fe^{3+}	712.5	4.9
Kaolinite		Native: No Fe detected		
		With Adsorbed Fe^{3+}:		
		adsorbed Fe^{3+}	711.4	3.8
Illite	(Fe^{2+}, Fe^{3+})	Native:		
		composite experimental peak	712.6	6.4
		lattice Fe^{2+}	710.4	3.9
		lattice Fe^{3+}	712.6	4.0
		With Adsorbed Fe^{3+}:		
		composite experimental peak	712.5	6.5
		lattice Fe^{2+}	710.4	3.9
		lattice Fe^{3+}	712.6	3.8
		adsorbed Fe^{3+}	711.5	3.7

$Cr(NO_3)_3$ solution at pH values of 2, 3, 4, 6, 8, and 10. Si 2p and Al 2p binding energies were found to be pH invariant. Cr 2p binding energies varied only slightly between pH values 2-4, but remained constant at pH values, 6, 8, and 10 (Table 4-21) (28). Above pH 6, the binding energy for chromium in the clay samples was identical with that of $Cr(OH)_3$, indicating precipitation had occurred. The binding energies of adsorbed Cr^{3+} below pH 6 was significantly (1.0 eV) lower than Cr^{3+} substituted in an octahedral lattice site (Kämmerite). These results were similar to those obtained from Fe^{3+} adsorption (21).

Table 4-21. XPS Data for the Reaction of $CrCl_3$ with Hydrite® R (From Koppelman *et al.*, 1979).

pH	Cr $2p_{3/2}$ E_b (eV±0.1)	Intensity Ratios	
		Cr $2p_{3/2}$/Si 2p	Al 2p/Si 2p
1	N.D.	–	0.931
3	577.5	0.0158	0.905
3.5	577.4	0.0563	0.909
4	577.3	0.0883	0.948
4.5	577.3	0.1485	0.975
4.6	577.2	0.1637	0.932
4.7	577.1	0.1700	0.930
4.8	576.9	0.1697	0.976
5	576.9	0.1689	0.965
6	576.9	0.1680	0.929

Adsorption of $Co(H_2O)_6{}^{2+}$ on chlorite has been investigated at pH values of 3 and 7 using XPS (25). It was observed that the binding energy of the adsorbed Co^{2+} species was independent of pH, but was 0.5 eV lower than Co^{2+} substituted in an octahedral site in lusakite (Table 4-22, Fig. 4-19). It was suggested that the degree of reduction in the adsorbed metal ion binding energy was dependent upon the oxidation state of the adsorbed species. Furthermore, it was noted that the binding energy for adsorbed Co^{2+} was significantly different than that of $Co(OH)_2$ indicating precipitation had not occurred.

Table 4-22. XPS Binding Energy Results for Cobalt Species (From Koppelman and Dillard, 1978).

Species	Binding Energy (±0.1 eV) Co $2p_{3/2}$	ΔE $2p_{1/2}$–$2p_{3/2}$
Lusakite	782.6	16.1
Co^{2+} adsorbed on chlorite,		
pH 3	782.1	16.0
pH 7	782.0	16.0
$Co(NO_3)_2 \cdot 6\,H_2O$	781.6	15.9
$Co(OH)_2$	780.9	16.1
CoO	780.4	16.1
$Co(NH_3)_6{}^{3+}$ adsorbed on chlorite	782.1	16.0
$[Co(NH_3)_6](NO_3)_3$	782.1	15.1
CoOOH	779.9	15.0
Co_2O_3	779.2	15.2

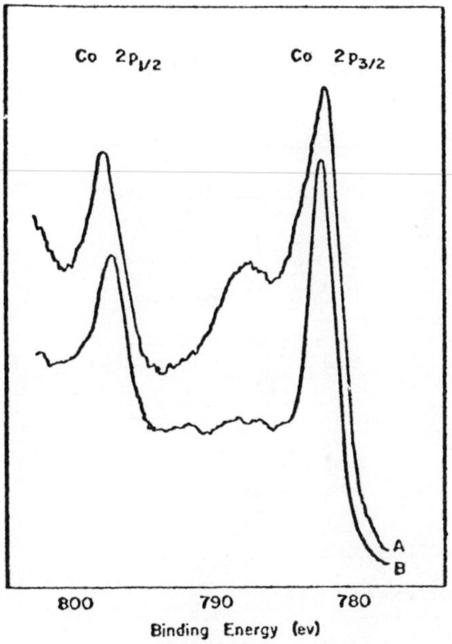

Figure 4-19. Co $2p_{1/2,\ 3/2}$ spectra for cobalt species adsorbed on chlorite (From Koppelman and Dillard, 1978).

The adsorption of cobalt (II) on Al_2O_3 and ZrO_2 was studied using XPS and electrophoretic mobility measurements (42, 43). A comparison of binding energies for adsorbed Co^{2+} with those for cobalt oxides and hydroxides revealed that cobalt adsorbed on alumina and zirconia exists as $Co(OH)_2$ (Table 4-23, Fig. 4-20). However, at 200° C, for cobalt adsorbed on the alumina surface, the cobalt photopeaks were similar to $CoAl_2O_4$, suggestive of surface transformation.

Table 4-23. XPS Line Positions for Cobalt Oxides and for Cobalt Adsorbed on Zirconium and Aluminum Oxides (From Tewari and Lee, 1975).

Compound	Binding Energy (eV) Co ($2p_{3/2}$)	Co ($3p_{1/2,3/2}$)
CoO	779.6 ± 0.2	61.0 ± 0.2
$Co(OH)_2$	780.7 ± 0.2*	60.6 ± 0.2
CoOOH	779.7 ± 0.2	61.3 ± 0.2
Co_2O_3	779.8 ± 0.2	
$CoO \cdot Al_2O_3$ ($CoAl_2O_4$	780.8 ± 0.2*	{61.3, 63.4} ± 0.2
Co adsorbed on ZrO_2 (30° C)	780.7 ± 0.2*	
Co adsorbed on Al_2O_3 (30° C)	780.7 ± 0.2*	
Co adsorbed on Al_2O_3 (200° C)	780.5 ± 0.3	{61.3, 63.4} ± 0.2

*Satellite peak observed 5.5 eV above the main cobalt $2p_{3/2}$ line.

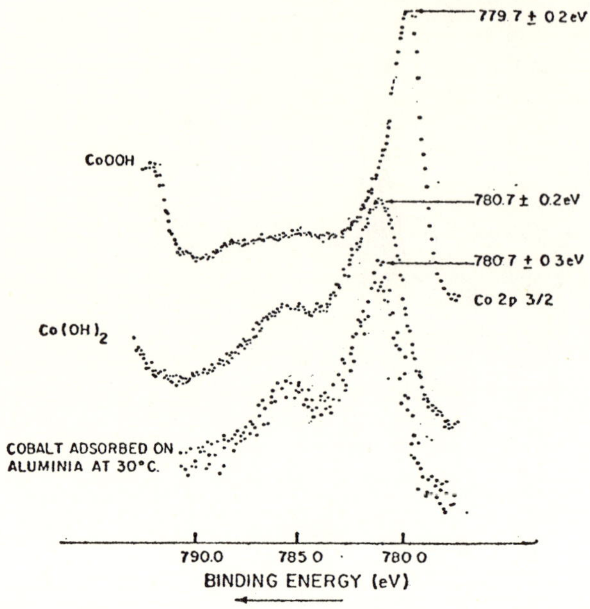

Figure 4-20. Co $2p_{3/2}$ photopeaks for cobalt oxides and cobalt adsorbed on alumina (From Tewari and Lee, 1975).

XPS measurements of cobalt adsorbed on hydrous manganese dioxide (a disordered birnessite, MnO_2) reveal strong evidence that Co(II) has been oxidized to Co(III) (Fig. 4-21) (33). This conclusion was reached, not through the examination and comparison of Co 2p binding energies, but by utilizing the position, shape, and intensity of the cobalt 2p satellite structure, and the energy of separation of the Co $2p_{1/2}$ and $2p_{3/2}$ levels. Due to spin state changes, absolute use of Co 2p binding energies alone for the assignment of cobalt oxidation state is generally not reliable. In this same study, manganese spectra were examined and are characteristic of Mn(IV) (Table 4-23).

Model calculations suggest that Co(II) cannot be oxidized by O_2 to Co(III) in bulk solution at seawater concentrations, but that oxidation can proceed in the presence of the strong electric potential at the MnO_2/solution interface (33). Ni(II), however, cannot be oxidized at this interface except at very high concentrations. These calculations suggest that the oxidation of Co(II) can explain the geochemical separation of cobalt from nickel.

XPS was used to examine the products of the reactions of kaolinite, chlorite, and illite with Cr(III) and Co(III) ammine complexes (21, 27). In the interaction of both chromium and cobalt hexammine complexes with chlorite, rapid and unanticipated rates of hydrolysis of the dissolved complexes were observed. XPS examination of the cobalt complex treated chlorite after both short (1 day) and long (1 week) interaction periods revealed that cobalt had been reduced to cobalt (II) (Fig. 4-22) (25). Relative rates of clay catalyzed hydrolysis of both chromium and cobalt hexammine complexes could be related to the amount of unoxidized (ferrous) iron in the surface region (25, 27). XPS atom ratio measurements for

N/Cr suggest that significant loss of coordinated ammine had occurred upon adsorption (27).

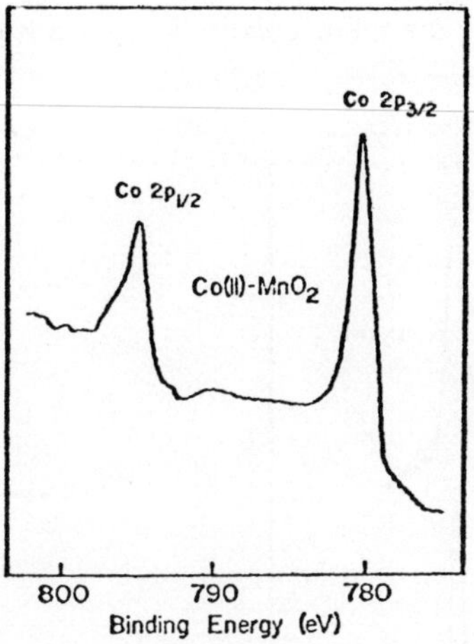

Figure 4-21. Cobalt 2p photopeaks for Co(II) adsorbed on MnO_2 (From Murray and Dillard, 1979).

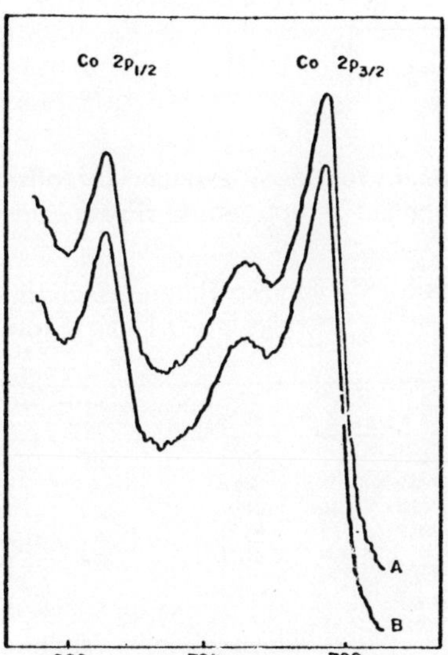

Figure 4-22. Co $2p_{1/2,\ 3/2}$ spectra of (A), $Co(NH_3)_6{}^{3+}$; (B), $Co(H_2O)_6{}^{3+}$, adsorbed on chlorite (From Koppelman and Dillard, 1978).

In a study of the adsorption of $Co(NH_3)_6{}^{3+}$ on a Y-type zeolite, XPS results indicated that $Co(NH_3)_6{}^{3+}$ is adsorbed as Co(III) (Fig. 4-23) (32). However, no examination of the Co/N atomic ratios was carried out to discover whether dissociation or decomposition of the complex occurred upon adsorption and subsequent heat treatment.

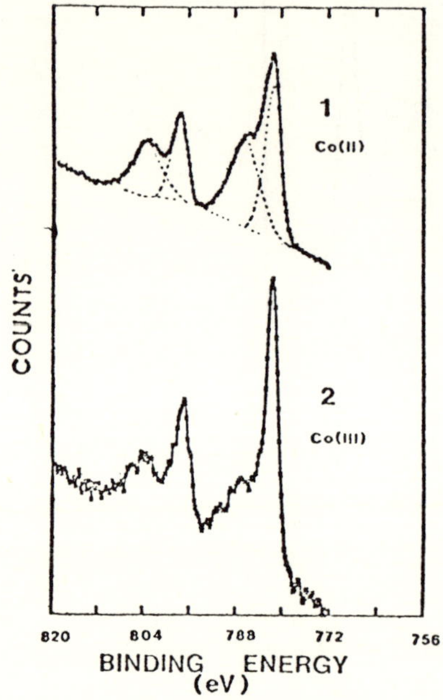

Figure 4-23. Co $2p_{1/2,\ 3/2}$ photopeaks for cobalt exchanged zeolites; (1) $Co^{2+}Y$; (2) $Co(NH_3)_6{}^{3+}$ exchanged Y-type zeolite (From Lunsford *et al.*, 1978).

Table 4-24. Binding Energy Values for Co and Mn Reference Oxides and Co (ads)–MnO_2. Precision of Binding Energies is ± 0.1 eV. (From Murray and Dillard, 1979).

Sample	pH	O $1s_{1/2}$	Co $2p_{1/2}$	Co $2p_{3/2}$	Δ_1**	Mn $2p_{3/2}$	Mn 3s(A) (high)	Mn 3s(B) (low)	Δ_2†
A	3.4	529.2	NM*	780.1	–	641.9	NM	NM	–
B	4.5	529.3	NM	780.2	–	642.0	NM	NM	–
C	5.2	529.3	795.4	780.2	15.2	641.9	88.2	83.7	4.5
D	7.0	529.3	795.3	780.2	15.1	641.8	88.2	83.7	4.5
MnO_2		529.2	–	–	–	641.9	88.4	83.7	4.7
Co_2O_3			794.4	779.2	15.2				
CoOOH			794.9	779.9	15.0				
$Co(OH)_2$			797.0	780.9	16.1				

*NM – not measured due to low photopeak intensity

**Δ_1 BE(Co $2p_{1/2}$–Co $2p_{3/2}$)

† Δ_2 BE (Mn 3s (A)–Mn 3s (B)

Examination of the mode of bonding of the metal ions Cu(II) and Ni(II) to clay minerals using XPS has also been investigated (24). Comparison of the binding energy for the adsorbed Ni^{2+} species with that of Ni(II) substituted in an octahedral site in lizardite revealed a lowering (0.4 eV) of Ni 2p binding energy for the adsorbed nickel species (Table 4-25). It was noted that this result was consistent with the results obtained for Co(II) adsorption (24, 25). The binding energy of the adsorbed copper species did not, however, show the same (≃ 0.5 eV) reduction in Cu 2p binding energy relative to dioptase (24). It was suggested from solution pH data during the reaction of the clays with $Cu(NO_3)_2$ that the adsorbed Cu(II) species was $Cu(OH)^+$.

Table 4-25. Binding Energies for Nickel and Copper (From Koppelman and Dillard, 1977).

Compound	Binding Energy (±0.1 eV) $2p_{3/2}$ Level
Ni (II) adsorbed (chlorite)	856.6
Ni $(OH)_2$	856.0
NiO	854.4*
	854.0*
	853.4*
Ni_2O_3	855.7
Pimelite	857.0
Lizardite	857.0
Cu(II) adsorbed (chlorite)	935.5
$Cu(OH)_2$	934.4
CuO	933.2
Cu_2O	933.2
Dioptase	935.1

*Data from other workers

4-6. SUMMARY

The applicability of XPS to the study of the chemistry of mineral surfaces is evident. XPS provides the researcher with a spectroscopic tool that is unique in that it is able to probe the surface region directly rather than by inference. Furthermore, XPS can provide insight into bonding at mineral surfaces as well as monitoring interfacial reaction processes. The analytical implications of XPS are clear and obvious. XPS is a tool which should not be limited to use only by chemists and physicists, and the realm of its usage should be explored by all physical scientists.

REFERENCES

1. Adams, I., J.M. Thomas, and G.M. Bancroft, 1972. An ESCA study of silicate minerals.*Earth Planet. Sci. Lett.* **16**: 429-432.
2. Adams, J.M., S. Evans, P.I. Reid, J.M. Thomas, and J.M. Walters, 1977. Quantitative analysis of aluminosilicates and other solids by x-ray photoelectron spectroscopy. *Anal. Chem.* **49**: 2001-2007.
3. Adams, J.M. and S. Evans. 1979. Exchange and selective surface uptake of cations by layer silicates using x-ray photoelectron spectroscopy (XPS). *Clays Clay Miner.* **27**: 248-252.
4. Alvarez, R., C.S. Fadley, J.A. Silva, and G. Uehara. 1976. A study of silicate adsorption on gibbsite ($Al(OH)_3$) by x-ray photoelectron spectroscopy (XPS). *Soil Sci. Soc. Amer. J.* **40**: 615-617.
5. Anderson, P.R. and W.E. Swartz, Jr. 1974. X-ray photoelectron spectroscopy of some aluminosilicates. *Inorg. Chem.* **13**: 2293-2294.
6. Baird, R.J., C.S. Fadley, S.K. Kawamoto, M. Mehta, R. Alvarez, and J.A. Silva. 1976. Concentration profiles for irregular surfaces from x-ray photoelectron angular distributions. *Anal. Chem.* **48**: 843-846.
7. Bancroft, G.M., J.R. Brown, and W.S. Fyfe. 1977a. Quantitative x-ray photoelectron spectroscopy (ESCA): Studies of Ba^{2+} sorption on calcite. *Chem. Geol.* **19**: 131-144.
8. Bancroft, G.M., J.R. Brown, and W.S. Fyfe. 1977b. Calibration studies for quantitative x-ray photoelectron spectroscopy of ions. *Anal. Chem.* **49**: 1044-1048.
9. Bancroft, G.M., J.R. Brown, and W.S. Fyfe. 1979. Advances in, and applications of, x-ray photoelectron spectroscopy (ESCA) in mineralogy and geochemistry. *Chem. Geol.* **25**: 227-243.
10. Betteridge, D., J.C. Carver, and D.M. Hercules. 1973. Devaluation of the gold standard in x-ray photoelectron spectroscopy. *J. Electron Spectrosc. Relat. Phenom.* **2**: 327-334.
11. Burness, J.H., J.G. Dillard, and L.T. Taylor. 1975. An x-ray photoelectron spectroscopic study of cobalt (II) schiff base complexes and their oxygenation products. *J. Amer. Chem. Soc.* **97**: 6080-6088.
12. Carriere, B. and J.P. Deville. 1977. X-ray photoelectron study of some silicon-oxygen compounds. *J. Electron Spectrosc. Relat. Phenom.* **10**: 85-91.
13. Castle, J.E., L.B. Hazell, and R.H. West. 1979. Chemical shifts in Al-Si compounds by Zr L_α photoelectron spectrometry. *J. Electron Spectrosc. Relat. Phenom.* **16**: 97-106.
14. Counts, M.E., J.S.C. Jen, and J.P. Wightman. 1973. An electron spectroscopy for chemical analysis study of lead adsorbed on montmorillonite. *J. Phys. Chem.* **77**: 1924-1925.
15. Defossé, C., P. Canesson, P.G. Rouxhet, and B. Delmon. 1978. Surface characterization of silica-aluminas by photoelectron spectroscopy. *J. Catal.* **51**: 269-277.
16. Dianis, W.P., and J.E. Lester. 1973. External standards in x-ray photoelectron spectroscopy. *Anal. Chem.* **45**: 1416-1420.
17. Emerson, A.B. 1979. An XPS investigation of the effect of pH on chromium and cobalt adsorption on clay minerals. M.S. thesis, VPI & SU, Blacksburg, Va., USA.

18. Hercules, D.M. 1974. Electron spectroscopy . . . . . for chemical analysis. *J. Electron Spectrosc. Relat. Phenom.* **5:** 811-826.
19. Huntress, W.T., Jr., and L. Wilson. 1972. An ESCA study of lunar and terrestrial materials. *Earth Planet. Sci. Lett.* **15:** 59-64.
20. Jaegle, A., A. Kalt, G. Nanse, and J.C. Peruchetti. 1978. Contribution a l'etude de l'effet de charge sur echantillon isolant en spectroscopie de photoelectrons (XPS). *J. Electron Spectrosc. Relat. Phenom.* **13:** 175-186.
21. Koppelman, M.H. 1976. An x-ray photoelectron spectroscopic investigation of the adsorption of metal ions on marine clay minerals. Ph.D. thesis, VPI & SU, Blacksburg, Va., USA. 251 pp.
22. Koppelman, M.H. and J.G. Dillard. 1975. An ESCA study of sorbed metal ions on clay minerals. In: T.M. Church (ed.), Marine chemistry in the coastal environment, ACS Symposium Ser. #18, pp. 186-201.
23. Koppelman, M.H., and J.G. Dillard. 1977a. Unpublished data.
24. Koppelman, M.H., and J.G. Dillard. 1977b. A study of the adsorption of Ni(II) and Cu(II) by clay minerals. *Clays Clay Miner.* **25:** 457-462.
25. Koppelman, M.H., and J.G. Dillard. 1978. An x-ray photoelectron spectroscopic (XPS) study of cobalt adsorbed on the clay mineral chlorite. *J. Colloid Interface Sci.* **66:** 345-351.
26. Koppelman, M.H., and J.G. Dillard. 1979. The application of x-ray photoelectron spectroscopy (XPS or ESCA) to the study of mineral surface chemistry. *Proc. Int. Clay Conf. 1978* (Pub. 1979): 153-164.
27. Koppelman, M.H., and J.G. Dillard. 1980. Adsorption of $Cr(NH_3)_6{}^{3+}$ and $Cr(en)_3{}^{3+}$ on clay minerals. Characterization of chromium using x-ray photoelectron spectroscopy. *Clays Clay Miner.* **28:** 000.
28. Koppelman, M.H., J.G. Dillard, A.B. Emerson, and J.R. Furey. 1979. Unpublished data.
29. Koppelman, M.H., A.B. Emerson, and J.G. Dillard. 1980. On the nature of adsorbed Cr(III) on chlorite and kaolinite: An x-ray photoelectron spectroscopic study. *Clays Clay Miner.* **28:** 119-124.
30. Lagaly, G. and A. Weiss. 1969. Determination of the layer charge in mica-type layer silicates. *Proc. Int. Clay Conf. 1969* **1:** 61-80.
31. Lindsay, J.R., H.J. Rose, W.E. Swartz Jr., P.H. Watts Jr., and K.A. Rayburn. 1973. X-ray photoelectron spectra of aluminum oxides: structural effects on the chemical shift. *Appl. Spectros.* **27:** 1-4.
32. Lunsford, J.H., P.J. Hutta, M.J. Lin, and K.A. Whitehorst. 1978. Cobalt nitrosyl complexes in zeolites A, X, and Y. *Inorg. Chem.* **17:** 606-610.
33. Murray, J.W. and J.G. Dillard. 1979. The oxidation of cobalt (II) adsorbed on manganese dioxide. *Geochim. Cosmochim. Acta* **43:** 781-787.
34. Nicholls, C.J., D.S. Urch, and A.N.L. Kay. 1972. Determination of coordination number in some compounds of magnesium and aluminum: a comparison of x-ray photoelectron (ESCA) and x-ray emission spectroscopies. *J.C.S. Chem. Comm. 1972:* 1198-1199.
35. Petrovic, R., R.A. Berner, and M.B. Goldhaber. 1976. Rate control in dissolution of alkali feldspars – I. Study of residual grains by x-ray photoelectron spectroscopy. *Geochim. Cosmochim. Acta* **40:** 537-548.
36. Schultz, H.D., C.J. Vesely, and D.W. Langer. 1974. Electron binding energies for silicon materials occurring in respirable coal dust. *Appl. Spectrosc.* **28:** 374-375.

37. Seals, R.D., R. Alexander, L.T. Taylor, and J.G. Dillard. 1973. Core electron binding energy study of group IIb-VIIa compounds. *Inorg. Chem.* **12:** 2485-2487.
38. Siegbahn, K., C. Nordling, A. Fahlman, R. Nordberg, K. Hamrin, J. Hedman, G. Johansson, T. Bergmark, S.E. Karlsson, I. Lindgren, and B. Kindbert. 1967. ESCA, Atomic, Molecular and Solid State Structure Studied by Means of Electron Spectroscopy, Almqvist and Wiksells, Uppsala, p. 276.
39. Stucki, J.W. and C.B. Roth. 1977. Oxidation-reduction mechanism for structural iron in nontronite. *Soil Sci. Soc. Amer. J.* **41:** 808-814.
40. Stucki, J.W., C.B. Roth, and W.E. Baitinger. 1976. Analysis of iron-bearing clay minerals by electron spectroscopy for chemical analysis (ESCA). *Clays Clay Miner.* **24:** 289-292.
41. Swartz, W.E. Jr., P.H. Watts Jr., J.C. Watts, J.W. Brasch, and E.R. Lippincott. 1972. Comparison of internal mixing and vacuum deposition procedures for calibrating ESCA spectra. *Anal. Chem.* **44:** 2001-2005.
42. Tewari, P.H., and W.J. Lee. 1975. Adsorption of Co(II) at the oxide-water interface. *J. Colloid Interface Sci.* **52:** 77-88.
43. Tewari, P.H. and N.S. McIntyre. 1975. Characterization of adsorbed cobalt at the oxide-water interface. *AIChE. Symposium Ser.* **71:** 134-137.
44. Thomassin, J.H., J. Goni, P. Baillif, and J.C. Touray. 1976. Étude par spectrométrie ESCA des premiers stades de la lixiviation du chrysotile en milieu acide organique. *C.R. Acad. Sci., Paris, Sér. D* **283:** 131-134.
45. Thomassin, J.H., J. Goni, P. Baillif, J.C. Touray, and M.C. Jaurand. 1977. An XPS study of the dissolution kinetics of chrysotile in 0.1 N oxalic acid at different temperatures. *Phys. Chem. Miner.* **1:** 385-398.
46. Thomassin, J.H., J.C. Touray, and J. Tricket. 1976. Étude par spectrométrie ESCA des premiers stades d'alteration d'une obsidienne: le compostement relatif de l'aluminium et du silicium. *C.R. Acad. Sci., Paris, Sér. D* **282:** 1229-1232.
47. Urch, D.S. and S. Murphy. 1974. The relationship between bond lengths and orbital ionization energies for a series of aluminosilicates. *J. Electron Spectrosc. Relat. Phenom.* **5:** 167-171.
48. Yin, L.I., S. Ghose, and I. Adler. 1971. Core electron binding energy difference between bridging and non-bridging oxygen atoms in a silicate chain. *Science* **173:** 633-635.

Chapter 5

THE APPLICATION OF NMR TO THE STUDY OF CLAY MINERALS

J.J. Fripiat

Director of Research, C.N.R.S., France, and
Professor at the University of Louvain, Belgium

5-1. INTRODUCTION: FUNDAMENTALS OF NMR

Consider an electromagnetic radiation interacting with an isolated nucleus bearing an angular momentum I. This vector presents an orientation with respect to the laboratory frame of reference (Fig. 5-1). Under some conditions an interaction between the magnetic moment $\vec{\mu}$ associated with $\vec{I}$

$$\vec{\mu} = \gamma \hbar \vec{I} \qquad [5\text{-}1]$$

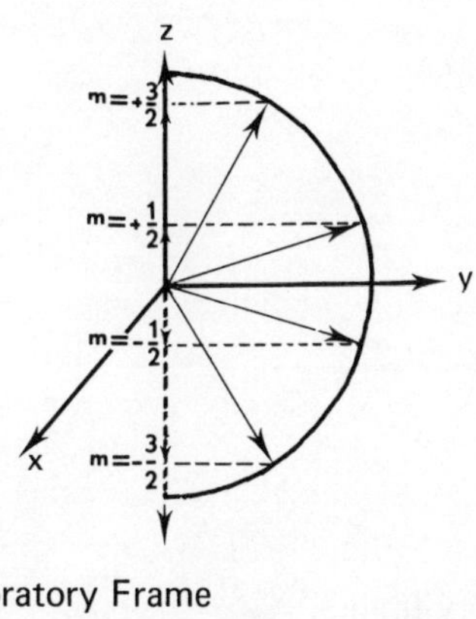

Figure 5-1. Laboratory Frame

(where γ is gyromagnetic ratio and $\hbar = h/2\pi$) and the oscillating magnetic field of the radiation can be obtained. Such an exchange of energy opens the door to a spectroscopic technique founded on the resonance between the energy levels defined by the interaction of a magnetic moment with a static magnetic field and the quantum of energy of an electromagnetic radiation. Purcell, Torrey and Pound (40) and Bloch, Hansen and Packard (5) were the first to realize this experiment in 1946 and this special kind of spectroscopy was called nuclear magnetic resonance spectroscopy (NMR). Resonance occurs if the difference $\Delta E = h\upsilon$ between two energy levels is equal to the frequency of the electromagnetic radiation multiplied by h.

J. W. Stucki and W. L. Banwart (eds.), Advanced Chemical Methods for Soil and Clay Minerals Research, 245–315.

This is the basic principle of any kind of spectroscopy, the uniqueness of NMR being that the energy levels are created by interacting $\vec{\mu}$ with a strong magnetic field $\vec{H}_o$.

Quantum physics dictates that the absolute value of $\vec{I}$ is $\hbar\sqrt{I(I+1)}$ where I is the spin quantum number. Table 5-1 contains the spin number I of various nuclei and the natural abundance of the isotope characterized by this quantum number. The absolute value of $\vec{\mu}$ is

$$\mu = \gamma\,\hbar\sqrt{I(I+1)} \qquad [5\text{-}2]$$

The projection of $\vec{I}$ on OZ (in Fig. 5-1) is $m\hbar$ where m is the magnetic quantum number and may take any one of the numerical values between + I, +(I−1), +(I−2), ...), . . .−(I−1), −I. This corresponds to the fact that $\vec{I}$ may take (2I+1) orientations with respect to OZ. This however is meaningless unless the reference frame is associated with some physical parameter. This parameter is the static magnetic field $\vec{H}_o$ which is made parallel to OZ. Several references discuss in detail the fundamentals of NMR (1, 3, 16, 31, 45).

5-1.1. The rotating frame

From classical mechanics we know that when a magnetic moment $\vec{M}$ interacts with a static field $\vec{H}_o$, the torque affecting $\vec{M}$ is given by the vectorial product $\vec{M}\wedge\vec{H}_o$ and that the rate at which $\vec{M}$ reorients (Fig. 5-2) is equal to the torque

$$\frac{1}{\gamma}\frac{d\vec{M}}{dt} = \vec{M}\wedge\vec{H}_o \qquad [5\text{-}3]$$

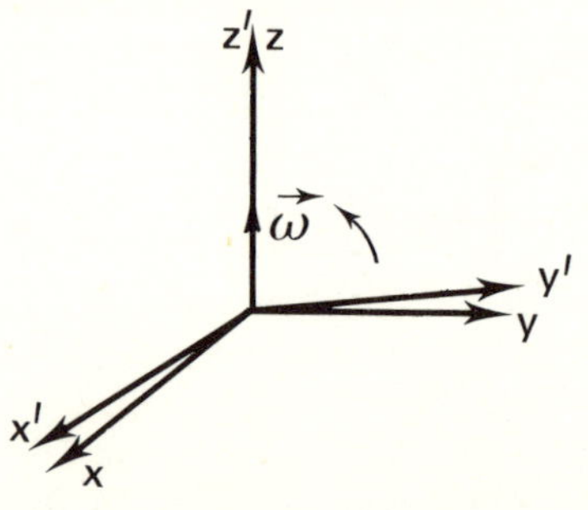

Figure 5-2. The rotating and laboratory frames.

This equation could be expressed with respect to a fixed frame of reference or with respect to a rotating frame as shown in Fig. 5-2. OZ' is parallel to OZ with the static magnetic field being along OZ // OZ'. OX' and OY' turn around OZ' at some angular speed, ω This motion may be represented by a vector $\vec{\omega}$ parallel to OZ' and directed toward the positive value of OZ if the motion is clockwise.

$$\frac{d\vec{M}}{dt} = \frac{dM_x}{dt}\vec{i} + \frac{dM_y}{dt}\vec{j} + \frac{dM_z}{dt}\vec{k} + M_x\frac{d\vec{i}}{dt} + M_y\frac{d\vec{j}}{dt} + M_z\frac{d\vec{k}}{dt} \qquad [5\text{-}4]$$

But since: $\frac{d\vec{i}}{dt} = \vec{\omega} \wedge \vec{i}$, $\frac{d\vec{j}}{dt} = \vec{\omega} \wedge \vec{j}$, and $\frac{d\vec{k}}{dt} = \vec{\omega} \wedge \vec{k}$; it follows that:

$$\frac{d\vec{M}}{dt} = \left(\frac{d\vec{M}}{dt}\right)_{rot} + \vec{\omega} \wedge \vec{M}$$

From eouation [5-4] and using equation [5-3],

$$\left(\frac{d\vec{M}}{dt}\right)_{rot} = \gamma \vec{M} [\vec{H}_o + \frac{\vec{\omega}}{\gamma}]. \qquad [5\text{-}5]$$

Thus in the rotating frame $\vec{M}$ is time independent if

$$\vec{\omega} = \vec{\omega}_o = -\gamma \vec{H}_o \qquad [5\text{-}6]$$

When this condition is fulfilled, $\vec{M}$ precesses around OZ in the laboratory frame with an angular speed W. If the magnetic moment $\vec{M}$ is the vectorial sum of a great number of individual spins

$$\vec{M} = \sum_i \vec{\mu}_i$$

each of them will precess at the same rate about OZ//OZ'//H_o (same γ) but without being in phase. This is the Larmor precession. Consider nuclei with spin number I = + 1/2 or − 1/2: the precession of the two sets of nuclei draw two cones as shown in Fig. 5-3.

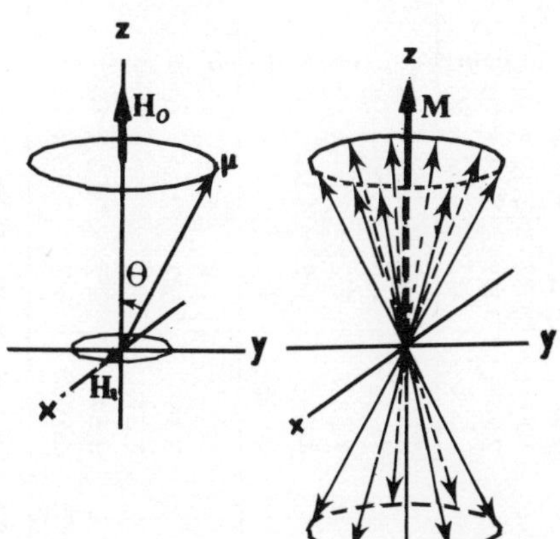

Figure 5-3. The precession of spins 1/2 and spins −1/2 (Farrar and Becker, 1971).

5-1.2. Resonance Experiment

The two sets of nuclei shown above, would keep precessing for ever if the static magnetic field is constant unless an additional torque is applied to force some of them to swing in the opposite "umbrella". This can be done for instance by applying an additional magnetic field H_1 along either the OX' or OY' direction in the rotating frame. In so doing the magnetic moment which so far was immobile in the rotating frame will precess either about OX' or OY'.

Table 5-1. NMR frequency table at 21.14 kG.

MHz	Nucleus	Natural Abundance %	Spin 1	Relative* Sensitivity	MHz	Nucleus	Natural Abundance %	Spin 1	Relative* Sensitivity
1.55	^{197}Au	100	3/2	2.53×10^{-5}	17.19	^{77}Se	7.50	1/2	6.93×10^{-3}
2.31	^{41}K	6.91	3/2	8.44×10^{-5}	17.88	^{29}Si	4.70	1/2	7.84×10^{-3}
2.83	^{103}Rh	100	1/2	3.12×10^{-5}	18.01	^{127}I	100	5/2	9.34×10^{-2}
2.92	^{57}Fe	2.245	1/2	3.38×10^{-5}	18.81	^{207}Pb	21.11	1/2	9.13×10^{-3}
3.64	^{107}Ag	51.35	1/2	6.62×10^{-5}	19.08	^{111}Cd	12.86	1/2	9.54×10^{-3}
3.68	^{105}Pd	22.23	5/2	7.94×10^{-4}	19.35	^{195}Pt	33.7	1/2	9.94×10^{-3}
3.70	^{183}W	14.28	1/2	0.98×10^{-5}	19.68	^{113}In	4.16	9/2	0.345
4.0	^{99}Ru	12.81	5/2	1.07×10^{-3}	19.72	^{115}In	95.84	9/2	0.347
4.19	^{109}Ag	48.65	1/2	1.01×10^{-4}	19.96	^{113}Cd	12.34	1/2	1.09×10^{-2}
4.20	^{39}K	93.08	3/2	5.08×10^{-4}	20.26	^{185}Re	37.07	5/2	0.133
4.41	^{89}Y	100	1/2	1.18×10^{-4}	20.47	^{187}Ro	62.93	5/2	0.137
5.07	^{47}Ti	7.75	5/2	2.09×10^{-3}	21.36	^{59}Co	100	7/2	0.281
5.08	^{49}Ti	5.51	7/2	3.76×10^{-3}	21.64	^{121}Sb	57.25	5/2	0.160
5.09	^{53}Cr	9.54	3/2	$9.03 \leqslant 10^{-4}$	21.00	^{69}Ga	60.2	3/2	0.91×10^{-2}
5.51	^{25}Mg	10.05	5/2	2.68×10^{-3}	21.87	^{45}Sc	100	7/2	0.301

*Sensitivity at constant field relative to protons

Table 5-1 (continued)

MHz	Nulceus	Natural Abundance %	Spin 1	Relative* Sensitivity	MHz	Nucleus	Natural Abundance %	Spin 1	Relative* Sensitivity
5.63	^{67}Zn	4.12	5/2	2.86×10^{-3}	22.00	^{93}Nb	100	9/2	0.482
5.75	^{143}Nd	12.20	7/2	5.49×10^{-3}	22.31	^{55}Mn	100	5/2	0.178
5.86	^{95}Mo	15.78	5/2	3.23×10^{-3}	22.55	^{79}Br	50.57	3/2	7.86×10^{-2}
5.92	^{201}Hg	13.24	3/2	$1.42 \times 10-3$	22.63	^{13}C	1.108	1/2	1.59×10^{-2}
5.99	^{97}Mo	9.60	5/2	3.44×10^{-3}	23.45	^{27}Al	100	5/2	0.206
6.06	^{43}Ca	0.13	7/2	6.40×10^{-2}	23.59	^{123}Te	0.89	1/2	1.80×10^{-2}
6.50	^{14}N	99.64	1	1.01×10^{-3}	23.66	^{51}V	99.76	7/2	0.382
6.90	^{33}S	0.74	3/2	3.26×10^{-3}	23.81	^{23}Na	100	3/2	9.26×10^{-2}
7.11	^{21}Ne	0.257	3/2	2.46×10^{-3}	23.86	^{63}Cu	89.09	3/2	9.31×10^{-2}
7.34	^{37}Cl	24.6	3/2	2.71×10^{-3}	24.31	^{81}Br	49.43	3/2	$9.85 \times 10-2$
7.38	^{131}Ke	21.24	3/2	2.75×10^{-3}	24.90	^{129}Xe	26.24	1/2	2.12×10^{-2}
8.01	^{61}Ni	1.25	3/2	3.53×10^{-3}	25.26	^{141}Pr	100	5/2	0.258
8.37	^{91}Zr	11.23	5/2	9.4×10^{-3}	25.56	^{65}Cu	30.91	3/2	1.14
8.69	^{85}Rb	72.8	5/2	1.05×10^{-2}	27.45	^{71}Ga	39.8	3/2	0.142
8.82	^{35}Cl	75.4	3/2	4.70×10^{-3}	28.43	^{125}Te	7.03	1/2	3.16×10^{-2}

Table 5-1 (continued)

MHz	Nucleus	Natural Abundance %	Spin 1	Relative* Sensitivity	MHz	Nucleus	Natural Abundance %	Spin 1	Relative* Sensitivity
8.94	^{135}Ba	6.59	3/2	4.90×10^{-3}	28.88	^{11}B	81.17	3/2	0.165
8.97	^{50}V	0.24	6	5.55×10^{-2}	29.45	^{87}Rb	27.2	3/2	0.175
9.12	^{15}N	0.365	1/2	1.04×10^{-3}	32.07	^{117}Sn	7.65	1/2	4.62×10^{-2}
9.57	^{10}B	18.83	3	1.99×10^{-2}	33.65	^{119}Sm	8.68	1/2	5.18×10^{-2}
10.00	^{137}Ba	11.32	3/2	$6.86 \leqslant 10^{-3}$	34.98	^{7}Li	92.57	3/2	0.294
10.8	^{181}Ta	100	7/2	3.60 x 10-2	36.44	^{31}P	100	1/2	6.63×10^{-2}
11.2	^{123}Sb	42.75	7/2	4.57×10^{-2}	39.00	^{87}Sr	7.02	9/2	2.69×10^{-3}
11.66	^{133}Cs	100	7/2	4.74×10^{-2}	68.56	^{3}He	$10^{-6} \times 10^{-7}$	1/2	0.442
12.20	^{17}O	3.7×10^{-2}	5/2	2.91×10^{-2}	84.57	^{19}F	100	1/2	0.833
12.65	^{9}Be	100	3/2	1.39×10^{-2}	90.00	^{1}H	99.98	1/2	1.00
13.24	^{6}Li	7.43	1	8.50×10^{-3}					
13.82	^{2}H	1.56×10^{-2}	1	9.65 x 10-3					
14.46	^{209}Bi	100	9/2	0.137					
15.41	^{75}As	100	3/2	2.51×10^{-2}					
16.1	^{199}Hg	16.86	1/2	5.67×10^{-3}					

Therefore the individual spins are going to jump from the upper to the lower umbrella if the example in Fig. 5-3 is considered. In order to obtain H_1 rotating with respect to the laboratory frame with the radial speed ω_o, the linearly polarized wave of an electromagnetic radiation of this frequency must be applied to the spins system.

The motion created by the oscillating magnetic field H_1 in the rotating frame can be predicted by equation [5-5].

$$\left(\frac{d\vec{M}}{dt}\right)_{rot} = \gamma \vec{M} \wedge \vec{H}_{eff} \qquad [5\text{-}7]$$

where

$$\vec{H}_{eff} = \vec{H}_o + \frac{\vec{\omega}}{\gamma} + \vec{H}_1 \qquad [5\text{-}8]$$

$\vec{H}_o$ and $\vec{\omega}$ are always parallel to OZ′ and OZ whereas $\vec{H}_1$ is either parallel to OX′ or OY′. Then, for instance with $\vec{H}_1 \parallel$ OX′.

$$\vec{H}_{eff} = (H_o + \frac{\omega}{\gamma})\, \vec{k}' + H_1 \vec{i}' \qquad [5\text{-}9]$$

where $\vec{k}'$ and $\vec{i}'$ are the unit vectors in the rotating frame. The precession occurring about $\vec{H}_{eff}$ is shown in Fig. 5-4. In the plane containing $\vec{H}_o$, $\vec{H}_{eff}$ and $\vec{H}_1$

$$\sin\theta = -\frac{\omega_1}{\gamma H_{eff}} \qquad [5\text{-}10]$$

$$H_{eff} = \sqrt{H_1^2 + (H_o + \frac{\omega}{\gamma})^2} \qquad [5\text{-}11]$$

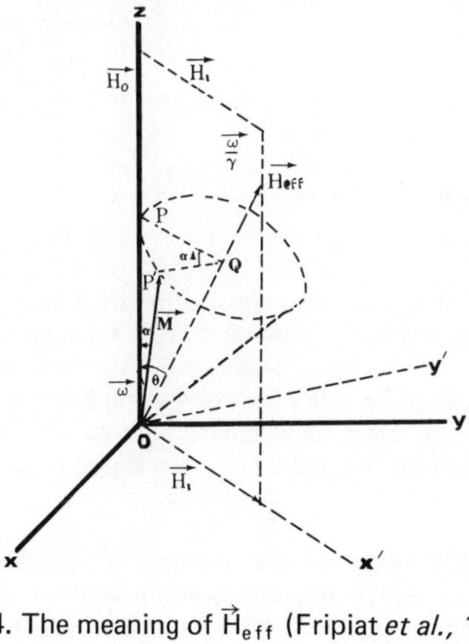

Figure 5-4. The meaning of $\vec{H}_{eff}$ (Fripiat *et al.,* 1971).

If $\omega = -\omega_o$, then $\vec{H}_{eff} = \vec{H}_1$ and $\vec{M}$ is now precessing about $\vec{H}_1$ which means that after time t_p of application of H_1, assuming $\vec{M}$ and $\vec{H}_o$ being parallel at time $t_p = 0$, $\vec{M}$ will have rotated by an angle $\omega_1 t_p$. The precession frequency about $\vec{H}_1$ being $\omega_1 = \gamma H_1$. The duration of $\vec{H}_1$ may be $\omega_1 t_p = \frac{\pi}{2}$ or $\omega_1 t_p = \pi$.

If $\omega_1 t_p = \pi$, $\vec{M} = -\vec{M}$, which means that for a system of spins 1/2, some of these spins which were in the upper umbrella of Fig. 5-3 at time $t_p = 0$ are in the lower umbrella at time $t_p = \pi/\omega_1$. Therefore the application of H_1 modifies the populations of the two energy levels. Similar reasoning may be applied to spins > 1/2 by considering more than two privileged orientations.

This raises the question of the correspondence between the conclusions obtained by the classical treatment exposed here and that derived from quantum mechanics.

5-1.3. Free Precession

Consider the laboratory frame of reference where we decompose the net magnetization into a component M_z which doesn't change with time as long as only $\vec{H}_o$ is present and a complex component M^+ in the x, y plane.

$$M^+ = M_x + i\,M_y = M_o \exp(i\,\omega_o t) \qquad [5\text{-}12]$$

At the thermodynamic equilibrium M^+ would be zero for a collection of identical spins which do not precess in phase. If the coil of a RF radio receiver is in the x, y plane, for instance along OY, no signal is obtained because there is no net magnetization in the x, y plane. On the contrary if field H_1 is applied for a short period t_p in the x, y plane $M_z(t_p) = M_o \cos(\omega_1 t_p)$ and $M^+(t_p) = M_o \sin(\omega_1 t_p)\exp(i\omega_o t)$.

Example for a "pulse":

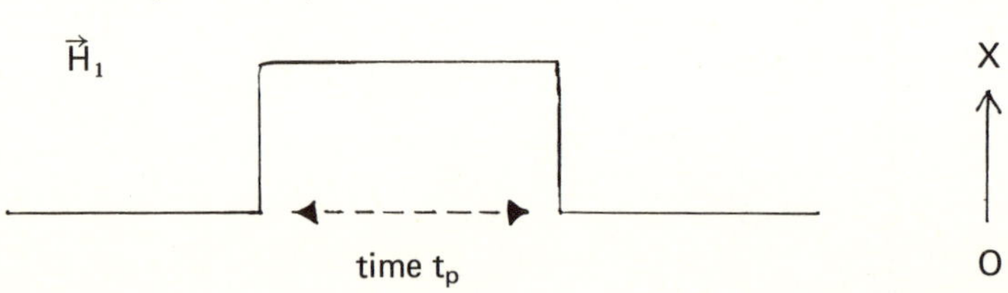

Thus during the pulse of application of H_1, the Z projection of the net magnetization decreases by $\cos(\omega_1 t_p)$ whereas in the x, y plane there is now a net magnetization $M_o \sin(\omega_1 t_p)$. After the pulse, this component will fade away because the individual magnetic moments μ are going out of phase having been exposed to a local magnetic field depending upon their local environments. This out-phasing process which is observed in the coil by the decrease of the signal (Fig. 5-5) can be described mathematically.

Assume that we represent the width of the distribution of the individual precession rates by a function $f(\omega)$ which has a maximum at the resonance frequency ω_o. After a time t following the pulse $M^+(t)$ is expressed as follows:

$$M^+(t) = M_o \sin(\omega_1 t_p) \int_{t=t_p}^{t>>t_p} f(\omega)\, e^{i\omega t}\, d\omega \qquad [5\text{-}13]$$

To understand the meaning of the integral, let us define u as $\omega - \omega_o$, f is a function of u and $e^{i\omega t} = e^{iut} \times e^{i\omega_o t}$

$$M^+(t) = M_o \sin(\omega_1 t_p) e^{i\omega_o t} \int_{t=t_p}^{t>>t_p} f(u)\, e^{iut}\, du \qquad [5\text{-}14]$$

If $t_p \to 0$, the integral is the Fourier transform G(t) of $f(\omega)$. However t_p must be such that $\omega_1 t_p = \frac{\pi}{2}$ in order for the observable signal in the coil to be maximum. This requires ω_1 (and H_1) to be very large as compared to the width of $f(\omega)$. The function $f(\omega)$ is simply the line shape which represents the actual distribution in frequency of local fields to which the nuclei are submitted. This distribution also involves the heterogeneity of H_o as well as the effects of all the magnetic species on the resonant nucleus (Fig. 5-6).

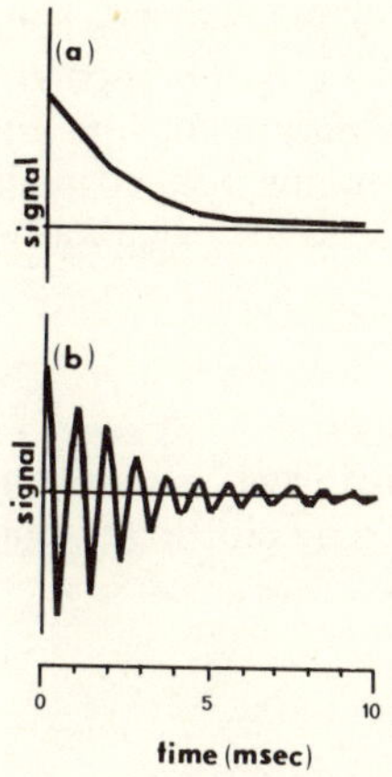

Figure 5-5. Free induction decay (FID) for rf precisely at the Larmor frequency (upper diagram) FID for rf off resonance (lower diagram). (Farrar and Becker, 1971).

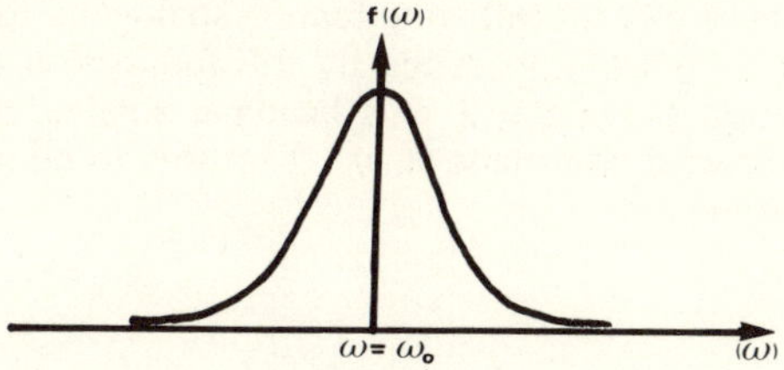

Figure 5-6. Distribution frequency.

What happens to magnetization in the rotating frame? Suppose $t_p = 0$ and $M(z) = M_o$ and $M^+ = 0$. After some time has elapsed $t_p = (\pi/2\,\omega_1)$, $M(z) = 0$ and $M^+ = M_o$ in the rotating frame. In that frame all spins are then in phase. They start to go out of phase, however, as indicated by the function

$$\int_{t = t_p}^{t >> t_p} f(u)\, e^{iut}\, du \qquad [5\text{-}15]$$

as time $t > t_p$ increases.

Suppose that $f(u) = [b/\pi]\;[1/(b^2 + u^2)]$, namely that the line shape is Lorentzian. In this instance $G(t)$ is defined as follows:

$$G(t) = \text{Fourier transform of } u = \exp(-bt) = \int_{t = t_p}^{t >> t_p} f(u)\, e^{iut}\, du \qquad [5\text{-}16]$$

In the laboratory frame the decay is modulated by the $e^{i\omega_o t}$ function and it can be further redressed to obtain the decay shown in Fig. 5-5. The constant b has the dimension of the inverse of time, $b = 1/T_2^*$. T_2^* is the apparent spin-spin relaxation time or it is the time characteristic for outphasing the collection of spins. T_2^* usually has a contribution $T_2^{-1}{}_{het}$ due to the applied magnetic field heterogeneity and a contribution of the local field due to the neighboring nuclei

$$T_2^{-1*} = T_2^{-1}{}_{het} + T_2^{-1} \qquad [5\text{-}17]$$

Later discussion will show how to cancel $T_2^{-1}{}_{het.}$ Because spins are precessing in a range of frequency covered by $f(\omega)$, interference effects such as shown in the lower half of Fig. 5-5 can be observed. T_2^* is easily obtained by plotting magnetization as a function of t.

5-2. THE BLOCH EQUATIONS

Assume that at thermal equilibrium, the magnetization along the OZ axis is

$$M_{oz} = \frac{\chi_o}{\mu_o} H_o \qquad [5\text{-}18]$$

where χ_o is the static magnetic susceptibility and μ_o the nuclear magneton (the nuclear moment $\mu = g\mu_o I$, where g is the splitting factor, analogous to the Lande factor in spectroscopy). After a "pulse", produced by the application of H_1 for a given time t_p, the magnetization along the Z axis becomes smaller because the nuclear moments are tipped toward the plane x, y. It returns to equilibrium by following a first-order rate process

$$\frac{d M_z}{dt} = -\frac{M_z - M_o}{T_1} \qquad [5\text{-}19]$$

with a time constant T_1. T_1 is a meaningful parameter of the system if the pulse has been long enough to tilt $\vec{M}$ with respect to $\vec{H}_o$ and OZ in such a way as to redistribute the spin population among the various magnetic energy levels of the system. This may be illustrated by considering an ensemble of spins 1/2 in a magnetic field $\vec{H}_o$ at equilibrium at temperature T. Let η_- be the number of nuclei per cm^3 in the upper energy state m = −1/2 and η_+ be the number in the lower energy state m = + 1/2. At equilibrium the number of transitions upward and downward are equal.

$$W_+\eta_+ = W_-\eta_- \qquad [5\text{-}20]$$

or

$$\frac{W_-}{W_+} = \frac{\eta_+}{\eta_-} = \exp\left(\frac{2\mu H_o}{kT}\right) \qquad [5\text{-}21]$$

W_- and W_+ are the corresponding transition probabilities considering the Boltzman distribution among the two levels. If μ H_o is << kT:

$$\frac{W_-}{W_+} = \frac{\eta_+}{\eta_-} \simeq 1 + (2\mu\, H_o/kT) \qquad [5\text{-}22]$$

(Note that for a proton placed in a field of 1 Tesla $\dfrac{\gamma\hbar H_o}{2k} \simeq 10^{-3}\ °K$)

By considering an average transition probability $W = \frac{1}{2}(W_+ + W_-)$ and an average population N

$$W_- = W[1 + \frac{\mu H_o}{kT}] \quad \eta_- = \frac{N}{2}\left(1 - \frac{\mu H_o}{kT}\right)$$

$$W_+ = W[1 - \frac{\mu H_o}{kT}] \quad \eta_+ = \frac{N}{2}\left(1 + \frac{\mu H_o}{kT}\right)$$

The total magnetization produced by the N nuclei of spin 1/2 at equilibrium is:

$$M_z(o) = \mu\frac{1}{2}(\eta_+ - \eta_-) = \frac{\mu^2 H_o N}{kT} = \frac{\gamma^2\hbar^2 H_o N}{4\,kT} = \frac{\chi_o}{\mu_o} H_o \qquad [5\text{-}23]$$

After the pulse η_- is greater than η_+ and M_z is less than $M_z(o)$. It is the return to equilibrium which obeys equation [5-19]. The return to equilibrium means that the energy absorbed by the spins system has to be released within the surroundings which is called in that particular case "the lattice". Thus T_1 is the spin-lattice relaxation time.

The time dependence of the magnetization is the same as that of $(\eta_+ - \eta_-)$. By defining this difference $\Delta\eta$, the variation $d\Delta\eta/dt$ is

$$\frac{d\Delta\eta}{dt} = 2\,W_-\eta_- - 2W_+\eta_+$$

since each time a transition occurs, $\Delta\eta$ changes by 2 units. When the system is not at the thermal equilibrium and $W_{-}\eta_{-} \neq W_{+}\eta_{+}$,

$$\frac{d\Delta\eta}{dt} = 2W(\Delta\eta_o - \Delta\eta) \qquad [5\text{-}24]$$

where $\Delta\eta_o$ is $(\eta_+ - \eta_-)$ at equilibrium. By integration of equation [5-24]

$$\Delta\eta_o - \Delta\eta = \text{const x} \exp(-2Wt) \qquad [5\text{-}25]$$

and $2W = T_1^{-1}$ as expected.

The projection of the magnetization in the x, y plane is different from zero after the pulse but the two components M_x and M_y will fade with the typical relaxation time T_2^* as explained previously. Neglecting for the moment $T_2^{-1}{}_{het}$ (equation [5-17]):

$$\frac{dM_x}{dt} = -\frac{M_x}{T_2} \qquad \frac{dM_y}{dt} = -\frac{M_y}{T_2} \qquad [5\text{-}26]$$

In addition the general equation describing the evolution of the magnetization vector is

$$\frac{d\vec{M}}{dt} = \gamma\, \vec{M} \wedge \vec{H}_o. \qquad [5\text{-}27]$$

It is now necessary to combine this equation with the equations accounting for the spin-spin and spin-lattice relaxation processes. Written for the laboratory frame

$$\frac{d\vec{M}}{dt} = \gamma\, \vec{M} \wedge \vec{H} - \vec{k}\left(\frac{M_z - M_o}{T_1}\right) - \vec{i}\,\frac{M_x}{T_2} - \vec{j}\,\frac{M_y}{T_2} \qquad [5\text{-}28]$$

Since $\vec{H} = \vec{H}_o + \vec{H}_1$ and that $H_{1x} = H_1 \cos\omega_1 t$ and $H_{1y} = H_1 \sin\omega_1 t$ it follows that

$$\left.\begin{aligned}
\frac{dM_z}{dt} &= -\gamma(M_x H_1 \sin\omega t + M_y H_1 \cos\omega t) - \frac{M_z - M_o}{T_1} \\
\frac{dM_x}{dt} &= \gamma(M_y H_o + M_z H_1 \sin\omega t) - \frac{M_x}{T_2} \\
\frac{dM_y}{dt} &= \gamma(M_z H_1 \cos\omega t - M_x H_o) - \frac{M_y}{T_2}
\end{aligned}\right\} \qquad [5\text{-}29]$$

After the pulse, $H_1 = 0$ and $t > t_p$, these equations become

$$\left.\begin{aligned}\frac{dM_z}{dt} &= -\frac{M_z - M_o}{T_1}\\ \frac{dM_x}{dt} &= \gamma M_y H_o - \frac{M_x}{T_2}\\ \frac{dM_y}{dt} &= -\gamma M_x H_o - \frac{M_y}{T_2}\ .\end{aligned}\right\} \qquad [5\text{-}30]$$

In the rotating frame,

$$\left(\frac{d\vec{M}}{dt}\right)_{rot} = \gamma \vec{M}_\Lambda \vec{H}_{eff} - \vec{k}\frac{M_z - M_o}{T_1} - \vec{i}'\frac{M'_x}{T_2} - \vec{j}'\frac{M'_y}{T_2} \qquad [5\text{-}31]$$

where H_{eff} is given by equation [5-8]. Calling $H_o + \frac{\omega}{\gamma} = \Delta\omega = \frac{\omega - \omega_o}{\gamma}$, it follows

$$\left.\begin{aligned}\frac{dM'_z}{dt} &= -\gamma M'_y H_1 - \frac{M_z - M_o}{T_1}\\ \frac{dM'_x}{dt} &= \gamma M'_y \Delta\omega - \frac{M'_x}{T_2}\\ \frac{dM'_y}{dt} &= \gamma M_z H_1 - \gamma M'_x \Delta\omega - \frac{M'_y}{T_2}\end{aligned}\right\} \qquad [5\text{-}32]$$

and after the pulse

$$\left.\begin{aligned}\frac{dM_z}{dt} = \frac{dM'_z}{dt} &= -\frac{M_z - M_o}{T_1}\\ \frac{dM'_x}{dt} &= \gamma M'_y \Delta\omega - \frac{M'_x}{T_2}\\ \text{and}\qquad \frac{dM'_y}{dt} &= -\gamma M'_x \Delta\omega - \frac{M'_y}{T_2}\end{aligned}\right\} \qquad [5\text{-}33]$$

A visual description of what happens in the rotating frame is shown in Fig. 5-7.

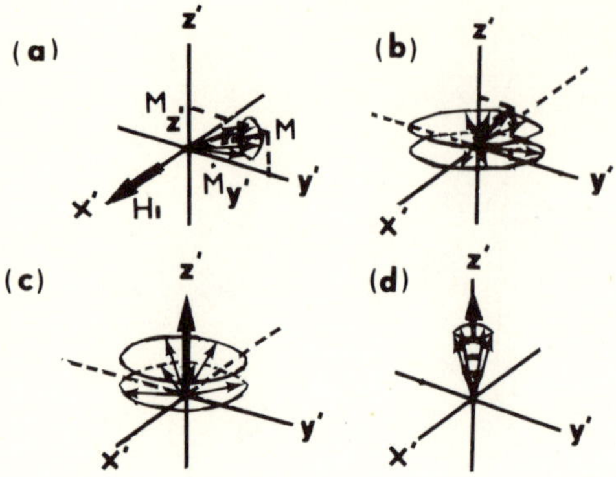

Figure 5-7. (a) Tipping of nuclear moments and macroscopic magnetization through an angle θ and establishment of $M_{y'}$. (b) Dephasing of nuclear moments by spin-spin relaxation and/or magnetic field inhomogeneity; reduction of $M_{y'}$. (c) Reduction of $M_{y'}$ to ~ 0. (d) Reestablishment of $M_{z'}$ at its equilibrium value, M_o. *N.B.* one umbrella only is considered for sake of simplicity. (Farrar and Becker, 1971).

5-2.1. T_1 Measurement

Fig. 5-7 shows clearly what happens after a pulse of duration $t_p = \frac{\pi}{2\omega_1}$ where $\omega_1 = -\gamma H_1$. Consider Fig. 5-3 as the starting point and suppose that the net magnetization resulting from summing up the independent spins in the two "umbrellas" is a vector $\vec{M}$ which is along the Z axis at time t = 0.

Then a pulse is applied for a duration t_p as shown in Fig. 5-7a. If t_p was rigorously $\frac{\pi}{2\omega_1}$, M should be exactly along Oy'. In Fig. 5-7b, out-phasing starts to occur, at time $t > t_p$ and finally the two umbrellas should be restored at time $t >> t_p$ (equilibrium state).

Suppose now that before the equilibrium is completely restored, a second $\frac{\pi}{2}$ pulse is applied when the system is in the state shown in Fig. 5-7c. This second pulse (a) rephases those individual spins which were running out of phase in plane Ox'y'; (b) converges them as one single vector and (c) tips this vector towards the "negative" end of the OZ axis. Those spins which were back in the initial (upper) umbrella will begin a new process as the one starting in Fig. 5-5a but the intensity of the signal generated in the coil will be less than that observed after the first pulse. Indeed a fraction of the spins which were "up" are now "down". The decrease of the signal amplitude will be $\frac{A_1}{A_2} = \exp(-\tau/T_1)$ where τ is the time between the two consecutive $\frac{\pi}{2}$ pulses.

The technique usually used to measure T_1 is that shown in Fig. 5-8. First a pulse with a duration $t_p = \frac{\pi}{\omega_1}$ is applied in order to turn the magnetization vector by 180°.

After time τ, the magnetization $M_Z(\tau)$ is sampled by using a $\frac{\pi}{2}$ pulse in order to orient it in the plane of the coil. The sampling is repeated many times changing

τ. The signal which was, say, negative for $\tau \simeq 0$ decreases progressively and becomes positive till it reaches, after $\tau \rightarrow \infty$ an amplitude practically equal to that observed for $\tau \rightarrow 0$. This procedure is illustrated in Fig. 5-8. Since $\frac{dM_z}{d\tau} = -\frac{M_z - M_o}{T_1}$

$$M_z(\tau) = M_o[1 - 2\exp(-\tau/T_1)] \qquad [5\text{-}34]$$

Theoretically $-M_o$ when τ is $0 = M_\infty$ when τ is ∞. For practical reasons the amplitude M_∞ is preferred to M_o because of uncertainty on the instrumental dead time.

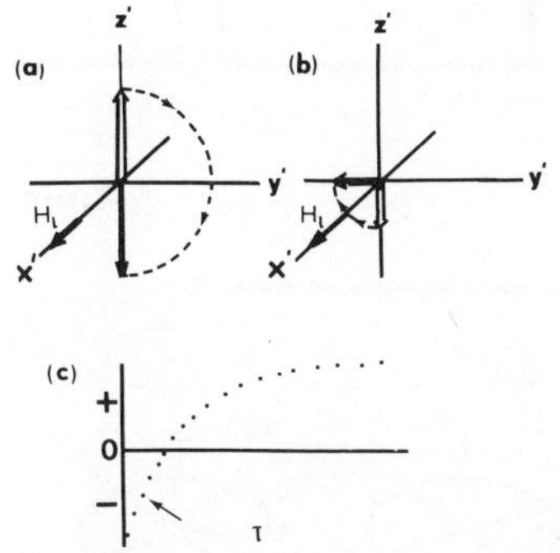

Figure 5-8. Determination of T_1 by 180°, τ, 90° sequences. (a) M is inverted by a 180° pulse at time 0. (b) After a time τ a 90° pulse rotates M to the y' (or −y') axis. (c) The initial amplitude of the FID after the 90° pulse, which is proportional to the value of M at time τ, is plotted as a function of τ. Note that each point results from a separate 180°, τ, 90° sequence. The point corresponding to (b) is indicated by the arrow. (Farrar and Becker, 1971).

5-2.2. T_2 Measurement by Spin-Echo

In order to obtain T_2 with accuracy and to get rid of the contribution of the static magnetic field heterogeneity within the sample, a series of two pulses at 90° and at 180° is carried out.

The first pulse brings the magnetization along oy'. Then outphasing begins. Those nuclear moments which are exposed to a weaker static field precess slower than those which are exposed to a stronger static field. There is thus a range of precession frequencies centered about υ_o, which is the rotation velocity of the rotating frame. After time τ a 180° pulse is applied.

Each magnetic moment still in the plane x'y' at that time is turned by 180° as shown in going from b to c in Fig. 5-9. Those rotating faster or those rotating slower continue and after time 2τ a signal (with a negative amplitude) will be observed called the echo (Fig. 5-9c). For running times longer than 2τ, the signal disappears again because of outphasing. It must be emphasized that only the contribution of the static field heterogeneity to T_2 is suppressed by the spin-echo technique. The contribution of a local field arising from another spin is going to change

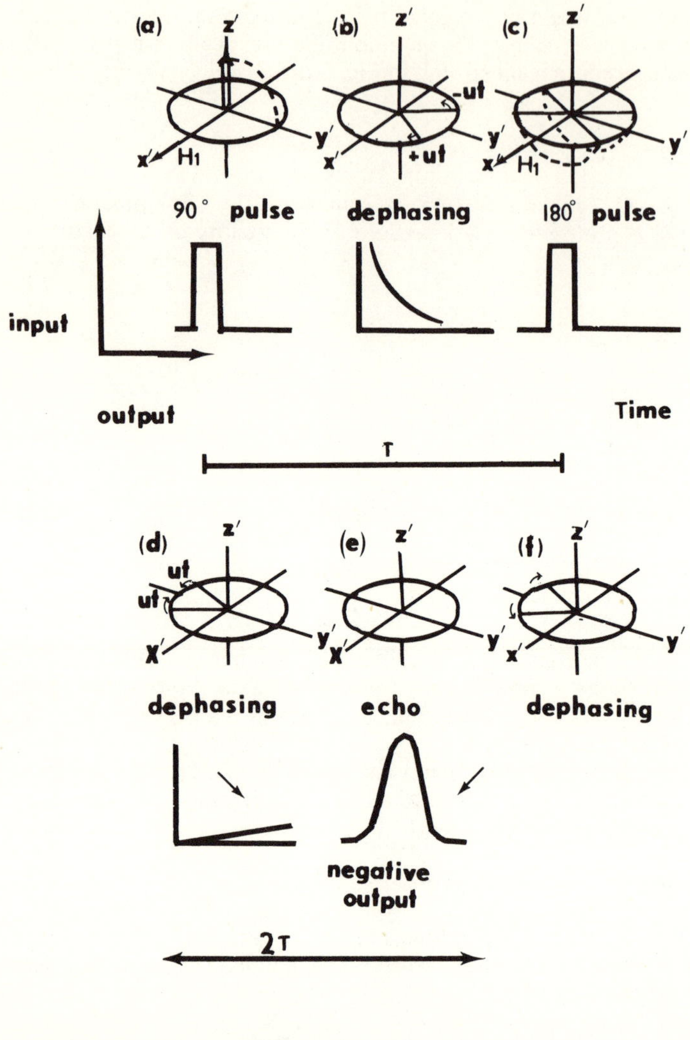

Figure 5-9. Spin echo: T_2 measurement.

sign after the first pulse since each spin is rotated by 180°. In the total magnetic field $\vec{H}_o + \vec{h}_{loc} + \vec{h}_{het}$, $\vec{h}_{loc}$ is changed into $-\vec{h}_{loc}$ and thus only h_{het} as a cause of out-phasing is suppressed. The echo amplitude is thus an exponential function of $-2\tau/T_2$ (not T_2^*). However the precise refocusing requires that *each* nucleus remain in a constant magnetic field during the time 2τ. Suppose that a spin diffuses with a diffusion coefficient D, given by the Einstein equation

$$D = 1^2/6\tau \qquad [5\text{-}35]$$

If the molecular diffusion is fast enough to translate the spin during time τ within an appreciable magnetic field gradient G, it can be shown that the echo amplitude is reduced

$$A(2\tau) \propto \exp\,[-(2\tau/T_2) - \frac{2}{3}\gamma^2\, G^2\, D\,\tau^3] \qquad [5\text{-}36]$$

The measurement of a long T_2 is particularly affected.

5-2.3. The Carr-Purcell Technique for Measuring T_2

A simple way to eliminate the effects of diffusion in the determination of T_2 is by using a $\pi/2$, τ, π, 2τ, π, 2τ, π, ...pulse sequence. The second π pulse will refocus an echo at time 4τ, the third at time 6τ etc. . . and the amplitude of the signal will be alternatively positive and negative. If τ between each π pulse is short, the field gradient sampled by the diffusing spin will be small and this cause of heterogeneity will not contribute appreciably to T_2.

5-3. LINE SHAPE

We have shown that the Fourier transform of the FID (free induction decay) is the function $f(\omega)$ which represents the width of the distribution function of the individual precession rates (Equations [5-13] and [5-14]). Since $\omega = \gamma H$ and H contains a term representing the local field h_{loc}, the static field heterogeneity is eliminated if $f(\omega) = f(h_{loc})$.

This paragraph discusses the kind of information that may be obtained from the line-shape of the NMR signal, in other words from the analysis of $f(h_{loc})$. As a first approximation, assume elimination of any source of line broadening other than that due to dipolar interaction. For an ensemble of identical spins interacting with each other, the Hamiltonian is

$$\mathcal{H} = \mathcal{H}_z + \mathcal{H}' \qquad [5\text{-}37]$$

where

$$\mathcal{H}_z = -\,\gamma\hbar H_o \sum_j I_z^j \qquad [5\text{-}38]$$

is the Zeeman contribution containing only the OZ components of the individual spins $\mu_z^j = \gamma\,\hbar I_z^j$ and where $\mathcal{H}'$ is the contribution of the spin interactions with each other. Suppose that the spins do not move and that they form a rigid lattice which means that the internuclei distances and the orientation of the internuclei vectors with respect to the laboratory frame are constant, then:

$$\mathcal{H}' = \sum_{i>j} \vec{\mu}_i\ \mathrm{grad}\,\frac{(\vec{\mu}_j\,\vec{r}_{ij})}{r_{ij}^3} \qquad [5\text{-}39]$$

where $\vec{\mu}_j\ \vec{r}_{ij}/r_{ij}^3$ is the potential experienced by spin $\vec{I}_i$ from spin $\vec{I}_j$ at a distance r_{ij}

$$\mathcal{H}' = -\gamma^2 \hbar^2 \sum_{i>j} \left(\frac{\vec{I}_i \vec{I}_j}{r_{ij}^3} - 3 \frac{(\vec{I}_i \vec{r}_{ij})(\vec{I}_j \vec{r}_{ij})}{r_{ij}^5} \right) \quad [5\text{-}40]$$

$\mathcal{H}'$ corresponds to the interaction of a nuclear moment with fields of the order of 1 gauss, whereas the Zeeman Hamiltonian ($\mathcal{H}_z$) corresponds to interactions with fields which are typically of the order of 10^4 gauss. $\mathcal{H}'$ may be considered as a perturbation of $\mathcal{H}_z$.

To simplify the problem, let us consider a pair of protons at a fixed distance as in H_2O (Fig. 5-10). $\mathcal{H}'$ can be developed as

$$\mathcal{H}' = (\gamma^2 \hbar^2 / r^3)\,(A+B+C+D+E+F) \quad [5\text{-}41]$$

if we express I_x and I_y in terms of the raising and lowering operators I^+ and I^- and transform the rectangular coordinates x, y, z in terms of spherical coordinates r, θ and φ.

$$\begin{aligned} A &= I_{z1} I_{z2}\,(1-3\cos^2\theta) \\ B &= -(1/4)\,(I_1^+ I_2^- + I_1^- I_2^+)\,(1-3\cos^2\theta) \\ C &= -\frac{3}{2}\,(I_1^+ I_{z2} + I_{z1} I_2^+)\cos\theta\sin\theta\exp(-i\varphi) \\ D &= C^* \qquad E = F^* \\ F &= -\frac{3}{4}\,I_1^+ I_2^+ \sin^2\theta \exp[-2i\varphi] \end{aligned} \quad [5\text{-}42]$$

where

$$I^+ = I_x + iI_y \qquad I^- = I_x - iI_y \quad [5\text{-}43]$$

In each of the A...F functions [5-42] there are two types of terms: (a) those which are functions of r, θ and φ, and thus of the location of the molecule, and (b) those which depend only on the spin operators.

If the pair of protons is moving, θ and φ become time dependent whereas r is constant as long as the molecule remains intact.

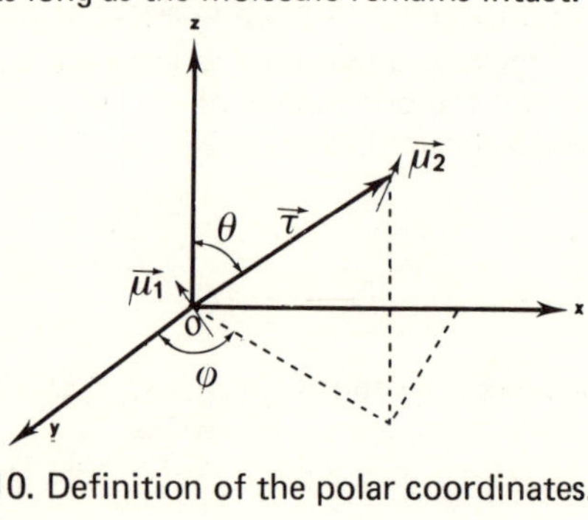

Figure 5-10. Definition of the polar coordinates.

5-3.1. The Proton-Proton Dipolar Interaction

It is important to observe the effect of $\mathcal{H}'$ on energy levels. For a pair of protons the Zeeman energy is:

$$E_z = <\mathcal{H}_z> = -\gamma\hbar H_o m_1 - \gamma\hbar H_o m_2 .$$

Since $m_1 = m_2 = \pm 1/2$, $E_z = -\gamma\hbar H_o$, O, $+\gamma\hbar H_o$. What pairs of states are connected by the perturbation?

The term A in [5-42] is proportional to I_{z1} and I_{z2}. It connects the two spins in the same energy level, e.g.:

$-1/2 \rightleftharpoons -1/2$

$-1/2 \rightleftharpoons +1/2$ or $+1/2 \rightleftharpoons -1/2$

$+1/2 \rightleftharpoons +1/2$

The term B is different because it contains the "raising" or "lowering" operators I^+ and I^-. Recall that these operators are so designated because of the effect they produce when they operate a wave function such as $U_{I,m}$:

$$\left.\begin{aligned} I^+ U_{I,m} &= \sqrt{I(I+1) - m(m+1)}\, U_{I,m+1} \\ I^- U_{I,m} &= \sqrt{I(I+1) - m(m-1)}\, U_{I,m-1} \end{aligned}\right\} \qquad [5\text{-}44]$$

I^+ or I^- turns $U_{I,m}$ into a function which m has been either raised or lowered by one unit. Therefore $<m'|I^{\pm}|m>$ vanishes unless $m' = m \pm 1$ whereas I_x or I_y interconnect states with either m+1 or m−1. Thus B flips the two spins in opposite directions:

$+1/2, -1/2 \rightarrow -1/2, +1/2.$

C and D flip one spin only

$+1/2, -1/2 \rightarrow -1/2, -1/2$

or

$-1/2, +1/2 \rightarrow +1/2, +1/2.$

Finally E and F flip either both spins up or both spins down:

$-1/2 -1/2 \rightarrow +1/2 + 1/2$

or

$+1/2 + 1/2 \rightarrow -1/2 - 1/2.$

These relationships are summarized in Fig. 5-11. The dipolar interaction Hamiltonian produces a second order energy shift of the Zeeman levels and it enables the alternating field to induce transitions of $\Delta m = m_1 + m_1 = 0$ or 2, while

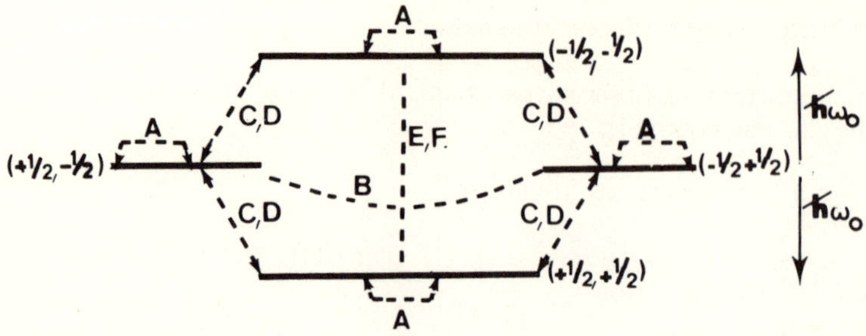

Figure 5-11. Zeeman energy levels for a pair of spins 1/2. (Fripiat *et al.*, 1971).

the normal rule is $\Delta m = \pm 1$ in absence of perturbation. The effect of terms C and F in [5-41] is therefore to give absorption near 0 and $2\omega_o$ but these peaks are very weak as compared with that at ω_o and therefore these terms may be disregarded. Finally:

$$\mathcal{H}' = (\gamma^2 \hbar^2 / r^3)\,(A+B) = (\gamma^2 \hbar^2 / r^3)\,(1-3\cos^2\theta)\times \quad [5\text{-}45]$$

$$[I_{1z}\,I_{2z} - \frac{1}{4}\,(I_1^+ I_2^- + I_1^- I_2^+)]$$

It can be shown, using the definition of the scalar product of vectors, that:

$$(3\,I_{1z}\,I_{2z} - \vec{I}_1\,\vec{I}_2) = 2\,[I_{1z}\,I_{2z} - \tfrac{1}{4}\,(I_1^+\,I_2^- + I_1^-\,I_2^+)]$$

and thus

$$\mathcal{H} = -\gamma\hbar\,H_o\,I_z + \frac{\gamma^2\hbar^2}{2r^3}\,(3\,I_{1z}\,I_{2z} - \vec{I}_1\cdot\vec{I}_2)\,(1-3\cos^2\theta) \quad [5\text{-}46]$$

where $I_z = m_1 + m_2 \equiv I_{z1} + I_{z2}$, where m_1 and m_2 are either + 1/2 or − 1/2. In addition I_z and $(3\,I_{1z} I_{2z} - \vec{I}_1\cdot\vec{I}_2)$ are two operators which commute. Recall that two operators, V and W for example, commute when VW = WV or VW − WV = 0. This property is represented generally by the symbol {V,W} = 0. Then W and V have the same eigen function. Therefore $\mathcal{H}$ and $\mathcal{H}'$ may have the same eigen function. Let U be such a function. The eigen values of the perturbation energy given by

$$\langle U^*|\mathcal{H}'|U\rangle = (\gamma^2\hbar^2/r^3)\,(1-3\cos^2\theta)\,[\langle U^*|I_{1z}\,I_{2z}|U\rangle - \frac{1}{4}\,\langle U^*|I_1^+ I_2^- + I_1^- I_2^+|U\rangle]. \quad [5\text{-}47]$$

are

$$+\frac{1}{4}\Delta E \text{ for } m_1 = \frac{1}{2},\ m_2 = \frac{1}{2}$$

$- \tfrac{1}{2}\,\Delta E$ for $m_1 = -\tfrac{1}{2}$, $m_2 = +\tfrac{1}{2}$ or $m_1 = +\tfrac{1}{2}$, $m_2 = -\tfrac{1}{2}$

$+ 1/4\,\Delta E$ for $m_1 = -1/2$, $m_2 = -1/2$

where $\Delta E = \gamma^2 \hbar^2 r^{-3}\,(1-3\cos^2\theta)$.

Therefore the Zeeman energy levels are displaced as follows:

$m_1 = 1/2, m_2 = 1/2 \quad [- \quad [-\gamma\hbar H_o + 1/4\,\Delta E]$

$$\left.\begin{array}{l} m_1 = -1/2, m_2 = +1/2 \\ m_2 = +1/2, m_2 = -1/2 \end{array}\right\} [0 - 1/2\,\Delta E]$$

$m_1 = -1/2, m_2 = -1/2 \quad [+\gamma\hbar\, H_o + 1/4\,\Delta E]$

The differences in energy in these levels are now either $-\gamma\hbar H_o + 1/4\,\Delta E + 1/2\Delta E = -\gamma\hbar H_o + 3/4\,\Delta E$ or $0 - 1/2\,\Delta E - \gamma\hbar H_o - 1/4\,\Delta E = -\gamma\hbar H_o - 3/4\,\Delta E$. Then two signals of equal intensity (same transition probability) appear at $\pm$ 3/4 ΔE from $-\gamma\hbar H_o$.

Example of application: the structure of water in the two-layer hydrate of Sodium vermiculite. Because a preferential orientation about the C crystal axis is readily obtained by sedimenting a suspension of clay particles, very useful NMR spectra can be obtained by using aggregated clay films.

These spectra can be used for instance for structural studies concerning water molecules within the interlamellar space.

NMR spectra of water adsorbed by clay minerals have been observed in several cases, the clearest example being perhaps that shown by Hougardy *et al.* (27) for the two-layer hydrate of a Na Llano vermiculite.

The experimental observations. Vermiculite is an expanding layer silicate that offers the unique ability to occlude water films one or two molecular diameters thick within an interlamellar space with a specific surface area of 800 m^2/g. Water adsorption outside the interlamellar space is negligible.

Studies on the Texas Llano vermiculite used here were initiated by Van Olphen (46, 47) who observed the step adsorption isotherm shown in Fig. 5-12. Such an isotherm is usually considered characteristic for a homogeneous surface. The unit cell formula of the sodium form of the Llano vermiculite is:

$$\{(Si_{5.22}Al_{2.72})^{IV}(Al_{1.32}Mg_{4.58})^{VI}O_{20}(OH)_4\}Na^+_{1.6}$$

The cation exchange capacity (CEC) is 2×10^{-3} eq/g. Therefore in the interlayer space the area available to each Na^+ is of the order of 30 Å^2. This means that if each interlamellar $Na^+_{1.6}$ cation is considered, on the average, to be the center of an octahedral hydration shell, the entire interlamellar space is occupied by coordinated water molecules forming a two-layer hydrate.

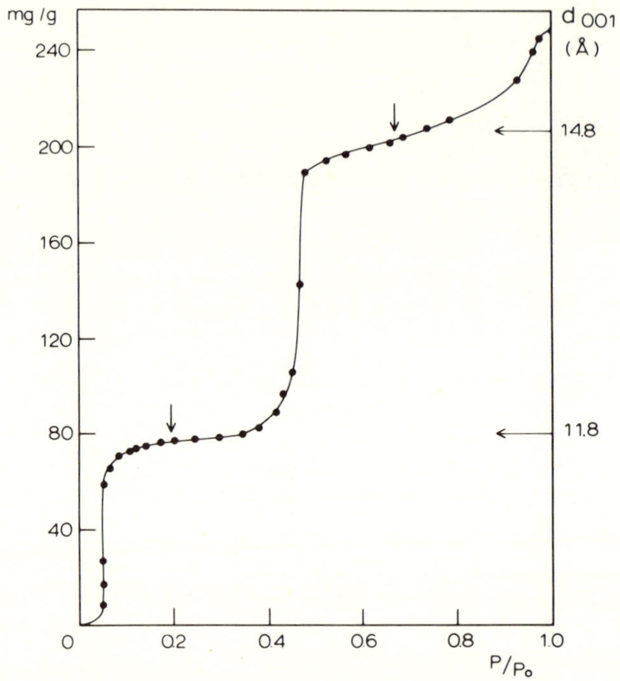

Figure 5-12. Water adsorption isotherm at 25°C obtained by Van Olphen (46) and X-ray d_{001} spacings for the Na Llano vermiculite. The two-layer hydrate is defined as the sample containing 203 mg water per g. (Hougardy *et al.*, 1976).

This is a crude representation and most probably there are some water molecules outside the hydration shell but few as compared to those directly coordinated to the cations. The 001 spacing of this two-layer hydrate is 14.8 Å. Such an arrangement therefore implies that the water molecules are largely the same type. Mathieson and Walker (33, 34), Bradley and Serratosa (6) and more recently Shirozu and Bailey (44) have studied by X-ray the structure of various such vermiculite hydrates.

In the two-layer hydrate of the Llano specimen, the Na^+ cations are midway in the interlamellar space. A randomized distribution of the sheets with respect to each other should prevail. Thus, the characteristic isotherm cannot actually be assigned to a tridimensional organization of the vermiculite sheets but merely to the predominant abundance of coordinated water.

The proton spectrum at room temperature is composed of a doublet and of a weak central line (40) (see Fig. 5-13). The linewidth and shape of this spectrum have been studied between 148° and 323° K for various orientations (δ) of the C^* axis with respect to the steady magnetic field H_o (Fig. 5-14).

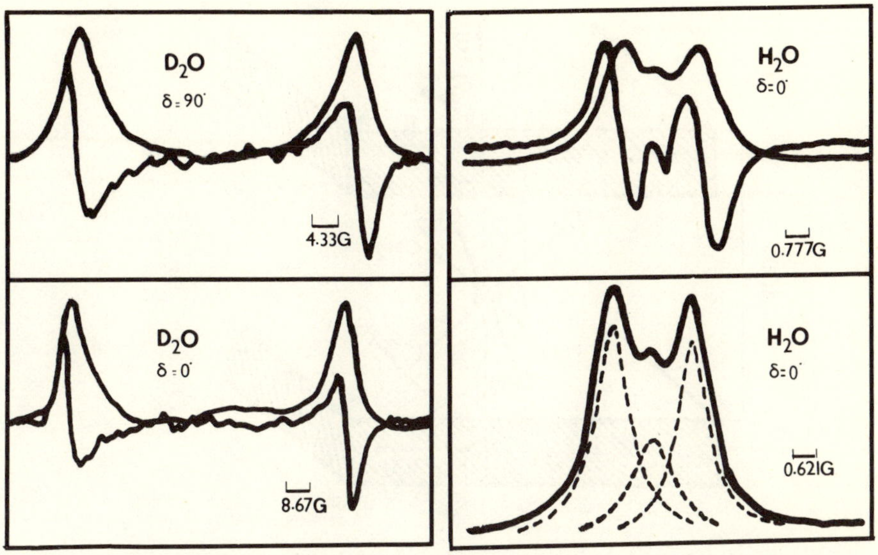

Figure 5-13. Experimental spectra at room temperature for D_2O at $\delta = 0°$ and $\delta = 90°$ and for H_2O at $\delta = 0°$ and the corresponding decomposed spectra showing doublet and central line (OH contribution has been substracted). Note that the frequency used for ^2H is 9.21 Mhz whereas it is 60 Mhz for ^1H in a field of 14 k. Gauss. (Hougardy *et al.*, 1976).

Above 270° K, the doublet is orientation dependent while below 220° K the doublet splitting increases and becomes orientation independent. In addition no central component emerges from the background.

From the evolution of the proton spectra it is thus adequate to distinguish two temperature regions (HT and LT) shown in Fig. 5-15, and separated by a transition domain extending from 220° to 270° K.

At first it should be emphasized that there is little evidence that the two-layer hydrate is destroyed at low temperature since Anderson and Hoekstra (2) have observed that a 16-17 Å spacing is maintained in hydrated montmorillonites cooled below −10°C and that Roby (43) has shown that d_{001} spacing of a Na fluorphlogopite undergoes only a moderate contraction from 15.11 to 15.01 Å as the temperature decreases from 273° to 123° K.

In addition, the formation of a separate ice phase outside the hydrated mineral phase should give rise to a broader line than that observed.

Thus the HT-LT transition is not likely to be provoked either by a change in the water content of the solid nor by a phase change of the water.

The deuteron WL spectra are composed solely of a doublet, as shown in Fig. 5-13 for two orientations.

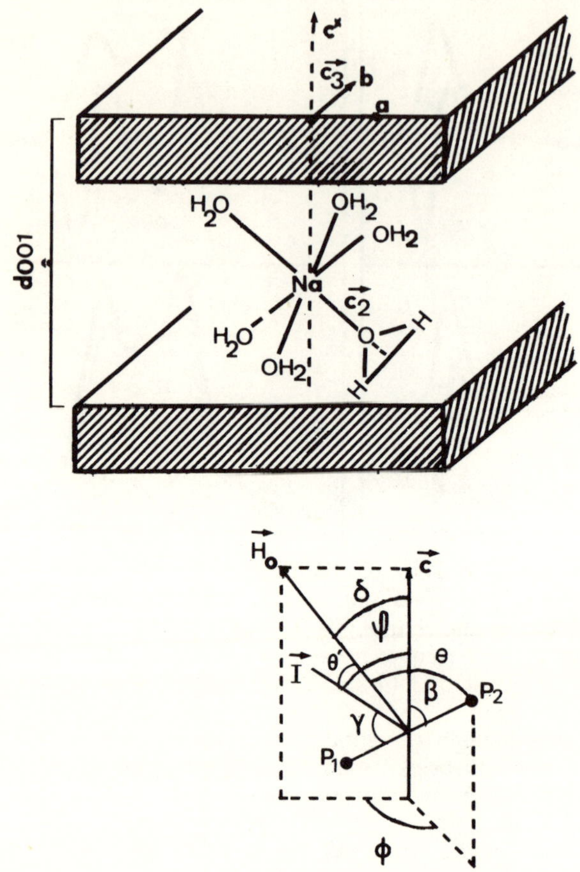

Figure 5-14. Schematic structure of vermiculite and definition of the various angular parameters used in the theoretical treatment. (Hougardy *et al.*, 1976).

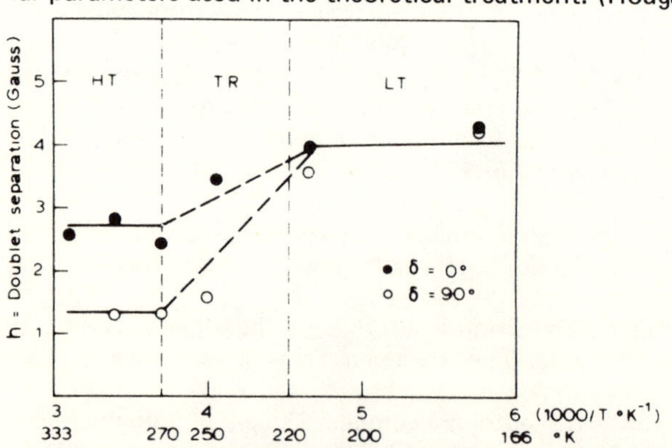

Figure 5-15. Variation of the doublet splitting of the proton spectra with temperature at two orientations δ of the C^* axis with respect to H_o. (Hougardy *et al.*, 1976).

Interpretation. As we have seen earlier, isolated pairs of protons such as those associated with water molecules should give rise to a doublet composed of two lines down field and up field, respectively, with a separation $2\alpha\ (3\cos^2\theta-1)$ gauss for each value of θ; using r = 1.58 Å, $\alpha = 3/4\ \gamma\hbar r^{-3} = 5.4$ G. If all the vectors $\vec{r}$ had the same orientation, the splitting should be in the ratio 1 to 2 for $\theta = 0$ and 90°, respectively. This is observed in the HT region in Fig. 5-15.

For a random distribution of vectors r, the doublet separation becomes insensitive to the position of the C* axis with respect to H_o. This situation is observed in the LT region.

The following quantitative analysis is based on the classical treatment proposed by Pake (36) and by Gutowsky and Pake (24).

Assume that the p-p vector of an isolated water molecule rotates about an axis $\vec{l}$ at a rate much greater than the frequency corresponding to the line splitting and therefore that θ fluctuates very rapidly. The angle γ between $\vec{r}$ and $\vec{l}$ remains constant as well as θ', the angle between $\vec{l}$ and $\vec{H}_o$.

The function $1/2\,(3\cos^2\theta-1)$ which is the second degree LEGENDRE polynomial $P_2(\cos\theta)$ must be averaged for a free rotation. This yields $P_2(\cos\theta) = 2P_2(\cos\theta') \times P_2(\cos\gamma) = \frac{1}{2}(3\cos^2\theta-1) \times (3\cos^2\gamma-1)$. The absorbance spectrum therefore consists of two lines, h gauss apart from each other:

$$h = \alpha(3\cos^2\theta'-1)(3\cos^2\gamma-1) \qquad [5\text{-}48]$$

in the case of a preferential orientation of $\vec{l}$ with respect to $\vec{H}_o$. Since δ (Fig. 5-14) is an experimental adjustable parameter, it is adequate to decompose $P_2(\cos\theta')$ into a function of ψ and δ. The following equation:

$$h = \frac{\alpha}{2}(3\cos^2\gamma-1)(3\cos^2\psi-1)(3\cos^2\delta-1) \qquad [5\text{-}49]$$

is obtained if $\vec{l}$ reorients about C* with a frequency higher than $\gamma\alpha/2\pi = 2.3\times10^4$ Hz. If the angle between the axis of rotation and proton-proton interatomic vector is 90° ($\vec{l}$ and the $(H_2O)C_2$ symmetry axis are parallel) $\vec{l}$ must be tilted by 65° with respect to C* in order to fit the HT experimental results shown in Fig. 5-15, α being 5.4 G. Moreover, as shown in Fig. 5-16, the variation predicted by Eq. [5-49] is obtained for all values of δ. In a pseudoregular octahedral hydration shell such as proposed by Shirozu and Bailey (44) for a magnesium vermiculite, the center of the cation is 1.16 Å from the planes containing the two sets of three water oxygens and the edge of the equilateral triangle formed by the molecules is 2.98 Å. The ψ angle in their case was 60°. Therefore an idealized network of octahedral shells in which the molecular rotational axis would be the coordinative bonds of the water oxygens to the central cation fits not only the geometrical requirements derived from the d_{001} spacing but also the NMR absorption data. If this hydration shell rotates around a threefold C_3 axis parallel to C* and passing through the center of an equilateral triangle, ψ is kept constant. Hence a diffusional rotation around this C_3 axis with a frequency higher than 2.3×10^4 Hz would allow the hydrated cation to diffuse in the interlamellar space while keeping an orientation dependent doublet.

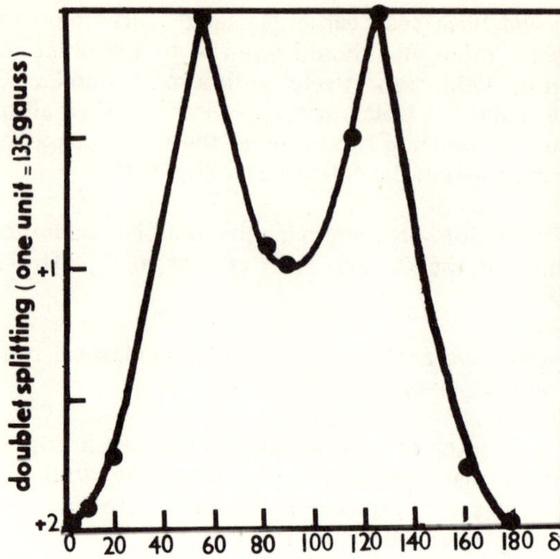

Figure 5-16. Variation of the proton doublet splitting with respect to δ. (Hougardy *et al.*, 1976).

In the low temperature domain, where doublet splitting becomes independent of the orientation of the clay film with respect to the magnetic field, the orientation of $\vec{I}$ is randomized.

5-3.2. Line Shape for Spins with Quantum Spin Number > 1/2. The Quadrupolar Interaction

A typical deuteron spectrum is obtained where water is completely replaced with heavy water by adsorption-desorption processes in a vermiculite oriented aggregate (Fig. 5-13). The spin quantum number of the ^2H nucleus is one. This changes the calculation of the perturbation experienced by the deuterons because the main interaction is no longer the dipolar type but it is due to the interaction between the nuclear quadrupole moment of the ^2H nucleus and the electrical field gradient at the nucleus. Figure 5-13 offers a good example of the benefit of replacing H_2O by D_2O in spite of the weaker sensitivity of the deuteron signal (Table 5-1).

While proton exchange between molecules affects the intramolecular proton interaction because the incoming proton may possibly have a different spin state, the coupling between the deuteron quadrupole moment and the electrostatic field gradient does not depend upon the spin state of the other deuteron partner. Since this coupling is very large compared to the dipole-dipole interaction, the exchange is not expected to be comparable to that for H_2O.

In H_2O if the proton exchange rate is of the order of the difference in frequency between the two peaks of the doublet, a central line must appear which progressively replaces the doublet as the exchange frequency increases. Such an increase may be obtained for instance by increasing the temperature. The absence of a central line in the D_2O spectra recorded under the same experimental conditions where such a line is present in the H_2O spectra favors assigning an exchange process to the origin of the central component.

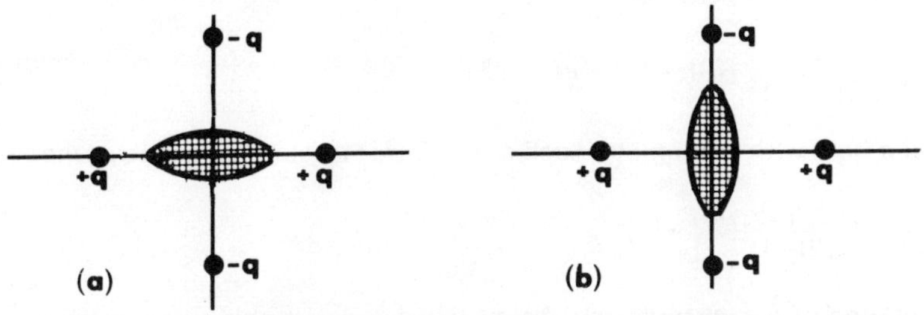

Figure 5-17. (a) A cigar-shaped nucleus in the field of four charges, +q on the x-axis; −q on the y-axis. The configuration of (b) is energetically more favorable because it puts the positive charge of the ends of the cigar closer to the negative charges −q. (Slichter, 1963).

Figure 5-17, taken from Slichter (45) is a model which accounts for the electrical effect on the energy required to reorient the nucleus. Suppose that the cigar-shaped nucleus is acted on by 4 charges. The situation described in b has a lower energy since it has put the tips of the positive nuclear charge closer to the negative external charge. The problem of evaluating the interaction energy E of a specified charge distribution (ρ) within the nucleus with an electrical potential V due to external sources is not easy since it implies one can solve the following equation:

$$E = \int \rho(r)\, V(r)\, dv \qquad [5\text{-}50]$$

where $\rho(r)$ is the charge density of the nucleus, $V(r)$ the electrostatic potential due to external sources and dv the volume element. It has been solved (Slichter (45)) by introducing the notion of a quadrupole moment of the nucleus, Q,

$$eQ = C\, I(2I-1) \qquad [5\text{-}51]$$

where C is a constant and where I is the total angular momentum of the nucleus. For I = 1/2, Q = 0: for I >1/2, Q >0. Then the Hamiltonian operator corresponding with the perturbation of the Zeeman energy levels is

$$\mathcal{H}_Q = \frac{eQ}{4I(2I-1)}\left[V_{zz}(3\, I_z^2 - I^2) + (V_{xx} - V_{yy})\,(I_x^2 - I_y^2)\right] \qquad [5\text{-}52]$$

where V_{zz}, V_{xx} and V_{yy} are the components of the electrical field gradient tensor.

Note that in the case of the dipolar interaction $\mathcal{H}_Q$ has the same role as $\mathcal{H}'$ for nuclei with I > 1/2. Usually two symbols are defined to make the equation more compact:

$$V_{zz} = eq = \text{field gradient}$$

and

$\eta = (V_{xx} - V_{yy})/V_{zz}$ = asymmetry parameter

Equation [5-52] then becomes

$$\mathcal{H}_Q = \frac{e^2 q Q}{4I(2I-1)} [(3I_z^2 - I^2) + \eta(I_x^2 - I_y^2)]. \quad [5\text{-}53]$$

In case of a field with a symmetry such as $\eta = 0$ (for instance an axial symmetry)

$$\mathcal{H}_Q = \frac{e^2 q Q}{4I(2I-1)} (3I_z^2 - I^2). \quad [5\text{-}54]$$

The calculation of the perturbation energy associated with this operator gives

$$E_Q = \frac{e^2 qQ}{8I(2I-1)} (3\cos^2\epsilon - 1)\,[3m^2 - I(I+1)] \quad [5\text{-}55]$$

where ϵ is the angle between the symmetry axis of the electrical field gradient and the z direction, or the H_o direction. Let us again consider D_2O in the two layer hydrate of Na vermiculite (5).

Application. Each of the deuteron nuclei in D_2O can be treated separately since the dipolar D-D interaction is much weaker than the interaction represented by E_Q. In other words D_2O may be considered as composed of two separate OD bonds. I for the deuteron nucleus is 1, and thus applying equation [5-55] where m = 1, 0 or −1, one gets three energy levels.

$$m = 1 \qquad -\gamma\hbar H_o + \frac{e^2 qQ}{8} (3\cos^2\epsilon - 1)$$

$$m = 0 \qquad 0 - \frac{e^2 qQ}{4} (3\cos^2\epsilon - 1)$$

$$m = -1 \qquad -\gamma\hbar H_o + \frac{e^2 qQ}{8} (3\cos^2\epsilon - 1)$$

and thus transitions corresponding to two peaks at

$$-\gamma\hbar H_o \pm 3/4\, e^2 qQ\, (3\cos^2\epsilon - 1)$$

These two peaks are shown in Fig. 5-13. Usually when quadrupolar interactions are considered $e^2 q Q$ is defined in terms of a quadrupole coupling constant QCC such as $e^2 q Q = h \times QCC$. To be symmetric with the equations used for H_2O, let $\alpha' = 3/8$ QCC. Then the ^2H doublet splitting is

$$h' = 2\alpha' (3\cos^2\epsilon - 1) \quad [5\text{-}56]$$

under the restriction that the deuteron experiences an axially symmetric electrical field gradient.

Starting from this relationship, the set of equations [5-48, 5-49] may be repeated, resulting in

$$h' = \alpha' (3 \cos^2 \gamma' - 1)(3 \cos^2 \psi - 1)(3 \cos^2 \delta - 1) \qquad [5\text{-}57]$$

where γ' is the angle between $\vec{I}$ (the rotation axis) and the symmetry axis of the electrical field gradient. It is assumed as before that the molecule experiences free rotation around $\vec{I}$. It is observed that the doublet splitting (h') decreases by a factor of 2 when δ goes from 0° to 90°. At 55° a single line is observed while at room temperature $h' = 98$ G for $\delta = 0$.

Neither the sign nor the absolute value of the quadrupole coupling constant are known. In hydrates with oxygen-oxygen distances between water molecules of about 2.8 Å, QCC = 250 kc (Reeves (41)), and $\alpha' = 143.7$ G. From equations [5-4] and [5-7], for a specified value of cos δ:

$$\frac{h'/\alpha'}{h/\alpha} = \frac{(3 \cos^2 \gamma' - 1)}{(3 \cos^2 \gamma - 1)} \qquad [5\text{-}58]$$

Since only one doublet is observed each of the OD bonds must be tilted with respect to $\vec{I}$ by the same angle. This implies $\gamma = 90°$, in agreement with the model previously suggested, see Fig. 5-14. It follows that $3 \cos^2 \gamma' - 1 = 1.37$ and $\gamma' = 27°$. The symmetry axis of the electrical field in D_2O is not oriented along the OD axis since the DOD angle is 104° in isolated molecules. A similar discrepancy has been observed by Woessner *et al.* (50, 51) by comparing the doublet splitting of H_2O and D_2O in oriented Na hectorite. Probably the perturbation produced by the rapid reorientation (Chiba (9)) and/or by neighboring cations on the D_2O molecules in the hydration shell effects the electrical field inside the molecule. Accordingly the electrical field gradient experienced by the deuteron nucleus and the value of the QCC should be modified.

<u>^{7}Li resonance in hectorite.</u> Another domain of interest for clay surfaces studies is the resonance study of exchangeable cations such as $^{23}Na^+$ and $^7Li^+$. $^{23}Na^+$ resonance in zeolite has been observed by H. Lechert and Henneke (30) both in the hydrated and dehydrated states. $^7Li^+$ resonance in hectorite has been investigated by J. Conard (11). These results complement the proton resonance data of adsorbed water. By applying equation [5-55] and recalling that the spin number I of $^7Li^+$ is 3/2, one gets the distribution of the Zeeman energy levels and of their shift due to the quadrupole splitting shown in Fig. 5-18.

Note that the central line is unaffected by the quadrupole splitting at least at first order. For $m = \pm 1/2$, $3m^2 - I(I+1) = 0$. The 1/2 to −1/2 transition is thus insensitive to the value of the electrical field gradient and thus to crystalline strains that would affect the symmetry of the electrical field. By contrast the difference in frequency between the central line and the two satellites is a function of QCC (equation [5-55]). In addition the magnitude of this difference may shed light on the symmetry of the crystal site occupied by Li^+. Fig. 5-19 shows the $^7Li^+$ NMR spectra obtained at 193° K, 273 and 293° K by J. Conard (11) for a Li hectorite at rather low water content (~ 5% by weight). An axially symmetric field gradient is responsible, at 293° K, for the near theoretical three-lines spectrum. As temperature decreases, the two side lines broaden and they shift away from the central line.

Finally at 190° K only the central line is observed. This transformation indicates either that the symmetry is progressively lost at the cation site or that the electrical field gradient increases as the temperature decreases. This would suggest a change in the arrangement of the water molecules or a change in the rate of their motions, affecting the net value of the electrical field gradient.

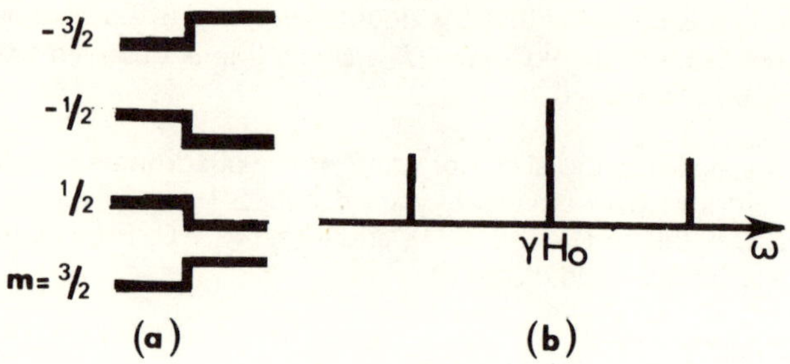

Figure 5-18. (a) Effect of a quadrupole coupling in first order. The shifts of all levels for I = 3/2 have the same magnitude. (b) Spectral absorption corresponding to the energy levels of (a). The central line is unaffected by the quadrupole coupling in first order. (Slichter, 1963).

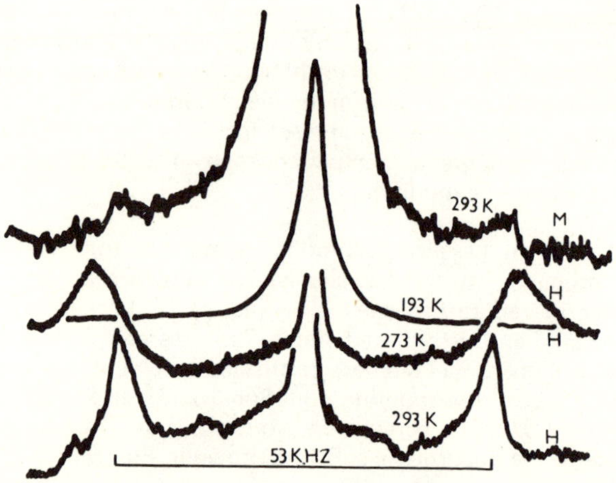

Figure 5-19. NMR ^{7}Li spectra of the Li hectorite (H) and montmorillonite (M). Fourier transform of the free precession at 34.9 MHz. A pure axial gradient, without protons' dipolar widening, is shown at room temperature. At 0° C some asymmetry is shown while the mean gradient grows up because of the thermal contraction of the hydrate. At −80° C no gradient can be measured probably because the dipolar field of protons is no longer averaged. In montmorillonite the width of the central line results from the high iron content (Conard, 1976, Magnetic Resonance in Colloid and Interface Science).

5-3.3. Polycrystalline Materials: The Second Moment

Polycrystalline materials have large surface areas, making them particularly adaptable for NMR studies of adsorption processes. Because a great many crystal grains have their individual crystal axes distributed randomly, the characteristic fine structure of the resonance spectrum is smeared out: the spectrum of the powder is the sum of the spectra of individual particles.

Consider dipolar interactions and suppose that the orientation of a dipolar pair is random: the fraction of pairs with θ contained in the interval $d\theta$ is $d(\cos\theta)$ (Fig. 5-20).

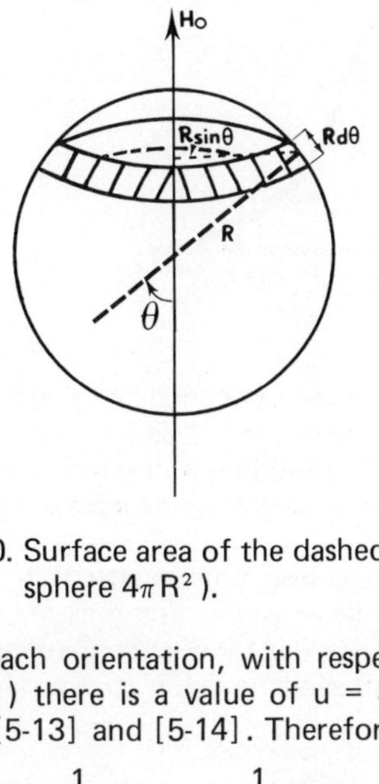

Figure 5-20. Surface area of the dashed zone: $2\pi R^2 \sin\theta \, d\theta$, (Surface area of the sphere $4\pi R^2$).

At each orientation, with respect to the magnetic field (equations [5-46] and [5-47]) there is a value of $u = \omega - \omega_o$ and thus a value of function f in equations [5-13] and [5-14]. Therefore

$$f(u)\, du = \frac{1}{2} \sin\theta \, d\theta = \frac{1}{2} d \cos\theta \qquad [5\text{-}59]$$

but

$$u \propto \alpha \, (3\cos^2\theta - 1) \qquad [5\text{-}60]$$

or

$$d\,u \propto \cos\theta \, d\theta \qquad [5\text{-}61]$$

Using [5-59] and [5-61] and expressing $\cos\theta$ as a function of u through equation [5-60], one gets:

$$f(u) \propto (1 \pm \frac{u}{\alpha})^{-1/2} \qquad [5\text{-}62]$$

where α has the same meaning as in equation [5-48]. The sign is + for $(2\alpha < u < \alpha)$ and – for $(\alpha < u < 2\alpha)$.

Instead of a spectrum composed of two lines, there is, as illustrated in Fig. 5-21 for gypsum, a function with two extremes occurring for $u/\alpha = \pm 1$, and a smoothed function with two maxima when the pair of protons is no longer isolated and when intermolecular interactions are considered. Such a transformation of a peak spectrum into a function with extremes is also observed for the powder spectrum obtained for ^{7}Li hectorite at 293°K in Fig. 5-19, whereas the transformation into a smoothed function is expected when intra molecular interactions are important.

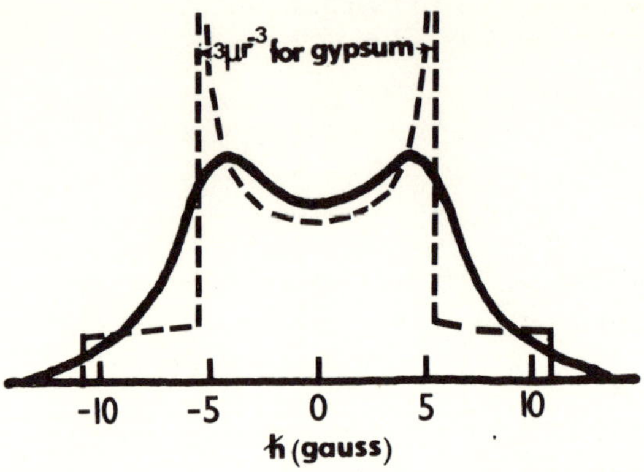

Figure 5-21. The broken line shows the calculated resonance line shape for the protons in polycrystalline gypsum $CaSO_4, 2H_2O$ taking into account nearest neighbor interactions only. The full line is obtained after taking into account the interaction of other neighbors. (Andrew, 1958).

When more than 2 protons are interacting, the equations to express the dipolar perturbation Hamiltonian become increasingly complicated, and thus a complicated line structure is predicted. Broadening by next neighbors removes the maxima and, when averaged over all orientations, a wide structureless band remains.

Although the line shape cannot be calculated, Van Vleck (48) has shown that the moments of the spectrum, specifically the second moment, can provide interesting information, i.e. the distances between interacting dipoles. By definition, the nth moment of the spectral function f (u) is

$$<u^n> = \frac{\int_0^\infty u^n \, f(u) \, du}{\int_0^\infty f(u) \, du} \, . \qquad [5\text{-}63]$$

For n = 2, equation [5-63] is called the second moment. $S_2 \equiv <u^2>$ is comparable to the mean-square of the line width

$$S_2 \propto \gamma^2 < h_{loc}>^2 .$$

For systems with one kind of spin,

$$S_2 = \frac{3}{2}\gamma^4 \hbar^2 I(I+1) \frac{1}{N} \sum_{j>k} \frac{(1-3\cos^2\theta_{jk})^2}{r^6_{jk}} \qquad [5\text{-}64]$$

This means that the summation is made by considering each spin (say spin j) as the origin of the r_{jk} vectors oriented with angle θ_{jk} with respect to H_o and by dividing by the number of spins over which the sum is taken. For a polycrystalline sample, the second moment is the average of the second moment of the individual grains. Since the isotropic average of $(3\cos^2\theta_{jk} - 1)^2$ is 4/5, equation [5-64] becomes

$$S_{2\ \mathrm{powder}} = \frac{6}{5}\gamma^4\hbar^2 I(I+1)\frac{1}{N}\sum_{j>k}\frac{1}{r^6_{jk}} \qquad [5\text{-}65]$$

If the system contains unlike spins, it is necessary to go back to equation [5-41] and to the meaning of the various operators it contains in order to calculate the local field. Recall that for like spins, only operators A and B need be considered. For unlike spins only A remains. B flips two spins + – into – +. If the two spins are identical, this lifts up the degeneracy, whereas for unlike spins, it produces second order energy shifts and gives rise to weak transitions. Since we are concerned with the width of the main transition, we can exclude these satellites.

Keeping the A operator for unlike spins, I, then S is a dipolar perturbation Hamiltonian of the type

$$\mathcal{H}'_{I,S} = \gamma_I\gamma_S\hbar^2 \sum_{k,1}\frac{(1-3\cos^2\theta_{k1})}{r^3_{kl}} I_{zk}S_{z1} \qquad [5\text{-}66]$$

where I_z and S_z are the z components of spin operators I and S, respectively. The corresponding term of the second moment is

$$S_{2I,S} = \frac{1}{3}\gamma_I^2\gamma_S^2\hbar S(S+1)\frac{1}{N}\sum_{j,k}\frac{(1-3\cos^2\theta_{jk})^2}{r^6_{jk}} \qquad [5\text{-}67]$$

averaging as

$$S_{2I,S\ \mathrm{powder}} = \frac{4}{15}\gamma_I^2\gamma_S^2\hbar S(S+1)\frac{I}{N}\sum_{j,k}\frac{1}{r^6_{jk}} \qquad [5\text{-}68]$$

for a powder.

Application – Kaolinite and Boehmite. The last equation is important for the calculation of the second moment of the structureless signal displayed by alumina such as boehmite or by a crystalline aluminosilicate, such as kaolinite.

Consider first the proton second moment (see Gastuche and Toussaint (21)) calculated for this clay by accounting for the proton-proton (I-I) and proton aluminum interaction (I-S) and the lattice parameters proposed by Brindley and Nakahira (7) for kaolinite. The calculated value is 4.55 gauss^2. The experimental "second moment" obtained for a kaolinite that is very low in paramagnetic impurities (Fe^{3+}), agrees reasonably with this value. If kaolinite is progressively dehydroxylated, the second moment should change, if the proton-proton distance was modified.

The time required for nucleation and growth of the dehydroxylation nuclei is very short, compared with the time required for a water molecule to diffuse inside the lattice toward a reaction interface. In other words, dehydroxylation proceeds by successive destruction of complete octahedral layers and the reaction probability is proportional to the amount of unreacted material. These conclusions agree with the kinetic studies which have shown that nucleation and growth of the nuclei affect complete crystal domains.

The next example of an application is that of boehmite since it shows an interesting temperature study of the second moment (S_2) of the structureless wide band displayed by this mineral (20). In the 250° K temperature region, shown in Fig. 5-22, the observed second moment is between 16.2 and 16.7 gauss^2, as compared to 18 ± 1 gauss^2 found by Holm *et al.* (26) in the same temperature range. In the high temperature region S_2 decreases progressively on heating and reaches approximately 12 gauss^2 at the temperature where the dehydroxylation process becomes noticeable. Below 400° C the variation of S_2 is reversible.

Using Van Vleck's (41) relationship for a polycrystalline sample and extending the calculation to the 28 unit cells, the following rigid lattice values were obtained on the basis of the structure parameters proposed by Fripiat *et al.* (17) (see Fig. 5-23 for the structure): (i) assuming a linear OH. . .O hydrogen bond: S_2 = 19.82 gauss^2. This result compares well with the value of 19.3 gauss^2 computed by Holm *et al.* (26) based on slightly different lattice parameters; (ii) assuming a OH. . .bond angle of 12°: S_2 = 14.5 gauss^2. According to Fripiat *et al.* (17), the intensity ratio of the two infra-red OH bending modes cannot be accounted for without allowing a deviation of about this magnitude from the linear OH---O bond: S_2 =

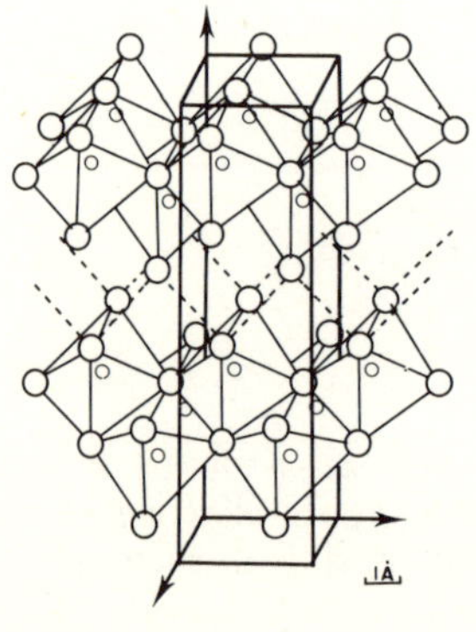

Figure 5-22. Temperature dependence of the second moment of the wide band. □, low temperature insert; ○, high temperature insert. The measurements were carried out at the indicated temperatures. (Fripiat *et al.*, 1967).

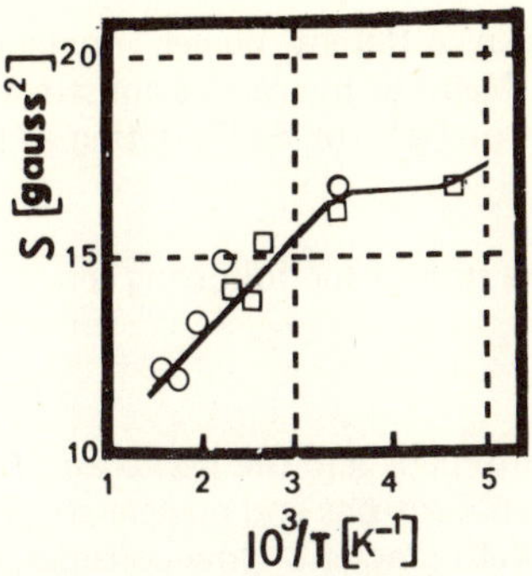

Figure 5-23. Structure of boehmite: small circle, aluminum; large circle, oxygen or hydroxyl; dashed line, hydrogen bonds. (Fripiat *et al.,* 1967).

It may be concluded that the second moment, obtained at low temperature, is close to that of the ideal hydrogenic structure of boehmite.

The decrease observed above 300°K could be attributed to the influence of the vibration modes or the spin motions in the lattice or both.

Pedersen (37) has shown that for isolated p-p vectors, the magnetic dipole interaction term ($3\cos^2\theta - 1$) is reduced by the bending vibrations by a factor: $f_R = [1 - 3/2(<\theta_b^2> + <\theta_c^2>)]$ where $<\theta_b^2>$ and $<\theta_c^2>$ are the square mean amplitudes of the two components of the p-p vector radial oscillation with respect to the reference Cartesian axes and θ, the angle between one of these axes and the steady magnetic field. For a polycrystalline specimen, $S_2 = f_R^2\ S_2$ rigid structure, where the p-p vector length is not appreciably modified. The square mean amplitude is derived from a relationship applied by Chiba (9).

$$<\theta_b^2> = \frac{h\upsilon_b}{f_b}\left\{\frac{1}{2} + (\exp\frac{h\upsilon_b}{RT} - 1)^{-1}\right\} \qquad [5\text{-}69]$$

where f_b is the restoring force constant and υ_b is the vibration frequency.

In boehmite, two adjacent protons in the chain contribute approximately 80% to the second moment: hence, the Pedersen's relationship may be considered as a good approximation. From the IR spectra $<\theta_{b,c}^2>$ can be calculated using equation [5-69]. At 300°K, $f_R^2 = 0.921$, whereas at 430°K, $f_R^2 = 0.892$. Between 300 and 430°K, the experimental second moment decreases by 19% from 16.5 gauss² to 13.4 gauss². Such a variation cannot be accounted for by the influence of the vibrational motions. It may be concluded that the narrowing of the wide line at increasing temperature involves the influence on the Zeeman Hamiltonian by spin motions of larger amplitudes than those of the vibration modes, for instance, a proton diffusion.

5-3.4. Interaction with Paramagnetic Centers

In addition to homodipolar interactions between identical magnetic nuclei, and to heteropolar interactions between different nuclei, nuclei-paramagnetic centers interactions are often important. Many of the earth's constituents like clays and micas contain rather large amounts of Fe^{3+} or Fe^{2+}, acting as paramagnetic centers.

The total Hamiltonian $\mathcal{H}$ is thus the sum of the following terms:

$$\mathcal{H} = \mathcal{H}_Z + \mathcal{H}_D + \mathcal{H}_P + \mathcal{H}_Q \quad [5\text{-}70]$$

where $\mathcal{H}_Z$ represents the Zeeman Hamiltonian, $\mathcal{H}_D$ the homo and hetero-dipolar contributions, $\mathcal{H}_P$ the interaction between electrons and nuclear spins, and $\mathcal{H}_Q$ the quadrupolar interaction with electrical field gradient. Now consider $\mathcal{H}_P$ that may originate from the interaction between a proton of a lattice OH group and the Fe^{3+} impurities, or between an electronic spin of an unpaired electron in a free radical and a proton belonging to that free radical.

The electronic magnetic moment

$$\vec{\mu}_p = \beta\, g \vec{S} \quad [5\text{-}71]$$

is about 10^3 times larger than the nuclear magnetic moment. However, under the usual conditions $\vec{\mu}_p$ reorients very rapidly in H_o with a correlation time τ_{ep} which is usually short with respect to the nucleus precession frequency and $\vec{\mu}_p$ must be time-averaged,

$$<\mu_p> = T_r|\rho,\mu_p|. \quad [5\text{-}72]$$

T_r is the sum of the diagonal elements of the matrix and ρ is the density matrix of the electronic states $\mathcal{H}_e = -g\beta S_Z H_o$ where S_Z is the Z component of $\vec{S}$, H_o being along the Z direction:

$$\rho = \exp(-\mathcal{H}_e/kT)/\sum_n \exp(-\mathcal{H}_e/kT) \quad [5\text{-}73]$$

At high temperatures

$$\overline{\mu}_{pz} = \frac{\beta^2 g^2 S(S+1) H_o}{3k(T-\theta)} \quad [5\text{-}74]$$

$$\overline{\mu}_{px} = \overline{\mu}_{py} = 0 \quad [5\text{-}75]$$

The lower the H_o and higher the temperature, the lower is $\overline{\mu}_{pz}$. The Curie temperature accounts for the weak interaction between the electronic spins. Usually θ is small and:

$$\overline{\mu}_{pz} \simeq \frac{\mu_p^2 H_o}{3kT} \quad [5\text{-}76]$$

As long as the nuclear (μ_n) and electron moments are far enough apart, their interaction is that of a pair of magnetic dipoles, the Hamiltonian being similar to $\mathcal{H}_D$ (see equation [5-40]) namely

$$\mathcal{H}_p = \frac{\vec{\mu}_p \vec{\mu}_n}{r^3} - 3 \frac{(\vec{\mu}_p \vec{r})(\vec{\mu}_n \vec{r})}{r^5} \qquad [5\text{-}77]$$

where r is the distance from the nucleus to the electron. This would be the only contribution to $\mathcal{H}_P$ as long as the electronic wave functions are zero at the nucleus (p or d state).

For the s state the electron wave-function is non zero at the nucleus and r is so small that the dipole approximation has broken down.

The effective energy which must then be taken into account is

$$E = -\vec{k}\,\overline{H}_Z\,\vec{\mu}_p \qquad [5\text{-}78]$$

where $\overline{H}_Z$ is the magnetic field in the Z direction due to the nucleus, and averaged over the electron orbital probability density. Slichter (45) has given a non-relativistic demonstration for $\overline{H}_Z$ showing that $\overline{H}_Z = \frac{8\pi}{3}\vec{\mu}_n\, u^2(O)$ where $u^2(O)$ is the square of the s electron wave function at the nucleus. Leaving the detail to the reader, the demonstration starts by considering the magnetic field generated by a current loop representing the nucleus charge turning in a circular path. The coupling Hamiltonian corresponding to the effective interaction energy, $E = -\frac{8\pi}{3}\vec{\mu}_p\,\vec{\mu}_n\, u^2(O)$, is conveniently represented as:

$$\mathcal{H} = -\frac{8\pi}{3}\vec{\mu}_p\,\vec{\mu}_n\,\delta(\vec{r}) \qquad [5\text{-}79]$$

where $\delta(\vec{r})$ is the Dirac function. This is the so-called Fermi contact interaction and it must be introduced into the total Hamiltonian.

5-4. RELAXATION MECHANISMS

How do the nuclei arrive at their thermal equilibrium? How does the spin system come to equilibrium with its environment, called the lattice? Different factors influence this process: a) the frequency distribution of molecular motions, and b) the spin-lattice interactions.

5-4.1. The Frequency Distribution of Molecular Motions

Molecular motions are distributed among a wide range of frequencies. Intuitively one feels that motions which are either very slow or very rapid with respect to the nuclear precession frequency will not appreciably affect the spin system. An example of a motion that is too rapid is an internal mode of vibration in a molecule (frequency in the range of $10^{13}-10^{14}$ Hz.).

Consider a motion which occurs, for instance, with a periodicity of about 10^{-8} sec. This means that a molecule remains in some state of motion for 10^{-8} sec. After this time, it suffers a collision which changes the state of motion for another 10^{-8} sec.

This complex motion can be analyzed using the Fourier method which allows one to know the various frequencies involved in it and their intensities.

The complex waveform, measured as a function of time is called the time domain function, while the spectrum of components with increasing periodicity is called the frequency domain function. The function f(t) which describes the complex motion can be usually represented by a Fourier series, i.e. an infinite series of sines and cosines:

$$f(t) = \sum_{n=0}^{\infty} A_n \cos(n\pi/T)t + \sum_{n=0}^{\infty} B_n \sin(n\pi/T)t \qquad [5\text{-}80]$$

A_n and B_n are the components of the spectrum. If f(t) is an even function of time (f(–t) = f(t)) all B_n = O. If f(t) is an odd function, the sine series only is needed. In equation [5-80] $\frac{n}{T}$ are the frequencies.

The Fourier theorem allows one to pass from the time domain to the frequency domain function. If $F(\omega)$ is the frequency domain function

$$f(t) = \frac{1}{2\pi} \int_{-\infty}^{+\infty} F(\omega) \exp(i\,\omega\,t)\, d\omega \qquad [5\text{-}81]$$

whereas the Fourier transform of f(t) is

$$F(\omega) = \int_{-\infty}^{+\infty} f(t) \exp(-i\,\omega\,t)dt. \qquad [5\text{-}82]$$

The Fourier sine or cosine transform is

$$2 \int_{o}^{\infty} f(t) \sin \omega t\, dt, \text{ or } 2 \int_{o}^{\infty} f(t) \cos \omega t\, dt.$$

To the time domain function f(t), there is a corresponding $F(\omega)$ the frequency domain function and thus the frequency domain function has some intensity in the region were $\omega = \omega_o$, the resonance frequency. It is this intensity which will affect the relaxation.

To clarify this relationship, consider the motion of a molecule carrying one or more nuclear spins. The molecular motion generates a fluctuating magnetic field in the space where the molecule is moving. The fluctuation of the magnetic field may have some component with an appreciable intensity at frequency ω_o and even at 2 ω_o. Thus the fluctuating magnetic field arising from the molecular motion behaves as the magnetic field H_1 generated for NMR measurement and it may be a source of relaxation under certain conditions. These conditions are related to the interaction of the fluctuating field with the magnetic moments.

5-4.2. Interaction between the Fluctuating Field and the Nuclear Magnetic Moment

Let us call $\vec{h}$ the fluctuating field

$$\vec{h} = \vec{i}\,h_x + \vec{j}\,h_y + \vec{k}\,h_z. \qquad [5\text{-}83]$$

Consider the interaction of $\vec{h}$ with some magnetization vector $\vec{M}$ originating from an ensemble of spins submitted to the fluctuating field. As previously discussed the torque acting on $\vec{M}$ is the vectorial product

$$\frac{d\vec{M}}{dt} = -\gamma \vec{h} \wedge M. \qquad \text{(see equ. [5-3])}$$

In the rotating frame of reference, the torque is conveniently expressed by

$$(-\gamma)x \begin{vmatrix} h_{x'}, M_{x'}, \vec{i'} \\ h_{y'}, M_{y'}, \vec{j'} \\ h_{z'}, M_{z'}, \vec{k'} \end{vmatrix} = \{\vec{i'}(h_{y'} M_{z'} - M_{y'} h_{z'}) + \vec{j'}(h_{z'} M_{x'} - M_{z'} h_{x'}) + \vec{k'}(h_{x'} M_{y'} - M_{x'} h_{y'})\}(-\gamma) \qquad [5\text{-}84]$$

Thus $h_{x'}$ is acting on $M_{y'}$ and $M_{z'}$; $h_{y'}$ is acting on $M_{z'}$ and $M_{x'}$; and $h_{z'}$ is acting on $M_{y'}$ and $M_{x'}$. Assume that we have displaced $\vec{M}$ from its equilibrium position by a R.F. pulse. In what way can $\vec{h}$ restore the initial state? $\vec{h}$ has contribution at all frequencies from $0 \to \infty$. Note that the zero frequency component, or the static value of $\vec{h}$, must be along the z' axis which remains by definition continuously parallel to the z axis of the laboratory frame.

On the other hand we see that $M_{y'}$ and $M_{x'}$ which are responsible for the T_2 relaxation process in the x, y plane are affected by $h_z = h_{z'}$ only. The situation is different for $h_{y'}$ and $h_{x'}$ since both of them interact with $M_{y'}$ or $M_{x'}$ *and* $M_{z'}$. Thus $h_{y'}$ and $h_{x'}$ will affect the T_2 *and* the T_1 relaxation mechanisms.

In other words the T_1 relaxation mechanism depends upon the magnitude of $h_{x'}$ and $h_{y'}$, whereas the T_2 relaxation mechanism is a function of the magnitude of the three components $h_{x'}$, $h_{y'}$ and $h_{z'}$. A consequence of this is that T_2 contains a frequency independent term whereas T_1 is necessarily frequency dependent since both $h_{y'}$ and $h_{x'}$ are moving with the rotating frame. Thus T_1 will be affected by frequency components (in the Fourier spectrum) which are at higher or lower frequency than the rotating frame.

As shown previously T_1^{-1} is fundamentally a direct function of the transition probability between the energy levels (equation [5-23]).

The magnitude of this probability is brought about by the interaction with the lattice, which is symbolized in equation [5-84] by the product of the components of the local field by the components of the magnetization vector of the considered spin system.

Consider an important fact that a local field may have various origins. It may be created by various kinds of nuclei bearing magnetic moments and/or by electronic spin. Consider for instance a proton moving within the field produced by a static paramagnetic center. Its own motion may modulate the field in the right range of frequency and be a source of relaxation.

For spins with quadrupole moments, a similar cause of modulation of the electrical field gradient may also be at the origin of relaxation. As the molecule reorients, the components of the quadrupole coupling tensor become random functions of time and provide another source for relaxation.

We may thus conclude that the spin-lattice relaxation time *and* the frequency dependent contribution to the spin-spin relaxation time are going to depend not only on the intensity of the Fourier component (in the right frequency domain) but also on the strength and on the nature of the coupling between the spin system and its neighbors.

The line shape contains information on the local field seen by the observed nucleus: the relaxation times are directly related to the molecular motions. We will consider how this information can be extracted from the experimental T_1 and T_2.

5-4.3. The Auto-Correlation Function

A function y(t) is said to be a random function of the time if a function p(y,t) can be defined which describes the probability for function y to have a specified value at time t. Then the average value of y is:

$$< y > = \int y\, p(y,t)\, dy. \qquad [5\text{-}85]$$

If a function f(y) is also a random function of the time its average value is

$$< f(y) > = \int f(y)\, p(y,t)\, dy. \qquad [5\text{-}86]$$

Generally speaking, in physical processes such as molecular motions, there frequently exists some correlation between the values of y at time t_1 and t_2. Let us call $p(y_1, t_1; y_2, t_2)$, the probability function which permits y to take the value y_1, at time t_1 *and* the value y_2 at time t_2, and $P(y_1, t_1; y_2 t_2)$ the conditional probability function which forces y to take the value y_2 at time t_2 if y was equal to y_1 at time t_1.

$$p(y_1, t_1; y_2, t_2) = P(y_1, t_1; y_2, t_2)\, p(y_1, t_1) \qquad [5\text{-}87]$$

By definition the auto-correlation function of the random function f(y) is

$$G(t_1, t_2) = <f(t_1) \times f(t_2)> = \iint p(y_1, t_1; y_2, t_2)\, f(y_1)\, f(y_2)\, dy_1\, dy_2 \qquad [5\text{-}88]$$

if $G(t_1, t_2) = G(t_2 - t_1) = G(\tau)$, then

$$G(\tau) = \iint p(y_1, y_2, \tau)\, f(y_1)\, f(y_2)\, dy_1\, dy_2 \qquad [5\text{-}89]$$

This means that the function G is independent of the origin of the time. This series of definitions and of equations is important for the calculation of the transition probability from state (1) to state (2).

5-4.4. Calculation of the Transition Probability Between Two States

Because of the molecular motions, the local magnetic field to which a magnetic nucleus is exposed is a function of time. The interaction energy is also between this nucleus and the field and therefore the Hamiltonian operator corresponding to this energy. (If we are dealing with stationary perturbation, the ensemble average of the Hamiltonian is equivalent to time average). Consider a system of spin which goes from one eigen state n to state m under the influence of a time dependent perturbation.

According to a classical relationship of quantum mechanics, the first derivative with respect to the time of the probability for a system which is at time t = 0 in state n, to be at time t in state m is

$$\frac{d\,Cm}{dt} = -\frac{i}{\hbar} Cn < \psi^{o}_{m} \mid A \mid \psi^{o}_{n} > F(t) \exp i(\omega)t \qquad [5\text{-}90]$$

where $\omega = (Em - En)/\hbar$, and where ψ_m and ψ_n are the time independent wave functions.

The time dependent Hamiltonian is decomposed into a time independent and a time dependent function A and F(t), respectively.

$$\mathcal{H}(t) = A\,F(t) \qquad [5\text{-}91]$$

We are looking for the transition probability which is, at time t′, Cm*Cm, or which is per unit time, $W_{n,m}$, defined as

$$W_{n,m} = Cm \frac{dCm^*}{dt} + Cm^* \frac{dCm^{\cdot}}{dt} \qquad [5\text{-}92]$$

Cm is obtained by integrating equation [5-90]

$$Cm = -\frac{i}{\hbar} Cn < \psi^{*}_{m} \mid A \mid \psi_{n} > \int_{o}^{t'} F(t') \exp(i\omega t')\,dt' \qquad [5\text{-}93]$$

Then

$$Cm^* \frac{dCm}{dt} = [+\frac{i}{\hbar} Cn^* < \psi_m \mid A \mid \psi^*_n > \int_{o}^{t'} F(t') \exp(-i\omega t')dt'] \times$$

$$[-\frac{i}{\hbar} Cn < \psi^*_m \mid A \mid \psi_n > F(t) \exp(i\omega t)]$$

$$Cm^* \frac{dCm}{dt} = \frac{1}{\hbar^2} Cn^* Cn \overline{< \psi_m \mid A \mid \psi_n >}^{2} \int_{o}^{t'} F^*(t') F(t) \exp(-i\omega(t' - t))dt'$$

$$Cm \frac{dCm^*}{dt} = \frac{1}{\hbar^2} Cn^* Cn \overline{< \psi_m \mid A \mid \psi_n >}^{2} \int_{o}^{t'} F(t) F^*(t') \exp(+i\omega(t' - t))\,dt'$$

$F^*(t')\ F(t)$ can be considered as an auto correlation function $G(t'-t) = G(\tau)$ if $\tau = t-t'$. At the steady state $G(\tau) = G(-\tau)$.

Combining the two equations above,

$$W_{n,m} = \frac{1}{\hbar^2}\ Cn^*Cn\ \overline{<\psi_m|A|\psi_n>^2}\ \{-\int_o^{-\tau} G(\tau) \exp i\omega\tau\ d\tau$$

$$+ \int_o^{\tau} G(\tau) \exp i\omega\tau\ d\tau\}$$

or

$$W_{n,m} = \frac{1}{\hbar^2}\ Cn^*Cn\ \overline{<\psi_m|A|\psi_n>^2} \int_{-\tau}^{\tau} G(\tau) \exp i\omega\tau\ d\tau\ . \qquad [5\text{-}94]$$

In this equation Cn^*Cn is unity at time $t = 0$. $\overline{<\psi_m|A|\psi_n>^2}$ is the observable corresponding to the time independent operator A. If we consider as usual the average value of $W_{n,m}$ over a length of time much longer than ω^{-1}, the integration limits are replaced by $-\infty$ and $+\infty$. $J(\omega)$ being the Fourier transform of $G(\tau)$, equation [5-94] becomes

$$\overline{W}_{n,m} = \frac{1}{\hbar^2} \overline{<\psi_m|A|\psi_n>^2}\ J(\omega) \qquad [5\text{-}95]$$

In the more general case where there spin functions are operated by several operators A, B etc. . .equation [5-95] may be generalized as

$$\overline{W}_{n,m} = \frac{1}{\hbar^2} \left\{ \overline{<\psi_m|A|\psi_n>^2} + \overline{<\psi_m|B|\psi_n>^2} + \ldots \right\} J(\omega)$$

Since $T_1^{-1} = \frac{1}{2} W_{n,m}$ then T_1 is obtained from that general equation which also contains the interaction term, { }, and the frequency domain function $J(\omega)$. In summary, equation [5-95] expresses in a quantitative manner the two types of parameters ruling T_1, namely the strength of the coupling and the intensity of the spectrum of frequency associated with the molecular motions.

Application to the motion of a water molecule in the liquid state. In order to provide material support to the above calculations, we will use a classical example, namely the relaxation in a liquid made from small molecules in which the main interaction is internal to the molecule. Water is a good example of the early application of the theory.

It is convenient in order to calculate $W_{n,m}$ in that case to introduce functions related to the sperical harmonics.

$$\left.\begin{aligned} Y_o &= r^{-3}\ (1-3\cos^2\theta) \\ Y_1 &= -3/2\, r^{-3} \sin\theta \cos\theta \exp(-i\varphi) \\ Y_2 &= -3/4\, r^{-3} \sin^2\theta \exp(-2i\varphi) \end{aligned}\right\} \qquad [5\text{-}96]$$

In the case of a rigid pair of protons, r is time independent whereas θ and φ become functions of the time with the consequence that Y_0, Y_1 and Y_2 become time dependent. Recall that we use r, θ, φ with the same definitions as in Fig. 5-10. Recall also that the operators A . . . F which are contributing to the perturbation Hamiltonian given by equation [5-41] and which are originating from the local field produced by one proton at the location occupied by its partner, contain the functions Y_0, Y_1 and Y_2 of the molecular coordinates. The time independent part of operators A. . .F in equation [5-41] are going to take the place of operator A in equation [5-94] whereas the correlation function is obtained from Y_0, Y_1 and Y_2. In calculating the time independent part of equation [5-94], only the operators which change the energy states have to be considered. From the scheme represented in Fig. 5-11, it follows that operator C and its conjugated D, and E and its conjugated F fulfill this requirement. (Thus the terms which are neglected in equation [5-45] are of interest here).

To carry out the calculation all spins near that undergoing the transition have to be considered. As far as operator D is concerned, considering a $m' \rightarrow m'-1$ transition, it can be shown that $\langle m'|D|m'-1\rangle^2 = (\frac{3}{2}\hbar^2\gamma^2)^2 \frac{1}{3} I(I+1)(I+m')(I-m'+1)$ whereas for operator F

$$\langle m'|F|m'-1\rangle^2 = 2(\frac{3}{4}\hbar^2\gamma^2)^2 \frac{2}{3} I(I+1)(I+m')(I-m'+1).$$

The average probability for the $m' \rightarrow m' - 1$ transition is

$$\frac{3}{4}\hbar^4\gamma^4 \; I(I+1)(I+m')(I-m'+1)$$

with $m' = 1/2$. If we add this term to the probability for the $m' \rightarrow m' + 1$ transition, we obtain:

$$\langle C+D+E+F\rangle = \frac{3}{4}\hbar^4\gamma^4 \; I(I+1) \qquad [5\text{-}97]$$

This sum represents the term { } in equation [5-94].

To calculate the components of the Fourier spectrum, consider a water molecule as a hard sphere rotating about one axis and diffusing simultaneously. This is the rotatory Brownian motion which has a correlation function $G_\rho(\tau) = \langle Y_\rho^*(t+\tau)\, Y_\rho(t)\rangle$ that can be defined using equation [5-89] and the spherical harmonics in equation [5-96]. The average $\langle\,\rangle$ is over an ensemble of nuclei. $G_i(\tau)$ can then be approximated by:

$$G_\rho(\tau) = G_\rho(o)\exp(-\tau/\tau_c) \qquad [5\text{-}98]$$

where

$$G_\rho(o) = \langle Y_\rho^*(o)\, Y_\rho(o)\rangle \qquad [5\text{-}99]$$

for $\tau = o$. These spherical harmonics are integrated by using a spherical element $\sin\theta \, d\theta \, d\varphi$. ρ represents subscripts 0, 1 or 2 of functions Y in equation [5-96].

$$G_\rho(o) = \langle Y_\rho^2 \rangle = \int_{\varphi=0}^{2\pi} \int_{\theta=0}^{\pi} Y_\rho^2 \sin\theta \, d\theta \, d\varphi \qquad [5\text{-}100]$$

with the results for ρ = 0, 1 and 2

$$G_0(o) = \frac{4}{5} r^{-6}, \; G_1(o) = \frac{2}{15} r^{-6}, \; G_2(o) = \frac{8}{15} r^{-6} \qquad [5\text{-}101]$$

By introducing equation [5-98] into [5-99], the time-dependent function becomes:

$$G_\rho(\tau) = G_\rho(o) \int_{-\infty}^{+\infty} \exp(-\tau/\tau_c) \exp(i\omega\tau) d\tau \qquad [5\text{-}102]$$

The corresponding Fourier transforms are:

$$\left.\begin{aligned} J(\omega=o) &= \frac{8}{5} r^{-6} \tau_c \\ J(\omega=\omega_o) &= \frac{4}{15} r^{-6} \tau_c/(1+\omega_o^2 \tau_c^2) \\ J(\omega=2\omega_o) &= \frac{16}{15} r^{-6} \tau_c/(1+4\omega_o^2 \tau_c^2) \end{aligned}\right\} \qquad [5\text{-}103]$$

In equations [5-98], [5-100] and [5-103], τ_c is the so-called correlation time which is actually a characteristic of the motion. In order to understand its meaning, consider a case where τ/τ_c is very small (τ_c very large): the interactions within the spin system are not going to change appreciably during the observation time τ. On the contrary if $\tau >> \tau_c$, the interactions at the end of the observation are noticeably modified. Thus τ_c may be considered as the life-time of a specified configuration.

Finally, from equations [5-94], [5-97] and [5-103] follows the classical equation of Kubo and Tomita (28) which is valid for water molecules in Brownian motions:

$$T_1^{-1} = \frac{3\gamma^4 \hbar^2}{10 r^6} \left[\frac{\tau_c}{1+\omega_o^2\tau_c^2} + \frac{4\tau_c}{1+4\omega_o^2\tau_c^2} \right] \qquad [5\text{-}104]$$

T_1 passes through a minimum for $\omega_o\tau_c$ = 0.616. In the region where $\omega_o\tau_c << 1$:

$$T_1^{-1} = \frac{3\gamma^4 \hbar^2}{10 r^6} \times 5\tau_c$$

and T_1 is independent of the frequency used in the experiments. In the region where $\omega_o\tau_c >> 1$

$$T_1^{-1} = \frac{3\gamma^4\hbar^2}{10\,r^6} \times \frac{2}{\omega_0^2\tau_c}.$$

T_1 depends upon the square of the frequency.

At constant frequency but variable temperature, T_1 varies with respect to 1/T, as shown in Fig. 5-24, according to the classical theory for random Brownian motion.

$$\tau_c = \tau_c(o)\exp H^*/RT \qquad [5\text{-}105]$$

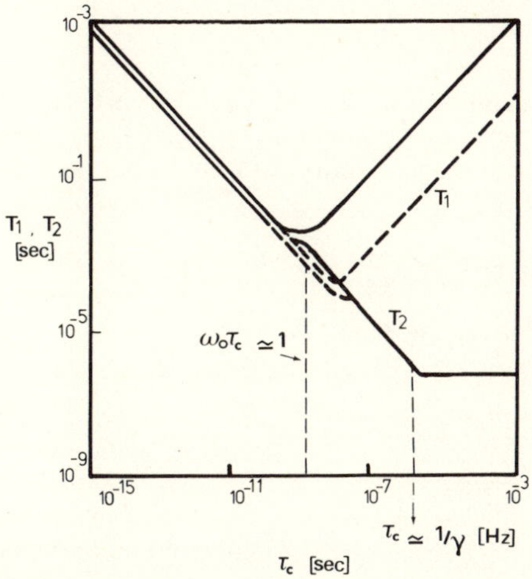

Figure 5-24. Variation of T_1 and T_2 with respect to the correlation time τ_c for a water molecule in Brownian motion in the liquid phase.

where H^* is the activation energy of this motion. Then in the $\omega_0\tau_c \ll 1$ domain, $T_1 \propto \exp(-H^*/RT)$, whereas $T_1 \propto \exp(+H^*/RT)$ in the low temperature domain. The slope of T_1 with respect to 1/T gives H^* and $\tau_c(o)$ can be obtained from the temperature at which the minimum of T_1 is observed. The diffusion coefficient is (Einstein (15)):

$$D = \frac{\langle \ell^2 \rangle}{6\tau_c}$$

where $\langle \ell^2 \rangle$ is the average quadratic jump distance: one may thus obtain the diffusion coefficient from the correlation time.

5-4.5. The Physical Meaning of the Relaxation Times

We have shown earlier (equation [5-95]) the relationship between the transition probability between two energy states and T_1. An illustration of this calculation has been carried out using the isolated water molecule as a model. Another way to calculate the relaxation times can be proposed which shows another aspect of their physical meaning. From equations [5-84], the derivatives of $M_{x'}$, $M_{y'}$, and M_z with respect to the time are

$$\left.\begin{aligned} \frac{dM_{x'}}{dt} &= \gamma(M_{y'}h_{z'} - h_{y'}M_{z'}) \\ \frac{dM_{y'}}{dt} &= \gamma(M_{z'}h_{x'} - h_{z'}M_{x'}) \\ \frac{dM_{z'}}{dt} &= \gamma(M_{x'}h_{y'} - h_{x'}M_{y'}). \end{aligned}\right\} \qquad [5\text{-}106]$$

We will attempt to calculate the evolution of $M_{x'}$, $M_{y'}$ and $M_{z'}$ after a short period of time following a situation, at time $t = 0$, where $M_{z'} = M_{z'}(0)$, $M_{x'} = M_{y'} = 0$. In other words, at the initial time, the magnetization has no component in the y',z' plane.

After a short period of time t_1

$$M_{x'}(t_1) = -\gamma\, M_{z'}(0) \int_0^{t_1} h_{y'}(t)\, dt \qquad [5\text{-}107]$$

From equations [5-106]

$$M_{x'}(t_1) - M_{x'}(o) = \gamma \int_0^{t_1} M_{y'}h_{z'}\, dt - \gamma \int_o^{t_1} M_{z'}\, h_{y'}\, dt$$

and at time t_1, $M_{z'}(t_1) \simeq M_{z'}(o)$ and $M_{x'}(t_1) \simeq o \simeq M_{y'}(t_1)$.

An equation similar to [5-107] can be written for $M_{y'}$

$$M_{y'}(t_1) = \gamma\, M_{z'}(o) \int_o^{t_1} h_{x'}(t)\, dt \qquad [5\text{-}108]$$

This is a first approximation and one may carry out a second approximation by considering that $M_{z'}(t_1)$ is actually slightly different from $M_{z'}(o)$. Consider a period of time T following t_1: at time T

$$M_{z'}(T) - M_{z'}(o) = \gamma \int_o^{T} (M_{x'}h_{y'} - h_{x'}M_{y'})t_1\, dt_1 \qquad [5\text{-}109]$$

We may replace in this equation $M_{x'}(t_1)$ and $M_{y'}(t_1)$ by their values in [5-107] and [5-108]:

$$M_{z'}(T) - M_{z'}(o) = (-1)\, M_{z'}(o)\, \gamma^2 \int_o^{T} dt_1 \int_o^{t_1} [h_{y'}(t)h_{y'}(t_1) + h_{x'}(t)\, h_{x'}(t_1)]\, dt. \qquad [5\text{-}109a]$$

Since $h_{x'}$, $h_{y'}$ and $h_{z'}$ are modulated by molecular motions, there is some correlation between the values of $h_{y'}$ at time t and at time t_1. According to the definition of the correlation function we may then define an auto-correlation function such as $G_1(t_1 t) = \langle h_{y'}(t_1) h_{y'}(t) \rangle$ in agreement with equation [5-88]. Let us call G(y′) and G(x′) the two functions that are contained in the integrand of equation [5-109]. Since we are interested in the evolution with respect to the time of the components of the magnetization in the laboratory frame and not in the rotating frame, equation [5-109] must be transformed into its equivalent in the laboratory frame. If h_x and h_y represent the x and y components of the local field in the laboratory frame respectively:

$$h_{x'} = h_x \cos \omega_o t + h_y \sin \omega_o t$$

$$h_{y'} = -h_x \sin \omega_o t + h_y \cos \omega_o t$$

whereas, $h_{z'} = h_z$. Then the integrand in equation [5-109a] becomes

$$\cos \omega_o (t_1 - t) [h_x(t_1) h_x(t) + h_y(t_1) h_y(t)] -$$

$$\sin \omega_o (t_1 - t) [h_x(t_1) h_y(t) - h_y(t_1) h_x(t)]$$

If there is no correlation between h_x and h_y, only the cosine term remains and:

$$M_z(T) - M_z(o) = (-1) M_z(o) \gamma^2 \int_o^T dt \int_o^{t_1} [G_x(t_1, t) + G_Y(t_1, t)]$$

$$\cos \omega_o (t_1 - t)\, dt.$$

For a Brownian motion:

$$G_x(t_1, t) = \langle h_x(t_1) h_x(t) \rangle = \overline{h_x^2}(o) \exp(-\tau/\tau_c)$$

$$G_y(t_1, t) = \langle h_y(t_1) h_y(t) \rangle = \overline{h_y^2}(o) \exp(-\tau/\tau_c) \quad [5\text{-}110]$$

and thus

$$\frac{M_z(T) - M_z(o)}{M_z(o)} = -\frac{T}{T_1} \quad [5\text{-}111]$$

where

$$T_1^{-1} = \gamma^2 \left[\overline{h_x^2}(o) + \overline{h_y^2}(o)\right] \int_o^\tau \cos \omega_o \tau \exp(-\tau/\tau_c) d\tau. \quad [5\text{-}112]$$

In this manner, T_1 appears as the Fourier transform of the auto-correlation function associated with the mean quadratic value of the components x and y of the local magnetic field.

We could also obtain the time variation of M_y and M_z by a parallel demonstration. For instance:

$$M_{x'}(T) - M_{x'}(o) = -\gamma^2 M_{x'}(o) \int_o^T dt \int_o^t [h_{y'}(t)\, h_{y'}(t_1) + h_z(t_1) h_z(t)]\, dt. \qquad [5\text{-}113]$$

The average value of the integrand is $\frac{1}{2}[h_x(t)h_x(t_1) + h_y(t)h_y(t_1)] \cos \omega_o (t_1 - t) + h_z(t_1)h_z(t)$ and

$$\frac{M_x(T) - M_x(o)}{M_x(o)} = -\frac{T}{T_2} \qquad [5\text{-}114]$$

where

$$T_2^{-1} = \gamma^2 \{\frac{1}{2}[\bar{h}_x^2(o) + \bar{h}_y^2(o)] \int_o^T \cos(\omega_o\tau)\exp(-\tau/\tau_c)d\tau + \bar{h}_z^2(o) \int_o^T \exp(-\tau/\tau_c)d\tau\}. \qquad [5\text{-}115]$$

The last integral may be easily approximated by τ_c, since for all values of $\tau >> \tau_c$:

$$\int_o^T \exp(-\tau/\tau_c)\, d\tau = -\tau_c [\exp + \tau/\tau_c) - 1] \qquad [5\text{-}116]$$

On the other hand: $\int \cos \omega_o \tau \exp(-\tau/\tau_c)d\tau$ is $\tau_c/(1 + \omega_o^2 \tau_c^2)$.

There are two interesting points in equations [5-112] and [5-115]. As discussed earlier (5-4.2) T_2 contains a frequency independent term (the one concerned with $\bar{h}_z^2$). This term is called the secular contribution. It can be shown easily that equation [5-115] predicts that

$$T_2^{-2} \propto \tau_c \qquad [5\text{-}117]$$

Therefore the variations of T_1 and T_2 with respect to temperature are quite different since T_2 decreases in a monotonous manner whereas T_1 passes through a minimum, see Fig. 5-24. Also considering the values of T_1^{-1} and T_2^{-1} derived from equation [5-116], one sees that

$$T_2^{-1} = \gamma^2 \bar{h}_z^2(o)\tau_c + \frac{1}{2} T_1^{-1} \qquad [5\text{-}118]$$

In equation [5-117] and above it was shown that the line width is a function of T_2^{-1}: the smaller T_2, the broader is the line. From equation [5-118], it is observed that the band width contains a non-secular term, T_1^{-1}, which is a function of the probability of transition between the magnetic energy levels. As the probability increases, T_1^{-1} increases and its contribution to the band width decreases.

Also from equations [5-112] and [5-115], put under the form

$$T_1^{-1} = \gamma^2 (\bar{h}_x^2(o) + \bar{h}_y^2(o)) \frac{\tau_c}{1+\omega_o^2\tau_c^2} \qquad [5\text{-}119]$$

$$T_2^{-1} = \gamma^2 [\frac{1}{2}(\bar{h}_x^2(o) + \bar{h}_y^2(o)) \frac{\tau_c}{1+\omega_o^2\tau_c^2} + \bar{h}_z^2(o)\tau_c] \qquad [5\text{-}120]$$

one sees that for $\omega_o^2 \tau_c^2 << 1$

$$\left.\begin{aligned} T_1^{-1} &= \gamma^2 (\bar{h}_x^2(o) + \bar{h}_y^2(o))\,\tau_c \\ T_2^{-1} &= \gamma^2 \left[\frac{1}{2}(\bar{h}_x^2(o) + \bar{h}_y^2(o)) + \bar{h}_z^2(o)\right]\tau_c \\ &\text{or} \\ T_1^{-1} &= T_2^{-1} \end{aligned}\right\} \quad [5\text{-}121]$$

In conclusion, the contributions of the time dependent h_x, h_y and h_z to the relaxation process are different. Thus h_z has the effect of accelerating or of slowing down the precession of the magnetization and effects T_2 only. If h_x and h_y have an important contribution in the Fourier spectrum at frequency ω_o, they have noticeable values in the rotating frame and may then effect the longitudinal (along H_o) or the transverse (in the a,b plane) component of the magnetization. This "classical" demonstration accounts for the existence of the J(o) and $J(\omega_o)$ frequency domain functions (equation [5-103]). The demonstration based on quantum mechanics, in addition, shows that $J(2\omega_o)$ cannot be neglected.

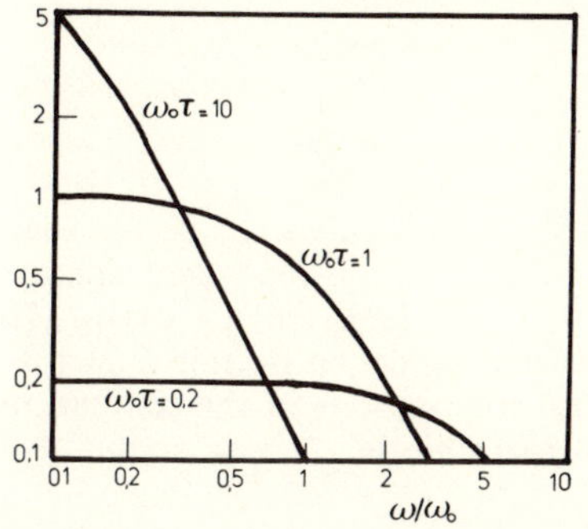

Figure 5-25. Log of a power function against the log of ω/ω_o at three values of $\omega_o\tau$.

Fig. 5-25 represents the variation of the domain frequency function $J(\omega)$ with respect to ω/ω_o for three values of τ_c: $\tau_c^{-1} = 5\omega_o$, $\tau_c^{-1} = \omega_o$ and $\tau_c^{-1} = 0.1$ ω_o respectively. When $\omega_o\tau_c << 1$ or $\tau_c^{-1} >> \omega_o$, the spin-lattice relaxation rate T_1^{-1} becomes very small because the power spectrum $J(\omega)$ (another "nickname" for the domain frequency function) spreads over a large frequency domain. The spin-lattice relaxation rate T_1^{-1} is also small when $\omega_o\tau_c >> 1$ or $\tau_c^{-1} << \omega_o$ because the power spectrum has weaker components in the region $\omega = \omega_o$ and *a fortiori* at $\omega = 2\omega_o$.

5-4.6. The Hetero Dipolar Relaxation

So far we have considered that the system contains one kind of spin only. However, we have studied the shape of the line observed at the resonance of one kind of spin and the influence of another species of nuclei. Such interactions are very general in solids or in the adsorbed state and before studying the application of relaxation techniques to clay minerals, one should consider spin systems with two interacting nuclei I and S that may or may not occur within the same molecule. If they occur in the same molecule, the distance between them is constant.

Suppose the distance r_{ij} between the spins is no longer constant because spins are moving with respect to each other. In equation [5-37], the perturbation Hamiltonian $\mathcal{H}'$ is time-dependent and using equation [5-40] it follows

$$\mathcal{H}'_{IS} = -\gamma_S \gamma_I \hbar^2 \Sigma\left(\frac{\vec{I}_i \vec{S}_j}{r^3_{ij}} - 3\,\frac{(\vec{I}_i \vec{r}_{ij})(\vec{S}_j \vec{r}_{ij})}{r^5_{ij}}\right) \qquad [5\text{-}122]$$

This equation is resolved into 6 terms as $\mathcal{H}'$, namely for two spins I and S

$$\left.\begin{aligned}
(a+b)_{IS} &= \{ I_Z S_Z - \tfrac{1}{4}(I_+S_- + I_-S_+) \} \times Y_0 \\
c_{IS} &= (I_+S_Z + I_Z S_+) \times Y_1 \\
d_{IS} &= (I_-S_Z + I_Z S_-) \times Y_1^* \\
e_{IS} &= (I_+S_+) \times Y_2 \\
f_{IS} &= (I_-S_-) \times Y_2^*
\end{aligned}\right\} \qquad [5\text{-}123]$$

(compare to equations [5-42] to [5-45] where $\vec{I}_1$ is $\vec{I}$ and $\vec{I}_2$ is $\vec{S}$). The Y functions are those defined by equations [5-96] : they are time-dependent since r, θ and φ are functions of the time. Finally: $\mathcal{H}'_{IS} = \gamma_S \gamma_I\ \hbar^2\,(a+b + c+d + e+f)_{IS}$. This perturbation Hamiltonian represents the local field produced by spin S at the location of spin I. This fluctuation is random since the motions of the spin are random. The Zeeman Hamiltonian for the system of two spins is

$$\mathcal{H}_Z = -\gamma_I\,\hbar\,H_o\,I_Z - \gamma_S\,\hbar\,H_o\,S_z \qquad [5\text{-}124]$$

which is the analog of equation [5-38], the total Hamiltonian being $\mathcal{H}' + \mathcal{H}_Z$. For ensembles of i nuclei $\vec{I}_i$ and j nuclei $\vec{S}_j$, in random motion with respect to each other we may average the Y functions as it was proposed by equations [5-98] and [5-99] and thus define an auto correlation function $G_i(\tau)$ such as

$$G_i(\tau) = \langle Y_i^*\,(o)\,Y_i(o)\rangle \exp(-\tau/\tau_c) \qquad [5\text{-}125]$$

Consequently the definition of the power function is unchanged (see equation [5-102]).

$$J_i(\omega) = \int_{-\infty}^{+\infty} G_i(\tau)\exp(-i\omega\tau)\,d\tau \qquad [5\text{-}126]$$

We now want to represent the evolution of the magnetization with time. Along the Z axis we may start from an equation which is similar to equation [5-111] and write in agreement with Abragam (1):

$$\frac{d<I_Z>}{dt} = -\frac{1}{T_{1II}}(<I_Z> - I_o) - \frac{1}{T_{1IS}}(<S_Z> - S_o) \qquad [5\text{-}127]$$

where $<I_Z>$ and $<S_Z>$ represent the contributions of spins $\vec{I}_i$ and $\vec{S}_j$ to the magnetization along the Z axis respectively. T_{1II} and T_{1IS} are different since for T_{1II} the interaction is between like spins whereas for T_{1IS} the interaction is between unlike spins. Another equation describing the variation of $<S_Z>$ is obtained by permuting I and S in equation [5-127].

Since the local field acting on I has contributions from like and unlike spins, the power function defined by equation [5-126] is a function of ω_I and ω_S, the resonance frequency of spins I and S respectively. The complete calculation gives:

$$T_{1II}^{-1} = \gamma_I^2 \gamma_S^2 \hbar^2 S(S+1) \left\{ \frac{1}{12} J_o(\omega_I - \omega_S) + \frac{3}{2} J_1(\omega_I) + \frac{3}{4} J_2(\omega_I + \omega_S) \right\} \qquad [5\text{-}128]$$

$$T_{1IS}^{-1} = \gamma_I^2 \gamma_S^2 \hbar^2 I(I+1) \left\{ -\frac{1}{12} J_o(\omega_I - \omega_S) + \frac{3}{4} J_2(\omega_I + \omega_S) \right\}$$

whereas

$$T_{2I}^{-1} = \gamma_I^2 \gamma_S^2 \hbar^2 S(S+1) \left\{ \frac{1}{6} J_o(o) + \frac{1}{24} J_o(\omega_I - \omega_S) + \frac{3}{4} J_1(\omega_I) + \frac{3}{2} J_1(\omega_S) + \frac{3}{8} J_2(\omega_I + \omega_S) \right\} \qquad [5\text{-}129]$$

Therefore a RF field at frequency ω_S has some action on spin I:

Consider now the situation where the modulation of the local field seen by nuclei I is not due to the motion of neighboring spins I and S but to a rapid fluctuation of spin S, the respective positions of spins S and I being unchanged during this fluctuation. This situation is frequently observed when, upon lowering the temperature, the correlation time characteristic of the motion of I with respect to S becomes higher than the period of the fluctuation of S. The fluctuation of S originates in a mechanism different than its interaction with I. This interaction is thus considered as negligible and the theoretical treatment is then very different from that described above. The system of spins S is considered as making part of the "lattice" with which it is continuously in equilibrium because of its fast relaxation rate. Keeping the same symbols as above

$$\frac{d<I_Z>}{dt} = -\frac{1}{T_{1I}}(< I_Z > - I_o) \qquad [5\text{-}130]$$

The perturbation Hamiltonian is the same as in equation [5-123]:

$$\mathcal{H}' = (\gamma_S \gamma_I \hbar^2 / r_{IS}^3)(a+b+c+d+e+f)_{IS} \qquad [5\text{-}131]$$

$$(a+b)_{IS} = \{ I_Z S_Z - \frac{1}{4} (I_+ S_- + I_- S_+) \} x Y_o$$

$$c_{IS} = (I_+ S_Z + I_Z S_+) \times Y_1$$

$$d_{IS} = (I_- S_Z + I_Z S_-) \times Y_1^*$$ see equation [5-123]

$$e_{IS} = I_+ S_+ \times Y_2$$

$$f_{IS} = I_- S_- \times Y_2^*$$

but where Y_0, Y_1 and Y_2 are now time independent; r_{IS} may be an average distance between spins I and S.

S_Z, S_+ and S_- are, on the contrary, time dependent operators. To take this particularity into account, one must substitute in equations [5-123] S_+, S_- and S_Z by the following values:

$$\left.\begin{aligned} S_+ &= S_+(t)e^{-i\omega_S t} \\ S_- &= S_-(t)e^{-i\omega_S t} \\ S_Z &= S_Z(t) \end{aligned}\right\} \qquad [5\text{-}132]$$

As a consequence, the operators a → f in equations [5-123] are again time dependent operators through S_+ and S_- but not through the Y_i. The probability for spin I to pass from one to another magnetic energy level under the influence of the fluctuation of the field produced by S is calculated by the theory of the time dependent perturbation. In addition one postulates that the fluctuation of spin S may be described by the following correlation function

$$\begin{aligned} &< S_z(t+\tau)\, S_z(t) > = 1/3\, S(S+1) \exp(-\tau/T_{1S}) \\ &< S_+(t+\tau)\, S_-(t) > = 2/3\, S(S+1) \exp(-\tau/T_{2S}) \end{aligned} \qquad [5\text{-}133]$$

where T_{1S} and T_{2S} are the relaxation times of the longitudinal and of the transverse components of S respectively. Using equations [5-127], [5-132] and [5-133] equation [5-130] can be solved and

$$T_{1I}^{-1} = \frac{2}{3} S(S+1)(\gamma_I^2 \gamma_S^2 \hbar^2 / r_{IS}^6) \{ \frac{1}{4}(1-3\cos^2\theta)^2 \frac{T_{2S}}{1+(\omega_S+\omega_I)^2 T_{2S}^2} + \frac{3}{4}\sin^4\theta \times \frac{T_{2S}}{1+(\omega_S-\omega_I)^2 T_{2S}^2} + \frac{3}{2}\sin^2\theta\cos^2\theta \frac{T_{1S}}{1+\omega_I^2 T_{1S}^2} \} \qquad [5\text{-}134]$$

Using the same type of calculation, the solution of

$$\frac{d < I_x >}{dt} = - \frac{I_x}{T_{2I}} \qquad [5\text{-}135]$$

is

$$T_{2I}^{-1} = \frac{1}{2} J_o(o) + J_I(\omega_I). \qquad [5\text{-}136]$$

For T_{1I} as well as for T_{2I}, the correlation time of the power functions are now T_{1S} and T_{2S} instead of being the correlation time of the motion as it was in the first situation. If, as commonly encountered, T_{1S} and T_{2S} are temperature independent, then T_{1I} and T_{2I} are also temperature independent as opposed to the behavior arising from the temperature dependence of τ_c in the first situation. One may thus expect to observe a transition in going from a temperature region where $\tau_c < T_{1S}$ to a temperature region where $\tau_c > T_{1S}$. If τ_c is *smaller* than T_{1S} at "high" temperature and larger than T_{1S} at "low" temperature, the observed T_{1I} would be temperature dependent at high temperature and it would become temperature independent beyond the transition.

5-4.7. Application – The Dynamics of the Water Molecule in the Interlamellar Space of Na Vermiculite

One of the most interesting characteristics of the expanding clay minerals is to accept one or two (and sometimes three) monolayers of water between the clay sheets.

If these intercalates are characterized by a discrete series of rational X-ray reflections, they may be described as well defined mono- or bi-layer hydrates. From the standpoint of the physical and chemical properties of water such hydrates are particularly valuable since water molecules are forming a continuous network of defined thickness between the sheets in which either the exchangeable cation or the surface oxygen are at the origin of structural perturbations because of the formation of cationic hydrates or of hydrogen bonds.

The word "structure" when applied to these layers of water molecules has a meaning which bears some similarity with that used for liquids, and more exactly for viscous liquids. To a large extent the molecular properties in an adsorbed phase are intermediate between those of a solid and of those of a liquid. The ordering no longer has the static aspect found in a solid where the structural atoms have a fixed average position about which they vibrate.

In adsorbed water layers, as we have seen earlier, there may be some preferential orientation and water protons exchange position in a particular way, defining what could be considered dynamic structure.

The experimental procedure and the origin of relaxation in the two layer hydrate of Na vermiculite. The relaxation time T_1 of the protons was measured with a pulse spectrometer operating at 60 MHz using a $\pi/2,\tau,\pi/2$ pulse sequence.

The following equation was used in order to separate the contribution of the lattice OH from that of the protons of hydration water. Let $M(\tau t)$ be the measured magnetization:

$$M(\tau,t) \propto C_{OH}\ [1-\exp(-\tau/T_1')]\ \exp(-t/2T_2')^2 + C_{H_2O}\ [1-\exp(-\tau/T_1)]\ \exp(-t/T_2)^2 \cos\delta\omega t$$

where τ is the time interval between two $\pi/2$ pulses, t the time at which $M(\tau,t)$ is measured after the second $\pi/2$ pulse. T_1' and T_2' are respectively the longitudinal and transverse relaxation times of the lattice OH. They were measured on a dehydrated sample and were found temperature independent. C_{OH} is the lattice OH content deduced from the chemical composition while C_{H_2O} is the hydration water content obtained from the adsorption isotherm. $\delta\omega$ is the doublet splitting, and T_2 the corresponding transverse relaxation time, both measured on the wide line spectrum. The field heterogeneity within the magnet is neglected.

Since the Llano vermiculite contains 1700 ppm Fe, it may be expected that the proton longitudinal relaxation rate is primarily affected by the paramagnetic centers (14). It is, therefore, very important to have information concerning both their location and their own relaxation properties.

By comparing EPR spectra obtained on oriented vermiculite aggregates with those obtained for a very well crystallized phlogopite, Olivier *et al.* (35) have been able to characterize four different crystal field symmetries acting on the Fe^{3+} cations: two octahedral species submitted to either an axial (0_a) or to a C_{2v} symmetry (0_b) have been detected and two tetrahedral species: T_b' with a C_{2v} symmetry which is far more abundant than an axially symmetrical (T_a) species.

The presence of tetrahedral Fe^{3+} indicates that the distance of closest approach between protons in the interlamellar space and the paramagnetic center is of the order of 3 Å whereas if the Fe^{3+} cations were located in the octahedral layer only, this distance would be about 6 Å. Since the distance of closest approach intervenes in the relaxation rate to the third power, the location of a noticeable fraction of the Fe^{3+} cations in the tetrahedral layer increases the proton relaxation rate by one order of magnitude.

By the saturation procedure, it was also found (46) that $10^{-7} \lesssim T_{1e} \lesssim 10^{-5}$ sec. (T_{1e} here is equivalent to T_{1S} in the general theory developed in Section 5-4.6.)

Therefore in the expression for the proton relaxation rate all terms but that corresponding to the $S_z I_\pm$ operator (equ. [5-123]) can be neglected since the others contain a denominator of the order of $\omega_s T_{2e} = 10^2$, ω_s (S = Fe) being 2.4×10^{11} rad sec^{-1}

In the sample of Llano vermiculite studied here, the average distance between paramagnetic centers is 46 Å assuming that the Fe^{3+} cations are randomly spread within the layer-silicate lattice.

Relaxation Analysis. The model suggested by the study of the proton or deuteron line shape will be the starting point for the relaxation analysis.

(a) The interlamellar cation is octahedrally coordinated to six water molecules spinning very rapidly (as compared to the rigid line width) about their C_2

symmetry axis directed towards the corners of a regular octahedron. The C_3 axis of this octahedron runs parallel to the crystallographic C* axis. For intermolecular relaxation processes each water molecule will be replaced by a fictitious proton situated on the C_2 axis, midway on the p-p vector since this vector is perpendicular to C_2 ($\gamma = 90°$). (see Fig. [5-14]).

(b) Cations diffuse in the interlamellar space with a diffusion coefficient of the order of 10^{-8} $cm^2\ sec^{-1}$ at room temperature and an activation energy of about 10 kcal $mole^{-1}$ (28). It is suggested that the orientation of the spinning axis remains well defined above 270°K and that the cation diffuses with the hydration shell rotating about the C_3 axis.

(c) In this motion of the hydration shell, a fictitious proton, for instance at P_2 in Fig. 5-14, is moving with respect to a fixed paramagnetic center located at P_1. P_1 is inside the vermiculite lattice while P_2 is in the interlamellar space. The motion of the hydration shell changes the length (r) and the orientation (θ) of the relaxation vector P_1P_2 with respect to H_o.

(d) There may be also a small proportion of molecules diffusing independently of those belonging to the hydration shell. These "free" molecules and/or their protons may exchange place with those inside the shell. This exchange may occur inside the shell without outside assistance but it also has been suggested that the "outside" molecules may relay this exchange, forming transient H_3O+ species.

Although these molecules are indispensable in insuring the exchange process, their relative concentration in the presently studied hydrate is considered very small as compared with those forming the hydration shell. Because of their low concentration, they do not constitute an observable separate population.

(e) Because paramagnetic centers are randomly distributed on both sides of the silicate sheets the reorientation and the change in length of the relaxation vector could be considered as quasi-isotropic.

In summary it is suggested that $T_1^{-1} = T_1^{-1}{}_{inter} + T_1^{-1}{}_{intra}$ with $T_1^{-1}{}_{inter} >> T_1^{-1}{}_{intra}$; $T_1^{-1}{}_{inter}$ is contributed by the interaction of the proton with the paramagnetic centers whereas $T_1^{-1}{}_{intra}$ is the intramolecular proton-proton interaction. The summation of two contributions for T_1^{-1} means that there are two relaxation pathways with independent probability.

Let us call τ_{ch} and τ_{cd} the correlation times associated with the motions of the hydration shell as a whole and of the "free" water molecules or protons, respectively. In the region where either τ_{ch}^{-1} or τ_{cd}^{-1} are much higher than T_{1e}^{-1} thus in the domain where the longitudinal relaxation time is temperature dependent.

$$T_1^{-1} = S(S+1)\gamma_p^2\gamma_S^2\,\hbar^2 \sin^2\theta \cos^2\theta\, r^{-6} \frac{3\tau_{c1}}{1+\omega_p^2\tau_{c1}^2} \quad [5\text{-}137]$$

In equation [5-137]

$$\tau_{c1}^{-1} = \tau_{ch}^{-1} + \tau_{cd}^{-1} + T_{1e}^{-1} \qquad [5\text{-}138]$$

and γ_p and γ_S are the gyromagnetic ratios of the proton and of the electronic spins, respectively. If τ_{ch} and τ_{cd} are sufficiently different, the observed relaxation rate is the sum of three contributions:

$$T_1^{-1} = T_{1h}^{-1} + T_{1d}^{-1} + T_{1p}^{-1} \qquad [5\text{-}139]$$

T_{1p}^{-1} being the "paramagnetic" contribution, e.g., that obtained when $T_{1e}^{-1} > \tau_{ch}^{-1}$ or τ_{cd}^{-1}. T_{1p}^{-1} should be practically temperature independent.

The experimental results obtained for the spin-lattice relaxation rate T_1 are shown in Figs. 5-26 and 5-27 for three orientations of the crystal C* axis with respect to H_o (angle δ in Fig. 5-14). T_1 is very weakly orientation dependent and there are two temperature regions characterized by the variation of T_1. A shallow minimum is observed at $T^{-1} = 4.3 \times 10^{-3}\ °K^{-1}$ and below $T^{-1} = 6 \times 10^{-3}\ °K^{-1}$ starts the region where $T_1^{-1} = T_{1p}^{-1}$. The motion modulated T_1 will be studied in the next section.

If we neglect the orientation dependence and if we consider

$$< \cos^2\theta \sin^2\theta > = \frac{4}{15}$$

then T_1^{-1} is obtained by integrating eq. [5-137] on a sphere element $4\pi r^2\,dr$ within the following limits: $r^* < r < \infty$; r^* is the shortest distance of approach between a proton and a paramagnetic center. N_{Fe} being the number of Fe^{+3} cations per cm^3, it follows that

$$T_1^{-1} = C\,\frac{\omega_p \tau_{c1}}{1+\omega_p^2\,\tau_{c1}^2} \qquad [5\text{-}140]$$

where

$$C = \frac{8\pi}{15}\,S(S+1)\gamma_p^2\gamma_S^2\,\hbar^2\,\frac{N_{Fe}}{\omega r^{*3}} \qquad [5\text{-}141]$$

For $S = 5/2$, $N_{Fe} = 10^{19}$ cations/cm^3, $\omega_p = 3.77 \times 10^8$ rad sec^{-1}, $\gamma_{Fe} = 1.78 \times 10^7$ rad sec^{-1} G^{-1}, $\gamma_p = 2.67 \times 10^4$ rad sec^{-1} G^{-1}

$$C = 0.9 \times 10^5 / r^{*3}\ (\text{Å})\ \text{sec}^{-1} \qquad [5\text{-}142]$$

The Proton and Molecular Motions. Regardless of the nature of the motion (rotational diffusion of the hydration shell or diffusion outside the hydration shell), the protons encounter a large variety of situations, such as remaining for some time in a deeper potential energy well on the surface, etc. It seems therefore adequate to assume a distribution function for the correlation times (42).

This is also suggested by the shallow character of the minimum observed at $T^{-1} = 4.3 \times 10^{-3}\ °K^{-1}$. Using the classical log normal distribution:

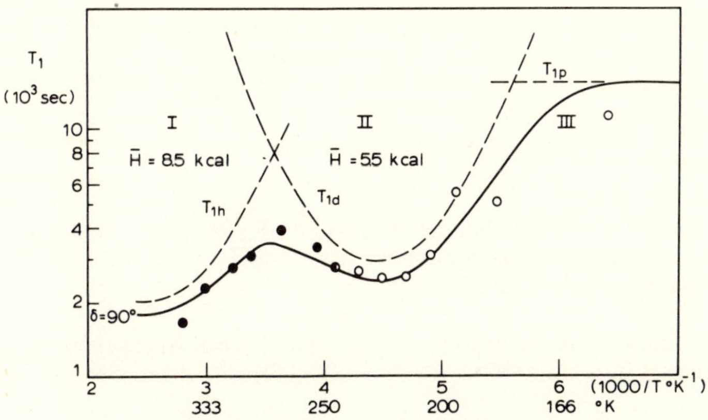

Figure 5-26. Variation of the proton T_1 with respect to 1/T at $\delta = 90°$. The solid line is obtained using Eq. [5-11] and the fitting parameters introduced in Eq. [5-19]. (Hougardy *et al.*, 1976).

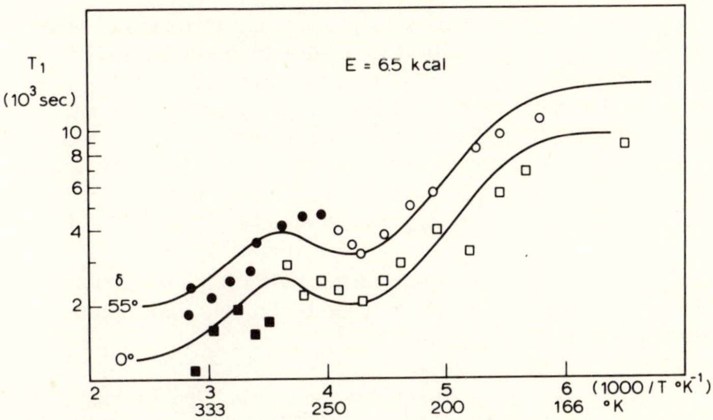

Figure 5-27. Variation of the proton T_1 with respect to 1/T at $\delta = 0°$ and 55°. (Hougardy *et al.*, 1976).

$$P(\tau)d\tau = \beta^{-1}\pi^{-1/2}\exp(-z/\beta)^2\ dz \qquad [5\text{-}143]$$

where

$$z = \ln(\tau/\tau_m) \text{ and } \tau_m = \tau_o \exp(\bar{H}/RT). \qquad [5\text{-}144]$$

Then $T_1^{-1} = CF_1$

where $$F_1 = \int_o^\infty P(\tau_{c1}) \frac{\omega_p \tau_{c1}}{1+(\omega_p\tau_{c1})^2} d\tau_{c1}. \qquad [5\text{-}145]$$

A T_1^{-1} maximum is observed for $\omega_p\tau_{c1} = 1$, r_{c1} is associated with τ_{ch} or τ_{cd}. Since a noticeable fraction of Fe^{3+} cations are in the lattice tetrahedral layer, the distance of closest approach of any water proton may be as low as 3 Å and thus C

$= 3 \times 10^4$ sec^{-1}. In the absence of a distribution of relaxation times, the maximum value of F_1 is 0.5. In the presence of such a distribution, $F_1 < 0.5$. In the absence of motion, e.g., when in equation [5-140] τ_{ch}^{-1} and τ_{cd}^{-1} are smaller than T_{1e}^{-1}, F_1 resumes to

$$\omega_p T_{1e}/[1+\omega_p^2 T_{1e}^2] \simeq (\omega_p T_{1e})^{-1}.$$

For all orientations in the low temperature region, T_1 levels out at about $T_{1p} = 15 \times 10^{-3}$ sec. For $r^* = 3$ Å, this is obtained assuming $T_{1e} = 1.20 \times 10^{-7}$, and this value is within the range predicted earlier.

In the temperature region where τ_{ch}^{-1} and τ_{cd}^{-1} are larger than T_{1e}, $(T_{1h}^{-1} + T_{1d}^{-1})$ is calculated by substracting $T_{1p}^{-1} = (1000/15)$sec^{-1} from the observed T_1^{-1}.

A well marked but somewhat shallow minimum is observed between 4.3 and 4.5×10^{-3} °K^{-1}, e.g., at about the same temperature as in X zeolites at a similar resonance frequency. In this case, it is generally assumed that a rotational diffusion motion is responsible for the variation of T_1^{-1} in this region.

T_{1d}^{-1} was then obtained by substracting T_{1p}^{-1} from the observed T_1^{-1} on the low temperature side of the minimum, where the contribution of T_{1h}^{-1} is negligible. The β and $\overline{H}_d$ parameters giving the best fit between $(T_1^{-1} - T_{1p}^{-1})$ and Eq. [5-144] were determined as described elsewhere (13). The results obtained were $\beta = 1.5$ and $H_d = 5.5$ kcal mole^{-1}. From the experimental data obtained for $\delta = 55°$, it follows $H_d = 6.5$ kcal mole^{-1} and $\beta = 1.5$. The average activation enthalpy for diffusion is thus about 6 ± 0.5 kcal mole^{-1}. T_{1d}^{-1} was then calculated on the left hand side of the minimum and T_{1h}^{-1} was obtained as $T_{1h}^{-1} = T_1^{-1} - T_{1d}^{-1} - T_{1p}^{-1}$.

The rotation of the complete cation hydration shell about C_3 parallel to C* suggested by the wide-line study may be tentatively considered as the relaxation mechanism operating T_{1h}^{-1}.

The variation of T_{1h}, shown by the dotted line in Fig. 5-26, can be adequately fitted by introducing $\beta = 3$, $\overline{H} = 8.5$ kcal, and $\tau_o = 6.1 \times 10^{-14}$ sec into Eq. [5-144]. The distribution of the correlation times is broader than that proposed for T_{1d} and the activation enthalpy is somewhat higher. At room temperature, the corresponding correlation time (τ_{ch}) is 10^{-7} sec. τ_{ch} could be related to the rotational diffusion of the hydrated cation by Eq. [5-35]. Assuming $(1^2) = 30$ Å^2, the corresponding diffusion coefficient is 0.5×10^{-8} cm^2 sec^{-1} at 298°K.

The Na^+ autodiffusion coefficients measured by Calvet (8) in homoionic montmorillonite at 293°K by the radioactive tracer technique are between 10^{-8} and 10^{-7} cm^2 sec^{-1} for water contents ranging from 5 to 12.5 weight percent with corresponding activation energies between 7 and 5 kcal. The heterodiffusion coefficients obtained for vermiculites (29) are also in the same range.

5-5. REVIEW OF SOME PROBLEMS: ORDER AND DISORDER IN ADSORBED WATER MOLECULES

The notion of time and the concept of structure have been apparently opposed for a long time although in describing or computing a structure, X-ray crystallographers frequently assign to any atom in a well ordered arrangement an ellipsoidal volume inside the border of which its vibrational motions are confined. Neutron scattering, by making use of the coherent elastic and incoherent inelastic processes, has bridged more efficiently the gap between the static and dynamic approach, but the notion of disorder, as opposed to the notion of order, is not often considered to include a time dependent perturbation. Let us then discuss the extent to which this point of view is valid when the organization of small molecules inside the interlamellar space of layer lattice minerals is considered.

The experimental results to be reviewed were obtained mostly by using NMR spectroscopy but other types of spectroscopy could be applied, if the time domain of each is clearly delineated. Classical infrared spectroscopy involves motions with frequencies between $10^{12} - 10^{14}$ Hz. The Fourier transform far infrared interferometer shows vibrational modes between 10 and 100 cm^{-1}. Vibrational modes of the charge balancing cations located in the interlamellar space of clay minerals appear within this domain. The number of the modes of vibration with respect to the surface, their frequency, and the band shape could be exploited to reveal the local environment and, eventually, the degree of ordering on a time scale of the order of 10^{-11} sec, corresponding to a frequency of about 10^{11} Hz. Research in this area is still in its infancy but many interesting developments should follow in the near future.

Electron paramagnetic resonance has been used by McBride, Pinnavia and Mortland (32) especially in the case of paramagnetic cations, such as Cu^{2+} balancing the lattice charge. Recently, M. Gutierrez-Le Brun and J.M. Gaite (23) have studied, in a more quantitative manner, the line shape of the EPR signal of hydrated copper. The time domain of EPR being in the range of $10^{10} - 10^{11}$ hz, the symmetry character of the Hamiltonian associated with the EPR signal must be dependent on a specified configuration around the paramagnetic center, for instance of water molecules in the Cu II hydration shell, which has a life-time of this order of magnitude. Nuclear magnetic resonance is another spectroscopic technique which operates in a broader time domain between approximately 10^{-6} to 10^{-10} sec if the relaxation processes are experimentally determined.

As with IR and EPR, NMR is sensitive to the order existing at rather small distances from the "probe", for which resonance is measured. Diameters of the order of 10 to 100 Å are typical of the spatial extension of the order, or of the disorder, at which these spectroscopic techniques are sensitive. Thus it is much more restricted than the long range statistical order which is measured by classical X-ray diffraction studies. In addition, configurations with life-times in these ranges are those influencing what could be considered as the resolution of these spectroscopic methods whereas X-ray techniques integrate a much longer time even if modern fast detectors and sources with high fluxes are used.

The aim of this discussion is to show the meaning of order and disorder for water molecules in the interlamellar space at the time scale of NMR. In addition, the predominant influence of the ordering at that scale on the thermodynamic and chemical properties of water molecules will be shown.

5-5.1. Materials: Vermiculite, Hectorite and Hydrated Halloysite

It is generally accepted that the nature of exchangeable cations has a strong influence on the ordering of water molecules in the interlamellar space. A textural factor appears to be important also, namely the extension of the layers in the a and b directions.

Experimental data of Prost (39) and of Bergaya *et al.* (4) have indeed shown that clays made from minute particles such as hectorite and montmorillonite have microporous volumes which fill up at low relative vapor pressure of the adsorbate and more precisely in the range of relative vapor pressure required for the adsorption within the interlamellar space. Therefore such a material contains in addition to molecules in interlamellar pores, molecules in this microporous volume which are probably not as well organized as the former. It should be noted that NMR permits a rough evaluation of the relative distribution of water molecules among these two populations. This is possible because the spin-lattice relaxation times are not the same in these two different environments. Under certain conditions, the rate at which the M_z magnetization is restored is the sum of two exponentials multiplied by the relative content in the two populations X_L and X_C with relaxation times $T_{1\,L}$ and $T_{1\,C}$ respectively:

$$\frac{M_z}{M_z(o)} = X_L \exp(-\frac{t}{T_{1\,L}}) + X_C \exp\left(-\frac{t}{T_{1\,C}}\right) \text{where } X_L + X_C = 1$$

In layer lattice silicates such as vermiculite the size of the sheets is such that the volume developed by the micropores is negligible. To take the influence of the microporous volume into account, the experimental data obtained for the two-layer hydrate of Na vermiculite (27) will be compared with those obtained for the one-layer hydrate of Li hectorite (19). An additional factor which must be outlined is the fact that the cation exchange capacity (CEC) of the Llano vermiculite dictates that the number of charge balancing cations, Na+, is in a 1/6 ratio with respect to the number of adsorbed water molecules. This, in addition to the value of the X-ray basal spacing constitutes a presumption in favor of an octahedral arrangement of water molecules.

Partially because there are water molecules outside the interlamellar space and partially because the CEC is smaller, there are within the interlamellar space of one-layer hydrate of Li hectorite molecules of second rank, e.g., not directly linked to Li^+.

Finally the third material to be examined is the hydrated halloysite (12). Here there is no exchangeable cation within the interlamellar space and the only ordering factor is through the hydrogen bonds between water molecules and surface oxygens or hydroxyls according to a study of Hendricks and Jefferson (25). Fig. 5-28 shows a representation of this structure.

By this choice of three different hydrates, we are examining three different situations in which the water molecule ordering is expected to decrease within the following sequence: two-layer hydrate of sodium vermiculite, one-layer hydrate of Li hectorite and finally the one-layer hydrate of halloysite. This expectation is founded 1) on the higher energy of interaction for a water molecule in a cation hydration shell than in a hydrogen bond, and 2) on the difference in the lattice charge of these two minerals as far as Li hectorite and Na vermiculite are concerned.

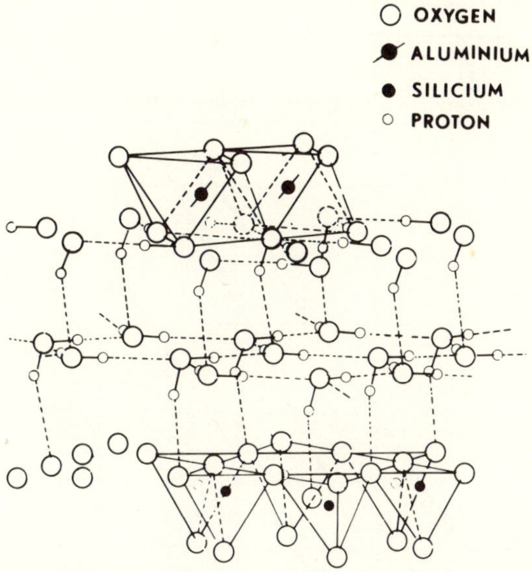

Figure 5-28. Schematic structure of endellite (hydrated halloysite) (according to Hendricks and Jefferson, 1938).

Eleven X-ray *00ℓ* orders of reflection were easily observed for the one-layer hydrate of Li-hectorite but the adsorption isotherm did not show steps. On the contrary the adsorption isotherm obtained by Van Olphen (46, 47) (see Fig. 5-12) for Na vermiculite shows very well defined steps and more than 15 00ℓ reflections are observable. The hydrated halloysite usually gives a very clean reflection at 10 Å. It is extremely sensitive to dehydration and must be kept permanently under saturated water vapor pressure. The monotonous character of the water adsorption isotherm in Li hectorite, as opposed to that observed in vermiculite must be related to water in the micropore volume, outside the interlamellar space.

5-5.2. The Water Molecules Arrangement as Deduced from the NMR Signals

The structure of water in the two layer hydrate of Na vermiculite has been discussed (5-3.1) and (5-3.2). In summary, as shown in Fig. 5-14, Na^+ is at the center of an octahedron of 6 water molecules. The value of the doublet splitting requires a double motion to occur: a molecular rotation about the C_2 axis and a rotation of the hydration shell about the C_3 axis. The relaxation analysis (see 5-4.7), founded on the T_1 measurements shown in Fig. 5-26 suggests that these two motions are at the origin of two relaxation mechanisms with correlation times shown in Fig. 5-29. As expected the correlation time of the hydration shell

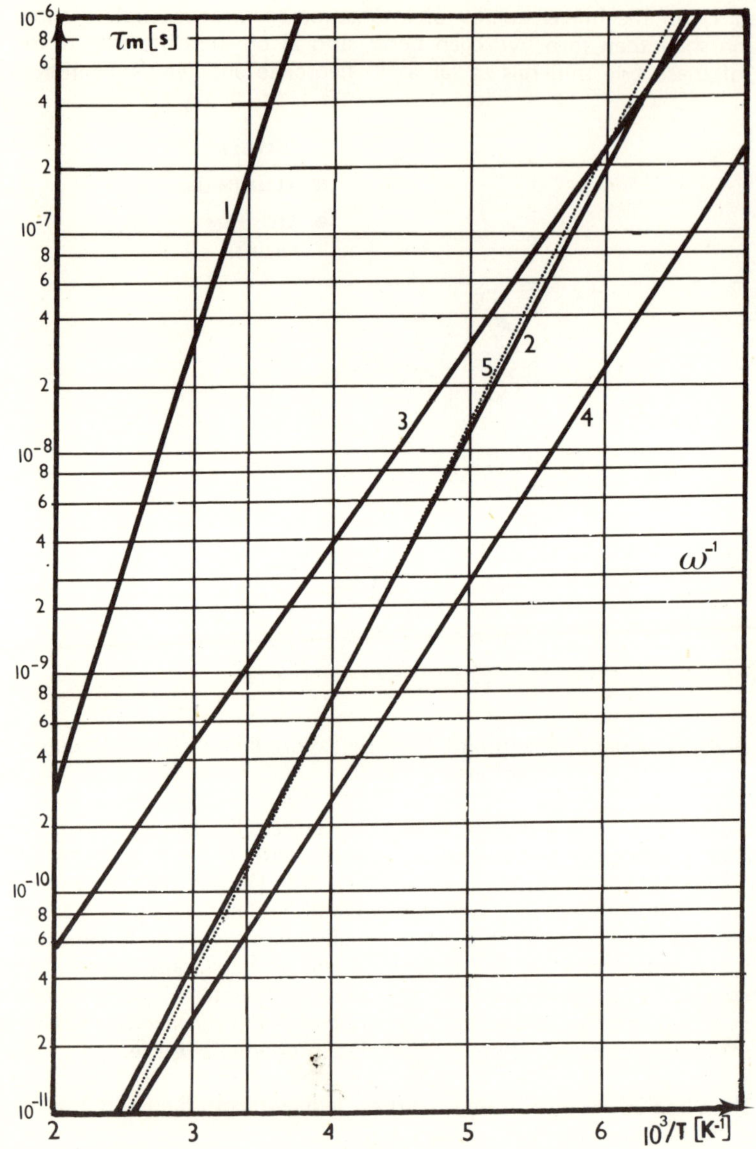

Figure 5-29. Correlation times: 1) rotation about the C_3 axis of the Na hydration shell in vermiculite; 2) rotation about the C_2 axis in Na vermiculite; 3) rotation about the C_3 axis of the Li hydration shell in hectorite; 4) rotation about the C_2 axis in Li hectorite; 5) observed in Na montmorillonite (one-layer) and Ca montmorillonite (two-layer hydrate).

rotation (#1 in Fig. 5-29) is much longer than that of the molecular rotation about C_2.

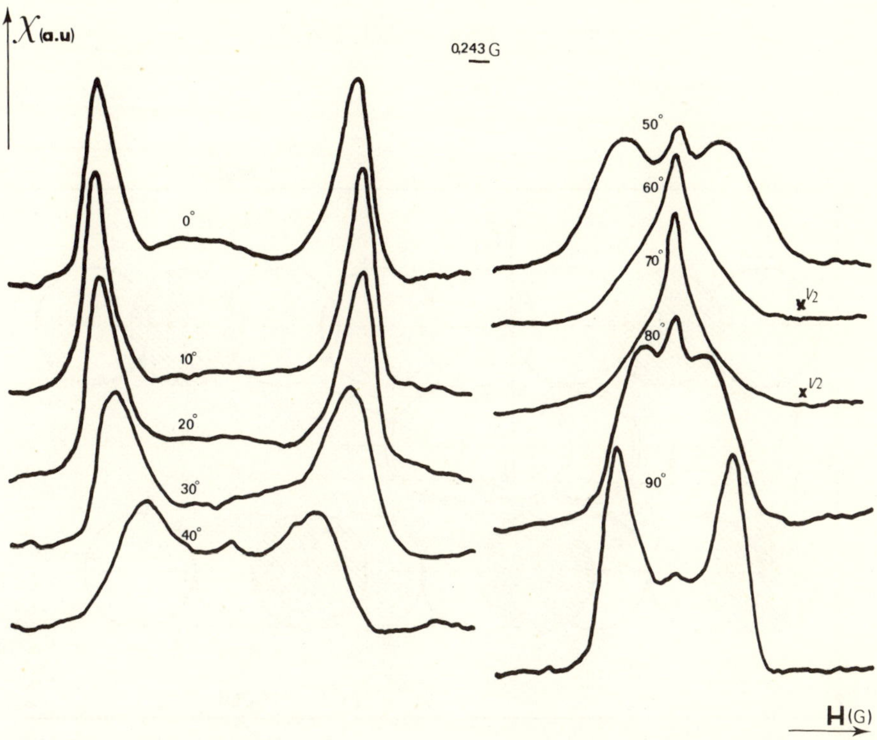

Figure 5-30. Proton NMR spectra observed at 14 MHz for the one-layer hydrate of Li hectorite.

In the case of Li hectorite, Fig. 5-30 shows the variation of the proton doublet with the angle between the normal to the film and H_o (angle δ in Fig. 5-14). The doublet obeys equation [5-49], $\vec{I}$ being tilted by 70° with respect to C*. Again there are two motions similar to those observed for the two layer-hydrate of Na vermiculite: one around C_2 and the other around a three-fold axis as schematically represented in Fig. [5-31]. However by considering more carefully Fig. 5-30, one sees clearly that the collapse of the doublet into one single line occurs in a larger angular domain than predicted by equation [5-49]. This is probably due to the misorientation of the clay sheets in the oriented aggregate used for these films. A discrepancy between experimental and theoretical misalignment of ± 20° of clay particles with respect to each other can be predicted. This value has been recently verified by an EPR study of copper in the one layer hydrate of the same hectorite (23). Proton relaxation in hectorite does not occur appreciably through paramagnetic impurities because of the low content of Fe (< 100 ppm). It is thus mainly through homodipolar interaction that protons relax.

Fig. 5-32 shows the two spin-lattice relaxation times observed by using a $\pi/2$, τ, $\pi/2$ sequence. Because in hectorite about 50% of the lattice OH are replaced by fluorine, the contribution of the lattice OH is negligible and the magnetization obeys a law similar to that shown in section 5-4.7 if $C_{OH} \simeq 0$ namely:

$$M(\tau,T) \propto C_{H_2O}\ \{1\text{-}\exp(-\tau/T_1)\}\exp(-t/T_2)\cos\delta\omega t$$

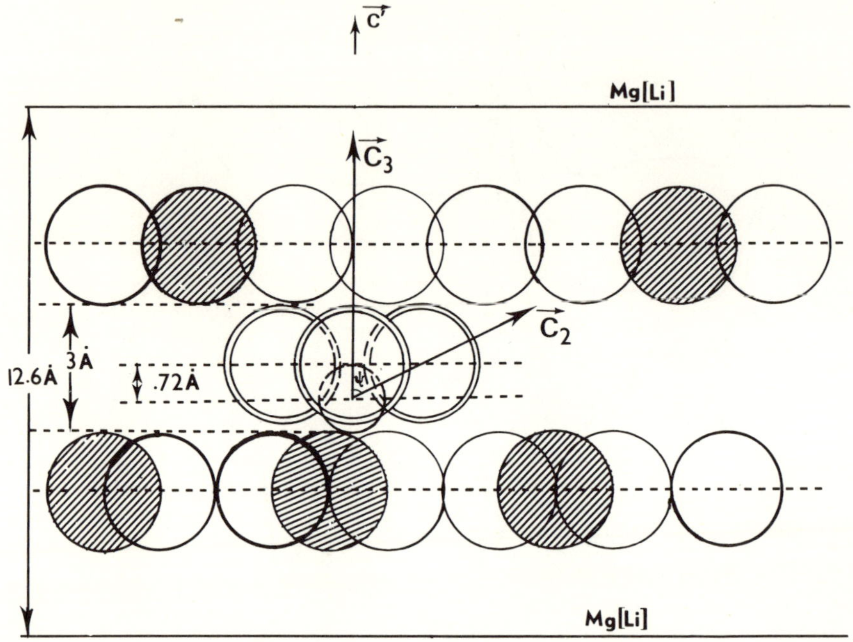

Figure 5-31. The ordering of water molecules in the one-layer hydrate of Li-hectorite. Projection of the surface oxygen atoms and of the oxygen atoms of water on plane 110. The position of C_3 passing through the center of an idealized hexagonal cavity has no special meaning because of diffusion.

where t is the running time after the second impulsion.

Two exponential decays are observed allowing the definition of two populations of water molecules, P_c and P_L, characterized by spin-lattice relaxation times T_{1C} and T_{1L} respectively. P_L increases from approximately 20% at room temperature to ~ 50% at ~ 200°K. The definitions of P_L and P_C may be represented as follows:

$$P_L = \frac{M_z(O)_L}{M_z(O)_L + M_z(O)_C} \text{ and } P_C = 1 - P_L.$$

P_L may be assigned to water molecules outside the interlamellar space because a longer relaxation time suggests the molecule is in a freer environment. The water molecules outside the interlamellar space would be in the micropores which were mentioned in 5-5.1.

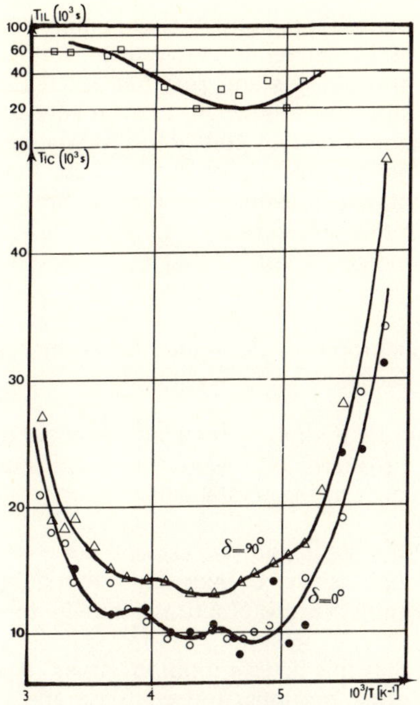

Figure 5-32. The T_{1C} and T_{1L} spin-lattice relaxation times observed for the one-layer hydrate of Li hectorite. T_{1C} is shown for two values of angle δ. The determination of T_{1L} was not accurate enough to obtain the angular dependence.

The relaxation time T_{1C} (see Fig. 5-32) is slightly orientation dependent because of the orientational dependence of the Y_i functions. In particular the spin-lattice relaxation rate in a pair of protons rotating about an axis perpendicular to the proton-proton vector has been shown to be (see Pfeifer (38)):

$$T_1^{-1} = \frac{9\gamma^4\hbar^2}{32\, r^6}\left\{ (1\text{-}\cos^4\theta')\,\frac{\tau_c}{1+(\omega_o\tau_c)^2} + (1+6\cos^2\theta' + \cos^4\theta')\,\frac{\tau_c}{1+4\omega_o^2\tau_c^2} \right\}$$

where θ' is the angle of the rotation axis $\vec{I}$ (e.g. the symmetry axis C_2) and $\vec{H}_o$. If $\vec{I}$ reorients about the crystal axis C*, one must consider the time average of $\cos^2\theta'$ and $\cos^4\theta'$.. These time-averaged functions could be calculated exactly if there was no misfit in the orientation of the clay particles but because of the rather broad distribution law for the orientation with respect to the "average" C* axis it seems impossible to account accurately for the orientation dependence of T_1, observed experimentally. Woessner (49) had proposed another model for the calculation of T_1. He considered a pair of protons at both ends of a vector that reorients about one axis. In that case T_1 is a function of γ only (see Fig. 5-14). For the one layer hydrate of Li hectorite as well as for the two-layer hydrate of Na vermiculite, γ is probably = 90°, and the model of Woessner doesn't fit the experimental data reported here.

Now if we consider T_{1C} in Fig. 5-31, it appears that it might be composed from two contributions in a comparable manner as observed for vermiculite. These two contributions, rotation about the C_3 axis and rotation about C_2, would be closer in magnitude than in vermiculite and this would explain why they collapse into a broad minimum. Fig. 5-33 shows the two contributions whereas the correlation times associated with these motions are shown in Fig. 5-29. Recently from neutron scattering, Conard (10) got two rotational relaxation times at about 4×10^{-11} and 4×10^{-12} sec at room temperature for a Li hectorite at lower water content (about 3 H_2O/Li). This reinforces the idea that the observed NMR T_1 is made from two contributions.

If we summarize the main conclusions of the study of the water molecules in the interlamellar space we may say that:

1) the cation imposes through its hydration shell a preferential organization of the water molecules. Their C_2 axis is directed toward the cation and C_2 is tilted by an almost constant angle with respect to the crystal axis.

2) From the dynamic point of view in the two cases, there is a fast rotation of the proton-proton vector orthogonal to the C_2 axis and a reorientation of the water molecules about an axially symmetric axis practically parallel to the C* axis.

It is worthwhile to point out that this double rotation "saves" a great deal of activation energy as compared to other motions. Indeed Giese and Fripiat (22) have shown that the combination of motions involving the C_3 and C_2 axes decreases the activation energy for molecular rotation from > 50 kcal mole^{-1} to $<$ 20 kcal mole^{-1}, in agreement with experimental activation energies of a similar order or smaller than 10 kcal mole^{-1} (see Fig. 5-29).

The mechanical analog of these two motions would be that of a ball bearing in which as the bearing (sphere of hydration-hydration shell) rotates the individual steel balls (water molecules) also rotate in a coupled fashion.

Finally, let us examine what NMR reveals about the ideal monolayer of water, namely that in the interlayer space of halloysite (12). The qualification of ideal is used because here there is no cation balancing the lattice charge, the net charge being negligible.

Considering Fig. 5-28 one might eventually expect a preferential orientation and thus the observation of a doublet but this is not the case. The line shape for the ^1H as well as for the ^2H signal is Lorentzian between 170° and 300°K. There is no preferential orientation at the time scale of the resonance, or more exactly of an eventual doublet splitting ($< 10^{-5}$ sec).

The reason being that water molecules in the monolayer have double orthogonal motions: one about C_2, the other one about an axis perpendicular to C_2. A rotation of 180° about C_2 restores, (see Fig. 5-28) the water network exactly as it is. On the contrary, the configuration obtained after a rotation of 180° about an axis perpendicular to C_2, is the mirror image of that shown in Fig. 5-28.

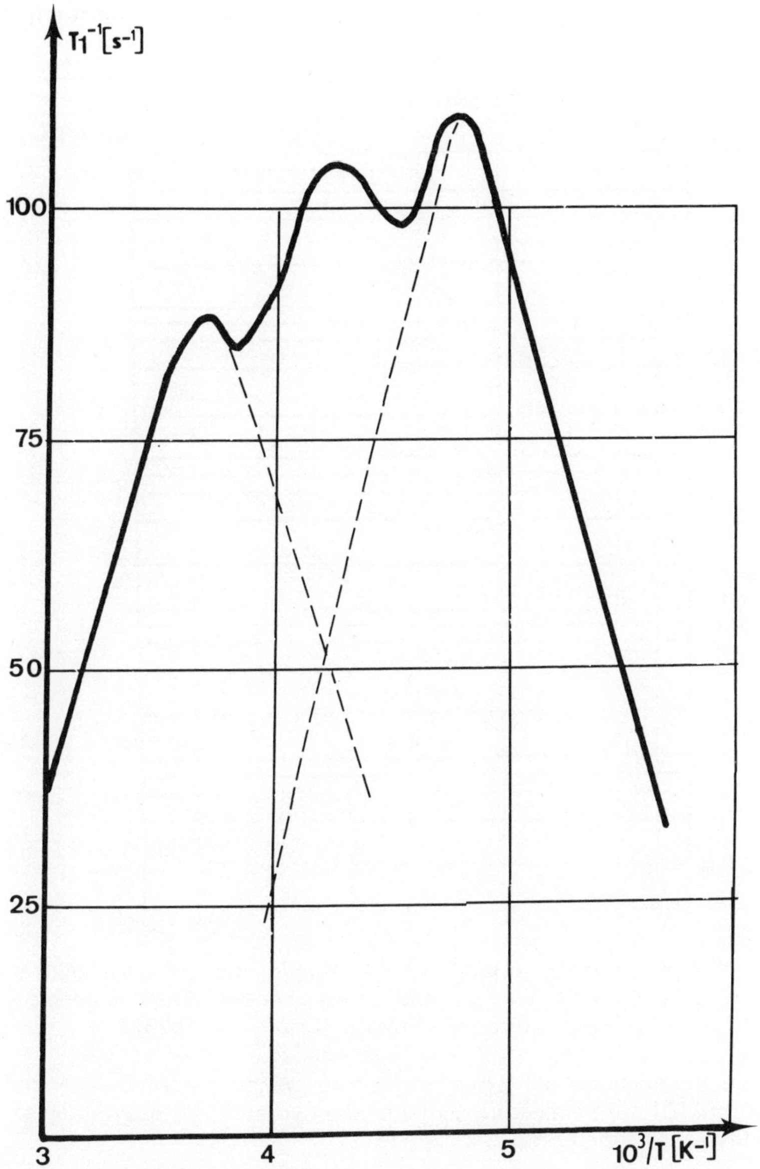

Figure 5-33. The two-contributions to T_{1C} in the one-layer hydrate of Li hectorite.

The two correlation times are shown in Fig. 5-34. The tumbling about the axis perpendicular to C_2 is much slower than that about C_2. The probability of changing the arrangement shown in Fig. 5-28 in its mirror image is much smaller than to flip the proton by 180° in the plane of the monolayer. An interesting consequence of this is to consider the molar heat capacity of water in the interlamellar space shown in Fig. 5-34. This molar heat capacity is obtained from the following equation:

$$C_p(H) = L\, C_p(D) + W\, C_p(HW).$$

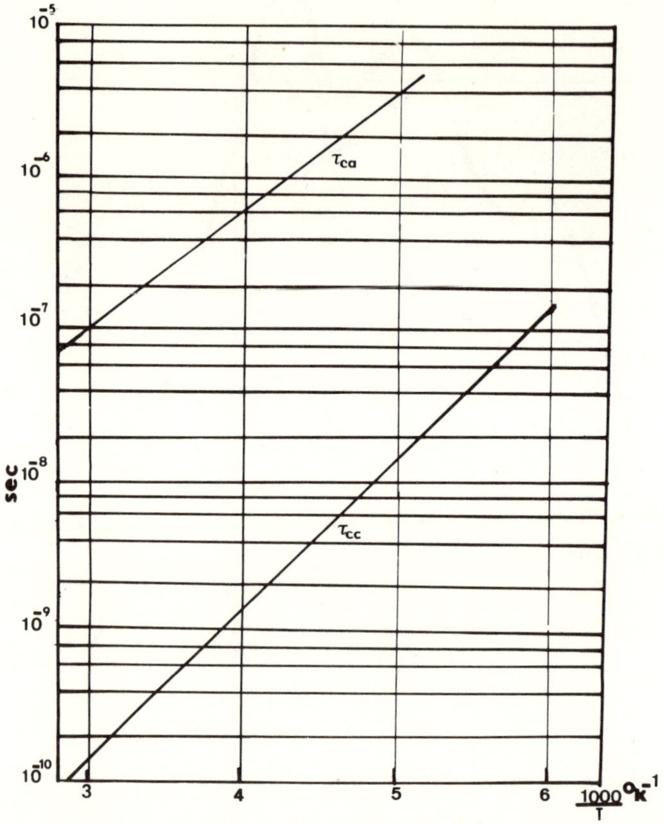

Figure 5-34. The correlation times of the 180° flipping motion (τ_{CC}) about the C_2 axis and of the tumbling motion (τ_{CA}) about an axis perpendicular to the C_2 axis in hydrated halloysite. (Cruz *et al.*, 1976).

where L is the kaolinite-like lattice content and W the hydration water content. $C_p(H)$ and $C_p(D)$ are the heat capacities of the hydrated and dehydrated samples respectively which are measured separately.

The graph in Fig. 5-35 shows that at about 140° K the heat capacity of water in the monolayer is equal to that of ice whereas at 260° K, it is approaching that of liquid water. Actually it can be shown that at 140° K, the tumbling motion is no longer efficient for relaxation and that the rotation about C_2 is probably transformed into a torsional vibration. At this temperature τ_{cc} equals T_2 of an immobile water molecule.

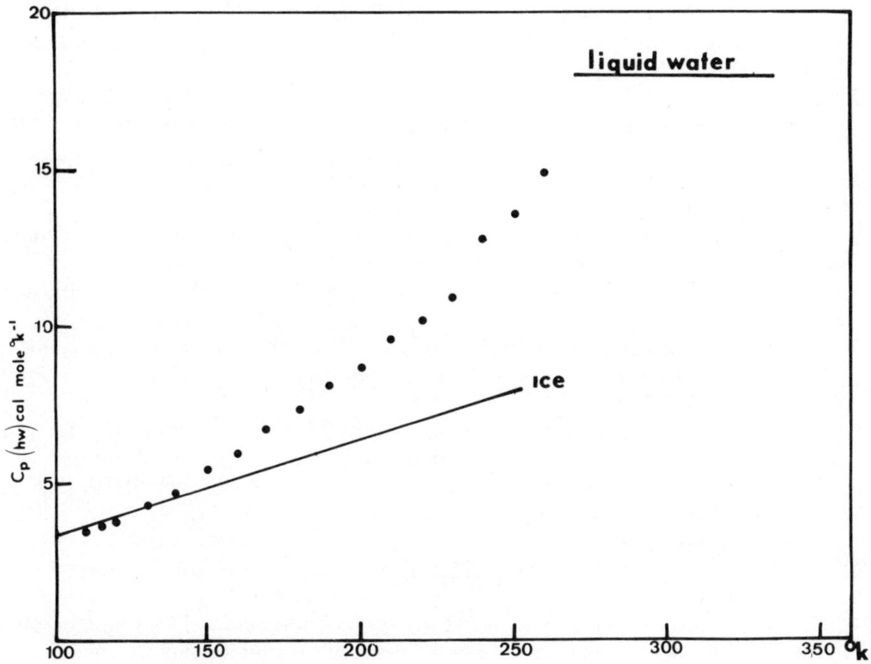

Figure 5-35. The molar heat capacity of hydration water in hydrated halloysite *vs* temperature. (Cruz *et al.,* 1978).

REFERENCES

1. Abragam, A. 1961. The principles of nuclear magnetism. Oxford Clarendon Press. 599p.
2. Anderson, D.M., and P. Hoekstra. 1965. C.R.R.E.L. Hanover (NH) Research Report, 192.
3. Andrew, E.R. 1958. Nuclear Magnetic Resonance. Cambridge University Press. 267p.
4. Bergaya-Annabi, F., M.I. Cruz, L. Gatineau and J.J. Fripiat. 1979. Quelques donnees sur la capacite calorifique et les proprietes de l'eau dans divers systemes proeux. *Clay Minerals* **14(3)**:161-172.
5. Bloch, F., W.W. Hansen, and M.E. Packard. 1946. Nuclear induction. *Phys. Rev.* **69**:127.
6. Bradley, W.F., and J.M. Serratosa. 1960. A discussion of the water content of vermiculite. 7th Natl. Conf. on Clays and Clay Min. 260-270.
7. Brindley, G.W., and M. Nakahira. 1958. Further consideration of the crystal structure of kaolinite. *Min. Mag.* **31**:781-786.
8. Calvet, R. Doctoral Thesis. 1972. Faculte des Sciences. Paris.
9. Chiba, T. 1963. Deuterium magnetic resonance study of barium chlorate monohydrate. *J. Chem. Phys.* **39**:947-953.
10. Conard, J. Personal communication.
11. Conard, J. 1976. Magnetic resonance in colloid and interface science. ACS Symposium Series 34. pp. 85-93.

12. Cruz, M.I., M. Letellier and J.J. Fripiat. 1978. NMR study of absorbed water. III. Molecular motions in the monolayer hydrate of halloysite. *J. Chemical Physics* **69**:2018-2027.
13. Cruz, M.I., W.E.E. Stone and J.J. Fripiat. 1972. The methanol-silica gel system. II. The molecular diffusion and proton exchange from pulse proton magnetic resonance data. *J. Phys. Chem.* **76**:3078-3088.
14. Deininger, D., and A. Gutsze. 1973. Preprint 237. Nicholas Copernicus University. Torun (Poland).
15. Einstein, A. 1956. Investigations on the theory of the Brownian movement. Dover publication. R. Furth. Tr. by A.D. Cowper (Eds.) 119p.
16. Farrar, T.C., and E.D. Becker. 1971. Pulse and Fourier transform NMR. Academic Press. 115p.
17. Fripiat, J.J., H. Bosmans and P.G. Rouxhet. 1967. Proton mobility in solids. I. Hydrogenic vibration modes and proton delocalization in Boehmite. *J. Phys. Chem.* **71**:1097-1111.
18. Fripiat, J.J., J. Chaussidon and A. Jelli. 1971. Chimie-physique des phenomenes de surface. Masson. Paris. 387p.
19. Fripiat, J.J., M. Kadi-Hanifi, J. Conard and W.E.E. Stone. 1979. Surface sciences. 2nd Intern. Symposium on the Application of NMR to Surface Chemistry. Menton.
20. Fripiat, J.J., and R. Touillaux. 1969. Proton mobility in solids. *Trans. Farad. Soc.* **65**:1236-1247.
21. Gastuche, M.C., F. Toussaint, J.J. Fripiat, R. Touillaux and M. Van Meersche. 1963. Study of intermediate stages in the kaolin→metakaolin transformation. *Clay Minerals Bulletin* **5**:227-236.
22. Giese, R., J.J. Fripiat. 1979. *J. Colloid and Interface Science* **71**:441.
23. Gutierrez-Le Brun, M., and J.M. Gaite. (in press 1979). 2nd Intern. Symp. on NMR in colloid and surface sciences. Menton.
24. Gutowsky, H.S., and G.E. Pake. 1950. Structural investigation by means of nuclear magnetism. II. Hinder rotation in solids. *J. Chem. Phys.* **18**:162-170.
25. Hendricks, S.B., and M.E. Jefferson. 1938. Structures of kaolin and talc pyrophyllite hydrates and their bearing on water sorption of the clays. *Am. Mineral* **23**: 863-875.
26. Holm, C.H., C.R. Adam and J.A. Ibers. 1958. The hydrogen bond in Boehmite. *J. Phys. Chem.* **62**:992-994.
27. Hougardy, J., W.E.E. Stone, and J.J. Fripiat. 1976. NMR study of absorbed water. I. Molecular orientation and protonic motions in the two-layer hydrate of a Na vermiculite. *J. Chem. Phys.* **64(9)**:3840-3851.
28. Kubo, R., and K. Tomita. 1954. A general theory of magnetic resonance absorption. *J. Phys. Soc. Japan* **9**:888-919.
29. Lai, T.M., and M.M. Mortland. 1968. Cationic diffusion in clay minerals. II. Orientation effects. *Clays and Clay Min.* **16**:129-136.
30. Lechert, H., and H.W. Henneke. 1977. Molecular sieve II. ACS Symposium Series 40, 53.
31. Legrand, A.P. 1976. Lecture notes to students in material sciences. Ecole de Chimie et de Physique. Paris.
32. McBride, M.E., T.J. Pinnavaia, and M.M. Mortland. 1975. Electron spin relaxation and the mobility of manganese (II). Exchange ions in smectites. *Am. Miner.* **60**:66-72.
33. Mathieson, A. McI. and G.F. Walker. 1954. Crystal structure of magnesium vermiculite. *Am. Min.* **29**:231-255, note 29.

34. Mathieson, A. McI., G.F. Walker. 1952. The structure of vermiculite. *Clay Min. Bull.* **1**:272-276.
35. Olivier, D., P. Lauginie and J.J. Fripiat. 1976. Relationship between the longitudinal relaxation rates of water protons and of well defined paramagnetic centers at low temperature in hydrated vermiculite. *Chemical Physics Letters* **40**:131-133.
36. Pake, G.E. 1948. Nuclear resonance. Absorption in hydrated crystals: Fine structure of the proton line. *J. Chem. Phys.* **16**:327-336.
37. Pedersen, B. 1964. NMR in hydrate crystals: Correction for vibrational motion. *J. Chem. Phys.* **41**:122-132.
38. Pfeifer, H. 1973. Nuclear magnetic resonance and relaxation of molecules absorbed on solids. *Adv. Nucl. Magn. Reson.* **55**.
39. Prost, R. 1975. Etude des interactions eau-orglie et des mecanismes de l'hydration des smectites. Doctoral Thesis. Paris.
40. Purcell, E.M., H.C. Torrey and R.V. Pound. 1946. Resonance absorption by nuclear magnetic moments in a solid. *Phys. Rev.* **69**:37-38.
41. Reeves, L.W. 1969. The study of water in hydrate crystals by nuclear magnetic resonance. *In* Progress in NMR spectroscopy, Vol. 4. Emsly, Feenay and Sutcliffe (eds.). Pergamon, New York. pp. 193-233.
42. Resing, H.A. 1967. Nuclear magnetic resonance relaxation of molecules adsorbed on surfaces. *Adv. Mol. Relaxation Process* **1**:109-154.
43. Roby, C. 1968. These 3eme Cycle. Thesis. Universite de Grenoble, France.
44. Shirozu, H., and S.W. Bailey. 1966. Crystal structure of a two-layer Mg-vermiculite. *Am. Min.* **51**:1124-1143.
45. Slichter, C.P. 1963. Principles of magnetic resonance, with examples from solid state physics. Harper and Row, New York. 246p.
46. Van Olphen, H. 1965. Thermodynamics of interlayer adsorption of water in clays. I. Sodium vermiculite. *J. Colloid Sci.* **20**:822-837.
47. Van Olphen, H. 1969. Thermodynamics of interlayer adsorption of water in clays. II. Magnesium vermiculite. *Proc. Int. Clay Conf.* Tokyo. Israel Universities Press, Jerusalem. **1**:649-657.
48. Van Vleck, J.H. 1948. The dipolar broadening of magnetic resonance lines in crystals. *Phys. Rev.* **74**:1168-1183.
49. Woessner, D.E. 1962. Spin relaxation processes in a two-proton system undergoing anisotropic orientation. *J. Chem. Phys.* **36**:1-4.
50. Woessner, D.E., B.S. Snowden and G.H. Meyer. 1970. A tetrahedral model for pulsed nuclear magnetic resonance transverse relaxation: Application to the clay water system. *J. Coll. Interface Sci.* **34**:43-52.
51. Woessner, D.E. 1974. Proton exchange effects on pulsed NMR signals from preferentially oriented water molecules. *J. Magn. Res.* **16**:483-501.

CHAPTER 6

DISTRIBUTION OF IONS IN THE OCTAHEDRAL SHEET OF MICAS

W.E.E. Stone
Section de Physico-Chimie Minérale (M.R.A.C.-Tervuren),
Place Croix du Sud 1, B-1348 Louvain-la-Neuve (Belgium)

and

J. Sanz
C.S.I.C., Serrano 115 Dpdo, Madrid 6, Spain

6.1. INTRODUCTION

In this chapter we will discuss a practical example of the use of NMR in the study of natural samples. It will be shown how NMR can provide information regarding the distribution of Fe^{2+}, F^- and OH^- ions within the octahedral sheet of micas (7, 9). Various octahedral associations are differentiated by carefully examining the H^+ and F^- NMR absorption signals as a function of frequency and orientation of the mica crystal in the applied magnetic field. Comparison of this method with x-ray and infrared spectroscopy will also be discussed.

The samples studied are phlogopites and biotites. Their ideal formula is $Si_3Al(R^{2+})_3\ O_{10}(OH,F)_2K$. The elementary layer of these trioctahedral micas consists of a central octahedral sheet between two tetrahedral sheets. The octahedra are formed by 4 oxygens and 2 hydroxyl groups (Fig. 6-1); two different cationic sites are possible according to whether the OH are in cis (M_2) or trans (M_1) positions. The OH groups are coordinated to one M_1 cation in the π symmetry plane and to 2 M_2 cations situated on both sides of the plane. The cations in these sites are essentially Mg^{2+} in phlogopites but can be replaced by Fe^{2+} ions up to a few percent in the case of phlogopites and up to 20% by weight in the case of biotites. Other cations such as Al^{3+}, Fe^{3+}, Ti^{3+} and octahedral vacancies are also present but at much lower concentration. The OH groups can also be isomorphously replaced by F^- ions to various degrees.

It is well known that the nature of the isomorphous substituents within the octahedral sheet is a determining factor in the vermiculitization process of these minerals; in particular when the fluorine content increases, the ease with which the K^+ in the interlayer region can be replaced by Na^+ decreases. On the contrary, when the content in Fe^{2+} increases the vermiculitization process also increases. It is therefore of interest in the study of alteration to know how these 2 ions, F^- and Fe^{2+}, are distributed in the lattice.

J. W. Stucki and W. L. Banwart (eds.), Advanced Chemical Methods for Soil and Clay Minerals Research, 317–329.

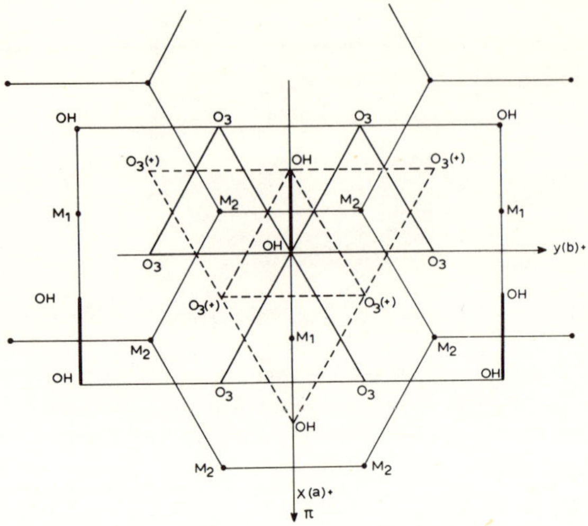

Figure 6-1. Projection on the ab plane of the octahedral layer of 1 M micas showing the position of hydroxyl groups and M_1 and M_2 ion sites.

In the experiments described here, the H^+ and F^- NMR absorption spectra were obtained at room temperature with a continuous wave spectrometer. Samples of about 1 cm^3 volume were formed by superposing platelets cut out of large plates of natural micas. The orientations of the a and b axes in the natural plates had been determined in order to obtain spectra at known angles with respect to the applied field $\overline{H}_o$. Twinned regions and crystals showing different polytypes were discarded.

6-2. INFLUENCE OF THE FE^{2+} IONS

Fe^{2+} being paramagnetic has an electronic spin, S, which interacts with the nuclear spin, I. The magnetic part of this S-I interaction, in a diamagnetic insulator can be written briefly as

$$\mathcal{H}_p = g\beta\hbar\gamma \, (\mathcal{H}_C + \mathcal{H}'_D) \qquad [6\text{-}1]$$

where g is the Landé factor, β the Bohr magneton, $\mathcal{H}_C$ the isotropic contact term and $\mathcal{H}'_D$ the anisotropic dipolar-dipolar interaction. As the magnetic moment of electrons, μ_p, is a thousand times larger than that of a nucleus, μ, one expects very large perturbations in the nuclear magnetic signal. It is anticipated that the polarized electrons create a local field, H_L, proportional to the applied field $\overline{H}_o$, at the sites of the nuclear spins. The local field that a nucleus will "feel" depends on its environment and distance from the electron. The result is that the nucleus will have its resonance frequency shifted away from the Larmor frequency, ω_o. At room temperature the electron spin usually reorients very quickly in the magnetic field so that the electron magnetic moment, μ_p, "seen" by the nucleus will only be an averaged value given by

$$\overline{\mu}_p = (\mu_p)^2 \frac{\overline{H}_o}{3k\,(T\text{-}\theta)} \qquad [6\text{-}2]$$

where θ is a constant introduced to take into account small interactions between electron spins. At room temperature and for usual values of $\overline{H}_o$, $\overline{\mu}_p$ is now only a few times larger than μ and consequently the shifts will be of the order of a few Gauss instead of 1000 Gauss as is true for a static μ_p. Equation [6-2] implies that the magnetic moment of S is now orientation independent (isotropic g) and that the electron relaxation time is shorter than the Larmor period. Equation [6-2] also shows that the interaction with a paramagnetic ion decreases linearly with the applied field. By working at different frequencies (8, 14, 56 and 60 MHz), we were able to discriminate between paramagnetic and diamagnetic influences on the NMR line.

When studying the effect of paramagnetic ions on the NMR line, we can somewhat arbitrarily distinguish between effects at long and short distances, i.e. separate the influence of paramagnetic ions which are either far or close to a particular nuclear site. In the first case, the local field created by the paramagnetic ions will be, at the most, of the order of magnitude of the local field created by the nuclei-nuclei interactions alone; whereas in the second case, it will be larger. Far away ions will therefore contribute an extra broadening to the line while nearby ions will shift the resonance line away from the average Larmor field.

The problem of evaluating the paramagnetic contribution to the width of a NMR line is a very difficult problem, especially in the case of low symmetry crystals where preferential distribution of ions in the various sites is possible. Moreover, when non-ellipsoidal samples are used, as here, the demagnetizing field will vary from point to point within the crystal and contribute to the width of the line.

For these reasons, only the shift of nearby nuclei will be considered in this chapter (i.e., up to S-I distance of 5-6 Å). The magnetic shift, ΔH, due to the anisotropic dipolar (I-S) interaction of equation [6-1] can be written

$$\Delta H = \frac{\overline{\mu}_p}{r^3} \,(3\cos^2\psi - 1) \qquad [6\text{-}3]$$

where $\overline{\mu}_p$ is given by equation [6-2] and r and ψ are the ion-nucleus polar coordinates relative to the direction of the magnetic field, $\overline{H}_o$. Using structural models (obtained by neutron diffraction experiments, for example (1)) and values for $\overline{\mu}_p$ obtained by susceptibility measurements (4), it is possible to calculate ΔH for different orientations and given magnetic fields. These can then be compared with experimental values (referenced with respect to the main central line).

6-3. H^+ SPECTRA OF PHLOGOPITES

Experimentally it is found that the observed shift in the position of the side-lines for H^+ depends on the sample orientation in the field, and that their position relative to the central line varies linearly with $\overline{H}_o$ (Fig. 6-2). It has also been observed that the intensity of these H^+ side-lines increases with iron content.

In Fig. 6-3 is given the calculated (Equation [6-3]) angular variation of shifts around the b-axis for nearest H^+ neighbors (r = 2.78 Å). When turning around the b-axis the 2 M_2 sites are equivalent and therefore a maximum of 2 side-lines is

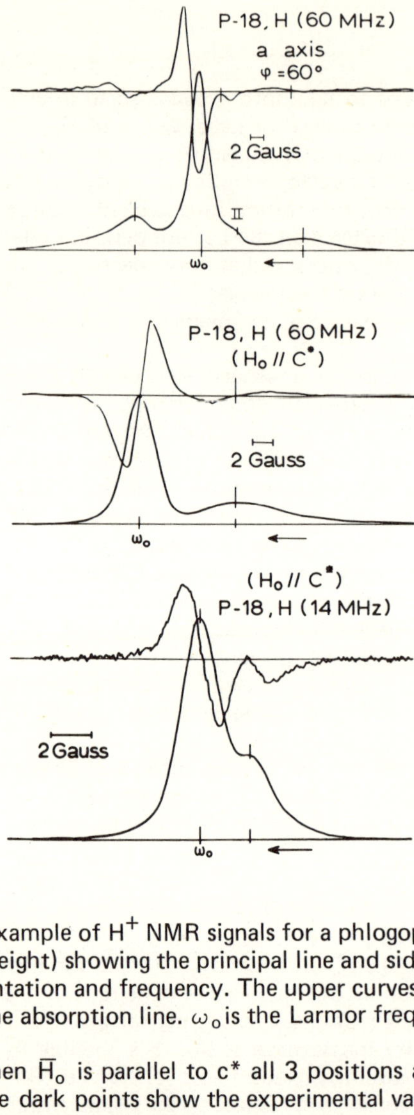

Figure 6-2. Example of H^+ NMR signals for a phlogopite sample (2.6% iron by weight) showing the principal line and side-lines as a function of orientation and frequency. The upper curves are the first derivative of the absorption line. ω_0 is the Larmor frequency.

expected. When $\overline{H}_0$ is parallel to c* all 3 positions are identical and only 1 line is expected. The dark points show the experimental values for a sample with 2.6% by weight of iron. The agreement is quite satisfactory. In the calculation, only Fe^{2+} ions are considered as they are the dominant paramagnetic ion (as shown by Mössbauer (8) and Chemical Analysis (5)). At about 30° orientation (see Fig. 6-5), the separation between lines is sufficiently large to allow a relative evaluation of the site occupied by Fe^{2+} ion. It is found that the line corresponding to OH groups adjacent to an ion in an M_2 site has an area approximately twice that of the line associated with the M_1 site. Therefore, as also shown by Mössbauer experiments (8), it seems that on the average the Fe^{2+} ions are randomly distributed between the two possible sites. In Mössbauer experiments, because the values of the two crystalline fields corresponding to M_1 and M_2 sites are close, the spectral resolution

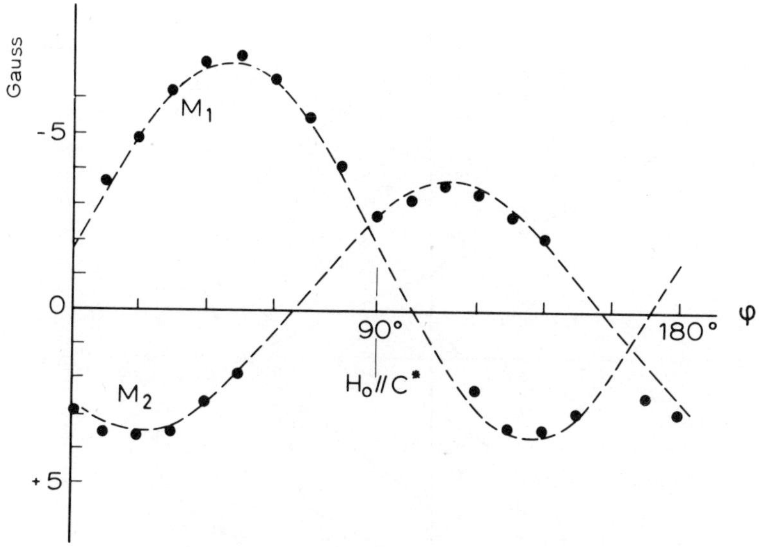

Figure 6-3. Calculated angular variation of shifts around the b-axis for nearest neighbor H^+, and experimental points at 14 MHz (•).

is poor. This is not the case in NMR where resolution can be optimized by choosing the appropriate orientation of the crystal in the applied field. Finally, for spectra run at high frequency, side-lines corresponding to second neighbor H^+ (Fig. 6-2) have been observed close to the main line.

6-4. H^+ SPECTRA OF BIOTITES

When one considers the H^+ spectra of biotites the situation is more complicated than for phlogopites due to the fact that the concentration of Fe^{2+} is much larger and the M_1 and M_2 sites may be occupied simultaneously by one, two or three Fe^{2+} ions around the same OH. Therefore, only low field spectra can be interpreted easily. Using the same model as in Section 6-3 and considering first neighbors only, the line shifts for the five possible associations of H^+ with the Fe^{2+} ions can be calculated (this has been done for rotations around the b axis and assuming that local fields are additive). They are shown in Fig. 6-4: I_1 corresponds to $2Mg^{2+}$ and $1Fe^{2+}$ (M_1); I_2, $2Mg^{2+}$ and $1Fe^{2+}$ (M_2); $II_{1,2}$, $1Mg^{2+}$ and $2Fe^{2+}$ ($M_1 + M_2$); II_2, $1Mg^{2+}$ and $2Fe^{2+}$ (M_2); and III, $3Fe^{2+}$. The vertical axis corresponds to zero shift, i.e. associations of an OH with $3Mg^{2+}$. Around 70°, two lines should be observed as shown in Fig. 6-5. Another interesting orientation is 30° where the 5 lines corresponding to the various associations can be readily observed. In this position one can see (Fig. 6-5) first, the relative increase of associations with one Fe^{2+} as one goes, for example, from sample P-18 (2.6% Fe^{2+}) to B-8 (9.8% Fe^{2+}) and with two Fe^{2+} from B-8 to B-10 (13.8% Fe^{2+}). Secondly, the intensities of I_2 and $II_{1,2}$ are twice that of I_1 and II_2 respectively, again showing the random distribution of iron on the M_1 and M_2 sites.

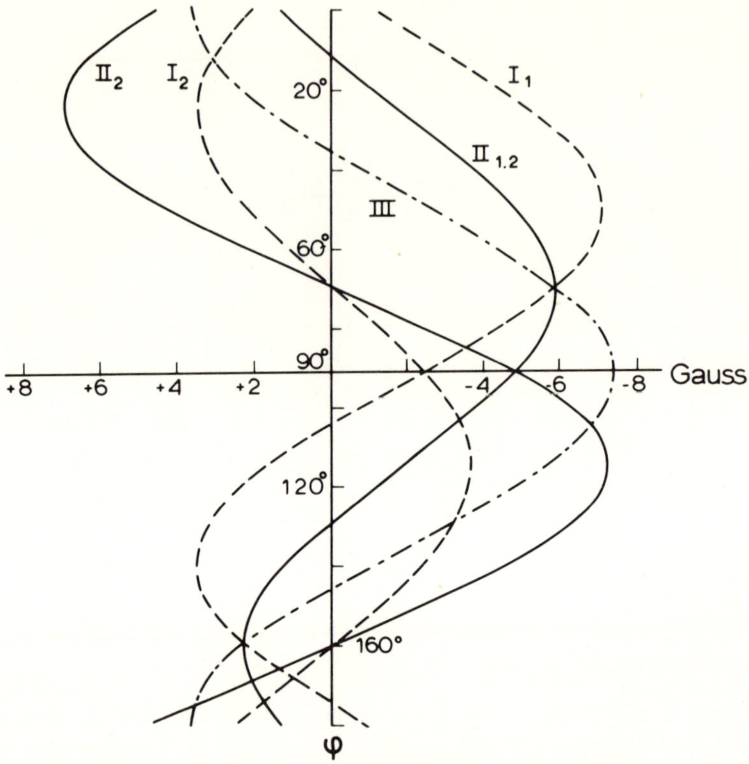

Figure 6-4. Calculated angular variation of shifts around the b-axis for the five possible associations of H^+ with Fe^{2+} situated in M_1 and M_2 sites immediately next to an OH group. [See text for definitions of I_1, I_2, etc.]

It seems therefore that the interaction model described above is correct (covalent bonding, g anisotropy and exchange interaction of little importance) and that H^+ can be used as an internal probe in determining the site distributions of the ions.

6-5. F^- SPECTRA

When the spectra of F^- ions (which isomorphously substitute for the OH) are examined, several interesting features are found. Concerning the paramagnetic influence on the F^- line it is found, first, that the width of the line is very much less dependent on the Fe^{2+} content than in the case of H^+; and second, when the position of the side lines is compared with that of calculated values only side-lines due to second or third neighbors are observed. Side lines corresponding to first neighbors have never been detected. The intensity of the side lines are, moreover, very much less dependent on the Fe^{2+} content than in the H^+ case. These various experimental facts indicate that most of the F^- ions are at a larger distance from the iron than are the H^+, and unlike the OH^- the F^- are not directly coordinated to the Fe^{2+}. This conclusion is corroborated by the following experimental data.

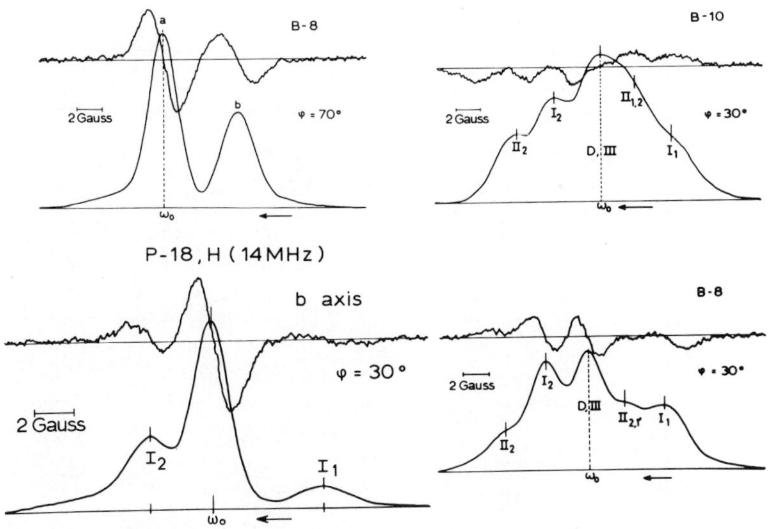

Figure 6-5. H^+ spectra of biotites B-8 (9,8% iron) and B-10 (13,8% iron) at 14 MHz for 2 orientations around the b axis. Also shown is the H^+ spectrum of phlogopite P-18 (2,6% iron) at 14 MHz.

At low field the F^- signal consists of doublets for all orientations. Recall that a doublet in NMR clearly indicates a diamagnetic interaction between a pair of nuclei with spin 1/2 (which is the case for H^+ and F^-). Also, such an interaction between two nuclei can be observed only if the interaction of each nucleus with its surroundings is less important than the interaction with its partner in the pair. The doublet separation, h, is given by the Pake expression (3):

$$h = \frac{a\mu}{r_3} (3 \cos^2 \theta - 1) \qquad [6\text{-}4]$$

where a is a constant equal to 3 for 2 identical spins 1/2 and 2 for 2 different spins 1/2.

Once again, it is possible to calculate (Equation [6-4]) the doublet separation for two ions situated in the same octahedron, M_2, F-F and F-H where r equals 2.64 and 3.45 Å, respectively. The result of this calculation around the b axis is given in Fig. 6-6 where the F-F distance has been taken from a fluor-phlogopite structure (2). The dark circles correspond to experimental points with two clearly distinguishable doublet separations from 90° to 180°. The inner doublet corresponds to F-H and the outer doublet to F-F. It can be seen that at 130° the two doublet separations are quite different and that this orientation is convenient for monitoring the number of associations of both types. In Fig. 6-7 the F^- spectra for an angle of 130° are given for different phlogopites where the F^- content increases according to $P_{21} < P_{12} < P_{15} < P_{18}$. By taking the area under the F-F doublet as a measure of the number of F pairs, it is possible to compare this figure

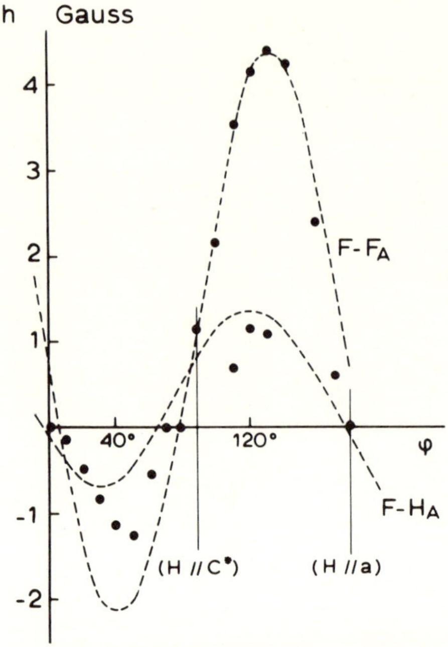

Figure 6-6. Calculated angular variation of the F doublet separation for F-F and F-H, h, around the b axis (— —), and experimental points with two clearly distinguishable doublets from 90° to 180°, •.

with the one obtained by taking the mineralogical formula and calculating the probability of finding F-F pairs. This is done in Fig. 6-8. It can be seen that the number of pairs found increases much faster than expected. This is interpreted as evidence for the existence of homogeneous, F^--rich domains in which Fe^{2+} ions are excluded and the F-F distance is equal to that in a pure fluor-phlogopite. The pictures obtained from the F^- and OH^- spectra are therefore quite complementary. NMR shows that the OH^- and F^- ions are highly differentiated with respect to cationic association. The chemical environment of F^- is homogeneous, whereas OH^- seems on the whole to be in a more heterogeneous environment (also see Section 6-6).
of cations around the anions does not necessarily follow a regular periodic pattern, and therefore cannot be detected by x-ray methods. This short range order is significant because it could play an important role in the vermiculitization of micas by modifying the cohesion between layers.

6-6. CORRELATION WITH I.R. RESULTS

Now consider two examples of how some of the results obtained by NMR can be correlated with results obtained by infrared spectroscopy. In the OH stretching region of the infrared spectrum phlogopites and biotites show a broad complex band from 3750 to 3450 cm^{-1}. This complex spectrum reflects the diversity of environments of the hydroxyl groups, since isolated OH^- groups all in

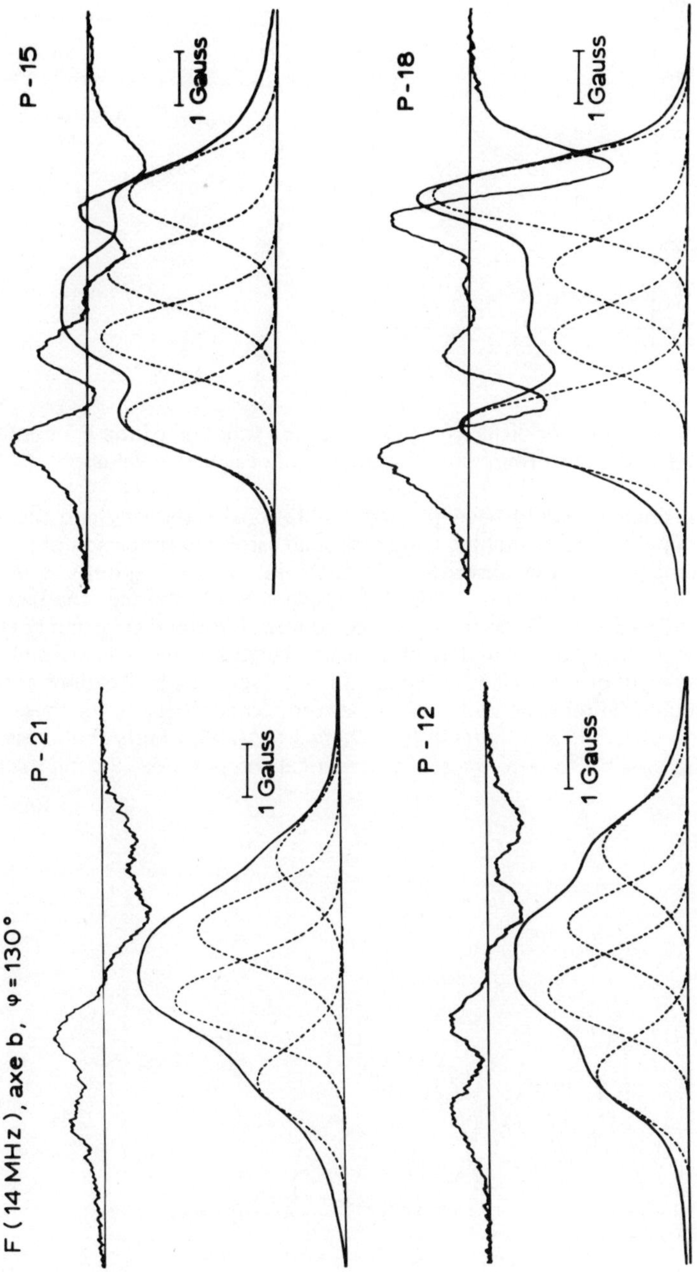

Figure 6-7. F spectra for different phlogopites with increasing content in F^-: P-21 < P-12 < P-15 < P-18.

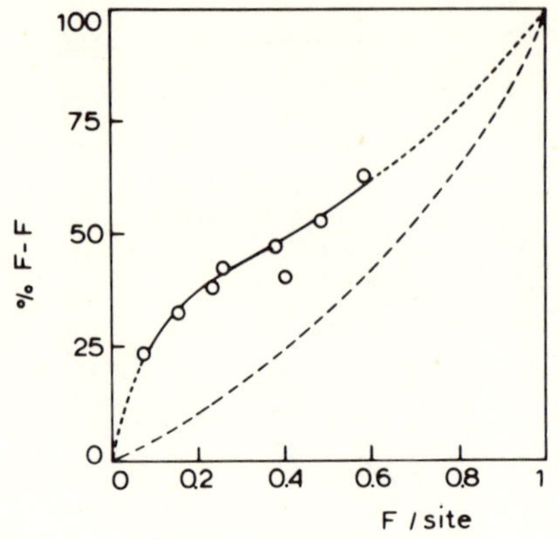

Figure 6-8. % F-F found for different samples (0) as a function of the F^- content. The broken line represents the statistically-calculated values.

the same environment would have the same vibrational frequency and therefore give rise to a single sharp stretching band. Through careful examination of samples of known composition, the absorption frequencies for OH^- groups in various environments have been identified (10, 11). Using this information, the band between 3750 and 3450 cm^{-1} can be attributed to three hydroxyl environments (see Fig. 6-9): the N type, associated with 3 divalent cations; the I type, due to 2 divalent and one tri- or tetravalent cation; and the V type, due to 2 cations and one vacancy. It should be remembered that the absorption coefficients for these three types are quite different and increase from N to I to V. With fairly sound assumptions it is quite possible to decompose, by numerical computation, the mica spectra

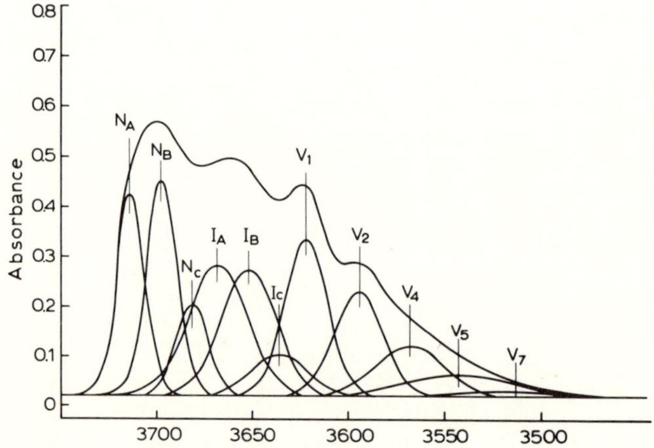

Figure 6-9. Typical example of the I.R. spectrum of a phlogopite specimen (N, I and V are defined in the text).

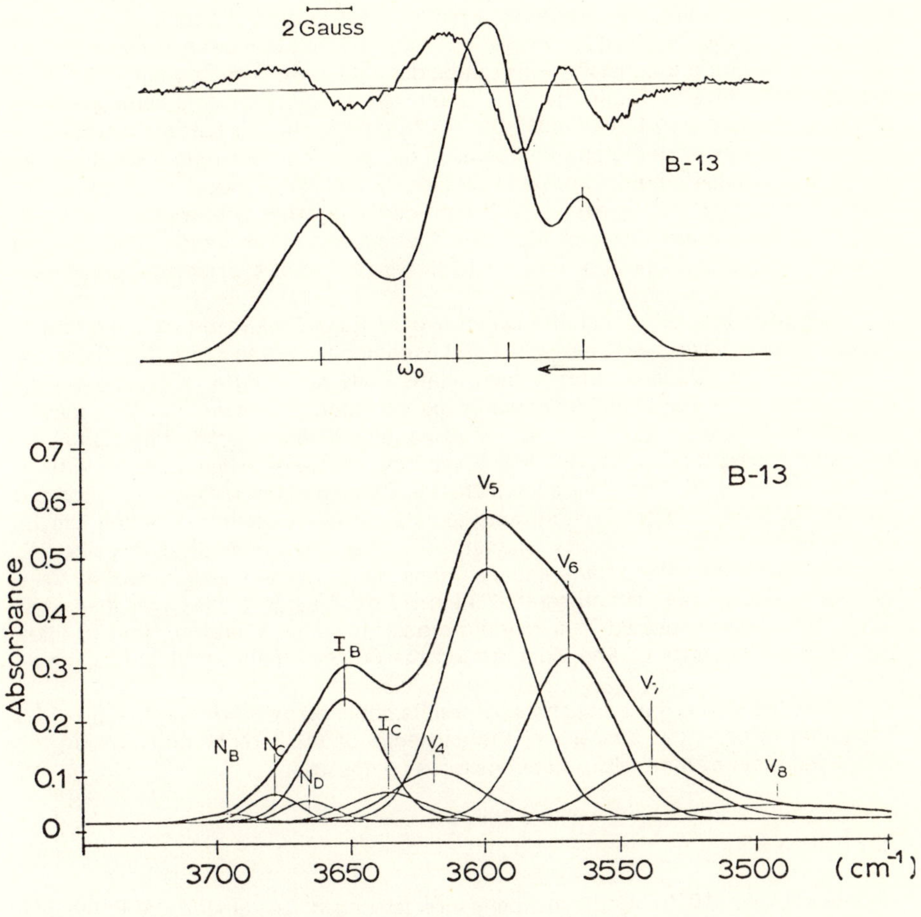

Figure 6-10. H^+ NMR (upper) and IR (lower) spectra of an Al-rich biotite. The NMR line is taken at 14 MHz with $H_o//c^*$. The important side-line at higher frequency is associated with a Al^{3+} Fe^{2+} vacancy site. In the IR spectra, I_c,V_4 and V_5 correspond to (Fe^{2+} Fe^{2+} Al^{3+}), (Al^{3+} Mg^{2+} vacancy) and Al^{3+} Fe^{3+} vacancy) sites, respectively.

into N, I and V components and therefore try to use the hydroxyls as a probe for cationic distributions. The way cations are distributed may then be studied by plotting the intensity ratio of appropriate bands against the ratio of the corresponding cationic composition determined by chemical analysis. This interpretation of I.R. spectra takes into account vacancies and different cations such as R^{2+}, R^{3+} and R^{4+}, it does not, however, consider the presence of F^- ions which, according to NMR, leads to very selective ionic distribution within the octahedral sheet.

Take now the N_A component associated with $3Mg^{2+}$ and the N_B component associated with $2Mg^{2+}$ and $1Fe^{2+}$. If x is the number of Mg^{2+} ions and z the number of Fe^{2+} ions, then assuming a random distribution the ratio of intensities can be written $N_A/N_B = x/3z$, where it is assumed that the absorption coefficient within the N group is constant. This can be done for various components but when the Mg^{2+}/Fe^{2+} ratio found by I.R. is plotted against the same ratio given by chemical analysis a 1 : 1 relationship is not found (6). In this random distribution model the points on the graph behave as if the Mg^{2+} content were lower than the chemical formulae would indicate. However, if the NMR model of fluorine domains, from which iron is excluded, is considered together with a plot of the I.R. over chemical analysis ratios of Mg^{2+}/Fe^{2+} versus the F^- content a definite trend is found. A clear tendency for fluorine to segregare toward Mg-rich environments is observed.

Another interesting parallel between the I.R. and NMR results is the case of Al-rich biotites having well resolved I and V components (see Fig. 6-10). Since the Fe^{2+} content of these biotites is high, statistically speaking an intense band I_C corresponding to $Fe^{2+}Fe^{2+}Al^{3+}$ would be expected. However, the I.R. spectra suggest the contrary. Moreover, as the Al content of these samples increases, the V_4 band corresponding to $Al^{3+}Mg^{2+}$ vacancy decreases whereas the V_5 band corresponding to $Al^{3+}Fe^{2+}$ vacancy increases. These observations are corroborated by the NMR H^+ spectra on the same samples where a distinct side-line can be associated with a Fe^{2+} vacancy geometry. Further, the intensity of this side line increases with the Al content and can thus be attributed unambiguously to a $Al^{3+}Fe^{2+}$vacancy site. The integrated intensity of this side-line is larger than what would have been expected in a random model. It seems, therefore, that in these biotites ordering patterns also exist around the OH especially as related to the Al cation.

In conclusion, the comparison of results obtained by NMR and I.R. provides interesting information concerning the existence of short range order within the octahedral sheet of minerals of heterogeneous composition.

REFERENCES

1. Joswig, W. 1972. Neutronenbengungsmessungen an einem 1M-Phlogopit. *Neues Jahrb. Mineral. Monatsh.* 1-11.
2. McCauley, J.W., R.E. Newnham and G.V. Gibbs. 1973. Crystal structure analysis of synthetic fluorophlogopite. *Am. Mineral.* **58**: 249-254.
3. Pake, G.E. 1948. Nuclear resonance absorption in hydrated crystals: fine structure of the H^+ line. *J. Chem. Phys.* **16**: 327-336.
4. Pake, G.E. 1962. Paramagnetic Resonance. Benjamin, New York.
5. Rousseaux, J.M., P.G. Rouxhet, L. Vielvoye and A. Herbillon. 1973. The vermiculitization of trioctahedral micas. I. K level and its correlation with chemical composition. *Clay Miner.* **10**: 1-16.
6. Rausell-Colom, J.A., J. Sanz, M. Fernandez and J.M. Serratosa. 1979. Distribution of octahedral ions in phlogopites and biotites. *Proc. Int. Clay Conf. 1978 (Pub. 1979)*: 27-36.
7. Sanz, J. and W.E.E. Stone. 1977. NMR study of micas. I. Distribution of Fe^{2+} ions on the octahedral sites. *J. Chem. Phys.* **67**: 3739-3743.

8. Sanz, J., J. Meyers, L. Vielvoye and W.E.E. Stone. 1978. The location and content of iron in natural biotites and phlogopites: a comparison of several methods. *Clay Miner.* **13:** 45-52.
9. Sanz, J. and W.E.E. Stone. 1979. NMR study of micas. II. Distribution of Fe^{2+}, F^{-} and OH^{-} in the octahedral sheet of phlogopites. *Am. Mineral.* **64:** 119-126.
10. Vedder, W. 1964. Correlations between infrared spectrum and chemical composition of mica. *Am. Mineral.* **49:** 736-768.
11. Wilkins, R.W.T. 1967. The hydroxyl-stretching region of the biotite mica spectrum. *Miner. Mag.* **36:** 325-333.

Chapter 7

GENERAL THEORY AND EXPERIMENTAL ASPECTS OF ELECTRON SPIN RESONANCE

Jacques C. Vedrine

Institut de Recherches sur la Catalyse — CNRS,
2, Av. A. Einstein, F 69626 Villeurbanne Cedex, France

7-1. INTRODUCTION

There has been increasing interest during the past decades in using different physical methods to better characterize inorganic solids such as clay minerals and catalysts. One of these physical methods, called "electron spin resonance" (ESR) spectroscopy, has been widely developed recently in its applications to structure determination of clays using paramagnetic probes such as transition metal ions. This technique originates from the experience of Zavoisky (79) in 1945 and was mainly developed in the late sixties. In this chapter the general theory of ESR and some experimental methods will be described.

7-1.1. Fundamental principles

The fundamental principles of ESR have been covered in detail by various authors (4, 7, 17, 19, 21, 56, 57, 69, 76). A somewhat briefer treatment follows.

Any spinning or rotating charge behaves like a magnet with its poles along the axis of rotation. An electron spinning about itself has a rotational angular momentum $\vec{S}$, designated its spin, and subsequently a magnetic moment $\vec{\mu}_e$ which is proportional to and colinear with $\vec{S}$. The expression $\gamma = \vec{\mu}_e / \vec{S}$ is called the gyromagnetic ratio. Along a quantification axis, the spin vector, $\vec{S}$, can take the value $\pm$ 1/2 in unit multiples of $\hbar$. If a system containing unpaired electrons (i.e., spin $\neq$ 0) is placed into an external magnetic field $\vec{H}$, the energy of the system is given by:

$$E = -\vec{\mu}_e.\vec{H} = -g_e\beta\,\vec{S}\cdot\vec{H} = \pm\,(1/2)g_e\,\beta\,H \qquad [7\text{-}1]$$

Where $\cdot$ represents a scalar product or dot product, g_e is a constant designated the g-factor and will be described in section 7-2, and β is the Bohr magneton for the electron. All the electronic spin axes are oriented by the magnetic field either in the same direction (+) (parallel) or in the opposite direction (—) (anti-

J. W. Stucki and W. L. Banwart (eds.), Advanced Chemical Methods for Soil and Clay Minerals Research, 331–389

parallel) with respect to $\vec{H}$. In the absence of an applied magnetic field the electrons are oriented at random. Under an applied magnetic field, $\vec{H}$ there are then two populations of spins and a difference in energy, ΔE, equal to

$$\Delta E = g_e \beta H. \qquad [7\text{-}2]$$

At equilibrium, the ratio of populations of both states is given by the Maxwell-Boltzmann law,

$$n^+/n^- = \exp(-\Delta E/kT) \simeq 1 - \Delta E/kT \qquad [7\text{-}3]$$

where k is the Boltzmann constant, and n^+ and n^- are the population of electrons with spin +1/2 and −1/2, respectively. It is worth noting at this stage that at 80 K, and for a microwave frequency of 9 GHz, n^+ and n^- populations differ by less than .005, which is much less than in optical spectroscopy.

When such an electronic spin system is subjected to a magnetic field $\vec{H}$ and a photon wave perpendicular to the magnetic field with a frequency of υ, a spin flip of electrons from antiparallel to parallel direction occurs when the photon energy, $h\upsilon$, is such that the condition $\Delta E = g_e\beta H = h\upsilon$ is obtained [7-2]. This is the "resonance condition" and results in an energy absorption as shown in Fig. 7-1. It also implies that the very first feature to remember is that ESR spectroscopy applies only to systems with at least one unpaired electron, i.e. to paramagnetic compounds. Note also that the whole material is concerned; that is, the technique is primarily a bulk technique, although it could be applied to surface chemistry when studying adsorbed species.

The electrons return to their initial spin states according to Boltzmann equilibrium, releasing their energy, $h\upsilon$, which is dissipated into the structure. This is designated as "spin-lattice relaxation" and is characterized by a time constant T_{1e} of the exponential decay in energy. Return to initial state also occurs for the spin phase by energy exchange between spins without energy loss to the lattice, and is designated as "spin-spin relaxation," characterized by the time constant T_{2e}. A more complete discussion of this is given by J.J. Fripiat in Chapter 5.

7-1.2. Experimental

As shown above, spin flip will occur when the material is submitted to both a continuous magnetic field and a microwave applied perpendicular to the magnetic field. Consequently, a spectrometer will be composed of a magnet giving a continuous magnetic field whose intensity may be varied over a large range (typically 0 to 10,000 G), a resonance cavity where the sample is introduced and a microwave source designated a klystron (typically 9,000 MHz). The microwave irradiates the sample by means of a wave guide (Fig. 7-2).

The experiment consists of sweeping the magnetic field until the resonance condition ($h\upsilon = g\beta H$) is fulfilled, resulting in an energy absorption. The detection is performed using magic T properties with a crystal detector located in one of the T arms. As in the case of a Wheatstone bridge, when misequilibrium is created by energy absorption due to resonance in one of the T arms where the sample is placed (resonance cavity), current variation occurs in the crystal detector giving an absorption curve.

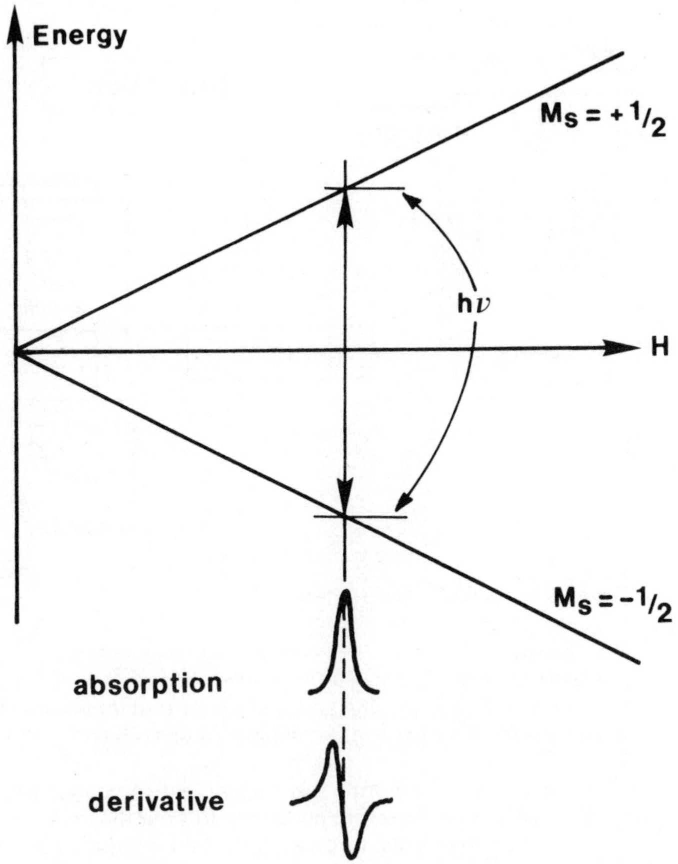

Figure 7-1. Energy level shift of the electron against an applied magnetic field.

In practice, absorption curves are weak, but their intensity may be sharply increased by modulating the magnetic field at low frequency (100 kHz). The result is a derivative curve.

A solid sample may be analyzed as a single crystal or as a powder. Single crystals may be attached to a sample holder within the resonance cavity and rotated in different directions. Powders are generally placed in silica tubes ($\simeq$ 30 mg), typically 5 or 6 mm o.d. The vertical detection region in the cavity is roughly 2 cm long with the sensitivity being the greatest at the center and decreasing uniformly (Gaussian type law) along the vertical axis. In some experiments, such as studying anisotropy of single crystals, the cavity and therefore the sample may be kept fixed while the magnet is rotated through 360°.

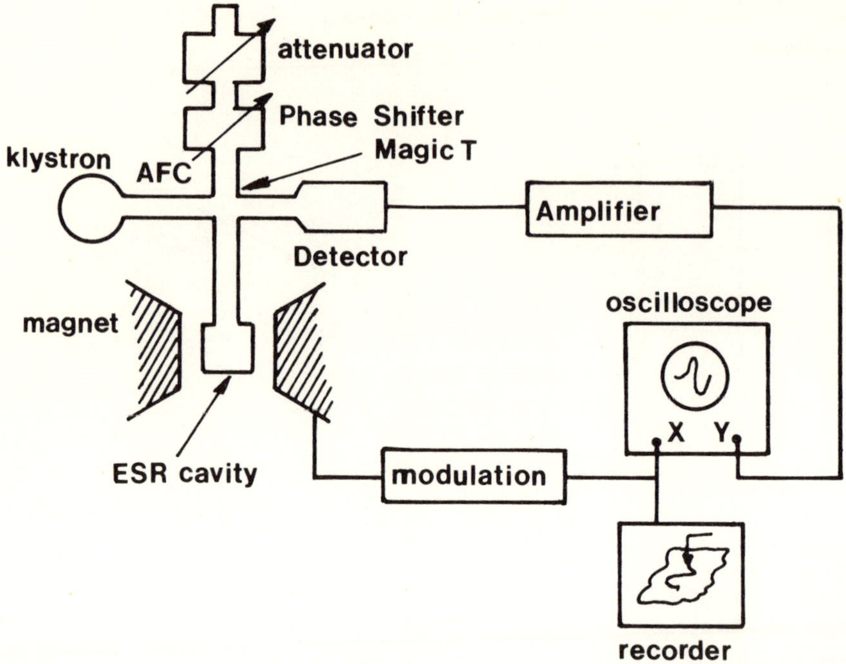

Figure 7-2. Scheme of an ESR spectrometer.

When outgassing a material is necessary, for instance when dipolar broadening of the signals due to paramagnetic oxygen occurs, the silica tube may be evacuated in an ordinary vacuum line and sealed off under vacuum. This is important when surface properties of the material have to be studied.

When introducing a sample into the magnet, the resonance frequency of the cavity is altered slightly. It is therefore necessary to tune the klystron frequency in accordance with that of the cavity in order to obtain stationary waves. Recording may be easily performed in a wide range of temperatures – typically 77 to 600 K. Additional accessories may be used to expand the temperature range down to 4.2 K (even 2.2 by pumping) or up to 1300 K.

As mentioned above, the spectrometer usually works with a 9,000 MHz klystron (λ = 3 cm, X band). However, for some experiments, other frequencies such as 23,000 MHz (K band) or 35,000 MHz (Q band, λ = 8 mm) are used but require special microwave bridges and different wave guides.

7-1.3. ESR Parameters

For a free electron, resonance occurs for a magnetic field intensity H according to the fundamental relationship

$$h\nu = g_e \beta_e H \qquad [7\text{-}3]$$

where g_e = 2.0023 and β_e is the Bohr magneton ($\beta_e = 9.273 \times 10^{-21}$ erg G^{-1}). The small deviation of g_e from 2.00 is due to relativistic correction. For such a free electron, a single resonance is always obtained at the same field position and the method should be then of very limited interest. Fortunately, however, the spin magnetic moment may be influenced by various interactions with the electron environment when the electron is involved in a molecule or a compound. These interactions result in a shift of the ESR line with respect to the free electron position and are thereby the principal factors responsible for the power of the technique. These interactions can be either electrical or magnetic.

The electron moves along a given orbit (molecular orbital for instance) resulting in a magnetic orbital moment ($\vec{\mu}_L$) and a kinetic or angular momentum ($\vec{p}_L$), designated by the quantum number L. In the same way, a nucleus rotating about itself has a nuclear magnetic moment ($\vec{\mu}_N$) and a kinetic or angular momentum designated its nuclear spin, $\vec{I}$. The magnetic interactions correspond to interactions between $\vec{\mu}_e$ and $\vec{\mu}_L$ or $\vec{\mu}_N$ or even $\vec{\mu}_e$ of other electrons (dipolar coupling). Electrical interactions involve interaction of the electron charge with neighboring charges (exchange interaction) or nuclei (quadrupolar interactions) or ions (crystal field influence). These different interactions are characterized by various parameters which will be considered in detail below. Each interaction corresponds to a given energy, and the Hamiltonian operator ($\mathcal{H}$) acting on the wave function (Ψ) of the electron is usually used to characterize the interaction. From the general Hamiltonian, only the part concerning the electron spin, designated the "spin Hamiltonian," is kept. In general, the Hamiltonian operator acts on the orbital wave function, resulting in eigenvalues that represent the energy of the system submitted to different types of interactions. The Hamiltonian is only a mathematical tool that is used for convenience in calculating eigenvalues, i.e. to represent the energies corresponding to the different interactions. Hence

$$\mathcal{H} \mid \Psi_k > = E \mid \Psi_k > \qquad [7\text{-}4]$$

and the eigenstates are given by

$$E_k = < k \,|\mathcal{H}|\, k> \qquad [7\text{-}5]$$

Where $|k>$ represents in short the wave function $|\Psi_k>$

The complete Hamiltonian for all interactions can be written

$$\mathcal{H} = \mathcal{H}_E + \mathcal{H}_V + \mathcal{H}_{LS} + \mathcal{H}_{SH} + \mathcal{H}_{LH} + \mathcal{H}_{SS} + \mathcal{H}_{SI} + \mathcal{H}_{IH} + \mathcal{H}_Q \qquad [7\text{-}6]$$

Where,

$\mathcal{H}_E$ = electronic interaction,
$\mathcal{H}_V$ = crystal field interaction, ($V = \sum_{i,j} Q_i/r_{ij}$; Q_{ij} = ionic charges)
$\mathcal{H}_{LS}$ = spin-orbit coupling,
$\mathcal{H}_{SH}$ = electron Zeeman effect,
$\mathcal{H}_{LH}$ = orbital Zeeman effect,
$\mathcal{H}_{SS}$ = electron spin-spin interaction,
$\mathcal{H}_{SI}$ = electron spin — nuclear spin interaction,

$\mathcal{H}_{IH}$ = nuclear Zeeman effect,
$\mathcal{H}_Q$ = quadrupolar interaction.

The relative energy domains of these interactions are given in Table 7-1. The largest of these terms is the crystal field interaction, and is determined by the expression

$$\mathcal{H}_E = \sum_i \left(\frac{p_i^2}{2m} - \frac{Ze^2}{r_i}\right) + \sum_{i<j} \frac{e^2}{r_{ij}} \qquad [7\text{-}7]$$

where $\mathcal{H}_E$ is the summation of the kinetic energy of each electron, the potential energy relative to the nuclei and the interelectronic replusive energies; p_i, the angular momentum of the electron; Z, the nuclear charge; e, the electronic charge; m, the electron mass; and r_i, the electron-nucleus distance.

Table 7-1. Energy domains in cm^{-1} of the various interactions influencing the spin magnetic moment of an electron in a matrix.

$\mathcal{H}_E = 10^5$	$\mathcal{H}_{SH} = 1$	$\mathcal{H}_{SI} = 10^{-2}$
$\mathcal{H}_V = 10^4$	$\mathcal{H}_{LH} = 1$	$\mathcal{H}_{IH} = 10^{-3}$
$\mathcal{H}_{LS} = 10^2$	$\mathcal{H}_{SS} = 1$	$\mathcal{H}_Q = 10^{-3}$

The term $\mathcal{H}_{LS} + \mathcal{H}_{SH} + \mathcal{H}_{LH}$ corresponds to the direct interaction of the spin with its orbit and plays a determining role in the calculation of the so-called g-factor (See Section 7-2). The $\mathcal{H}_{SI} + \mathcal{H}_{IH}$ term characterizes the interaction of the nuclei and will be considered below in the so-called "hyperfine structure" (Section 7-3). The $\mathcal{H}_{SS} + \mathcal{H}_Q$ term characterizes the interelectron interaction and will be analyzed in the so-called fine structure (Section 7-4). These are the three ESR parameters of main interest.

Before considering them in detail, it is appropriate to first describe relaxation phenomena, line widths and line shapes in ESR (9, 20, 41, 56).

7-1.4. Relaxation Processes and Line Widths

A general and comprehensive treatment of relaxation processes and line widths necessitates consideration of three main effects (20): (a) interaction of each electronic dipole with all other electronic and nuclear dipoles and surrounding diamagnetic molecules (spin-spin and spin-lattice interaction); (b) effect of vibrational, rotational and translational motion of these species and of the electron exchange between them (motional and exchange modulation); and (c) chemical reactions of the paramagnetic species.

These interactions can increase the linewidth by decreasing the life time of the spin state undergoing the observed transition, or by changing the magnetic field experienced by the spin at the instant of the transition. Line broadening may also be caused by unresolved nuclear hyperfine structure (see below) or by inhomogeneity in the paramagnetic species.

To enter into the details of relaxation mechanisms and linewidth is beyond the purpose of this Chapter, since more precise papers and books have been published on this subject (7, 55). Suffice it to give some simplified ideas, and to focus attention on the effects of spin-spin and spin-lattice relaxation on line width.

Spin-spin interaction. Spin-spin interaction corresponds to dipolar interaction between electrons or between the electron and the nucleus. The dipolar Hamiltonian can be written

$$\mathcal{H}_{\mathrm{dip.}} = -\gamma_1 \gamma_2 \hbar^2 \left[\frac{\vec{\mu}_1 \vec{\mu}_2}{r^3} - 3\,\frac{(\vec{\mu}_1 \vec{r})(\vec{\mu}_2 \vec{r})}{r^5}\right] \qquad [7\text{-}8]$$

where subscripts 1 and 2 correspond to the two partners, and $\vec{r}$ is the vector between them; and the gyromagnetic ratios are defined by $\gamma = \vec{\mu}/\vec{S}$. This interaction thus depends on the relative angles between $\vec{\mu}$ and $\vec{r}$ and on the distance r between the spins involved. The result is a distribution of resonances around a mean position. Therefore, a peak is an envelope of resonances, characterized by the linewidth. As it turns out, the concentration of spins greatly influences the linewidth – the more concentrated the paramagnetic centers, the greater the broadening. Hence, every effort should be made to sufficiently dilute the paramagnetic species in a diamagnetic matrix so as to avoid dipolar broadening. This is very important for clay mineral studies, since paramagnetic probes are often used. ESR methods are particularly efficient for dilute systems because when paramagnetic concentration increases, one first observes dipolar broadening followed by narrowing due to rapid electron exchange. The latter effect smears out all hyperfine or fine structure, thus precluding determination of the three ESR parameters mentioned above and thereby defeating the primary purpose of the method.

Spin-lattice interaction. The spin-lattice interaction is mainly related to spin-orbit coupling and orbit-lattice interaction. When both spin-spin and spin-lattice relaxations contribute to the ESR line width, one can write

$$\Delta H \simeq \frac{1}{T_{2e}} + \frac{1}{T_{1e}} \qquad [7\text{-}9]$$

where ΔH is the line width, and T_{2e} and T_{1e} are the spin-spin and spin-lattice relaxation times, respectively.

In general, $T_{2e} < T_{1e}$ and the linewidth depends mainly on spin-spin interactions. However, when T_{1e} becomes very short ($<10^{-7}$ sec), its effect on the lifetime of a species in a given energy level makes an important contribution to the linewidth. For free radicals or S-state ions, spin-orbit coupling is negligible and T_{1e} is large. By contrast, T_{1e} for transition ions or rare earth ions may be very small and broad lines are obtained. In some cases the ESR lines are broadened beyond detection. It has been shown that T_{1e} is then inversely proportional to temperature ($T_{1e} \propto T^{-n}$), the value of n depending on the relaxation mechanism. In such cases, cooling down of the sample increases T_{1e} and gives rise to detectable ESR lines. It is therefore necessary to record such spectra at liquid nitrogen or even at liquid helium temperature. Obviously, in order to do this it is also necessary to have the instrumental capability for a wide range of recording temperatures. Note also at this stage that ESR line intensities, which depend on paramagnetic susceptibility, are inversely proportional to the absolute temperature (Curie law), and the capability of recording spectra at low temperatures could be very interesting and informative as to the magnetic properties of the sample.

Saturation effects. Another aspect of relaxation that must be considered carefully is if the lattice relaxation time, T_{1e}, is too long after spin flip, electrons may have insufficient time to return to their initial state. The result is that n^+ and n^- populations become equal and the ESR signal intensity decreases, i.e. it is no longer proportional to the number of unpaired electrons or paramagnetic species. To avoid this effect, known as "saturation," the sample must be exposed to low microwave power. In order to determine the microwave intensities that avoid saturation, one plots the line intensity versus the square root of the microwave power (Fig. 7-3), then selects a value in the linear part of the curve, which is the unsaturated region.

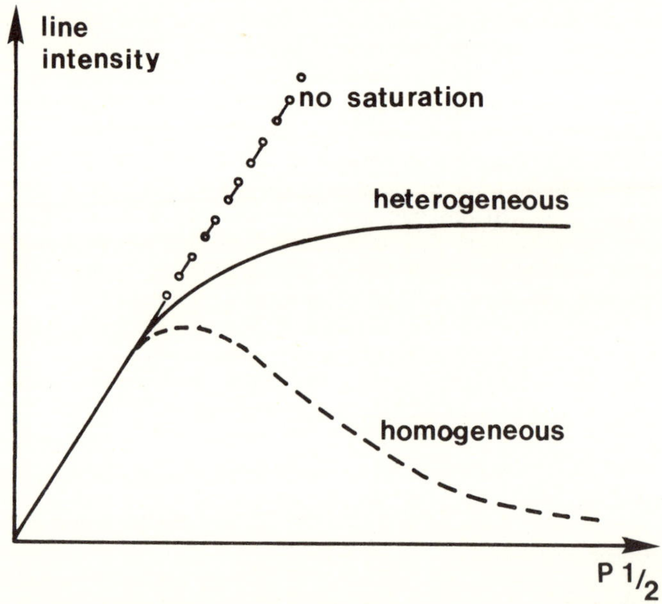

Figure 7-3. Saturation curves obtained by plotting line intensities against the square root of microwave power, P.

Homogeneous saturation is characteristic of a simple ESR line, such as one that arises from a single ESR parameter. Heterogeneous saturation occurs when a line arises from the overlapping of several individual lines. The result is line broadening.

Exchange interaction. Exchange interaction corresponds to exchanges between electrons that occur when there exists large overlap between the wave functions of these electrons. For fast exchange, only the mean position of the electron resonance will be recorded. If both electrons belong to similar ions or radicals, the line will be narrowed at the top and broadened at the bottom resulting in a Lorentzian line-shape. If interacting electrons belong to different ions or radicals, a Gaussian shape will be observed.

Effect of Relaxation on the ESR Spectra. Spin-lattice relaxation occurs because of various interactions between an electron spin and its surroundings (20, 77). These interactions enable the spin system to exchange energy with the translational, rotational and vibrational motions of its environment, i.e. the lattice or surrounding molecular structure. The rate at which the energy can be transferred is determined by the nature of the coupling between the spin system and the structure, and is characterized by T_{1e} which may vary from several seconds to 10^{-9} seconds or less.

A general picture of spin-lattice relaxation processes may be visualized as follows. The magnetic interaction between an electron spin and its environment causes the electron to "see" an additional magnetic field, h_L, that is superimposed on the external field but not necessarily parallel to it. This additional magnetic field fluctuates with the molecular motions of the system, and may be described by a Fourier series having a very broad frequency spectrum. Included in that frequency spectrum is a component at the transition frequency of the electron spin, which is then capable of inducing additional electron spin transitions.

If the magnetic field, h_L, associated with the spin-lattice relaxation fluctuates between the limits $\pm H_L$ with a time constant τ_c (τ_c is known as the correlation time for the molecular motions that cause the field to vary), then the theory of spin-lattice relaxation gives the following expression (3):

$$\frac{1}{T_{1e}} = \frac{2}{3}\left(\frac{g\beta H_L}{\hbar_L}\right)^2 \left(\frac{\tau_c}{1+(2\pi\upsilon_o\tau_c)^2}\right) \qquad [7\text{-}10]$$

where υ_o is the electron spin transition frequency. This equation shows that the spin-lattice relaxation time is very sensitive to the strength of the coupling between the spin system and the lattice. In Chapters 8 and 9 of this book are given detailed examples of motion effects as well as examples of spin labelling (17, 31, 73) using such a property. If T_{1e} becomes very short, the individual magnetic energy levels of the spin system are broadened in accordance with the uncertainty principle, $\Delta E \cdot \Delta t \geqslant \hbar$, where the lifetime in a given state (Δt) may be roughly equated to T_{1e}. Thus, ESR lines are broadened according to the relation

$$\Delta\upsilon_o \simeq \frac{1}{2\pi T_{1e}} \qquad [7\text{-}11]$$

This problem of line broadening due to an overefficient, spin-lattice relaxation mechanism is often encountered in compounds with a degenerate electronic ground state or with more than one unpaired electron. Frequently, the degeneracy of the ground state is lifted by interaction with nuclei within the structure, causing the energy levels of the spin transitions to be split. Thus, nuclear interaction plays a major role in determining the g-factor (see below). When this coupling is very strong, and if there are any appreciable motions of either the paramagnetic compound or the host structure, the spin-lattice relaxation will be very rapid and no resonance is observed. This has been proposed as the reason why paramagnetic compounds with degenerate ground states, such as several transition metal ions, have not been observed by ESR in many systems where they are known to be present.

The conclusion that can be drawn at the present time is that ESR spectroscopy can detect paramagnetic compounds, but if the relaxation mechanisms are too fast, ESR signals may not be seen. In other words, it is correct that when an ESR signal is observed, a paramagnetic compound is necessarily involved, while the absence of an ESR signal does not necessarily mean that paramagnetic compounds are absent in a material.

Another point is also worth emphasizing. When a strong magnetic dipole-dipole interaction takes place between different magnetic species there exists a very efficient relaxation path, provided there is a slight overlap in the magnetic energy levels. This cross-relaxation process is important for surface studies because O_2 from the air is paramagnetic with two unpaired electrons and can contaminate the surface very readily. This results in a dipolar broadening that masks the signal of the paramagnetic species under study. One way to determine if a paramagnetic species is present in the outermost layer of a material ($d \leqslant 10$ Å) is to record its ESR spectrum in vacuum and in air or O_2. The presence of an ESR signal in both cases clearly indicates that the paramagnetic species is in the bulk. If there is a signal in vacuum which disappears reversibly in air or O_2, then the paramagnetic species is in the surface layer.

7-1.5. Line Intensity

The line intensity is an important characteristic of an ESR spectrum since ESR spectroscopy can be a very reliable method for quantitative determination of spin concentrations. This arises from the fact that the spin populations directly determine susceptibility. The line intensity is proportional to the magnetic susceptibility, X_o, given by the relationship

$$X_o = \frac{1}{3\,k\,T} g^2\, \beta^2\, N_o J\,(J+1) \qquad [7\text{-}12]$$

where $\vec{J} = \vec{L} + \vec{S}$ and N_o is the number of unpaired spins.

To perform quantitative measurements one must first integrate the derivative curves point by point in order to get the absorption lines. Then the surface area under the integrated lines is determined and compared with a standard sample recorded under the same conditions (microwave power, modulation width, amount of material. . .). Standard samples are usually either diphenyl picryl hydrazyl free radical (DPPH) dissolved in known amounts of benzene or a pitch (C in KCl) from Varian (strong pitch equals 3×10^{14} spins per cm length).

7-2. G-FACTOR TENSOR

7-2.1. General Introduction

The g-factor (11, 18) was defined previously (Section 7-1.1) by the fundamental relationship

$$h\upsilon = \Delta E = g\beta H. \qquad [7\text{-}2]$$

It also corresponds to the Landé factor given by

$$g_L = 1 + \frac{J(J+1) + S(S+1) - L(L+1)}{2J(J+1)} \qquad [7\text{-}13]$$

where $\vec{J}$ is the kinetic momentum equal to the vectorial sum of the orbital ($\vec{p}_L$) and spin ($\vec{S}$) momenta. $\vec{p}_L$, the orbital angular momentum, is defined by the vectorial product

$$\vec{p}_L = m\,\vec{r}\,\vec{v} \qquad [7\text{-}14]$$

where m is the electron mass, $\vec{v}$ is its velocity, and $\vec{r}$ is the orbital radius. For a free electron or a S-state ion, L = O and subsequently g_L = 2. In fact, g_L = 2.0023 because of relativistic correction. For the rare earth ions, g-factor values are given by equation [7-13]. This is not observed to be true for transition ions. This "discrepancy" is explained by considering that the paramagnetic 3d levels of these transition ions may be influenced by very intense electrical fields due to ions or polar molecules attached to the central ion. This acts to break (more or less completely) the coupling between $\vec{L}$ and $\vec{S}$, since $\vec{L}$ is sensitive to electrical fields while $\vec{S}$ is not. There then exist 2L + 1 orbital sublevels and one can consider the orbital moment, $\vec{p}_L$, as blocked or "quenched" by the crystal field, or that it is non-orientable under the influence of a magnetic field. The paramagnetism then stems only from electron spins of magnetic moment

$$\mu = g_L\beta S = 2\beta\sqrt{S(S+1)} \qquad [7\text{-}15]$$

For transition metal ions, the L–S coupling is only partly quenched and the spin-orbit coupling interaction is represented by the operator

$$\mathcal{H}_{SO} = \lambda\,\vec{p}_L \cdot \vec{S} \qquad [7\text{-}16]$$

where λ is a constant designated the spin-orbit coupling constant. The spin Hamiltonian can then be written

$$\mathcal{H}_S = \beta\vec{H}\cdot\mathbf{g}\cdot\vec{S} = \lambda\vec{p}_L\vec{S} + \beta\vec{H}\,(\vec{p}_L + g_e\cdot\vec{S}) \qquad [7\text{-}17]$$

where **g** is a second order tensor defined by the latter relationship. The g_L (Landé) factor thus becomes meaningless. The **g** tensor can be represented by the third order matrix operator

$$\begin{bmatrix} g_{xx} & g_{xy} & g_{xy} \\ g_{yx} & g_{yy} & g_{yz} \\ g_{zx} & g_{zy} & g_{zz} \end{bmatrix} \qquad [7\text{-}18]$$

When the x, y, and z axes are chosen to be those of the **g** tensor, one has a diagonal tensor, viz.

$$\begin{bmatrix} g_{xx} & 0 & 0 \\ 0 & g_{yy} & 0 \\ 0 & 0 & g_{zz} \end{bmatrix} \qquad [7\text{-}19]$$

Equation [7-17] may also be represented by defining the "effective" magnetic field H_{eff} by

$$\vec{H}_{eff} = \mathbf{g}\cdot\vec{H}_o \qquad [7\text{-}20]$$

This effective field is in general not colinear with the actual field, $\vec{H}_o$, since g-factor values depend on the orientation. One can write

$$\vec{H}_{eff} = \mathbf{g}.\vec{H}_o = (\vec{i}g_{xx}\vec{i} + \vec{j}g_{yy}\vec{j} + \vec{k}g_{zz}\vec{k}) \cdot (\vec{i}H_x + \vec{j}H_y + \vec{k}H_z) \qquad [7\text{-}21a]$$

and

$$\vec{H}_{eff} = \vec{i}\,g_{xx}H_x + \vec{j}\,g_{yy}H_y + \vec{k}\,g_{zz}H_z \qquad [7\text{-}21b]$$

If $\vec{H}$ is expressed in polar coordinates, θ and ϕ, with respect to the x, y, and z axes defined as the main axes of the **g** tensor, one can write further

$$\frac{\vec{H}_{eff}}{H_o} = \vec{i}\,g_{xx}\frac{H_x}{H_o} + \vec{j}\,g_{yy}\frac{H_y}{H_o} + \vec{k}\,g_{zz}\frac{H_z}{H_o} \qquad [7\text{-}22a]$$

or

$$g = \text{mod.}\,\{\vec{i}\,g_{xx}\sin\theta\cos\phi + \vec{j}\,g_{yy}\sin\theta\sin\phi + \vec{k}\,g_{zz}\cos\theta\} \qquad [7\text{-}22b]$$

resulting in

$$g^2 = g^2{}_{xx}\sin^2\theta\cos^2\phi + g^2{}_{yy}\sin^2\theta\sin^2\phi + g^2{}_{zz}\cos^2\theta. \qquad [7\text{-}23]$$

Thus the g-value is given by the modulus of the $\vec{H}_{eff}$ vector whose extremity describes an ellipsoid with semi axes g_{xx}, g_{yy} and g_{zz}. As a matter of fact, the components of the $\vec{H}_{eff}/H_o$ vector along the x, y, and z axes are $g_{xx}\eta_x$, $g_{yy}\eta_y$ and $g_{zz}\eta_z$ where η_x, η_y and η_z are the director cosines of $\vec{H}_o$ along the x, y, and z axes which fulfill the ellipsoid formula

$$\frac{x^2}{g_{xx}^2} + \frac{y^2}{g_{yy}^2} + \frac{z^2}{g_{zz}^2} = 1. \qquad [7\text{-}24]$$

Since

$$\frac{g_{xx}^2\,\eta_x^2}{g_{xx}^2} + \frac{g_{yy}^2\,\eta_y^2}{g_{yy}^2} + \frac{g_{zz}^2\,\eta_z^2}{g_{zz}^2} = 1 \qquad [7\text{-}25a]$$

or

$$\eta_x^2 + \eta_y^2 + \eta_z^2 = 1. \qquad [7\text{-}25b]$$

For clarity, a representation of the g-factor is given in Fig. 7-4.

The important point to note at this stage is that equation [7-23] shows that g-factor values depend on the orientation of g-factor coordinates relative to the magnetic field. Thus, anisotropy could occur leading to complex ESR spectra.

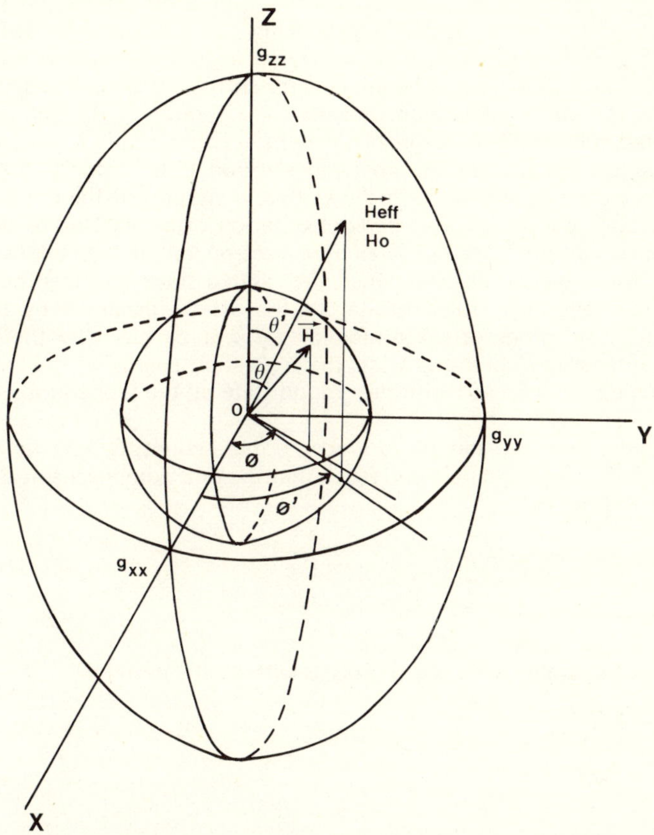

Figure 7-4. Ellipsoidal representation of variations in the g-tensor value in the x, y, z coordinates.

Values for the g-factor tensor may be calculated theoretically, considering the Hamiltonian (equation [7-17]) as a perturbation of the spin Hamiltonian and using perturbation theory. Considering the spin-orbit coupling interaction as a perturbation, it can be shown that

$$g_{ij} = g_e \delta_{ij} - 2\lambda \cdot \sum_{n \neq 0} \frac{\langle 0|L_i|n\rangle\langle n|L_j|0\rangle}{E_n - E_o} \qquad [7\text{-}26]$$

where i, j correspond to two axes of the coordinate system, δ_{ij} is the Kronecker symbol, ($\delta_{ij} = 1$ for $i = j$ and $\delta_{ij} = 0$ for $i \neq j$). 0 represents the ground state and n the different excited states.

7-2.2. Uses of g-factor Values

Equation [7-26] is important and a number of interesting, qualitative features may be extracted from it: (a) The g-factor shifts ($g-g_e$) depend on the magnetic field orientation because the alignment of the electron spins along H by the Zeeman interaction determines, via the spin orbit interaction (L-S), which component of L will be active in producing the g-factor shift. Different components of L differ indeed in their ability to couple the various states with the ground state and thus give different g-factor shifts. (b) Since the spin-dependent part of the spin-orbit interaction involves the operator S, the contribution of a given excited state to the g-factor shift may have either sign (+ or −), depending on whether or not the spin of the unpaired electron is the same as the spin of the excited electron. Equation [7-26] shows that excitation of the unpaired electron gives a negative g-shift ($g<g_e$) whereas excitation of an electron of opposite spin gives a positive g-shift ($g>g_e$). The excited electron will have an opposite spin with respect to the unpaired electron when the excited state is formed by exciting an electron from an inner, filled orbital to the orbital occupied by the unpaired electron. (c) The order of magnitude of Δg will be given by the ratio of the spin-orbit interaction constant, λ, to the excitation energy of the lowest excited states which can be admixed with the ground state by the L operator.

Consider now an example to illustrate these concepts. Let the z-axis be the direction of the magnetic field and recall that the quantum mechanical form of the L operator is given by

$$\ell_z = -i\,(dx/dy - dy/dx) \qquad [7\text{-}27]$$

where i is $\sqrt{-1}$.

The results of operating on s and p orbitals with ℓ_z are given by

$$\ell_z|s\rangle = 0 \qquad [7\text{-}28a]$$

$$\ell_z|p_z\rangle = 0 \qquad [7\text{-}28b]$$

$$\ell_z|p_x\rangle = i|p_y\rangle \qquad [7\text{-}28c]$$

$$\ell_z|p_y\rangle = -i|p_x\rangle. \qquad [7\text{-}28d]$$

The paramagnetic NO_2 molecule is known to be bent 134° in the ground state with the unpaired electron occupying a nitrogen hybrid orbital directed along the external bisector of the ONO angle. When the magnetic field is perpendicular to the NO_2 plane, two excited states may be involved. Excited state 1 is produced by exciting an electron from one of the NO bonding orbitals into the non-bonding, ground-state orbital. Because of the Pauli exclusion principle, the spins must be opposite, resulting in a positive value for Δg_1. The second excited state is just the reverse, the unpaired electron is excited into an NO σ antibonding orbital. This gives a negative value for Δg_2. Although a detailed calculation of energy levels and the wave function are required for a definite answer, the antibonding orbital is known to be further above than the NO σ orbital is below the non-bonding, ground-state orbital. Therefore, the first excited state must be dominant and Δg_2 is expected to be positive, and g_2 can be assigned a value of 2.0062.

If the direction of the magnetic field is changed so that $\vec{H}$ lies along the third axis of the NO_2 molecule, i.e. in the plane perpendicular to the bisector axis, L can excite the unpaired electron from its sp orbital into a nitrogen orbital along the second axis which lies perpendicular to the plane. This is believed to be a low-lying excited state because it differs in energy from the ground state only by virtue of the bending of the NO_2 molecule. The g shift is then expected to be negative since the Nitrogen orbital is empty. This prediction is consistent with the observed value of 1.9910 ($\Delta g_3 = -0.0113$).

At least three kinds of information can be obtained once the g-values are known. First, an idea as to the type of molecular motion can be deduced. This is possible because a rotation around a given axis averages out the g-factor components from the other two axes, resulting in one value for g and one for $g_\perp$ (g means parallel to the rotation axis and $g_\perp$ is the average of the other two components, $\frac{g_1 + g_2}{2}$. For a complete rotation around all axes, or a random motion, the three g values are averaged to an isotropic g-value, i.e. $g_{ave} = \frac{g_1 + g_2 + g_3}{3}$.

Second, paramagnetic species can be identified. Once the g-values have been extracted from an experimental spectrum (see below), the paramagnetic species may be immediately identified in many cases simply by comparing the observed g-value with either theoretical or known values for paramagnetic species. The g-values are given by the equation [7-26] which shows that the g shift with respect to g_e depends on the coupling between the ground state and different excited states. When the excited state arises by the transition from a half-filled orbital to an empty orbital the terms $(E_n - E_o)$ and λ are positive. This gives rise to a g-value lower than g_e. On the other hand, when the electron is excited from a filled orbital into a half-filled orbital, an "electron hole" results at the lower energy level, and $(E_n - E_o)$ is negative yielding a g-value greater than g_e. λ is assigned a negative sign by convention.

Two well known examples illustrate these qualitative predictions. In NO (nitrous oxide) molecules, which are paramagnetic, there is one electron in the antibonding π^* orbitals; while in O_2^- (superoxide) there are three electrons. Upon adsorption of these paramagnetic species onto an oxide surface, the surface crystal field splits the π^* electron energy level thus removing its degeneracy to give two π^* orbitals separated by an energy of Δ. So, for NO the lone electron is excited from the lower, half-filled π^* orbital to the higher, empty π^* orbital. Whereas, for O_2^-, one electron from the lower, filled orbital is excited into the higher, half-filled orbital. Consequently, the g-values are less than g_e (in a 1.9 to 2.0 range) for NO and larger than g_e for O_2^- (2.038, 2.008 and 2.002 for instance for O_2^- adsorbed on Al_2O_3), as expected.

Third, the symmetry of the ion environment can be characterized. Equation [7-26] clearly indicates that g values are dependent on the spin-orbit coupling between the ground state and the different excited states. However, the crystal field to which the paramagnetic ion is submitted greatly influences the magnitude of separation in orbital levels. This results in a variation in g values, which may then be correlated to perturbations in crystal field symmetry.

7-2.3. Transition metal ions

The properties of the ground-state level of free transition metal ions with d^n electronic configurations (where n = 1 to 9) are reported in Table 7-2.

Table 7-2. Characteristics of the ground-state levels of free transition metal ions.

Number of d electrons	quantum numbers S	L	J	Spectroscopic terms $(^{2S+1}L_J)$	orbital degeneracy
1	1/2	2	3/2	$^2D_{3/2}$	5
2	1	3	2	$^3F_2$	7
3	3/2	3	3/2	$^4F_{3/2}$	7
4	2	2	0	$^5D_0$	5
5	5/2	0	5/2	$^6S_{5/2}$	1
6	2	2	4	$^5D_4$	5
7	3/2	3	9/2	$^4F_{9/2}$	7
8	1	3	4	$^3F_4$	7
9	1/2	2	5/2	$^2D_{5/2}$	5

The orbital degeneracy of the free ion is 1, 5, or 7, but when placed in a crystal structure, or when the ion is coordinated to polar molecules, this orbital degeneracy is partially removed by the crystal field.

Two theorems play an important role in the understanding of the orbital energy levels of transition metal ions: (a) The Jahn-Teller theorem (50) states that in any orbitally-degenerate ground state there will be a distortion to remove the degeneracy, except in linear molecules and in systems having Kramer's doublets; and (b) Kramer's theorem states that any system containing an odd number of electrons will show at least two-fold degeneracy in the absence of a magnetic field. It follows from (b) that ions having an even number of electrons exhibit a complete splitting of energy levels, i.e. no ground-state degeneracy. In most cases the energy difference of this splitting is very large ($>1\ cm^{-1}$) so no ESR signal is expected. In ions with an odd number of electrons, levels with Kramer's degeneracy are present and an ESR signal is expected since the applied magnetic field splits this degeneracy.

Indeed the spin-orbit constant, ξ, which measures the energy of the interaction between the spin and the orbital angular momentum of the electron, is a property of the electron configuration. It may be calculated from the expression

$$\xi = \frac{Z_{eff}e^2}{2m^2 c^2 <r^3>} \qquad [7\text{-}29]$$

where e, m, and c bear their usual significance. ξ is thus a positive quantity but depends on the effective nuclear charge, Z_{eff}, and on the average distance of the electron from the nucleus, $<r>$. ξ is also related to λ, the spin-orbit interaction constant, by the expression

$$\lambda = \pm \xi/2S \qquad [7\text{-}30]$$

where S is the spin multiplicity.

In the literature, the intensity of crystal fields is often referred to as being either weak, medium or strong. The strong field is defined as one with sufficient strength to pair electrons such that they occupy lower levels. Medium and weak fields obviously correspond to lesser strengths usually not high enough to pair electrons.

According to Hund's law the d electrons are distributed in the five d orbitals in such a way as to maximize S. The result, then, is that d^1 and d^6 configurations split in the same manner when exposed to a given crystal field. The same considerations prevail for d^2 and d^7, d^3 and d^8, and d^4 and d^9. To illustrate this principle more clearly, examples for each configuration are given below and in reference 24.

d^1 Ions (Mo^{5+}, Ti^{3+}, V^{4+}, W^{5+}, Cr^{5+}, etc.). Coordination compounds of these ions usually exhibit either octahedral (O_h) or tetrahedral (T_d) symmetry, which splits the five-fold degenerate state of the free ion into two states of different energy; namely, t_{2g} (triply degenerate) and e_g (doubly degenerate) (30, 39, 42, 46, 59, 68). Under the effect of Jahn-Teller distortion or through spin-orbit coupling the degeneracy of the ground state is finally removed. Furthermore, when a component with a crystal field of lower symmetry is superimposed on the O_h or T_d symmetry further splitting of the t_{2g} and e_g states will occur. Consider first the effects of these distortions when the ion is in octahedral coordination. The splitting that removes five-fold degeneracy as the ion is placed in an octahedral crystal field is shown in Fig. 7-5.

Two types of distortion can occur: (a) for elongated octahedral or square pyramidal complexes, the unpaired electron occupies the doubly degenerate level, (which corresponds to the d_{yz} and d_{xz} orbitals), which is the lowest level and no ESR spectrum would be observed; and, (b) for compressed octahedral or square pyramidal complexes the non degenerate t_{2g} (d_{xy}) level lies lowest. The g-values are then given by (see Appendix 7-1)

$$g_{xx} = g_{yy} = g_{\perp} = g_e - 2\lambda/\delta \qquad [7\text{-}31a]$$

$$g_{zz} = g_{\parallel} = g_e - 8\lambda/\Delta \qquad [7\text{-}31b]$$

where δ and Δ are defined in Fig. 7-5.

An example of a $3d^1$ ion in tetragonally distorted octahedral symmetry is given by the ESR spectrum of Ti^{3+} in CaO with $g_{\parallel} = 1.9427$ and $g_{\perp} = 1.9380$. The experimental results of $g_{\perp} < g_{\parallel} < g_e$ fit the predicted results of equations [7-31]. For comparison, note that Ti^{3+} ion in anatase (TiO_2) has $g_{\perp} = 1.99$ and $g_{\parallel} = 1.959$ while in rutile $g_{\perp} = 1.975$ and $g_{\parallel} = 1.9360$.

When the ion is in a purely tetrahedral crystal field the five-fold degeneracy is split into a two-fold degenerate e_g ground state and an upper, three-fold degenerate t_{2g} state. A tetragonal distortion may lift the ground state degeneracy as shown in Fig. 7-6.

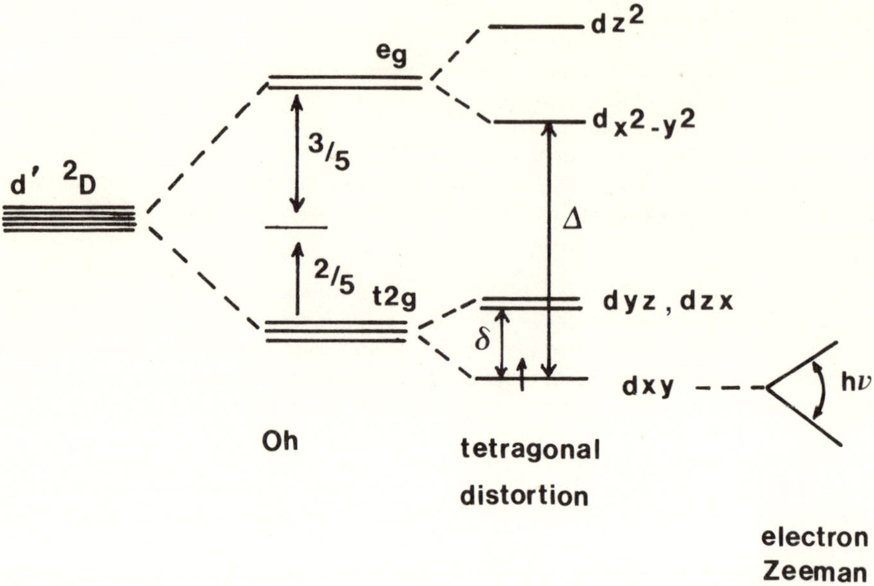

Figure 7-5. Energy levels of a d^1 ion placed in an octahedral field with a tetragonal distortion (octahedral compression).

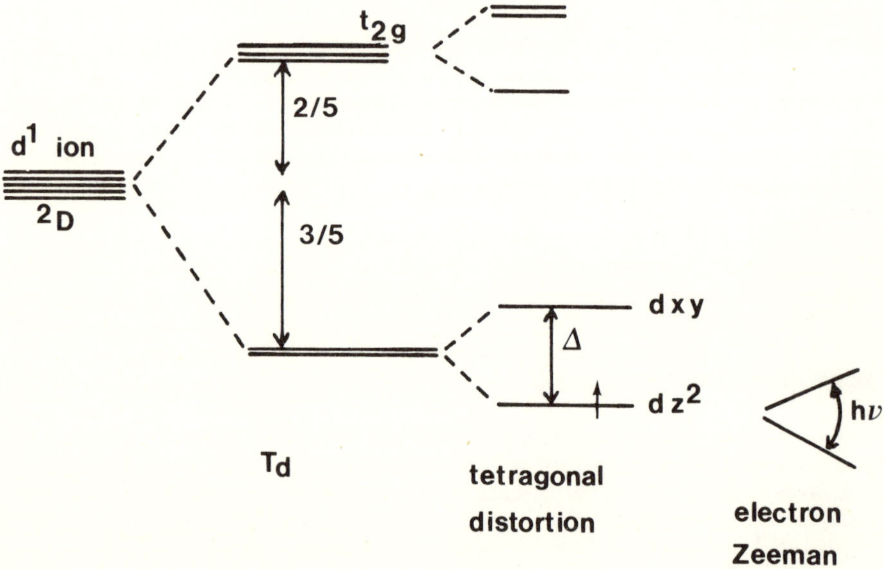

Figure 7-6. Energy levels of a d^1 ion placed in tetrahedral symmetry with a tetragonal distortion (compression).

However, the ground state could be either the d_{z^2} or the d_{xy} orbital depending on the distortion.

For a *compressed* tetrahedron the d_{z^2} level lies lower and one can show that g-values are given by

$$g_{zz} = g_{\parallel} = g_e \qquad [7\text{-}32a]$$

$$g_{xx} = g_{yy} = g_{\perp} = g_e - 6\lambda/\Delta. \qquad [7\text{-}32b]$$

where Δ is defined in Fig. 7-6. This is the case found for Cr^{5+} in $CrO_4{}^{3-}$ with $g_{\parallel}$ = 1.9936 and $g_{\perp}$ = 1.9498.

For an *elongated* tetrahedron the d_{xy} level lies lower and the theoretical values of the g-factor would be

$$g_{zz} = g_{\parallel} = g_e - 8\lambda/\Delta \qquad [7\text{-}33a]$$

$$g_{xx} = g_{yy} = g_{\perp} = g_e - 2\lambda/E \qquad [7\text{-}33b]$$

where Δ is given in Fig. 7-6 and E is the energy difference between the t_{2g} and e_g states.

In summary, from equations [7-32] and [7-33] it is clear that d^1 ions in a tetrahedral site compressed along a four-fold axis results in $g_{\parallel} > g_{\perp}$; while if the tetrahedron is elongated along the same axis the result is $g_{\parallel} < g_{\perp}$. Furthermore, in both cases g-values would be less than g_e. It is worth noting also that the splitting of the e_g ground state by tetragonal distortion is generally small. This could therefore result in short relaxation times and ESR spectra could only be observable at the temperature of liquid nitrogen or below. Finally, these considerations indicate that ESR spectra are very sensitive to the symmetry of the environment surrounding the d^1 ion. This sensitivity, in turn, renders ESR an important tool for obtaining information about this symmetry.

d^2 ions (V^{3+}, Cr^{4+}) (58, 80). A Jahn-Teller distortion will split the ground state (t_{2g}) in O_h symmetry into a lower, non-degenerate a_{2g} and a two-fold degenerate level (e_g) if the distortion is trigonal. Spin-orbit coupling on the lowest level produces a non-degenerate state in the lower level where $M_s = 0$ and a higher ($\simeq 10\ cm^{-1}$) doubly-degenerate level with $M_s = \pm 1$. All other levels are much higher in energy (> several hundred cm^{-1}). This large, zero-field splitting will be discussed in detail in Section 7-5, and presents an ESR signal that is easily detected.

If symmetry is lower than D_{3h} or D_{4h}, the energy levels are further split at zero field as will be discussed in Section 7-5 and illustrated in Fig. 7-15. Values for g are given by

$$g_{\parallel} = g_e - \frac{63\,\lambda^2}{8\,\Delta^2} \qquad [7\text{-}34a]$$

$$g_{\perp} = g_e - \frac{9\lambda^2}{2\Delta^2} \qquad [7\text{-}34b]$$

where Δ is the splitting of t_{2g}, i.e. between a_{2g} and e_g.

d^3 ions (V^{2+}, Cr^{3+}, Mn^{4+}). (29, 37, 45, 48). In an octahedral crystal field the seven-fold degeneracy (see Table 7-2) is removed into three levels, the ground state being a non-degenerate level. Often the excited states are lying so high in energy that the spin-lattice relaxation time becomes long, and a nearly isotropic g value is obtained. An ESR spectrum could then be observed at room temperature with $g_{iso} = g_e - 8\lambda/\Delta$, where Δ is the separation between the ground state and excited states. Only the $\Delta M_s = +1/2 \Leftrightarrow -1/2$ transition is observable for this configuration because large distortions mask the $+1/2 \Leftrightarrow +3/2$ and $-1/2 \Leftrightarrow -3/2$ transitions.

Cr^{3+} ion in MgO gives a symmetrical signal at g = 1.98 and is often used as a standard reference for g-value determinations. In hydrated zeolites, Cr^{3+} gives about the same symmetrical line as in MgO. However, upon dehydration of the zeolite the signal decreases and disappears completely when high dehydration is attained. It is restored upon rehydration. The disappearance of the ESR spectrum after dehydration is attributed to the migration of Cr^{3+} towards specific cationic sites of the zeolite cavities (designated I and II) which have a tetragonal or trigonal symmetry. In such an environment no ESR spectrum can be detected for d^3 ions.

d^4 ions (24). In O_h symmetry the ground state is e_g and spin-orbit coupling splits this state into five levels (orbital and spin levels) which are very close in energy. Some of the doubly-degenerate levels are populated, which allows the Jahn-Teller theorem to be applicable and an axial field will split this state into a_1 and b_1 levels. In zero-field, the $M_s = 0$ level lies lowest (Section 7-5) and since it is not a Kramer's doublet and levels split by spin-orbit coupling are high in energy, one can expect the ESR signal to be very weak. One expects the magnitudes of these energy levels to be given by the relations

$$E|+2\rangle = 2g\beta H + 2D \qquad [7\text{-}35a]$$

$$E|+1\rangle = g\beta H - D \qquad [7\text{-}35b]$$

$$E|0\rangle = -2D \qquad [7\text{-}35c]$$

$$E|-1\rangle = -g\beta H - D \qquad [7\text{-}35d]$$

$$E|-2\rangle = -2g\beta H + 2D \qquad [7\text{-}35e]$$

D is defined in Section 7-5.

For a B_{1g} ground state (D_{4h}) one has

$$g_{\parallel} = g_e - \frac{8\lambda}{E_{d_{x^2-y^2}} - E_{d_{xy}}} \qquad [7\text{-}36]$$

$$g_{\perp} = g_e - \frac{2\lambda}{E_{d_{x^2-y^2}} - E_{d_{xz,yz}}}; \qquad [7\text{-}37]$$

for an A_{1g} ground state

$$g_{\parallel} = g_e \quad [7\text{-}38]$$

$$g_{\perp} = g_e - \frac{6\lambda}{E_{d_{z^2}} - E_{d_{xz,yz}}}; \quad [7\text{-}39]$$

and in D_{3d} symmetry

$$g_{iso} = g_e - \frac{4\lambda}{\Delta}. \quad [7\text{-}40]$$

Very few reports of ESR spectra for d^4 ions have appeared due to the short spin-lattice relaxation times.

d^5 ions (Mn^{2+}, Fe^{3+}, Cr^+) (13, 14, 15, 25, 33, 37, 60, 65, 66, 78). The ground state of both O_h (octahedral) and T_d (tetrahedral) symmetries is A_1. Six-degenerate levels are present ($\Gamma6$ and $\Gamma8$) and since $\Gamma8$ is four-fold degenerate, then it may be split by a Jahn-Teller distortion. There is no zero-field splitting and energy levels are at $\pm 5/2$, $\pm 3/2$ and $\pm 1/2$ $g\beta H$, and a single resonance at $g \simeq 2$ occurs.

If there is an axial distortion, a zero-field splitting is expected and three Kramer's doublets appear with energy levels given by equations [7-41] (see also Fig. 7-14):

$$E|\pm 5/2\rangle = \pm \frac{5}{2} g\beta H + \frac{10D}{3} \quad [7\text{-}41a]$$

$$E|\pm 3/2\rangle = \pm \frac{3}{2} g\beta H - \frac{2D}{3} \quad [7\text{-}41b]$$

$$E|\pm 1/2\rangle = \pm \frac{1}{2} g\beta H - \frac{8D}{3} \quad [7\text{-}41c]$$

This case is important for clay mineral studies since Fe^{3+} and Mn^{2+} ions with $S = 5/2$ are often present as impurities in variable concentrations.

The transition probabilities of the three Kramer's doublets have been calculated (1, 16) as a function of D and the apparent g values (g_{app}) for the three doublets. This means that g_{app} is used in place of g in equations [7-41] for the calculation of transition energies, i.e. g_{app} includes the fine interaction and differs from $g \simeq g_e$. It is also related to the spin-orbit coupling interaction as described above. Values for g_{app} may vary in a range from less than 1.5 up to 10 for Fe^{3+} ion, which makes detailed interpretation of spectra very difficult, especially for powdered samples, because of this huge broadening. This is treated in more detail in Chapter 8 and also by Olivier *et al.* (53) and Harvey (26).

d^6 ions (Fe^{2+}) (24). In an O_h field the ground state is t_{2g}. This allows the Jahn-Teller theorem to be applicable and hence to split the state into b_{2g} and e_g. Spin-orbit coupling subsequently splits these levels and the $M_s = 0$ level lies lowest at zero-field. Since the lowest level is not a Kramer's doublet no ESR spectrum is expected.

d^7 ions (Fe^+, Co^{2+}, Ni^{3+}) (38, 67, 70). In an O_h field the ground state is t_{2g} but the lowest level is a Kramer doublet under the effect of spin-orbit coupling. Jahn-Teller distortions are therefore not expected. Axial fields split the ground state into a_{2g} and e_g with Kramer's doublets in either level. ESR spectra may thus be expected, but spin-orbit coupling may mix these states and result in short spin-lattice relaxation times. An example of this case is the $Co(H_2O)_6^{2+}$ complex which is not detected by ESR. However, by dehydration in a zeolite matrix at 200° C, a signal is observed at low temperature (77 K). This is due to Co^{2+} ions located in sites II of the zeolite framework and coordinated to lattice oxygens in a compressed tetrahedron. In practice, the ground state is a_{2g} and the system behaves as in the case of d^3 ions in an octahedral field.

In low spin configuration (S = 1/2) spectra are readily observed, provided that the symmetry is low. For instance, hexacyanomethyl Co(II) complexes were observed by Lunsford *et al.* (38, 70) in zeolites at room temperature as a consequence of long relaxation time. In a square pyramidal symmetry the unpaired electron would occupy the d_{z^2} orbital, while in a trigonal pyramidal coordination the ground state would be the $d_{x^2-y^2}$ or d_{xy} orbitals. The g-factor values for C_{4v} symmetry (d_{z^2} ground state) are then given by the expressions

$$g_\parallel = g_e \quad [7\text{-}42a]$$

$$g_\perp = g_e + 6\lambda/(E_{d_{z^2}} - E_{d_{xy}}); \quad [7\text{-}42b]$$

and for trigonal symmetry ($d_{x^2-y^2}$ ground state),

$$g_\parallel = g_e + 8\lambda/(E_{d_{x^2-y^2}} - E_{d_{xy}}) \quad [7\text{-}42c]$$

$$g_\perp = g_e + 2\lambda/(E_{d_{x^2-y^2}} - E_{d_{xy\text{-}yz}}). \quad [7\text{-}42d]$$

It then turns out that ESR spectroscopy provides a relatively easy means to discriminate low and high spin complexes. Moreover, the g-factor values enable one to determine the environment symmetry involved. Co ions also may be unambiguously identified because of their characteristic hyperfine pattern as shall be seen in section 7-3, since I = 7/2 produces 8 hyperfine lines. An example is found in the low spin Co(II) complex obtained by adsorbing methyl isocyanide at room temperature into a Co(II) Y-type zeolite dehydrated at 500° C. The resulting ESR spectrum (38) is characterized by $g_\parallel$ = 2.003 and $g_\perp$ = 2.172, which is superimposed by considerable hyperfine and superhyperfine structure due to Co (I = 7/2) and ^{13}C (I = 1/2), respectively. This shows that the Co(II) environment has square pyramidal symmetry and the ground state of Co ions is its d_{z^2} orbital.

d^8 ions (S = 1; Ni^{2+}, Cu^{3+}) (48). In an O_h field the ground state is a_2 and a Jahn-Teller effect is possible since spin-orbit coupling produces a three-fold degenerate level. As in the case of d^3 ions the spectra are usually readily observed. Spin-orbit coupling does not split the M_s = 0 and M_s = ± 1 levels at zero field, resulting in isotropic g and hyperfine values. When the symmetry is reduced, the zero field splitting becomes large and since the lowest level is not a Kramer's doublet an ESR spectrum is not easily observed. Hence, $g = g_e + 4\lambda/\Delta$ is expected to be isotropic. Ni^{2+} ions are indeed very difficult to observe by ESR since symmetry distortions are common. Only O_h symmetry yields ESR absorption.

d^9 ions ($S = 1/2$; Cu^{2+}, Ni^+, Co°, Rh°) (12, 27, 34, 43, 47, 48, 50, 61, 63). This configuration can be considered as having one hole in the orbital levels, and may be treated as a d^1 configuration, except that g values are larger than g_e rather than smaller as is true for d^1 ions. In O_h complexes g values are the same as for the d^4 ion with the signs changed, and for tetrahedral complexes g-values are the same as for d^1 ions except δ and Δ are now negative.

The crystal field splitting is described in Figs. 7-5 and 7-6. As can be seen in these figures, the ground state is e_g in O_h symmetry. Jahn-Teller distortion is possible since spin-orbit coupling produces a four-fold degeneracy and is often very large. In lower symmetries the ground state is a Kramer's doublet and spectra are readily observed, even at room temperature.

An example of this case is Cu^{2+}, which has been often introduced into different clay minerals. A detailed analysis of copper complexes in clays is given in Chapter 8.

In summary, transition metal ions may be studied by ESR provided they are paramagnetic; however, because of either ground state degeneracy or too efficient relaxation, many are undetected by ESR. When paramagnetic ions are detected, however, ESR spectroscopy provides a wealth of information concerning the symmetry of the environment and any distortions due to alterations of the lattice (e.g., defects or other nuclei such as F and OH.) It then becomes the most powerful method available for this purpose. As a final precaution it should be emphasized again that, in order to avoid misinterpretations of ESR spectra, it is absolutely necessary that its limitations, as described above, receive ample consideration.

7-3. HYPERFINE INTERACTION.

As indicated in Section 7-1, hyperfine interaction corresponds to the interaction between electrons and nuclei (2, 4, 7, 18). One can write:

$$\mathcal{H}_{SI} = \sum_{i=1}^{n} \vec{S} \cdot \mathbf{A}_i \cdot \vec{I}_i \qquad [7\text{-}43]$$

where $\mathbf{A}_i$ is the hyperfine tensor and i corresponds to the different nuclei involved. The nuclear Zeeman interaction is given by

$$\mathcal{H}_{IH} = \sum_{i=1}^{n} \gamma_{N_i} \vec{I}_i \vec{H},$$

where γ_{N_i} is the nuclear gyromagnetic ratio of nucleus i. This term is generally rather small and can be neglected in first order approximations.

As in the case of the g-factor tensor, the hyperfine tensor characterizes the anisotropy of the interaction of nuclei with p and d orbitals, and should be analyzed carefully during clay mineral studies. The tensor **A** is usually decomposed into isotropic (**a**) and anisotropic (**b**) parts, and is then written as

$$\mathbf{A} = a\,\mathbf{u} + \mathbf{b}, \qquad [7\text{-}44]$$

where **u** is the tensor unity. Some characteristic properties of nuclei are given in Appendix 7-2.

7-3.1. Isotropic Hyperfine Interaction

Let us first consider the interaction with a single nucleus of non-zero nuclear spin, I. The spin Hamiltonian may then be written

$$\mathcal{H}_S = g_o\, \beta \vec{S}\cdot\vec{H} + ha\, \vec{S}\cdot\vec{I} - g_N \beta_N\, \vec{I}\cdot\vec{H} \qquad [7\text{-}45]$$

where a is the so-called isotropic hyperfine coupling, expressed in frequency units (hertz). Note that because of isotropic interaction, g-factor and hyperfine tensors have been replaced by scalar values g_o and a.

The term **a** was introduced by Fermi who demonstrated that its magnitude is related to the spin density, $|\Psi(0)|^2$, of the unpaired electron at the nucleus and is given by the relation

$$a = \frac{-8\pi}{3} g_e g_N \beta\, \beta_N |\Psi(0)|^2 \qquad [7\text{-}46]$$

It follows that an isotropic hyperfine interaction requires a non-zero spin density at the nucleus, which theoretically precludes all p or d orbitals since their wave functions exhibit nodes ($|\Psi(0)|^2 = 0$) at the nucleus.

In the absence of a magnetic field, $\vec{I}$ and $\vec{S}$ are coupled resulting in a momentum $\vec{F}$. For S=1/2, the electronic level is split into two levels with a separation equal to

$$\Delta E = \frac{1}{2} ha\,(2I + 1) \qquad [7\text{-}47]$$

When the material is submitted to an external magnetic field, $\vec{H}$, the $\vec{I}\cdot\vec{S}$ coupling is broken and $\vec{I}$ and $\vec{S}$ orient independently along the direction of the field. This is known as the Back-Goudsmit effect. The eigen values of the energy solution of the spin Hamiltonian can be expressed as

$$E\,|M_S, M_I> = g_o \beta M_S H + ha M_S M_I - g_N \beta_N M_I H \qquad [7\text{-}48]$$

where M_S and M_I are the magnetic quantum numbers of electron and nucleus, respectively, along $\vec{H}$.

When the microwave and magnetic fields are applied perpendicularly, as is usually done in ESR as described above, the selection rules correspond to $\Delta M_S = \pm 1$ and $\Delta M_I = 0$. These rules are reversed for NMR transitions. It follows as shown in Fig. 7-7 that (2I + 1) transitions may occur resulting in (2I + 1) "hyperfine" lines of equal intensity.

As the microwave energy is kept constant and the magnetic field is swept, the hyperfine lines are obtained as in the case of Fig. 7-7 (I=1) for magnetic field values such as

$$\Delta E = g_o \beta H_1 + ha$$

$$\Delta E = g_o \beta H_2 \qquad [7\text{-}49]$$

$$\Delta E = g_o \beta H_3 - ha,$$

i.e. if H_o is the magnetic field for the central line (no hyperfine coupling), the lines are obtained for $H_o - \frac{ha}{g_o\beta}$, H_o and $H_o + \frac{ha}{g_o\beta}$.

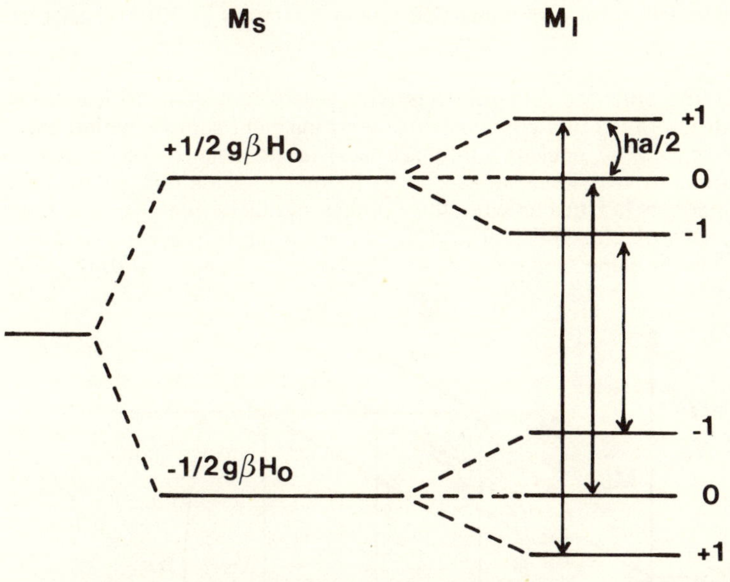

Figure 7-7. ESR transition corresponding to an isotropic hyperfine interaction of one electron with one nucleus of spin I = 1.

It follows that the splitting between two successive lines equals $ha/g_o\beta$. Such a splitting makes possible the experimental determination of the hyperfine coupling constant, a. This also shows that the Zeeman nuclear term, $g_N \beta_N M_I H$, does not modify the hyperfine splitting, so its contribution is generally neglected. Low intensity, forbidden transitions may also occur with $\Delta M_I \neq 0$ and the Zeeman nuclear term must then be considered.

Equation [7-48] is only valid if $\vec{I}$ and $\vec{S}$ are completely decoupled (strong field approximation). In the absence of complete decoupling, the general solution of Hamiltonian equation [7-45] is given by the so-called Breit and Rabi (4) equation

$$E|I, \pm 1/2\rangle = -\frac{\Delta E}{2(2I+1)} - g_N \beta_N M_F H \pm \frac{\Delta E}{2} \left(1 + \frac{4M_F x}{2I+1} + x^2\right)^{1/2} \quad [7\text{-}50]$$

where M_F is the quantum number of $\vec{F}$ (the resultant of $\vec{S}$ and $\vec{I}$), x equals $(g-g_n)\beta H_o/\Delta E$, and ΔE was defined in equation [7-47]. The relationship in equation [7-50] is important since it shows that the variations in nuclear energy levels with magnetic field are not linear but slightly curved, which means that hyperfine separations between successive hyperfine lines are not all equal but depend on the value of M_I. The significance of these differences is greatest when the hyperfine splitting is large, as in the cases of vanadyl and manganese ions, for instance. To obtain the actual value of the hyperfine coupling constant, a, one must introduce magnetic field values for each hyperfine line into relation [7-50] and calculate the value for a.

When the unpaired spin interacts with several non-zero nuclear spins, the hyperfine structure is more complex. In the strong field approximation each electronic level is split by nucleus i into $(2I_i + 1)$ levels. Each is then split again by nucleus j into $(2I_j + 1)$ levels and so on. It follows that the number of hyperfine lines will increase drastically with the number of nuclei present, and follows a multiplicity rule of $(2I_i+1)\cdot(2I_j+1)\cdot\ldots\cdot(2I_k+1)$. However, if hyperfine coupling with given nuclei is the same, one observes an overlapping of levels as shown in Fig. 7-8.

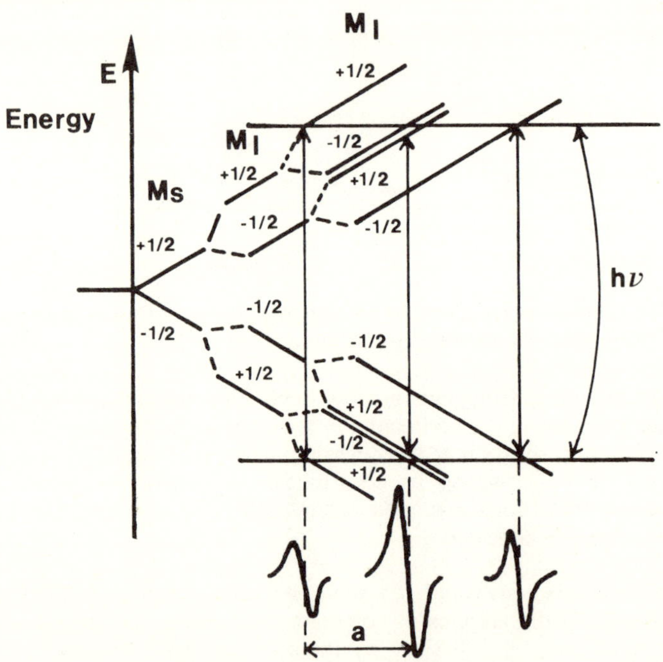

Figure 7-8. Hyperfine splitting of each electronic level by equal coupling with 2 nuclei of nuclear spin of one.

If n nuclei of nuclear spin I are involved, it can thus be easily shown that the number of hyperfine lines is equal to (2nI + 1). Their relative intensities can be determined by the Pascal triangle rule, and also equal the coefficients of the binomial expansion of $(1 + x)^n$.

If there are two or more groups of nuclei with different hyperfine splittings, the overall pattern is more complex than before but can be constructed by extending the energy level diagram shown in Fig. 7-8. Two main conclusions can then be deduced: (a) when hyperfine lines are observed with equal intensity, one can identify the nuclei with unequal coupling since the number of lines is given by $(2I_i+1).(2I_j + 1)...(2I_k + 1)$; when hyperfine lines are observed with unequal intensity, from their relative intensities and their number one may deduce the number and the nature of nuclei concerned by applying the above principles. And, (b) from the hyperfine coupling constant, one can obtain the unpaired spin density at each nucleus of a molecule (23), i.e. a kind of map of unpaired spin density along a molecule, using the relation

$$a = A_o\ C_{ns}^2. \qquad [7\text{-}51]$$

where A_o is the hyperfine splitting if the electron spends all its time on the nucleus (the values of A_o are known for each nucleus) and C_{ns}^2 is the spin density on orbital ns (See Appendix 7-3).

Note that the g-value is measured by the magnetic field at the center of the hyperfine pattern as a first-order approximation. For high values of a (>50G) a second order correction has to be introduced downfield (see Sections 7-3.1, 7-4.3 and equations 7-50 and 7-85).

7-3.2. Configuration Interaction

As indicated above, isotropic coupling at the nucleus level should be absent for orbitals such as p, d, and f which present a node at the nucleus. However, in spite of this expectation, isotropic coupling does indeed exist and has been observed (e.g., d^5 Mn^{2+} ion where a ≃ 100G). This phenomenon is attributed to configuration interaction, which is the interaction between the fundamental ($3s^2 3d^5$ for Mn^{2+}) and excited ($3s^1 3d^5 4s^1$) electronic levels. In the case of a molecule, configuration interaction corresponds to a mixture of an excited σ orbital with the fundamental π orbital, which confers s character onto the resulting orbital. In the case of hydrocarbon radicals such an interaction may allow mixture of a 2p excited state with a σ-type orbital.

It follows that the unpaired spin density on a proton, or the proton hyperfine splitting, is proportional to the spin density, ρ_c, at the carbon atom next to the C bonded to the proton. McConnell (4, 31) has thus proposed a relation

$$a_H = Q_{CH}^{H}\ \rho_c \qquad [7\text{-}52]$$

where Q_{CH}^{H} is a constant equal to −28 G for all aliphatic hydrocarbons and −22.5 G for aromatics. The negative sign of Q indicates that the spin density on the proton is negative while that of the neighboring carbon is positive.

7-3.3. Anisotropic Hyperfine Interaction

Dipolar interaction was discussed in Section 7-1.4, and from the dipolar Hamiltonian (equation [7-8]) recall that

$$E \propto \left[\frac{\vec{\mu}_1 \cdot \vec{\mu}_2}{r^3} - 3\,\frac{(\vec{\mu}_1 \cdot \vec{r})(\vec{\mu}_2 \cdot \vec{r})}{r^5}\right] \qquad [7\text{-}53]$$

and the Hamiltonian can be written:

$$\mathcal{H}_{IS} = -g_e g_N \beta_e \beta_N \left[\frac{\vec{I} \cdot \vec{S}}{r^3} - \frac{3(\vec{I} \cdot \vec{r})(\vec{S} \cdot \vec{r})}{r^5}\right]. \qquad [7\text{-}54]$$

Now designate by Φ the angle between the vector $\vec{r}$ connecting an electron and a nucleus, and the main axis of the anisotropic hyperfine tensor **b** (defined in equation [7-44]). If Θ is the angle between the main axis of tensor **b** and the applied magnetic field, the dipolar magnetic interaction is given by

$$E|M_S, M_I\rangle = \frac{g_e g_N \beta_e \beta_N}{r^3}(1 - 3\cos^2\Phi)\frac{1}{2}(1 - 3\cos^2\Phi)M_S M_I. \qquad [7\text{-}55]$$

Because of the motion of the electron along its orbital, the $(1-3\cos^2\Phi)/r^3$ term may be replaced by a mean value calculated along the orbital, i.e. $\langle\frac{1-3\cos^2\Phi}{r^3}\rangle$. If the unpaired electron spends all of its time on the orbital, the anisotropic hyperfine constant, b_o, is defined as

$$B_o = \frac{1}{2}\frac{g_e g_N \beta_e \beta_N}{h}\left[\left\langle\frac{1-3\cos^2\Phi}{r^3}\right\rangle\right] \qquad [7\text{-}56]$$

It thus becomes $b = B_o\,(1 - 3\cos^2\Theta)$. The spin Hamiltonian can be written

$$\mathcal{H}_S = g\beta\vec{S}\cdot\vec{H} + hb\vec{I}\cdot\vec{S} + g_N\beta_N\vec{I}\cdot\vec{H} \qquad [7\text{-}57]$$

which in strong field approximation results in the energy relation

$$E|M_S, M_I\rangle = g\beta M_S H + hbM_I M_S + g_N\beta_N M_I H. \qquad [7\text{-}58]$$

For an isotropic g-factor tensor, the hyperfine spectra obtained for usual selection rules ($\Delta M_I=0$, $\Delta M_S = \pm 1$) contain 2I+1 lines at magnetic field positions of

$$H_{M_I} = H_o - \frac{M_I b\, h}{g_o\beta}. \qquad [7\text{-}59]$$

The hyperfine pattern is symmetrically distributed about the position $H_o = h\upsilon/g_o\beta$ with a central line at that position if I is odd-numbered and no central line for even numbers of I.

It is worthwhile to note that B_o values can be calculated from the relation [7-56]. Some of these values are given in Appendix 7-2. For instance, for a 2p orbital, one can show that $<\cos^2\Phi> = \frac{3}{5}$ leading to

$$B_o = \frac{2}{5}\frac{g_e g_N \beta \beta_N}{h} <r^{-3}> \qquad [7\text{-}60]$$

in frequency units. When $\vec{H}$ is oriented parallel to the p orbital axis, the term $1-3\cos^2\Theta = -2$; therefore, $b_\parallel = 2B_o$. If $\vec{H}$ is oriented perpendicularly, $1-3\cos^2\Theta = 1$; therefore, $b_\perp = B_o$. It follows from this fact and from equation [7-59] that hyperfine lines for each M_I value will vary in field position when the orientation of $\vec{H}$ changes. This is the reason for the term "anisotropic". As for isotropic coupling one can show that the hyperfine constant, $b = b_\perp = -b_\parallel/2$, is given by $b = B_o C^2_{np}$: i.e., is proportional to the unpaired spin density on the p, d and f orbitals involved. Obviously, $\Sigma C^2_x = 1$ (x being all orbitals s, p, d, f...).

7-3.4. Total Hyperfine Interaction

Now consider the general case where both isotropic and anisotropic interactions occur. The total hyperfine tensor is written:

$$\mathbf{A} = a\,\mathbf{u} + \mathbf{b} = \begin{bmatrix} A_{xx} & A_{xy} & A_{xz} \\ A_{yx} & A_{yy} & A_{yz} \\ A_{zx} & A_{zy} & A_{zz} \end{bmatrix} = a\,\mathbf{u} + \begin{bmatrix} b_{xx} & b_{xy} & b_{xz} \\ b_{yx} & b_{yy} & b_{yz} \\ b_{zx} & b_{zy} & b_{zz} \end{bmatrix} \qquad [7\text{-}61]$$

and the spin Hamiltonian is

$$\mathcal{H}_S = \beta\vec{H}\cdot g\cdot\vec{S} + h\vec{I}\cdot\mathbf{A}\cdot\vec{S}. \qquad [7\text{-}62]$$

Determination of the unpaired spin density within a molecule or a transition metal ion complex is one of the most striking features of ESR spectroscopy since it allows one to draw a kind of spin density map all along a compound. Determination of the s and/or p character of a given orbital bearing the unpaired electron is also of great interest. Again consider the paramagnetic NO_2 radical as an example. The unpaired electron is located in the sp orbital of the N atom. The hyperfine pattern, resolved as will be presented below in Section 7-4, gives

$$a/h = 146.5 M_c/s \qquad [7\text{-}63a]$$

$$B/h = 19.2 M_c/s \qquad [7\text{-}63b]$$

when the experimental values of $A_\parallel = a + b_\parallel$, $A_\perp = a - b_\parallel/2$ and $b_\parallel - 2b$ are given. From $b = B_o\, C^2_{np}$ and $a = A_o\, C^2_{ns}$ one then gets

$$C^2_{2s} = 0.095$$

$$C^2_{2p} = 0.405$$

and $C^2_{2s} + C^2_{2p} = 0.50$ [7-65]

See Appendix 7-2 for values for A_o and B_o.

This means that the unpaired electron spends half of its time in the sp orbital of N and the other half on the oxygen atoms. Moreover, the unpaired electron orbital is roughly 20% s character and 80% p character. It can be shown that this hybridization, due to the bending of the ONO bond, is related to the bond angle α by

$$\alpha = 2 \cos^{-1} (\lambda^2 + 2)^{1/2} \quad [7\text{-}65]$$

in C_2 symmetry where $\lambda^2 = (C_{2p}/C_{2s})^2$. Using this expression for λ^2 and the above values for C^2_{2p} and C^2_{2s}, one can easily calculate the value of α, which is 132.8Å. The actual value is 134°. The agreement is quite good and illustrates the usefulness of ESR parameters.

In a molecule with C_{3v} symmetry one must use the relationship

$$\alpha = \cos^{-1} \left[\frac{1.5}{\lambda^2+3} - \frac{1}{2}\right]. \quad [7\text{-}66]$$

The bond angles for various hybridizations are given in table 7-3.

Table 7-3. Variation of bond angle with hybridization of sp^n in C_{3v} and C_{2v} molecules.

	1	2	3	4	5	6	7	8	9	10
(C_2v)	109°28′	120°	126°52′	131°44′	135°34′	138°34′	141°04′	143°08′	144°54′	146°26′
(C_3v)	101°32′	106°37′	109°28′	111°21′	112°39′	113°35′	114°20′	114°54′	115°23′	115°46′

If the g-factor is isotropic, or its main axes are coincident with those of the hyperfine tensor, the overall pattern should arise as the sum of all the g and hyperfine factors. For a given orientation, the number and relative intensities of the hyperfine lines allow, as shown above, identification of the nuclei involved in interactions with the unpaired electron. The g-factor values are determined from the center of the hyperfine pattern corresponding to $h\upsilon = g\beta H_o$. Often, the magnetic field H_o and the frequency υ are only approximately known. One then uses a double resonance cavity which allows simultaneous recording of a standard with the sample (common standards are DPPH, g = 2.0036; Varian Pitch, g = 2.0028). In such cases, one has

$$h\upsilon = g\beta H_o = g_{ref}\, \beta H_{ref} \quad [7\text{-}67a]$$

$$g = \frac{g_{ref}\, H_{ref}}{H_o} \quad [7\text{-}67b]$$

When the reference signal is close to the signal of the sample being studied, one can approximate the equation [7-67b] with

$$\frac{\Delta g}{g_{ref}} = -\frac{\Delta H}{H_{ref}} \qquad [7\text{-}68]$$

where $\Delta H = H_o - H_{ref.}$ and $\Delta g = g - g_{ref.}$

If the g-factor and hyperfine tensors have no colinear axes, the problem is much more complex since the main values of the tensors are not obtained for the same orientations of the magnetic field with respect to the crystal. For single crystals, values for ESR parameters may be easily obtained by rotating the sample with respect to the magnetic field. But large difficulties exist in the case of powders as will be discussed in Section 7-4.2.

7-3.5. Hyperfine and Superhyperfine Interactions

The superhyperfine interaction corresponds to the interaction of nuclei other than the nucleus directly involved with the unpaired electron. It obviously arises from the overlap of the molecular orbital with the ligand atoms, and corresponds to a splitting of each first hyperfine line (27, 42, 46).

An example of superhyperfine interaction is found when tungsten ion is introduced into a rutile-type, SnO_2 material by impregnating SnO_2 with an ammonia solution of WO_3, and then calcining the sample at high temperature (1000°C) in air. W^{5+} ions are thus incorporated into the SnO_2 matrix. In the SnO_2 rutile structure there are two equivalent sites into which Sn has been substituted. These are obtained by rotating 90° about the crystallographic y axis ([00l] direction), and each Sn is surrounded by an elongated octahedron of oxygen ions. In addition, Sn is substituted into 4 other interstitial sites that are equivalent. Each of these Sn ions is surrounded by a flattened oxygen octahedron. The ESR spectra may be obtained at either liquid nitrogen or room temperature, and give the parameters: g_{xx} = 1.671, g_{yy} = 1.500 and g_{zz} = 1.732, A_{xx} = 62, A_{yy} = 80 and A_{zz} = 120G for the ^{183}W hyperfine structure (I = 1/2 for ^{183}W with 14% natural abundance); $^1A_{xx}$ = 430, $^1A_{yy}$ = 520 and $^1A_{zz}$ = 450G for the superhyperfine structure of the two equivalent Sn nuclei designated (I = 1/2 16% natural abundance for ^{117}Sn and ^{119}Sn); and $^2A_{xx}$ = 78, $^2A_{yy}$ = 80 and $^2A_{zz}$ = 64G for the superhyperfine structure of the 4 equivalent Sn nuclei designated 2.

W^{5+} occupies a substitutional site and the next-nearest neighbors can be divided into three groups: a, b and c along the y (2Sn), x (4Sn) and z (4Sn) axes, respectively. The W-Sn distance for group (a) is 3.2Å which is smaller than the 3.7Å distance for group (b). One can then expect the superhyperfine splitting due to the 2 Sn nuclei of group (a) to be much larger than that of the 4 in group (b) and thus corresponds to the designation 1 in the previous paragraph while designation 2 corresponds to group (b). Interaction with group (c) is too weak to detect since the unpaired electron is in a $d_{x^2-y^2} - \lambda d_{z^2}$ orbital which will not "see" the Sn atoms of this group.

This example demonstrates how ESR may be used to locate paramagnetic ions in a matrix, and to identify the molecular orbital bearing the unpaired electron. Further, interactions of the unpaired electron with nuclei of nearest neighbors may be characterized, and thereby the location of the electron within the lattice can be identified (66, 71).

7-4. ANALYSIS OF ESR SPECTRA

It is now important to consider how values for the ESR parameters are determined from the ESR spectrum of a clay mineral sample (5, 10, 11, 32, 34, 36, 56). The method will depend on whether the sample is a single crystal whose crystallographic axes can be determined from x-ray diffraction; or a powder, which presents a random orientation of particles with respect to the magnetic field.

7-4.1. Single Crystals

Since the crystallographic axes of a single crystal are readily determined, one may record its ESR spectrum simply by rotating the sample with respect to the magnetic field. For the small cylindrical cavity (mainly used in Q band) the sample is kept fixed in the cavity and the magnet itself is rotated. One determines the hyperfine coupling constant from the hyperfine line splitting as described above, and g-factor values are given by the centers of the hyperfine patterns. It is worth noting that anisotropy varies in the case of hyperfine interaction from $2B_o$ to B_o; or, in other words, the extreme positions of the lines as a function of orientation occur with the main axes colinear with $\vec{H}$. This allows determination of the orientation of the g-factor and hyperfine tensor axes with respect to the magnetic field.

7-4.2. Powder Samples (5, 10, 11, 34, 36, 40, 49, 62)

Powder samples are a very common yet very complex case which must be considered in detail to obtain values for the ESR parameters (5, 10, 11, 34, 36, 40, 49, 62). If we assume that the polycrystallites are randomly distributed, then the probability of finding a crystallite at an angle Θ with respect to the magnetic field is proportional to the area of an angular ring of width dΘ and circumference $2\pi r$ cos Θ. This will cause the number of polycrystallites, dN, to vary as

$$dN = \frac{N_o}{2} \sin \Theta \, d \Theta \qquad [7\text{-}69]$$

where N_o is the total number of polycrystallites and $0 \leqslant \Theta \leqslant \pi/2$. The absorption line intensity in this domain is

$$\frac{dN}{dH} = \frac{dN}{d\Theta} \cdot \frac{d\Theta}{dH} \cdot \qquad [7\text{-}70]$$

For axial symmetry, and assuming no hyperfine coupling, one has

$$h\upsilon_o = (g_{\parallel}^2 \cos^2 \Theta + g_{\perp}^2 \sin^2 \Theta)^{1/2} \beta H \qquad [7\text{-}71]$$

and

$$g_{ave} = \frac{2g_{\perp} + g_{\parallel},}{3} \qquad [7\text{-}72]$$

leading to

$$\frac{dN}{dH} = \frac{N_o}{2} g_o^2 \frac{H_o^2}{H^3} (g_\parallel^2 - g_\perp^2)\left[\left(\frac{g_o H_o}{H}\right)^2 - g_\perp{}^2\right]^{-1/2}. \quad [7\text{-}73]$$

The ESR line shape should then be given by

$$S(H) = \frac{dN}{N} \cdot \frac{1}{dH} = H_\perp^2 H_\parallel (H_\perp^2 - H_\parallel^2)^{-1/2} H^{-2} (H_\perp^2 - H^2)^{-1/2} \quad [7\text{-}74]$$

with $\int_0^\infty S(H)dH = 1$, and where $H_\perp$ and $H_\parallel$ are the values of H for $\Theta = \frac{\pi}{2}$ and 0, respectively. The calculated line shapes are shown in Fig. 7-9, which assumes a Dirac-type delta function for individual lines since the spectrum turns out to be due to an overlap of all lines corresponding to the whole angular range. Individual lines, indeed, are of insufficient width to be treated independently, but may be represented by either a Lorentzian or a Gaussian law F(H-Ho). Thus, the overall ESR spectrum is represented by

$$I(H) = \int_{H_\perp}^{H_\parallel} F(H\text{-}H_o) S(H_o) dH_o. \quad [7\text{-}75]$$

where $F(H\text{-}H_o)$ represents either the Gaussian or Lorentzian function.

Lebedev (36) has calculated this integral according to the hypotheses: (a) crystallites are independent and randomly distributed; (b) no thermal motion occurs; (c) environmental symmetry is of axial type; (d) widths for individual lines, (ΔH_i), are Θ-independent; and, (e) anisotropy is small, i.e.

$$\Delta H_{an} = |H_\parallel - H_\perp| << H. \quad [7\text{-}76]$$

One can write, then

$$I(H) = \int_{H_\perp}^{H_\parallel} \left(\frac{H - H_o}{\Delta H_i}\right) (|H_o - H_\perp|)^{-1/2} dH_o \quad [7\text{-}77]$$

where for a Gaussian line

$$F = f\left(\frac{H-H_o}{\Delta H_i}\right) = (\pi^{1/2}\Delta H_i)^{-1} \exp.\left[-\frac{(H-H_o)^2}{\Delta H_i{}^2}\right] \quad [7\text{-}78a]$$

and for a Lorentzian line,

$$F = f\left(\frac{H-H_o}{\Delta H_i}\right) = (\pi \Delta H_i)^{-1}\left[1 + \frac{(H-H_o)^2}{\Delta H_i{}^2}\right]^{-1}. \quad [7\text{-}78b]$$

The different types of line shapes as a function of the anisotropy parameter, ΔH_{an}, are given in Fig. 7-10. The corresponding g values are also shown and are determined approximately from the inflexion points. Accurate determination necessitates computer simulation of the spectrum to the best fit with the experimental spectrum.

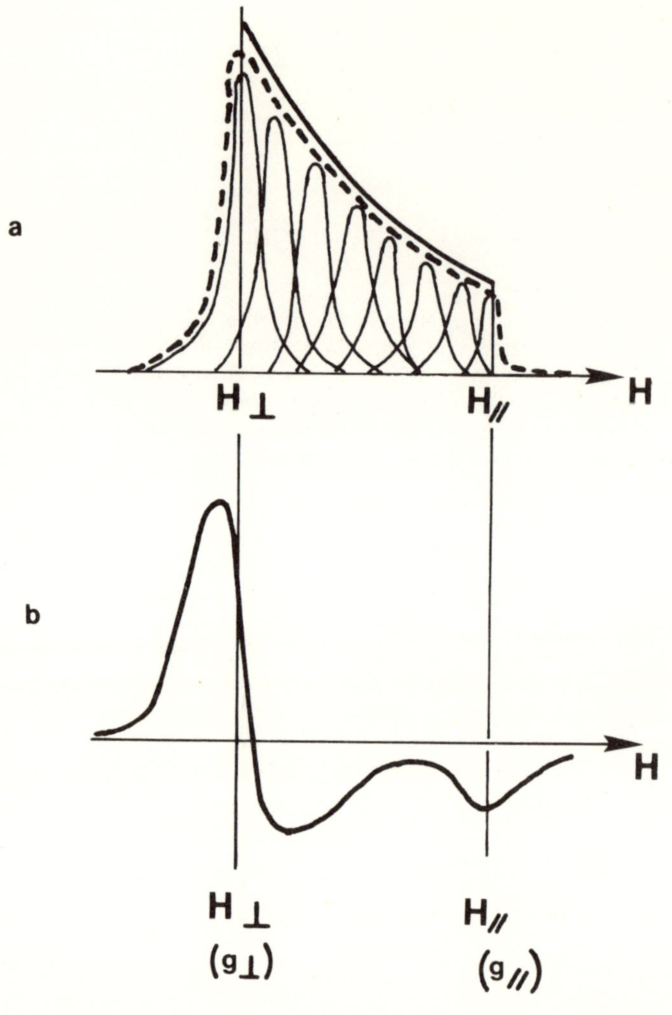

Figure 7-9. Typical ESR absorption (a) and derivative (b) spectra for powdered samples calculated from equation [7-35]. — — assumes a δ function for linewidths; —— assumes Lorentzian or Gaussian lineshapes for individual lines.

In the case of orthorhombic symmetry in ESR, the symmetry is lower than axial symmetry and consequently the calculation of the line shape is somewhat more difficult. The resulting spectra are similar to those shown in Fig. 7-11.

The three g values corresponding to the eigen values of the g-factor tensor are given by the inflexion points as shown in Fig. 7-11. Resolution of the spectrum depends on the degree of anisotropy ($g_1 - g_2$ or $g_2 - g_3$) relative to the individual linewidth. For poorly resolved spectra, computer simulation of the ESR spectrum is absolutely necessary if accurate values of the g-factor are expected.

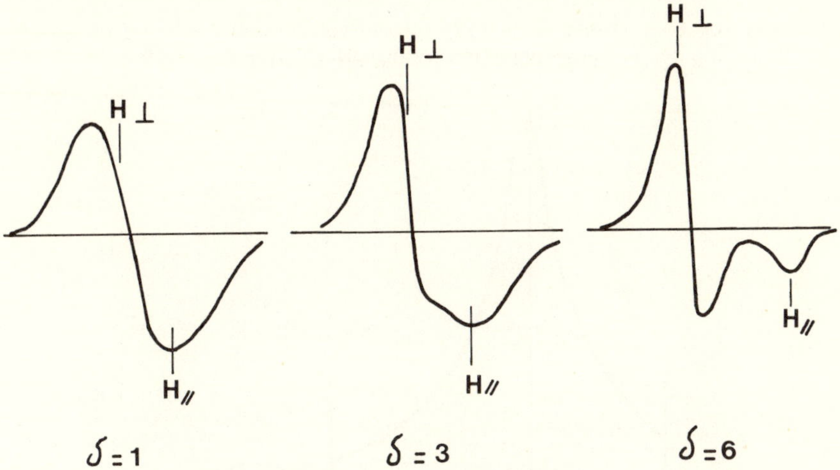

Figure 7-10. Changes in lineshapes for powder spectra as a function of the ratio of anisotropy to individual linewidth ($\Delta H_i:\delta$).

In the presence of hyperfine splitting, assumed to be axially symmetric, the hyperfine lines are easily shown to resonate at magnetic field values given by

$$H = \frac{g_o H_o}{g} - \frac{M_I K}{\beta_o g^2} \qquad [7\text{-}79]$$

where

$$K = (g_\parallel^2 A_\parallel^2 \cos^2\Theta + g_\perp^2 A_\perp^2 \sin^2\Theta)^{1/2} \qquad [7\text{-}80]$$

and

$$g^2 = g_\parallel^2 \cos^2\Theta + g_\perp \sin^2\Theta. \qquad [7\text{-}81]$$

The line shape is then given by the relation

$$\frac{dN}{dH} = \frac{1}{2} N_o \left(\frac{2\cos\Theta}{g^2} \left[\frac{(g_\parallel^2 - g_\perp^2) g_o H_o}{2g} + \frac{M_I}{\beta} \frac{g_\parallel^2 A_\parallel^2 - g_\perp^2 A_\perp^2}{2K} - \frac{K(g_\parallel^2 - g_\perp^2)^{-1}}{g^2} \right] \right) \qquad [7\text{-}82]$$

Extreme hyperfine lines described by the relations

$$H_\parallel = \frac{g_o H_o}{g_\parallel} - M_I A_\parallel / \beta g_\parallel \qquad [7\text{-}83a]$$

$$H_{\perp} = \frac{g_o H_o}{g_{\perp}} - \frac{M_I A_{\perp}}{\beta g_{\perp}} \qquad [7\text{-}83b]$$

are obtained for $\Theta = 0$ and $\Theta = \pi/2$, respectively. Typical spectra of powder samples, including g- and hyperfine-anisotropies, are given in Fig. 7-12.

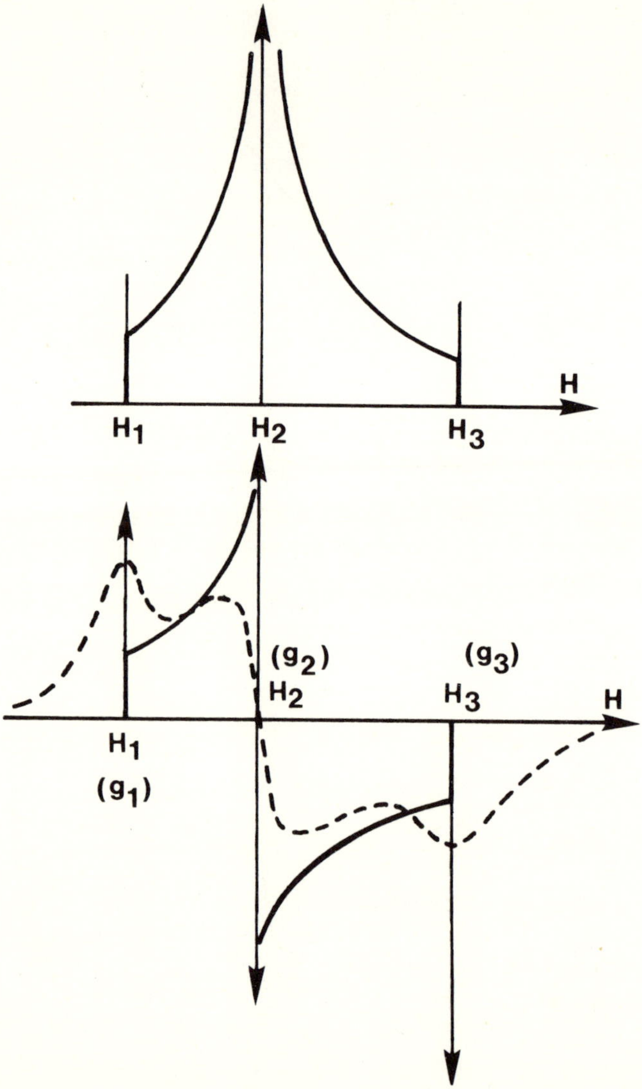

Figure 7-11. Calculated absorption (upper curve) and derivative (lower curve) lineshapes for orthorhombic-type symmetry assuming a δ function (—) and Gaussian or Lorentzian (— — —) individual lineshapes.

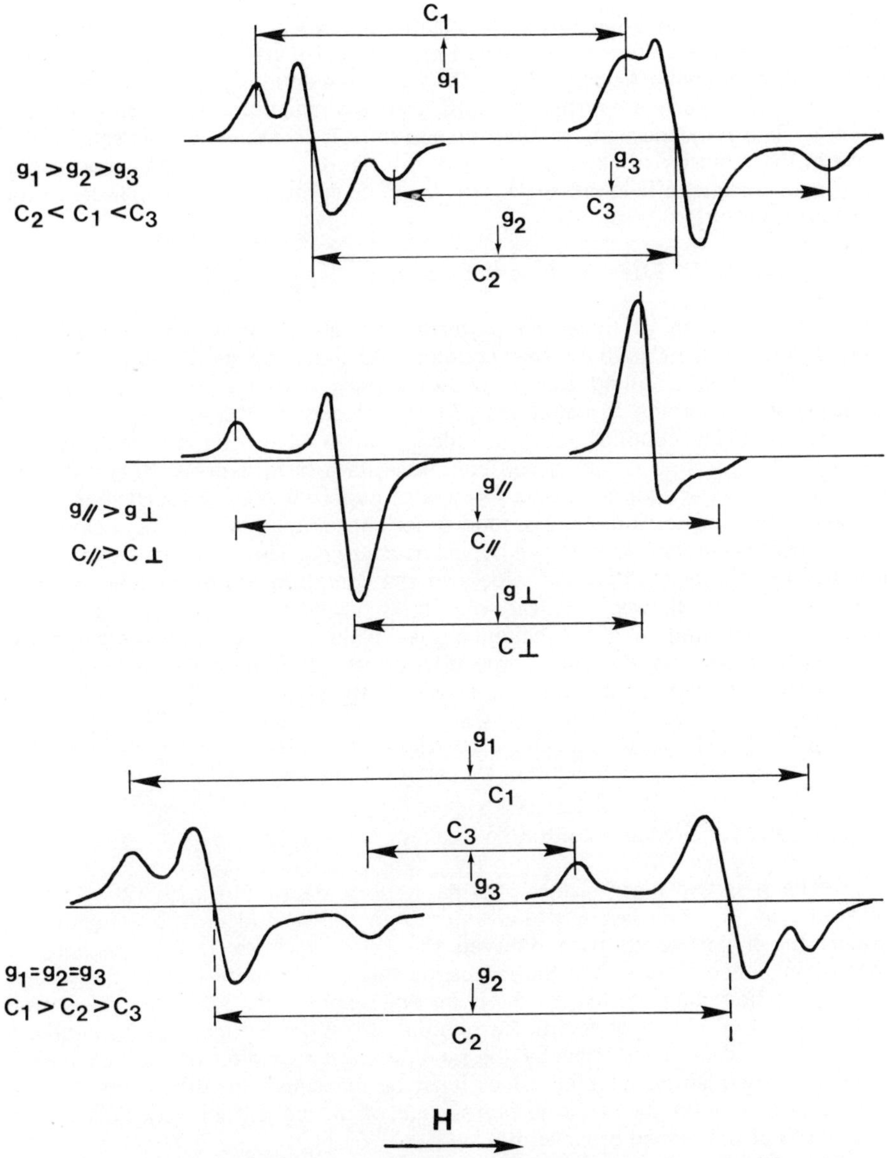

Figure 7-12. Typical ESR powder spectra for different g-factor values (g) and hyperfine tensor eigenvalues (A).

Values for the hyperfine coupling, A, should be expressed in units of frequency, Mc/s, rather than units of magnetic field. The conversion factor is

$$2.8\,H_G = H_{Mc/s} \qquad [7\text{-}84]$$

which stems from equation [7-3] and assumes that $g = g_e$. Any change in the value of g will alter the value of the conversion factor. Hence, it turns out that when expressed in magnetic field units (G), the hyperfine coupling may differ from the experimental hyperfine splitting depending on the magnitude to which g deviates from g_e. The frequency unit is more precise since it corresponds to an energy. But since in the majority of cases $g \simeq g_e$, scientists habitually express the experimental splitting in magnetic field units (G). This could be confusing to one unaware of the inherent assumption.

7-4.3. Second Order Effect in Hyperfine Patterns

Interpretation of hyperfine patterns must also consider the possibility of second order effects in hyperfine coupling. As has been shown above (Section 7-4.2), hyperfine coupling can be deduced from hyperfine splitting (equation [7-48]), and according to equations [7-83], calculation of a more realistic value for the hyperfine coupling must include g values when g is different from 2. Indeed, all hyperfine coupling considerations have been expressed heretofor assuming a first order approximation. For a large hyperfine coupling such an approximation is no longer valid and a second order approximation must be included to obtain more accurate values for the hyperfine coupling. The theory of second order interactions has been discussed widely in the literature. By evaluating the non-diagonal elements between nuclear spin states having the same value of I (total nuclear momentum), and by including these values in the equations for determining the energy levels. The energy differences between two states of a component M_I with identical values of I are given by the relation (4, 75)

$$\Delta E = g_{zz}\beta H + A_z M_I + \frac{1}{2}\frac{A_{xx}^2 + A_{yy}^2}{2g_{zz}\beta H_o}\left[I(I+1) - M_I^2\right] \qquad [7\text{-}85]$$

which is valid for circular permutation of axes.

The first two terms give the normal hyperfine structure with (2I + 1) lines separated by A_z. The last term corresponds to the second order treatment and affects the hyperfine splitting between the (2I + 1) lines. It follows that: (a) separation between hyperfine lines depends slightly on the values of the second order hyperfine components; (b) the hyperfine separation is greater for the central lines (M_I = 0 or 1/2) than for the extremities (M_I = I); (c) lines corresponding to the same M_I value are shifted by the same amount regardless of the sign on M_I; and, (d) subsequently, g-factor values must be corrected since the center of two hyperfine lines with the same absolute value of M_I are shifted downfield by the same value of the second order term.

7-5. FINE STRUCTURE

The discussion in the preceding sections has focussed on systems with only one unpaired electron, where the electron spin is 1/2. The majority of systems studied by ESR fall within this category. However, for biradicals and transition metal ions, the number of unpaired electrons may be more than 1, resulting in $S > 1/2$. In such cases it can be shown by following the same general approach as in preceding sections that the magnetic field splits the electron levels into 2S + 1 magnetic sublevels, characterized by their M_S value (11, 28, 44, 74). The selection rule $\Delta M_S = \pm 1$ is still valid and gives 2S transitions. For high environmental symmetry, the 2S transitions occur at the same magnetic field value, resulting in a single resonance line (Fig. 7-13a).

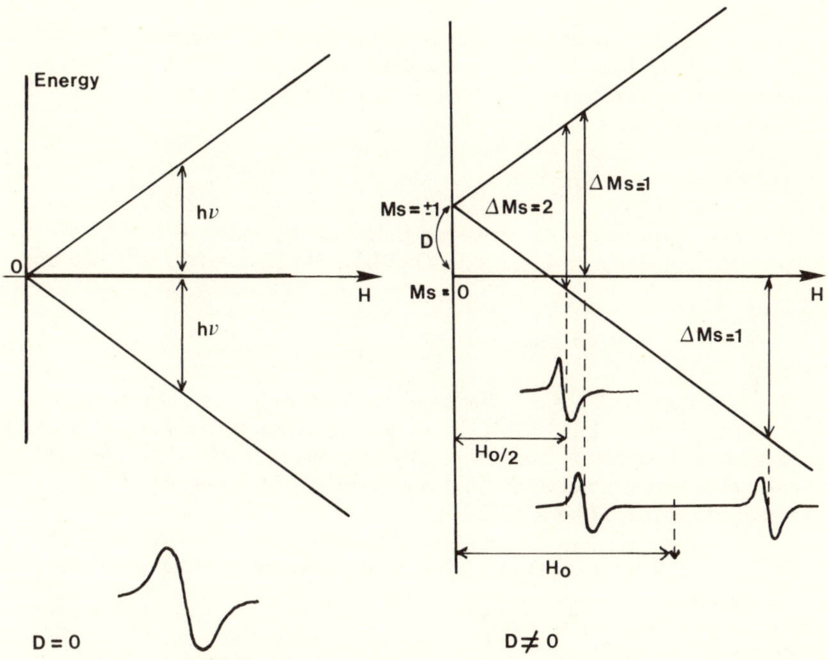

Figure 7-13. Electron energy levels for S = 1 and a crystal field environment with (a) high symmetry and (b) an axial component.

If the crystal field contains a component in a given direction, and in the absence of any magnetic field the fundamental energy level is split into as many sublevels as there are $|M_s|$ values (Fig. 7-13b). This splitting is designated the decomposition at zero-field and arises from the dipolar Hamiltonian, viz.

$$\mathcal{H}_d = \vec{S} \cdot \mathbf{D} \cdot \vec{S} \qquad [7\text{-}86]$$

where **D** is a second order tensor as g and A.

The **D** tensor is particularly important for transition metal ions since it corresponds to an indirect coupling of electron spins via spin-orbit coupling, and arises from a second order perturbation calculation. One can write:

$$\mathcal{H}_d = S\cdot\mathbf{D}\cdot S = D_{xx}S_x^2 + D_{yy}S_y^2 + D_{zz}S_z^2$$

or

$$\mathcal{H}_d = D[S_z^2 - S(S+1)/3] + E(S_x^2 - S_y^2) \qquad [7\text{-}87]$$

where $D = D_{zz} - (D_{xx}+D_{yy})/2$, and $E = (D_{xx}-D_{yy})/2$.

D is a dipolar tensor having a null trace in its axis system. The D and E constants are called fine structure parameters at magnetic field equal to zero. E represents deviations from axial symmetry. Recall that the trace of a tensor is the sum of its diagonal terms.

7-5.1. The Case of Axial Symmetry

For axial symmetry, the value of E is zero in equation [7-87]. When the magnetic field is parallel to the z axis, $D = D_{zz}$ and $D_{xx} = D_{yy} = 0$ and one can write

$$\mathcal{H}_S = g_\parallel \beta H S_z + D[S_z^2 - S(S+1)/3] \qquad [7\text{-}88]$$

For S = 1, the $\Delta M_S = \pm 1$ transitions are separated by 2D (energy units) or $2D/g_\parallel\beta$ (magnetic field units). This allows determination of the value of D and also of $g_\parallel$, which corresponds to the center of the fine structure. A $\Delta M_S = \pm 2$ transition is also observed at a magnetic field value which is half that for $\Delta M_S = \pm 1$. This is the so-called half-field line (Fig. 7-13b).

When H is perpendicular to the z axis, the energies of the allowed transitions ($\Delta M_S = \pm 1$) are

$$h\upsilon = g_\perp\beta H_\perp \pm D/2$$

$$\text{or} \quad H_\perp = (h\upsilon/g_\perp\beta) \pm (D/2g_\perp\beta). \qquad [7\text{-}89]$$

This means that the resonance lines are separated by D when using energy units, and by $D/g_\perp\beta$ when using magnetic field units. The center of the pattern corresponds to $g_\perp$ (Fig. 7-14).

7-5.2. The Case of Orthorhombic Symmetry

The value of E in the dipolar Hamiltonian (equation 7-87) is no longer zero with the departure from axial symmetry. One may show that for S = 1 and $\vec{H}$ parallel to the z axis, three energy levels are obtained: $(\pm\, g_{zz}^2\beta^2H^2 + E^2)^{1/2} + (1/3)D$ for $S_z = \pm 1$; and $-(2/3)D$ for $S_z = 0$ as shown in Fig. 7-15. The separation between lines equals 2D (energy units). For $\vec{H}$ perpendicular to the z axis, the transitions are separated by D + 3E and D − 3E.

7-5.3. Significance of D and E

The D and E terms can be expressed as a function of the spin-orbit coupling constant, ξ, and of environmental symmetry. General trends are given in Table 7-4 according to the discussion in sections 7-5.1 and 7-5.2.

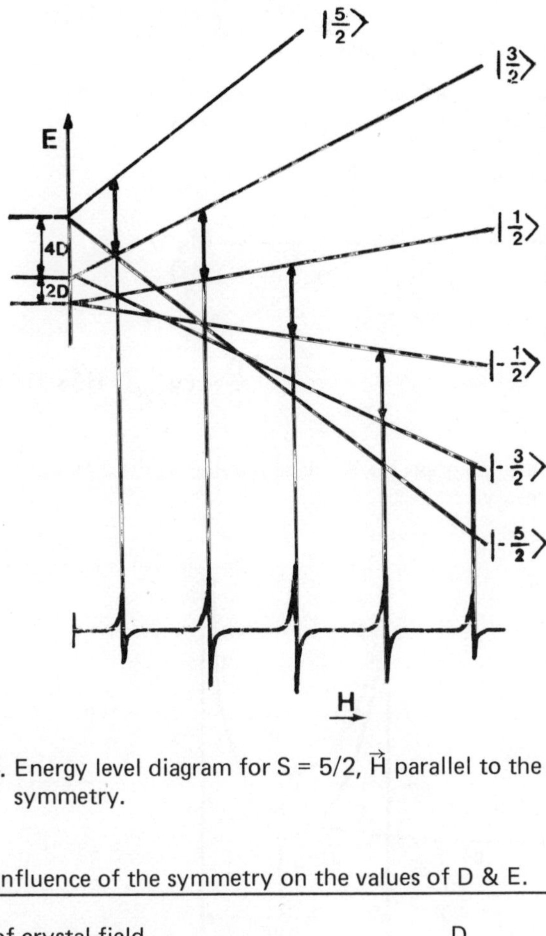

Figure 7-14. Energy level diagram for S = 5/2, $\vec{H}$ parallel to the z axis and axial symmetry.

Table 7-4. Influence of the symmetry on the values of D & E.

symmetry of crystal field	D	E
octahedral, tetrahedral	0	0
axial symmetry (trigonal, tetragonal . . .)	>0	0
lower symmetry (orthorhombic . . .)	>0	>0

It is worthwhile to note that if N unpaired electrons are present on the metal ion d orbitals, N lines of fine structure may be observed but are not detected in the majority of cases. This is due to line-broadening or low transition probabilities. Moreover, D values must also be less than $h\nu$ for a transition to be possible.

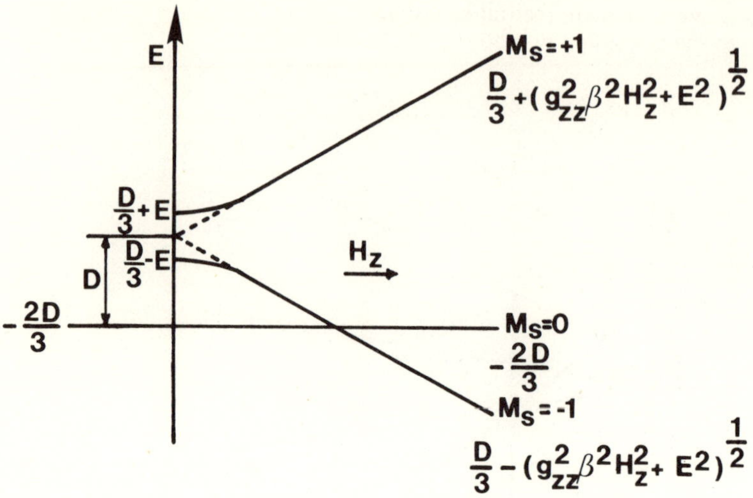

Figure 7-15. Energy level diagram for S = 1, $\vec{H}$ parallel to the z axis and orthorhombic symmetry.

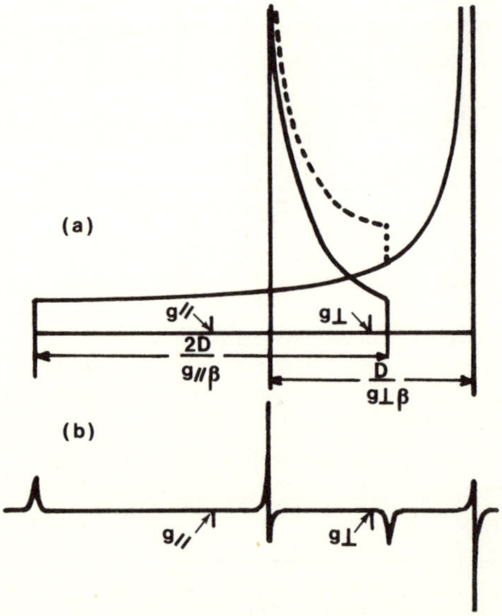

Figure 7-16. Theoretical absorption (a) and derivative (b) ESR spectra for S = 1 and axial symmetry.

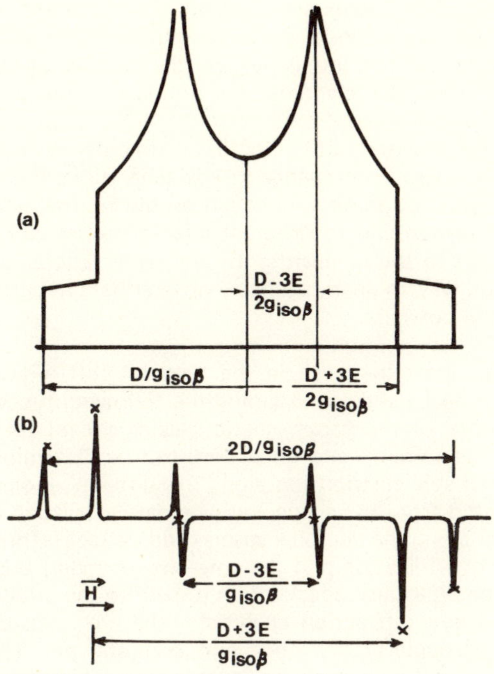

Figure 7-17. Theoretical absorption (a) and derivative (b) ESR spectra for S = 1, orthorhombic symmetry and isotropic g-factor.

7-5.4. Fine Structure of Powder Samples

This is the most striking and complex case of fine structure, and only a few examples have been clearly analyzed. The approach for calculating line shapes is similar to that given for g-factor and hyperfine tensors in Section 7-4. For clarity, Figs. 7-16 and 7-17 show the cases for axial and orthorhombic symmetries, respectively, where S=1 as discussed in Sections 7-5.1 and 7-5.2. When S>1 and the g-factor is anisotropic, the spectrum generally turns out to be rather complex and difficult to analyze.

7-6. SUMMARY

In summary, this chapter has considered all of the interactions involved in ESR, all interactions between the unpaired electron(s) and its (their) surroundings. The spin Hamiltonian includes terms representing all of these interactions so the ESR spectrum will, therefore, be very complex. Hopefully, these considerations will assist experimentalists in obtaining an idea concerning the main ESR parameters and in interpreting a complex ESR spectrum. Some general features of ESR are as follows: (a) parallel components are of much lower intensity than the perpendicular components; (b) repetition of a given ensemble clearly indicates nuclei-electron interaction; and (c) relative intensities and the number of hyperfine lines allow a determination as to the nature and number of nuclei involved.

When the sample is irradiated with a fluctuating microwave frequency, (e.g., X to Q band fluctuation), there is no change in line separations due to fine structure or hyperfine structure since these are unchanged by the frequency modifications. However, lines corresponding to different g-factor values will be shifted in magnetic field with respect to the g_e position by the same ratio as the frequency fluctuation. This is a good way to unambiguously differentiate hyperfine couplings from g-factor anisotropy in powders.

Current knowledge and experience in the field of ESR indicates that the technique has been most successful in obtaining the following types of chemical information: (a) the nature of the paramagnetic species and of its environment (nature of atoms, type of symmetry and distortion) can be determined with good precision; (b) the unpaired spin distribution along the different molecular orbitals (s, p, d...) can be calculated from hyperfine coupling data – one can then obtain a kind of spin density map along the complex species and extract information about overlapping of orbitals, hybridization and the covalent (or ionic) bond character; (c) the orientation of a paramagnetic species with respect to the principal axes of a crystal (determined by x-ray diffraction analysis) – the high sensitivity of ESR makes this application particularly attractive and accurate; and (d) the use of paramagnetic probes provides information about crystallographic arrangements and motions in a crystal – the Fe^{3+}, Mn^{2+} V^{4+} ions have been used with particularly good success. In references 11, 22, 51, 52, 54, 64 and 72, and in Chapters 8 and 9 of this book, examples of the applications of many of these principles of ESR to the understanding of clay mineral structures are explored.

APPENDIX 7-1

Let us take an example of a theoretical calculation of g-factor values for a d^1 ion in a tetragonally distorted O_h symmetry, as described in Fig. 7-5.

The d orbitals may be characterized by the corresponding m_L values: $|2\rangle, |1\rangle, |0\rangle, |-1\rangle$ and $|-2\rangle$. One can write:

$$e_g \begin{cases} d_{z^2} = |0\rangle \\ d_{x^2-y^2} = \frac{1}{\sqrt{2}}\, |2\rangle + |-2\rangle) \end{cases} \qquad [7\text{-}90]$$

and

$$t_{2g} \begin{cases} d_{xy} = -\frac{1}{i\sqrt{2}}\,(|2\rangle - |-2\rangle) \\ d_{yz} = -\frac{1}{i\sqrt{2}}\,(|1\rangle + |-1\rangle) \\ d_{xz} = -\frac{1}{\sqrt{2}}\,(|1\rangle - |-1\rangle) \end{cases} \qquad [7\text{-}91]$$

The ground state function is d_{xy} and relation 7-26 can be rewritten:

$$g_{ij} = g_e\delta_{ij} - 2\lambda \sum_{n \neq xy} \frac{\langle xy|L_i|n\rangle\langle n|L_j|xy\rangle}{E_n - E_{xy}} \qquad [7\text{-}92]$$

7-1.1. g_{zz} calculation

Let us calculate the matrix elements $\langle n|L_i|0\rangle$

from relation [7-91]: $|xy\rangle = \frac{1}{i\sqrt{2}}(|2\rangle - |-2\rangle)$. One can write: $L_z|xy\rangle = \frac{1}{i\sqrt{2}}$ $(L_z|2\rangle - L_z|-2\rangle)$; but $L_z|Y_\ell^m\rangle = m\hbar|Y_l^m$, resulting in:

$$L_z|2\rangle = L_z|Y_2^2\rangle = 2|Y_2^2\rangle$$
$$L_z|-2\rangle = L_z|Y_2^{-2}\rangle = 2|Y_2^{-2}\rangle \qquad [7\text{-}93]$$

or $$L_z|xy\rangle = -2i|x^2-y^2\rangle \qquad [7\text{-}94]$$

The element $\langle x^2-y^2|L_z|xy\rangle$ equals $\langle x^2-y^2|-2i|x^2-y^2\rangle = -2i$ where L_z is an hermitian operator. It then follows that

$$\langle x^2-y^2|L_z|xy\rangle^* = \langle xy|L_z|x^2-y^2\rangle = 2i \qquad [7\text{-}95]$$

and from relation [7-92]

$$g_{zz} = g_e - 2\lambda\,\frac{-4\,i^2}{E_{x^2-y^2} - E_{xy}} = g_e - \frac{8\lambda}{E_{x^2-y^2} - E_{xy}}. \qquad [7\text{-}96]$$

The summation on $n \neq xy$ gives only one term since other excited states — $|z^2\rangle$, $|yz\rangle$ and $|zx\rangle$ — lead to $\langle n|L_z|xy\rangle$ elements equal to zero because of orthogonality of wave functions.

7-1.2. g_{xx} and g_{yy} calculation

One uses the shift operators: $L^+ = L_x + iL_y$ and $L^- = L_x - iL_y$ and since

$$L^{\pm} Y_l^m = \{1(1+1) - m(m^{\pm} 1)\}^{1/2}\} Y_l^{m \pm 1} \quad [7\text{-}97]$$

$$L_x|xy\rangle = \frac{L^+ + L^-}{2}|xy\rangle, \quad [7\text{-}98]$$

and since $|xy\rangle = \frac{1}{i\sqrt{2}}(|2\rangle - |-2\rangle)$ one has:

$$L_x|xy\rangle = \frac{1}{2}[\frac{1}{i\sqrt{2}}\{L^+|2\rangle - L^+|-2\rangle + L^-|2\rangle - L^-|-2\rangle\}] \quad [7\text{-}99]$$

$$= -\frac{i}{\sqrt{2}}(|Y_2^1\rangle - |Y_2^{-1}\rangle) = +i\{\frac{|1\rangle - |-1\rangle}{\sqrt{2}}\} = i|xz\rangle$$

In the same way, $L_y|xy\rangle = i|yz)$. From relation [7-92] one then gets

$$g_{xx} = g_e - \frac{2\lambda}{E_{xz} - E_{xy}} \quad [7\text{-}100]$$

$$g_{yy} = g_e - \frac{2\lambda}{E_{yz} - E_{xy}} \quad [7\text{-}101]$$

APPENDIX 7-2

Table 7-5. Hyperfine coupling constants for some nuclei.

nucleus	Natural abundance (%)	nuclear spin (I)*	hyperfine coupling constants (G) A_o	B_o
^{1}H	99.98	1/2	508	–
^{2}D	0.02	1	78	–
^{7}Li	92.57	3/2	105	–
''B	81.17	3/2	720	18.9
^{13}C	1.11	1/2	1,110	32.5
^{14}N	99.64	1	550	17.0
^{17}O	0.04	5/2	−1,653	−51.5
^{19}F	100	1/2	17,110	541
^{23}Na	100	3/2	224	343
^{27}Al	100	5/2	981	21
^{29}Si	4.7	1/2	−1,218	−31
^{31}P	100	1/2	3,636	102
^{35}Cl	75.4	3/2	1,665	50
^{37}Cl	24.6	3/2	1,385	41.8
^{39}K	93.08	3/2	52	32
^{41}K	6.91	3/2	29	17

*in multiples of $\hbar = h/2\pi$

Table 7-6. Magnetic properties of Nuclei

Isotope	Natural abundance (%)	Magnetic moment, μN, in multiples of the nuclear magneton (eh/4Mc)	Nuclear Spin (I) in multiples of $h/2\pi$
1. ^{1}H	99.9844	2.79268	1/2
2. ^{2}H	$1.56\ 10^{-2}$	0.857386	1
3. ^{3}He	$10^{-5}\ 10^{-7}$	−2.1274	1/2
4. ^{6}Li	7.43	0.82192	1
5. ^{7}Li	92.57	3.2560	3/2
6. ^{9}Be	100	−1.1773	3/2
7. ^{10}B	18.83	1.8005	3
8. ^{11}B	81.17	2.6880	3/2
9. ^{13}C	1.108	0.70220	1/2
10. ^{14}N	99.635	0.40358	1
11. ^{15}N	0.365	−0.28304	1/2
12. ^{17}O	$3.7\ 10^{-2}$	−1.8930	5/2
13. ^{19}F	100	2.6273	1/2
14. ^{21}Ne	0.257	−0.66176	3/2
15. ^{23}Na	100	2.2161	3/2
16. ^{25}Mg	10.05	−0.85471	5/2
17. ^{27}Al	100	3.6385	5/2

Table 7-6. (continued)

Isotope	Natural abundance (%)	Magnetic moment, μN, in multiples of the nuclear magneton (eh/4Mc)	Nuclear Spin (I) in multiples of $h/2\pi$
18. ^{29}Si	4.70	−0.55477	1/2
19. ^{31}P	100	1.1305	1/2
20. ^{33}S	0.74	0.64274	3/2
21. ^{35}Cl	75.4	0.82091	3/2
22. ^{37}Cl	24.6	0.68330	3/2
23. ^{39}K	93.08	0.39094	3/2
24. ^{41}K	6.91	0.21488	3/2
25. ^{43}Ca	0.13	−1.3153	7/2
26. ^{45}Sc	100	4.7492	7/2
27. ^{47}Ti	7.75	−0.78711	5/2
28. ^{49}Ti	5.51	−1.1022	7/2
29. ^{50}V	0.24	3.3413	6
30. ^{51}V	99.76	5.1392	7/2
31. ^{53}Cr	9.54	−0.47354	3/2
32. ^{55}Mn	100	3.4611	5/2
33. ^{57}Fe	2.245	0.0903	1/2
34. ^{59}Co	100	4.6388	7/2
35. ^{61}Ni	1.25	0.746	3/2
36. ^{63}Cu	69.09	2.2206	3/2
37. ^{65}Cu	30.91	2.3790	3/2
38. ^{67}Zn	4.12	0.87354	5/2
39. ^{69}Ga	60.2	2.0108	3/2
40. ^{71}Ga	39.8	2.5549	3/2
41. ^{73}Ge	7.61	−0.87677	9/2
42. ^{75}As	100	1.4349	3/2
43. ^{77}Se	7.50	0.5325	1/2
44. ^{79}Br	50.57	2.0991	3/2
45. ^{81}Br	49.43	2.2626	3/2
46. ^{83}Kr	11.55	−0.96705	9/2
47. ^{85}Rb	72.8	1.3482	5/2
48. ^{87}Rb	27.2	2.7414	3/2
49. ^{87}Sr	7.02	−1.0893	9/2
50. ^{89}Y	100	−0.13682	1/2
51. ^{91}Zr	11.23	−1.298	5/2
52. ^{93}Nb	100	6.1435	9/2
53. ^{95}Mo	15.78	−0.9099	5/2
54. ^{97}Mo	9.60	−0.9290	5/2
55. ^{99}Tc[a]	–	5.6572	9/2
56. ^{99}Ru	12.81	−0.63	5/2
57. ^{101}Ru	16.98	−0.69	5/2
58. ^{103}Rh	100	−0.0879	1/2
59. ^{105}Pd	22.23	−0.57	5/2
60. ^{107}Ag	51.35	−0.1130	1/2

Table 7-6. (continued)

Isotope	Natural abundance (%)	Magnetic moment, μN, in multiples of the nuclear magneton (eh/4Mc)	Nuclear Spin (I) in multiples of h/2π
61. ^{109}Ag	48.65	−0.1299	1/2
62. ^{111}Cd	12.34	−0.6195	1/2
63. ^{113}In	4.16	5.4960	9/2
64. ^{115}In	95.84	5.5073	9/2
65. ^{115}Sn	0.35	−0.9132	1/2
66. ^{117}Sn	7.67	−0.9949	1/2
67. ^{119}Sn	8.68	−1.0409	1/2
68. ^{121}Sb	57.25	3.3417	5/2
69. ^{123}Sb	42.75	2.5334	7/2
70. ^{123}Te	0.89	−0.7319	1/2
71. ^{125}Te	7.03	−0.8824	1/2
72. ^{127}I	100	2.7937	5/2
73. ^{129}Xe	26.24	−0.77255	1/2
74. ^{131}Xe	21.24	0.68680	3/2
75. ^{133}Cs	100	2.5642	7/2
76. ^{135}Ba	6.59	0.83229	3/2
77. ^{137}Ba	11.32	0.93107	3/2
78. ^{139}La	99.911	2.7615	7/2
79. ^{141}Pr	100	3.92	5/2
80. ^{143}Nd	12.20	−1.25	7/2
81. ^{145}Nd	8.30	−0.78	7/2
82. ^{147}Sm	15.07	−0.68	7/2
83. ^{149}Sm	13.84	−0.55	7/2
84. ^{151}Eu	47.77	3.441	5/2
85. ^{153}Eu	52.23	1.521	5/2
86. ^{155}Gd	14.68	−0.25	3/2
87. ^{157}Gd	15.64	−0.34	3/2
88. ^{159}Tb	100	1.52	3/2
89. ^{161}Dy	18.73	−0.38	5/2
90. ^{163}Dy	24.97	−0.53	5/2
91. ^{165}Ho	100	3.31	7/2
92. ^{167}Er	22.82	0.48	7/2
93. ^{169}Tm	100	−0.229	1/2
94. ^{171}Yb	14.27	0.4926	1/2
95. ^{173}Yb	16.08	−0.677	5/2
96. ^{175}Lu	97.40	2.230	7/2
97. $^{176}Lu^{a}$	2.60	4.2	6
98. ^{177}Hf	18.39	0.61	7/2
99. ^{179}Hf	13.78	−0.47	9/2
100. ^{181}Ta	100	2.340	7/2
101. ^{183}W	14.28	0.115	1/2
102. ^{185}Re	37.07	3.1437	5/2

Table 7-6. (continued)

Isotope	Natural abundance (%)	Magnetic moment, μN, in multiples of the nuclear magneton (eh/4Mc)	Nuclear Spin (I) in multiples of $h/2\pi$
103. ^{187}Re	62.93	3.1760	5/2
104. ^{187}Os	1.64	0.12	1/2
105. ^{189}Os	16.1	0.6507	3/2
106. ^{191}Ir	38.5	0.16	3/2
107. ^{193}Ir	61.5	0.17	3/2
108. ^{195}Pt	33.7	0.6004	1/2
109. ^{197}Au	100	0.1439	3/2
110. ^{199}Hg	16.86	0.4979	1/2
111. ^{201}Hg	13.24	−0.5513	3/2
112. ^{203}Tl	29.52	1.5960	1/2
113. ^{205}Tl	70.48	1.6115	1/2
114. ^{207}Pb	21.11	0.5837	1/2
115. ^{209}Bi	100	4.0389	9/2
116. ^{209}Po[a]	–	–	1/2
117. ^{227}Ac[a]	–	1.1	3/2
118. ^{231}Pa[a]	–	1.96	3/2
119. ^{233}U[a]	–	0.54	5/2
120. ^{235}U[a]	0.71	0.35	7/2
121. ^{237}Np[a]	–	2.5	5/2
122. ^{239}Np[a]	–	–	1/2
123. ^{239}Pu[a]	–	0.4	1/2
124. ^{241}Pu[a]	–	1.4	5/2
125. ^{241}Am[a]	–	1.4	5/2
126. ^{243}Am[a]	–	1.4	5/2
127. ^{244}Cm[a]	–	–	7/2

[a] Radioactive.

APPENDIX 7-3.

Table 7-7. Calculated Values of Anisotropic (B_o) and Isotropic (A_o) Couplings for a Free Ion (from Goodman and Raynor, 1970).

Ion	Electronic configuration	$2B_o$ (G)	A_o (G)
^{1}H 0	$1s^1$	–	508
^{2}H 0	$1s^1$	–	78
^{3}He 0	$1s^2$	–	−2192
^{6}Li 0	$2s^1$	–	39
^{7}Li 0	$2s^1$	–	103
^{9}Be 0	$2s^2$	–	−128
^{10}B 0	$2p^1$	17.3	242
^{11}B 0	$2p^1$	38.1	723
^{13}C 0	$2p^2$	63.8	1119
^{14}N 0	$2p^3$	33.5	557
^{15}N 0	$2p^3$	−46.6	−781
^{17}O 0	$2p^4$	−102	−1659
^{19}F 0	$2p^5$	1085	17160
^{21}Ne 0	$2p^6$	−132	−9886
^{23}Na 0	$3s^1$	687	224
^{25}Mg 0	$3s^2$	−233	−119
^{27}Al 0	$3p^1$	43.2	983
^{29}Si 0	$3p^2$	−61.5	−1218
^{31}P 0	$3p^3$	202	3676
^{33}S 0	$3p^4$	56.6	975
^{35}Cl 0	$3p^5$	102	1672
^{37}Cl 0	$3p^5$	84.4	1391
^{39}K 0	$4s^1$	64.1	51.9
^{41}K 0	$4s^1$	35.2	28.6
^{43}Ca 0	$4s^2$	−182	−150
^{45}Sc −1	$3d^2 4s^2$	−27.97	276
^{45}Sc 0	$3d^4$	−20.21	–
^{45}Sc 0	$3d^1 4s^2$	−37.74	654.1
^{45}Sc +1	$3d^2$	−31.43	–
^{45}Sc +2	$3d^1$	−41.65	–
$^{47,49}Ti$ −1	$3d^3 4s^2$	9.96	−70.06
$^{47,49}Ti$ 0	$3d^4$	7.76	–
$^{47,49}Ti$ 0	$3d^2 4s^2$	12.36	−175.7
$^{47,49}Ti$ +1	$3d^3$	10.53	–
$^{47,49}Ti$ +2	$3d^2$	13.08	–
$^{47,49}Ti$ +3	$3d^1$	15.60	–
^{51}V −1	$3d^4 4s^2$	−63.23	364.3
^{51}V 0	$3d^5$	−52.53	–
^{51}V 0	$3d^3 4s^2$	−75.60	932.3
^{51}V +1	$3d^4$	−65.67	–
^{51}V +2	$3d^3$	−78.77	–
^{51}V +3	$3d^2$	−92.11	–
^{51}V +4	$3d^1$	−105.18	–

Table 7-7. (continued)

Ion	Electronic Configuration	$2B_o$ (G)	A_o (G)
$^{53}Cr-1$	$3d^5 4s^2$	17.88	−85.5
$^{53}Cr\,0$	$3d^6$	14.89	–
$^{53}Cr\,0$	$3d^4 4s^2$	20.58	−224.9
$^{53}Cr+1$	$3d^5$	18.29	–
$^{53}Cr+2$	$3d^4$	21.23	–
$^{53}Cr+3$	$3d^3$	24.36	–
$^{53}Cr+4$	$3d^2$	27.56	–
$^{53}Cr+5$	$3d^1$	30.66	–
$^{55}Mn-1$	$3d^6 4s^2$	−97.45	433.8
$^{55}Mn\,0$	$3d^7$	−84.15	–
$^{55}Mn\,0$	$3d^5 4s^2$	−112.0	1093
$^{55}Mn+1$	$3d^6$	−99.13	–
$^{55}Mn+2$	$3d^5$	−114.5	–
$^{55}Mn+3$	$3d^4$	−129.2	–
$^{55}Mn+4$	$3d^3$	−144.5	–
$^{55}Mn+5$	$3d^2$	−160.3	–
$^{55}Mn+6$	$3d^1$	−175.6	–
$^{57}Fe-1$	$3d^7 4s^2$	−15.57	63.2
$^{57}Fe\,0$	$3d^8$	−14.34	–
$^{57}Fe\,0$	$3d^6 4s^2$	−17.51	160.5
$^{57}Fe+1$	$3d^7$	−15.79	–
$^{57}Fe+2$	$3d^6$	−17.88	–
$^{57}Fe+3$	$3d^5$	−20.14	–
$^{59}Co-1$	$3d^8 4s^2$	−138.0	519.6
$^{59}Co\,0$	$3d^9$	−123.7	–
$^{59}Co\,0$	$3d^7 4s^2$	−153.0	1308
$^{59}Co+1$	$3d^8$	−139.4	–
$^{59}Co+2$	$3d^7$	−155.8	–
$^{59}Co+3$	$3d^6$	−172.9	–
$^{61}Ni-1$	$3d^9 4s^2$	−61.58	216.9
$^{61}Ni\,0$	$3d^{10}$	−55.41	–
$^{61}Ni\,0$	$3d^8 4s^2$	−67.59	539.7
$^{61}Ni+1$	$3d^9$	−61.16	–
$^{61}Ni+2$	$3d^8$	−68.69	–
$^{61}Ni+3$	$3d^7$	−75.47	–
$^{63}Cu-1$	$3d^{10} 4s^2$	−216.0	694.5
$^{63}Cu\,0$	$3d^{10} 4s^1$	−216.5	–
$^{63}Cu\,0$	$3d^9 4s^2$	−235.0	1767
$^{63}Cu+1$	$3d^{10}$	−217.4	–
$^{63}Cu+2$	$3d^9$	−238.1	–
$^{65}Cu-1$	$3d^{10} 4s^2$	−231.4	744.0
$^{65}Cu\,0$	$3d^{10} 4s^1$	−216.5	–
$^{65}Cu\,0$	$3d^9 4s^2$	−251.8	1893
$^{65}Cu+1$	$3d^{10}$	−232.9	–
$^{65}Cu+2$	$3d^9$	−250.1	–

Table 7-7. (continued)

Ion	Electronic Configuration	$2B_o$ (G)	A_o (G)
^{67}Zn 0	$3d^{10}$	64.5	454
^{69}Ga 0	$4p^1$	106	2667
^{71}Ga 0	$4p^1$	135	3389
^{73}Ge 0	$4p^2$	−25.2	−535
^{75}As 0	$4p^3$	179	3431
^{77}Se 0	$4p^4$	270	4816
^{79}Br 0	$4p^5$	459	7764
^{81}Br 0	$4p^5$	495	8370
^{83}Kr 0	$4p^6$	−87.2	−1439
^{85}Rb 0	$5s^1$	297	200
^{87}Rb 0	$5s^1$	1005	678
^{87}Sr 0	$5s^2$	−1715	−1656
^{89}Y 0	$4d^1 5s^2$	9.1	−235
^{89}Y +1	$4d^2$	7.81	—
^{89}Y +2	$4d^1$	9.98	—
^{91}Zr 0	$4d^2 5s^2$	24.3	−522
^{91}Zr +1	$4d^1 5s^1 5p^1$	—	−1548
^{91}Zr +1	$4d^3$	19.70	—
^{91}Zr +2	$4d^2$	26.86	—
^{91}Zr +3	$4d^1$	31.37	—
^{93}Nb 0	$4d^3 5s^2$	−83.3	1556
^{93}Nb +1	$4d^2 5s^1 5p^1$	—	4577
^{93}Nb +1	$4d^4$	−67.36	—
^{93}Nb +2	$4d^3$	−90.80	—
^{93}Nb +3	$4d^2$	−104.1	—
^{95}Mo 0	$4d^4 5s^2$	27.4	−462
^{95}Mo +1	$4d^3 5s^1 5p^1$	—	−1259
^{95}Mo +1	$4d^5$	25.96	—
^{95}Mo +2	$4d^4$	29.60	—
^{95}Mo +3	$4d^3$	33.37	—
^{97}Mo 0	$4d^4 5s^2$	28.4	−471
^{97}Mo +1	$4d^3 5s^1 5p^1$	—	−1285
^{97}Mo +1	$4d^5$	26.51	—
^{97}Mo +2	$4d^4$	30.22	—
^{97}Mo +3	$4d^3$	34.06	—
^{99}Tc 0	$4d^5 5s^2$	−116	1749
^{99}Tc +1	$4d^4 5s^1 5p^1$	—	5162
^{99}Ru 0	$4d^6 5s^2$	27.4	−376
^{99}Ru +1	$4d^5 5s^1 5p^1$	—	−1129
^{99}Ru +1	$4d^6$	23.95	—
^{99}Ru +2	$4d^5$	28.76	—
^{99}Ru +3	$4d^4$	31.84	—
^{101}Ru 0	$4d^6 5s^2$	30.4	−416
^{101}Ru +1	$4d^5 5s^1 5p^1$	—	−1236
^{101}Ru +2	$4d^6$	31.50	—

Table 7-7. (continued)

Ion	Electronic Configuration	$2B_o$ (G)	A_o (G)
$^{101}Ru^{+3}$	$4d^5$	34.92	–
$^{103}Rh\,0$	$4d^7\,5s^2$	22.7	–286
$^{103}Rh^{+1}$	$4d^6\,5s^1\,5p^1$	–	–849
$^{103}Rh^{+1}$	$4d^8$	19.60	–
$^{103}Rh^{+2}$	$4d^7$	23.85	–
$^{103}Rh^{+3}$	$4d^6$	26.10	–
$^{105}Pd\,0$	$4d^8\,5s^2$	34.1	–397
$^{105}Pd^{+1}$	$4d^7\,5s^1\,5p^1$	–	–1187
$^{105}Pd^{+1}$	$4d^9$	30.02	–
$^{105}Pd^{+2}$	$4d^8$	34.70	–
$^{105}Pd^{+3}$	$4d^7$	37.69	–
$^{107}Ag\,0$	$4d^9\,5s^2$	38.8	–419
$^{107}Ag^{+1}$	$4d^8\,5s^1\,5p^1$	–	–1256
$^{107}Ag^{+1}$	$4d^{10}$	36.21	–
$^{107}Ag^{+2}$	$4d^9$	39.20	–
$^{107}Ag^{+3}$	$4d^8$	42.31	–
$^{109}Ag\,0$	$4d^9\,5s^2$	44.5	–482
$^{109}Ag^{+1}$	$4d^8\,5s^1\,5p^1$	–	–1443
$^{109}Ag^{+1}$	$4d^{10}$	41.62	–
$^{109}Ag^{+2}$	$4d^9$	45.06	–
$^{109}Ag^{+3}$	$4d^8$	48.63	–
$^{111}Cd\,0$	$4d^{10}\,5s^2$	–229	–2326
$^{113}Cd\,0$	$4d^{10}\,5s^2$	–240	–2433
$^{113/115}In\,0$	$5p^1$	149	3417
$^{115}Sn\,0$	$5p^2$	–335	–6669
$^{117}Sn\,0$	$5p^2$	–367	–7268
$^{119}Sn\,0$	$5p^2$	–385	–7603
$^{121}Sb\,0$	$5p^3$	338	6089
$^{123}Sb\,0$	$5p^3$	183	3297
$^{123}Te\,0$	$5p^4$	–478	–8081
$^{125}Te\,0$	$5p^4$	–576	–9738
$^{127}I\,0$	$5p^5$	455	7320
$^{129}Xe\,0$	$5p^6$	–754	–11827
$^{131}Xe\,0$	$5p^6$	223	3504
$^{133}Cs\,0$	$6s^1$	466	351
$^{135}Ba\,0$	$6s^2$	442	472
$^{137}Ba\,0$	$6s^2$	494	527
$^{177}Hf\,0$	$5d^2\,6s^2$	–14.9	322
$^{177}Hf^{+}$	$5d^1\,6s^1\,6p^1$	–	965
	$5d^3$	–11.26	–
$^{179}Hf\,0$	$5d^2\,6s^2$	9.2	–199
$^{179}Hf^{+}$	$5d^1\,6s^1\,6p^1$	–	–578
	$5d^3$	6.74	–
$^{181}Ta\,0$	$5d^3\,6s^2$	–73.1	1416
$^{181}Ta^{+}$	$5d^2\,6s^1\,6p^1$	–	4166

Table 7-7. (continued)

Ion	Electronic configuration	$2B_o$ (G)	A_o (G)
	$5d^4$	−56.21	—
$183_{\mathrm{W}}0$	$5d^4\,6s^2$	−30.3	536
$183_{\mathrm{W}}+$	$5d^3\,6s^1\,6p^1$	—	1561
	$5d^5$	−23.51	—
$185_{\mathrm{Re}}0$	$5d^5\,6s^2$	−195	3188
$185_{\mathrm{Re}}+$	$5d^4\,6s^1\,6p^1$	—	9274
	$5d^6$	−154.8	—
$187_{\mathrm{Re}}0$	$5d^5\,6s^2$	−197	3221
$187_{\mathrm{Re}}+$	$5d^4\,6s^1\,6p^1$	—	9269
	$5d^6$	−156.4	—
$189_{\mathrm{Os}}0$	$5d^6\,6s^2$	−77.7	1183
$189_{\mathrm{Os}}+$	$5d^5\,6s^1\,6p^1$	—	3427
	$5d^1$	−62.33	—
$191_{\mathrm{Ir}}0$	$5d^7\,6s^2$	−21.9	396
$191_{\mathrm{Ir}}+$	$5d^6\,6s^1\,6p^1$	—	901
	$5d^8$	−17.62	—
$193_{\mathrm{Ir}}0$	$5d^7\,6s^2$	−23.1	419
$193_{\mathrm{Ir}}+$	$5d^6\,6s^1\,6p^1$	—	957
	$5d^8$	−18.71	—
$195_{\mathrm{Pt}}0$	$5d^8\,6s^2$	−278	3709
$195_{\mathrm{Pt}}+$	$5d^7\,6s^1\,6p^1$	—	10840
	$5d^9$	−225.7	—
$197_{\mathrm{Au}}0$	$5d^9\,6s^2$	−25.0	313
$197_{\mathrm{Au}}+$	$5d^8\,6s^1\,6p^1$	—	913
	$5d^{10}$	−20.24	—
$199_{\mathrm{Hg}}0$	$5d^{10}\,6s^2$	−286	3416
$199_{\mathrm{Hg}}+$	$5d^9\,6s^1\,6p^1$	—	9606
	$5d^{10}\,6s^1$	−249.6	—
$201_{\mathrm{Hg}}0$	$5d^{10}\,6s^2$	105	−1258
$201_{\mathrm{Hg}}+$	$5d^9\,6s^1\,6p^1$	—	−3680
	$5d^{10}\,6s^1$	92.12	—
$203_{\mathrm{Tl}}0$	$6p^1$	660	14893
$205_{\mathrm{Tl}}0$	$6p^1$	666	15040
$207_{\mathrm{Pb}}0$	$6p^2$	351	6868
$209_{\mathrm{Bi}}0$	$6p^3$	358	6394

REFERENCES

1. Aasa, R. 1970. Powder line shapes in the EPR spectra of high spin ferric complexes. *J. Phys. Chem.* **32**: 3919-3930.
2. Abragam, A. and M.H.L. Pryce. 1951. Theory of the nuclear hyperfine structure of paramagnetic resonance spectra in crystals. *Proc. Roy. Soc.* A **205**: 135-153.
3. Adrian, F.J. 1968. Guidelines for interpreting ESR spectra of paramagnetic species adsorbed on surfaces. *J. Colloid Interface Sci.* **26**: 317-354.
4. Ayscough, P.B. 1967. ESR in Chemistry. Methuen, London. 451 pp.
5. Blinder, S.M. 1960. Orientation dependence of magnetic hyperfine structure in free radicals. *J. Chem. Phys.* **33**: 748-752.
6. Buch, T., B. Clerjaud, B. Lambert and P. Kovacs. 1973. EPR study of $3d^5$ ions in mixed-polytype ZnS_2. *Phys. Rev. B,* **7**: 184-191.
7. Carrington, A. and A.D. McLachlan. 1967. Introduction to magnetic resonance. Harper and Row, New York. 266 pp.
8. Chao, C.C. and J.H. Lunsford. 1972. EPR study of Cu(II) ion pairs in Y-type zeolites. *J. Chem. Phys.* **57**: 2890-2898.
9. Che, M., J. Demarquay and C. Naccache. 1969. Mathematical method of determining shapes and widths of individual lines in powder EPR spectra. *J. Chem. Phys.* **51**: 5177-8.
10. Che, M., J. Vedrine, and C. Naccache. 1969. Le facteur g de l'électron: représentation physique et mesure par RPE. Méthodes d'analyse d'un spectre de poudres. *J. Chim. Phys.* **66**: 579-594.
11. Che, M., J. Fraissard and J.C. Vedrine. 1974. Application de la RPE et de la RMN à l'étude des silicates et des argiles. *Bull. Groupe Franc. Argiles* **25**: 1-53.
12. Clementz, D.M., T.J. Pinnavaia and M.M. Mortland. 1973. Stereochemistry of hydrated Cu(II) ions on the interlamellar surfaces of layer silicates. An ESR study. *J. Phys. Chem.* **77**: 196-200.
13. Cordischi, D., R.L. Nelson and A.J. Tench. 1969. Surface reactivity of MgO doped with Mn: an ESR and chemisorption study. *Trans. Faraday Soc.* **65**: 2740-2757.
14. Derouane, E.G., M. Mestdagh and L. Vielvoye. 1974. EPR study of the nature and removal of Fe(III) impurities in NH_4-exchanged NaY zeolite. *J. Catal.* **38**: 169-175.
15. Derouane, E.G. and V. Indovina. 1973. Mn ions as paramagnetic probes in MgO. *Bull. Soc. Chim. Belg.* **82**: 645-656.
16. Dowsing, R.D. and J.F. Gibson. 1969. ESR of high spin d^5 systems. *J. Chem. Phys.* **50**: 294-303.
17. Dugas, H. 1973. Introduction à la RPE et à la méthode du marqueur de spin – un article revue. *Can. J. Spectrosc.* **18**: 110-118.
18. Farach, H.A. and C.P. Poole. 1971. The spin hamiltonian for completely anisotropic g-factor and hyperfine coupling tensors. *Il nuovo cimento* **4**: 51-58.
19. Flockhard, B.D. 1973. ESR studies of adsorbed species. *In* M.W. Roberts and J.M. Thomas, (eds). Surface and defect properties of solids. Vol. 2. *The Chem. Soc. London.* pp. 69-96.
20. Freed, J.H. and G.K. Fraenkel. 1963. Theory of linewidths in ESR spectra. *J. Chem. Phys.* **39**: 326-348.
21. Freymann, R. and M. Soutif. 1960. La spectroscopie hertzienne appliquée à la chimie. Dunod, Paris. 185 pp.

22. Gaite, J.M. and J. Michoulier. 1970. Application de la RPE de l'ion Fe^{3+} à l'étude de la structure des feldspaths. *Bull. Soc. Fr. Minéral. Cristall.* **93**: 341-356.
23. McGarvey, B.R. 1974. Determination of electron distribution from hyperfine interaction. *In* D.R. Lide, Jr. and M.A. Paul, (eds). Critical evaluation of chemical and physical structural informations. National Academy of Sciences, Washington, D.C. pp. 415-435.
24. Goodman, B.A. and J.B. Raynor. 1970. ESR of transition metal complexes. *Adv. Inorganic Chem. and Radiochem.* **13**: 135-162.
25. Griscom, D.L. and R.E. Griscom. 1967. Paramagnetic resonance of Mn^{2+} in glasses and compounds of the Li Borate System. *J. Chem. Phys.* **47**: 2711-2722.
26. Griffith, J.C. 1964. Theory of the isotropic g-value of 4.27 found for some high-spin ferric ions. *Molec. Phys.* **8**: 213-216.
27. Guzy, C.M., J.B. Raynor and M.C.R. Symons. 1969. ESR spectrum of Cu-phthalocyanine. A reassessment of the bonding parameters. *J. Chem. Soc.* **A15**: 2299-2303.
28. Harvey, J.S.M., 1970. Relativistic contribution to the second-order crystal-field splitting in Eu^{2+}: $CaWO_4$. *Can. J. Phys.* **48**: 574-579.
29. Hemidy, J.F., F. Delavennat and D. Cornet. 1973. Etude en RPE de zéolithes chromeés. I. Cr(III). *J. Chim. Phys.* **70**: 1716-1720.
30. Hemidy, J.F. and D. Cornet. 1974. Etude en RPE de zéolithes chromeés. II. Cr(V). *J. Chim. Phys.* **71**: 739-745.
31. Hubbell, W.L. and H.M. McConnell. 1969. Motion of setroid spin labels in membranes. *Proc. National Acad. Sci.* **63**: 16-22.
32. Jones, M.T., M. Komarynsky and R.D. Rataiczak. 1971. An ESR line-shape analysis for determination of unresolved metal hyperfine splittings in ion pairs. Its application to the benzene anion radical. *J. Phys. Chem.* **75**: 2769-2773.
33. Kemp, R.C. 1972. ESR of Fe^{3+} in phlogopite. *J. Phys. C. Solid State Phys.* **5**: 3566-3572.
34. Kivelson, D. and R. Neiman. 1961. ESR studies on the bonding in Cu complexes. *J. Chem. Phys.* **35**: 149-155.
35. Kottis, P. and R. Lefebvre. 1964. Calculation of the ESR line-shape of randomly oriented molecules in a triplet state. Correlation of the spectrum with the zero-field splittings. Introduction of an orientation-dependent linewidth. *J. Chem. Phys.* **41**: 379-393.
36. Lebedev, Y.S. 1963. Calculation of EPR spectra with an electronic computer. II. asymmetrical lines.*Zh. strukt Khimii* **4**: 19-24.
37. Loveridge, D. and S. Parke. 1971. ESR of Fe^{3+}, Mn^{2+} and Cr^{3+} in glasses. *Phys. Chem. Glasses* **12**: 19-27.
38. Lunsford, J.H. and E.F. Van Sant. 1973. Formation and structure of penta- and hexa-coordinate Co(II) methyl isocyanide complexes in Y-type zeolites. *J. Chem. Soc., Faraday Trans II* **69**: 1028-1035.
39. Martini, G., M.F. Ottaviani and G.L. Seravalli. 1975. ESR study of vanadyl complexes adsorbed on synthetic zeolites. *J. Phys. Chem.* **79**: 1716-1720.
40. Maruani, J. 1967. Sur l'interprétation des spectres de RPE des radicaux libres polyorientés. *Cahiers Phys.* **202**: 1-147.
41. Marx, R. 1966. RPE et mouvements moléculaires. *J. Chim. Phys.* **63**: 128-135.
42. Meriaudeau, P., Y. Boudeville, P. de Montgolfier and M. Che. 1977. g, hyperfine and superhyperfine tensors of pentavalent tungsten in polycrystalline tin dioxide. *Phys. Rev. B* **16**: 30-36.

43. Meissier, J. and G. Marc. 1971. Etude par RPE de la structure de couches monomoléculaires de stéarate de cuivre. *J. Phys.* **32**: 799-804.
44. Mialhe, P., A. Briquet and B. Tribollet. 1971. Détermination du signe relatif de A et D à partir d'un spectre de RPE d'un échantillon à l'état de poudre cristalline. *J. Phys. Chem. Solids* **32**: 2639-2643.
45. Mikheikin, I.D., O.I. Brotikovskii and V.B. Kazanskii. 1972. ESR investigation of the sites where Cr^{3+} ions are localized in dehydrated type Y zeolite. *Kinet. Katal.* **13**: 481-482.
46. de Montgolfier, P., P. Meriaudeau, Y. Boudeville and M. Che. 1976. Hyperfine and superhyperfine interactions for Mo^{5+} in SnO_2. *Phys. Rev. B* **14**: 1788-1795.
47. Naccache, C. and Y. Ben Taarit. 1971. Oxidizing and acidic properties of Cu-exchanged Y zeolite. *J. Catal.* **22**: 171-181.
48. Naccache, C. and Y. Ben Taarit. 1973. Nature of NO and NO_2 adsorbed on Cr and Ni exchanged zeolites. ESR and IR study. *J. Chem. Soc., Faraday Trans. I* **69**: 1475-1486.
49. Neiman, R. and D. Kivelson. 1961. ESR line shapes in glasses of Cu complexes. *J. Chem. Phys.* **35**: 156-161.
50. O'Brien, M.C.M. 1964. The dynamic Jahn-Teller effect in octahedrally co-ordinated d^9 ions. *Proc. Roy. Soc. London.* **A 281**: 323-339.
51. Olivier, D., J.C. Vedrine and H. Pezerat. 1975. Application de la RPE à la localisation du Fe^{3+} dans les smectites. *Bull. Groupe Franç. Argiles* **27**: 153-165.
52. Olivier, D., J.C. Vedrine and H. Pezerat. 1975. RPE de Fe^{3+} dans les argiles altérés artificiellement et dans le milieu naturel. *Proc. Inter. Clay Conf. 1975* (Publ. 1976): 231-238.
53. Olivier, D., J.C. Vedrine and H. Pezerat. 1977. Application de la RPE à la localisation des substitutions isomorphiques dans les micas: localisation du Fe^{3+} dans les muscovites et les phlogopites. *J. Sol. Stat. Chem.* **20**: 267-279.
54. Papp, J., I.D. Mikheikin and V.B. Kazanskii. 1970. EPR study on the stable H atoms on γ-irradiated clinoptilolites containing La, Mg and K ions. *Kinet. Katal.* **11**: 671-2.
55. Pescia, J. 1966. La relaxation des spins électroniques avec le réseau. Théorie élémentaire et méthodes de mesure du temps T_1. *J. Phys.* **27**: 782-800.
56. Poole, Ch. 1967. ESR a comprehensive treatise on experimental techniques. *Interscience,* New York. Chaps. 18 and 20.
57. Poole, Ch. and H.A. Farach, (eds). 1972. The theory of magnetic resonance. Wiley – Interscience, New York. 452 pp.
58. Rei, D.K. 1962. Analysis of the EPR spectrum of V^{3+} ions in corundum (Al_2O_3) *Soviet. Phys. Solid State* **3**: 1606-1612.
59. Rei, D.K. 1962. Concerning the paramagnetic resonance spectrum of V^{4+} ions in rutile (TiO_2). *Soviet. Phys. Solid State* **3**: 1845-1847.
60. Rius, G. and A. Herve. 1972. EPR of Cr^+ in MgO. *Solid State Comm.* **11**: 795-797.
61. Rupert, J.P. 1973. ESR spectra of interlamellar Cu(II)-arene complexes on montmorillonite. *J. Phys. Chem.* **77**: 784-790.
62. Sands, R.H. 1955. Paramagnetic resonance absorption in glass. *Phys. Rev. B.* **99**: 1222-1226.
63. Sroubek, Z. and K.Z. Dansky. 1966. ESR of Cu^{2+} ion in $CdWO_4$, $ZnWO_4$ and $MgWO_4$. *J. Chem. Phys.* **44**: 3078-3083.

64. Taarit, Ben Y., M.V. Mathieu and C. Naccache. 1971. Acidic and oxidizing properties of rare earth exchanged Y zeolites. Advances in Chemistry Series. No. 102. Molecular Sieves II. *American Chemical Society.* pp. 362-372.
65. Thikhomirova, N.N., S.N. Dobryakov and I.V. Nikolaeva. 1972. The calculation of ESR spectrum of Mn^{2+} ions in polycrystalline samples. *Phys. Stat. Sol.* (a) **10:** 593-603.
66. Thikhomirova, N.N., I.V. Nikolaeva, V.V. Demkin, E.N. Rosolovskaya and K.V. Topchieva. 1973. ESR study of hydrated synthetic zeolites containing various cations. *J. Catal.* **29:** 105-111.
67. Tsay, F.D., H.B. Gray and J. Danon. 1971. EPR and optical spectra of pentacyanocobaltate (II). *J. Chem. Phys.* **54:** 3760-3768.
68. Tynan, E.C. and T.F.U. Yen. 1970. General purpose computer program for exact ESR spectrum calculations with application to V chelates. *J. Magn. Reson.* **3:** 327-335.
69. Ursu, I. 1968. La résonance paramagnétique électronique. Dunod, Paris. pp. 504.
70. Vansant, E.F. and J.H. Lunsford. 1972. ESR spectra of five or six coordinate Co(II) methyl isocyanide complexes in Co(II) Y zeolites. *J.C.S. Chem. Comm.* p. 830-832.
71. Vedrine, J.C., E.G. Derouane and Y. Ben Taarit. 1974. Temperature dependence of hyperfine coupling for Cu complexes in NaY zeolite. *J. Phys. Chem.* **78:** 531-535.
72. Vedrine, J.C. and E.G. Derouane. 1973. EPR studies of the formation, mobility and reactivity of surface species in heterogeneous catalysis. *Ind. Chim. belges* **38:** 375.
73. Volino, F. and A. Rousseau. 1972. Etude des mouvements moléculaires dans le cyclohexane plastique à l'aide d'une sonde radicalaire par RPE. *Molec. Crystals and liq. crystals* **16:** 247-262.
74. Wasserman, E., L.C. Snyder and W.A. Yager. 1964. ESR of the triplet state of randomly oriented molecules. *J. Chem. Phys.* **41:** 1763-1772.
75. Weil, J.A. 1971. The analysis of large hyperfine splitting in paramagnetic resonance spectroscopy. *J. Magn. Reson.* **4:** 394-399.
76. Wertz, J.E. and J.R. Bolton. 1972. Electron spin resonance elementary theory and practical applications. McGraw-Hill, New York. 497 pp.
77. Wilson, R. and D. Kivelson. 1966. ESR linewidths in solution. Experimental studies of anisotropic and spin-rotational effects in Cu complexes. *J. Chem. Phys.* **44:** 4445-4452.
78. Wickman, H.H., M.P. Klein and D.A. Shirley. 1965. EPR of Fe^{3+} in polycrystalline ferrichrome A. *J. Chem. Phys.* **42:** 2113-2117.
79. Zavoisky, E. 1945. *J. Phys. USSR* **9:** 211-245.
80. Zverev, G.M. and A.M. Prokhorov. 1961. EPR of the V^{3+} ion in corundum. *Soviet. Phys. JETP.* **13:** 714-715.

Chapter 8

APPLICATIONS OF ESR SPECTROSCOPY TO INORGANIC-CLAY SYSTEMS

Thomas J. Pinnavaia

Department of Chemistry
Michigan State University
East Lansing, Michigan 48824

8-1. INTRODUCTION

The surface chemistry and physical properties of clay minerals are often very much dependent on the nature of the metal ions which balance the negative charge of the oxygen framework. The most abundant metal ions normally found in clays (silicon, aluminum, magnesium, and alkali and alkaline earth metals) are diamagnetic, but paramagnetic ions such as Fe^{3+} may also be found to substitute for silicon, aluminum or magnesium in tetrahedral or octahedral positions. A variety of paramagnetic ions or metal complexes, such as VO^{2+} or $Cu(phen)_2{}^{2+}$, can become part of a clay structure by replacing the interlayer alkali or alkaline earth exchange cations. Thus, it is only natural that electron spin resonance spectroscopy (ESR), sometimes called electron paramagnetic resonance (EPR), should be a useful tool in studying the behavior of metal ions in clays.

The theory of ESR for metal ions has been well developed, and several excellent treatises and general texts are available, including those by McGarvey (28, 29), Wertz and Bolton (39), and Abragam and Bleaney (1). Since time and space do not allow for a treatment of general theory, interested students are referred to these and other references given in the text. The objective of this chapter will be to discuss some typical problems of interest to clay chemists and mineralogists which can be examined by ESR spectroscopy. Examples will include the use of the technique in examining orientation and mobility of surface-bound ions and ligand dissociation reactions of surface metal complexes. We will also discuss the chemical significance of ESR spectra which arise from paramagnetic centers in the oxygen framework.

8-2. SURFACE-BOUND METAL IONS.

8-2.1. Simple Hydrated Ions

Copper. The size and charge of interlamellar cations in smectites have an

J. W. Stucki and W. L. Banwart (eds.), Advanced Chemical Methods for Soil and Clay Minerals Research, 391–421.

important influence on the swelling properties of the minerals. At low degrees of swelling (1-3 monolayers of water) the interlayer cations are expected to be solvated by ordered layers of water molecules. The first ESR evidence to verify this solvation model for Cu^{2+} in smectites has been reported by Clementz *et al.* (14).

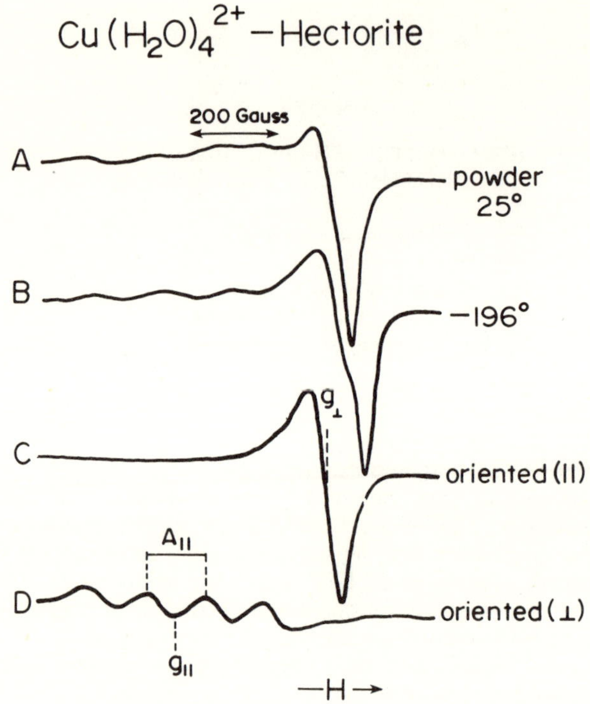

Figure 8-1. ESR spectra (first derivative curves) for Cu(II) Hectorite. Spectra A and B, respectively, are for randomly oriented powder samples at 300 and 77° K. Spectrum C is for an oriented film sample at 300° K with the silicate layers positioned parallel to H. In spectrum D the layers are positioned perpendicular to H (from Clementz *et al.*, 1973).

Fig. 8-1 illustrates ESR spectra for random powders and oriented film samples of Cu^{2+}-hectorite under conditions where a monolayer of H_2O occupies the interlayers (d_{001} = 12.4 Å). The powder spectra indicate that the hydrated Cu^{2+} has tetragonal symmetry ($g_{\parallel}$ = 2.34, A = .0165 cm^{-1}, $g_{\perp}$ = 2.08). When the magnetic field direction is oriented parallel (∥) to the silicate sheets, the electron spin is quantized in the direction ∥ to the silicate sheets, and only the $g_{\perp}$ component is observed. The $g_{\parallel}$ component is observed when the magnetic field is perpendicular ($_{\perp}$) to the silicate sheets. These results show that the copper ion is solvated by a highly oriented monolayer of water as illustrated in diagram A (Fig. 8-2).

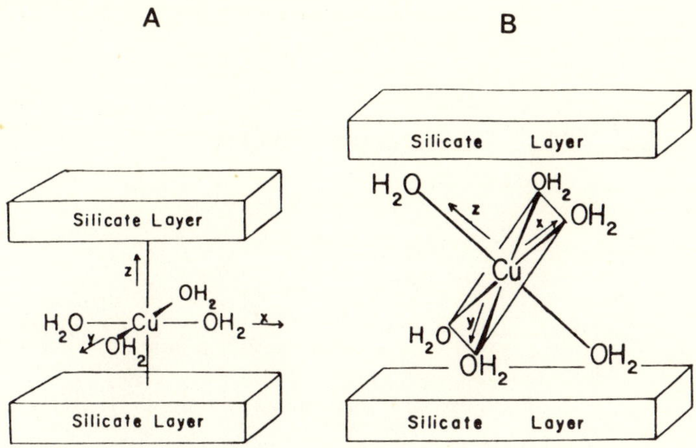

Figure 8-2. Schematic representation of the stereochemistry of hydrated Cu(II) under conditions where (A) one layer and (B) two layers of water occupy the interlamellar regions (from Clementz *et al.*, 1973).

When the Cu^{2+} ion is part of a two-water layer system, as in fully hydrated Cu^{2+}-vermiculite with d_{001} = 14.2 Å, spectra similar to those shown in A and B of Fig. 8-1 are obtained ($g_{\parallel}$ = 2.38, $A_{\parallel}$ = 0.0145 cm^{-1}, $g_{\perp}$ = 2.16). However, the spectra are orientation independent, i.e., both $g_{\parallel}$ and $g_{\perp}$ are observed for ∥ and ⊥ orientation. It may be concluded, therefore, that the tetragonally distorted $Cu(H_2O)_6{}^{2+}$ ion is highly oriented on the surface with the symmetry axis inclined with respect to the silicate sheets near an angle of 45°, as shown in diagram B (Fig. 8-2).

The spectra for Cu^{2+} doped into the three-water layer phase of Mg^{2+}-hectorite (d_{001} = 15.0 Å) are shown in Fig. 8-3 ($g_{\parallel}$ = 2.335, $g_{\perp}$ = 2.065, $A_{\parallel}$ = 0.0156 cm^{-1}, $A_{\perp}$ = 0.0022 cm^{-1}). Note that splitting of the $g_{\perp}$ line by ^{63}Cu and ^{65}Cu is resolved in this case, but that the splitting is not resolved in fully saturated Cu^{2+}-smectite or vermiculite. In the latter systems dipolar interactions between neighboring copper ions broaden the resonance somewhat and obscure the splitting. The orientation of the $Cu(H_2O)_6{}^{2+}$ ion that is deduced from the ESR spectra is shown in Fig. 8-4.

The ESR spectra of oriented film samples of Cu^{2+} doped into air-dry Na^{+}-hectorite (d_{001} ≃ 13.6 Å) (see Fig. 8-5) are quite different from those observed for the hydrated Cu^{2+} ions in the examples discussed above. It has been suggested that the broad anisotropic signal arises from binding of the ion in hexagonal cavities of silicate oxygens, perhaps near special, doubly-charged sites in the framework. Nevertheless, the specially bound ions are still available for reaction. As can be seen from spectra b in Fig. 8-5, the addition of pyridine to Cu^{2+}-doped Na^{+}-hectorite gives a new set of anisotropic lines corresponding to $Cu(py)_x{}^{2+}$ ($g_{\parallel}$ = 2.23, $g_{\perp}$ = 2.05, $A_{\parallel}$ = 0.0167 cm^{-1}; py = pyridine). This example illustrates that ESR can be a powerful tool for observing complex formation between exchange ions and adsorbate molecules in clays. We shall return to the application of ESR to metal complexes on clay surfaces in section 8-2.2.

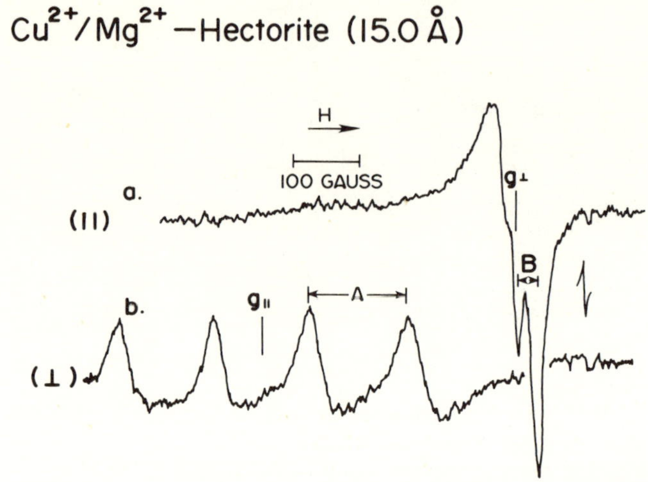

Figure 8-3. First derivative ESR spectra at 25° of oriented film samples of the d_{001} = 15.0 Å hydration state of Mg^{2+}-hectorite doped with Cu^{2+}: (a) silicate sheets parallel to the applied field, H; (b) silicate sheets perpendicular to H. Position of a standard pitch signal is shown with g = 2.0028 (from McBride *et al.*, 1975a).

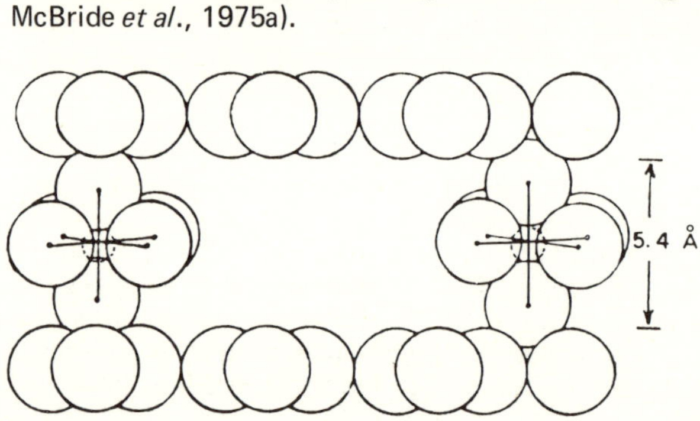

Figure 8-4. Orientation of $Cu(H_2O)_6^{2+}$ in the 15.0-Å state of hectorite. Open circles represent surface oxygen atoms of the silicate structure and the ligand water molecules of $Cu(H_2O)_6^{2+}$ (from McBride *et al.*, 1975a).

We have seen that the ESR data for Cu^{2+} solvated by one-, two-, and three-water layer systems show highly ordered orientations, indicative of low ion mobility in an authentic solid phase. But what happens when the interlayers are fully swollen with multiple layers of water? Fig. 8-6 shows the ESR spectra for fully wetted Cu^{2+}-hectorite ($d_{001} \simeq 21$ Å). A single isotropic resonance is observed, similar to the resonance that is found for $Cu(H_2O)_6^{2+}$ in dilute aqueous

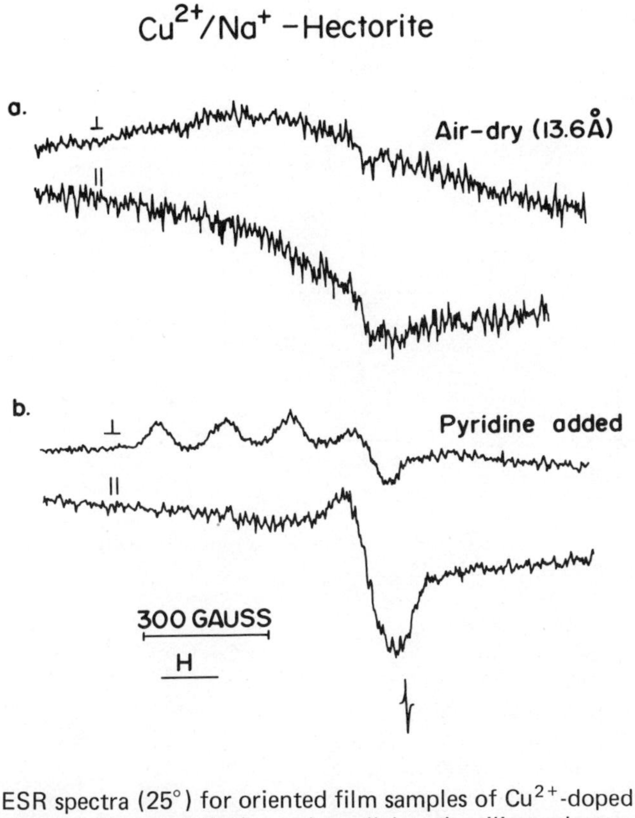

Figure 8-5. ESR spectra (25°) for oriented film samples of Cu^{2+}-doped Na^+-smectite with H perpendicular and parallel to the silicate sheets: (a) air-dried hectorite, d_{001} = 13.6 Å; (b) after exposure of air-dried hectorite to pyridine (from McBride *et al.*, 1975a).

solution. The averaging of $g_{\parallel}$ and $g_{\perp}$ and concomitant collapse of hyperfine splitting may arise by two different mechanisms: (1) rapid isotropic tumbling in a solution-like interlayer environment or (2) dynamic Jahn-Teller exchange in a rigid, ice-like environment. The second effect, which is responsible for the isotropic ESR line observed for $Cu(H_2O)_6{}^{2+}$ in frozen aqueous solution (18), occurs by coupling of vibration modes of the ligands. The vibrational coupling results in

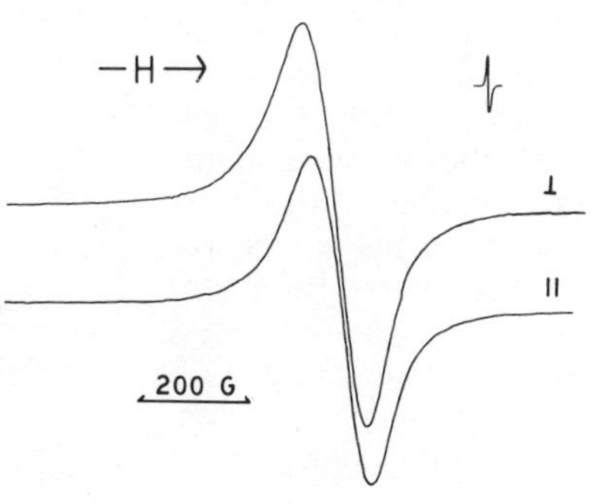

Figure 8-6. X-band spectra for an oriented film sample of Cu^{2+}-hectorite fully wetted with H_2O (d_{001} ≈ 21 Å).

the interchange of the z-axis with the x-y axes as shown in Fig. 8-7. We will later show, based on ESR studies of hydrated Mn^{2+} ions in smectite, that the ions do, in fact, tumble rapidly in a solution-like environment when the interlayers contain multiple layers of water.

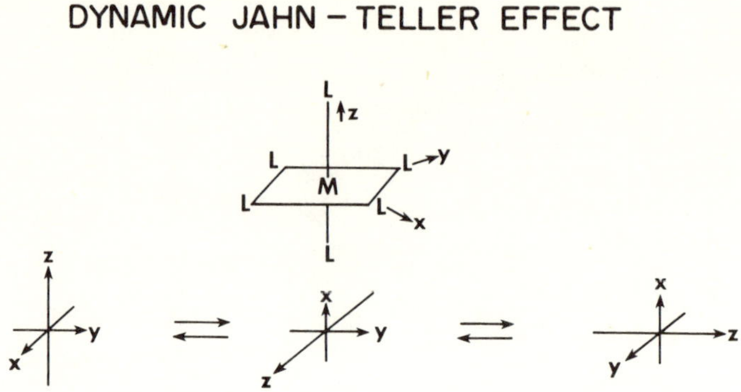

Figure 8-7. Dynamic Jahn-Teller effect for a tetragonally elongated Jahn-Teller ion such as $Cu(H_2O)_6{}^{2+}$.

The thermal dehydration (215°C) of Cu^{2+}-montmorillonite results in migration of Cu^{2+} into hexagonal cavities in the silicate framework (23). Some Cu^{2+} also migrates into vacant octahedral sites of the dioctahedral mineral to give a mineral with reduced charge. The properties of the reduced charged montmorillonite (RCM) are similar, but not identical, to those of RCM's prepared by thermal migration of Li^{+} into octahedral cavities (8). Fig. 8-8 illustrates the proposed pathway for migration of the interlayer cation to the octahedral sites. In the case of hectorite, all octahedral sites are already occupied by Mg^{2+} and Li^{+}, and Cu^{2+} exchange ions can only migrate to hexagonal cavities.

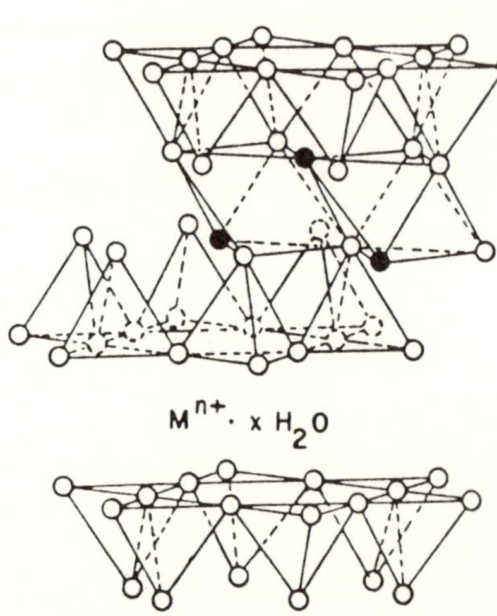

Figure 8-8. Proposed pathway for thermal migration of an exchange cation of suitable size from the interlayer region to octahedral sites in a reduced charge montmorillonite (RCM).

Fig. 8-9 illustrates the spectra of thermally dehydrated Cu^{2+}-montmorillonite and -hectorite. The heated montmorillonite shows almost no observable Cu^{2+}

signal, whereas hectorite shows a strong anisotropic signal. The loss of Cu^{2+} signal for the montmorillonite is attributed to spin interactions of the Cu^{2+} with structural Fe^{3+} which decreases the relaxation time and broadens the ESR signal (2). Hectorite contains little structural Fe^{3+}, whereas the montmorillonite contains ~ 0.32 Fe^{3+} per unit cell. Rehydration of the Cu^{2+}-montmorillonite (d_{001} = 9.7 Å) results in solvation of Cu^{2+} mainly on exchange sites at the external surfaces (*cf*, Fig. 8-9).

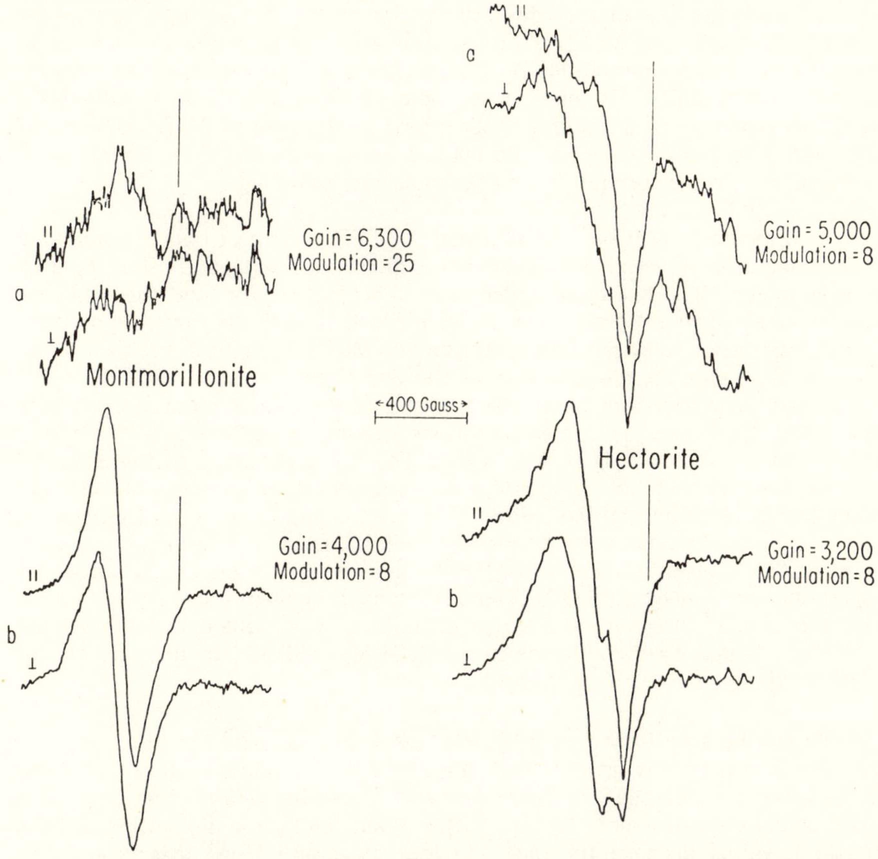

Figure 8-9. ESR spectra of Cu^{2+}-montmorillonite and -hectorite after being heated to 215° for 24 hr. (a) the dehydrated minerals in absence of moist air; (b) after rehydration over a free water surface for 24 hr. (from McBride and Mortland, 1974).

Cu^{2+} has also been used as an ESR probe of the external cation exchange sites in RCM prepared by Li^{+} migration (15) and in kaolinite (21). Orientation-dependent spectra for both types of clays indicate that the symmetry axis of the solvated ion is ⊥ to the silicate sheets under air-dry conditions. When the surfaces are fully hydrated, the ion gives an isotropic spectrum indicative of rapid tumbling or Jahn-Teller interchange of $g_{\parallel}$ and $g_{\perp}$ orientations. Based on the orientation dependence observed under air-dried conditions ($g_{\perp}$ is parallel to the silicate sheets), it has been suggested that the planar $Cu(H_2O)_4{}^{2+}$ ion occupies sites on external basal planes rather than at crystallite edges. The result is not unequivocal, however, because the ion may adopt an analogous orientation by complexing to the hydroxyl oxygens at edge positions to form Al-0-$Cu(H_2O)_x$ or related species.

Manganese. The hydrated $Mn(H_2O)_6{}^{2+}$ ion has a high spin d^5 configuration and a ground state orbital angular momentum (S-state) of zero. In most environments all three components of the g-tensor are equal, and isotropic spectra are observed. Although the Mn^{2+} is not as well suited as hydrated Cu^{2+} for ESR studies of metal ion orientations on clay surfaces, its line widths may be readily related to ion mobility. As seen earlier, rapid tumbling or a dynamic Jahn-Teller effect can give rise to an isotropic spectrum. In the case of $Mn(H_2O)_6{}^{2+}$, the Jahn-Teller effect does not apply, and the line width of the Mn^{2+} resonance can be related directly to the correlation time for molecular tumbling.

McBride *et al*. (25) have investigated the ESR spectra of hydrated Mn^{2+} as the exchange sites of vermiculite, nontronite, montmorillonite, and hectorite. Typical spectra for Mn^{2+}-hectorite under fully hydrated, air-dry, and thermally dehydrated (200°C) conditions are compared in Fig. 8-10 with the spectrum of Mn^{2+} in dilute methanol solution. The spectrum for Mn^{2+} in solution consists of six hyperfine lines due to the coupling of the S=5/2 electron spin with the I=5/2 nuclear spin. Each hyperfine component consists of three superimposed Lorentzian lines arising from five $\Delta M_S = 1$ transitions between $|\pm 5/2\rangle \Leftrightarrow |\pm 3/2\rangle$, $|\pm 3/2\rangle \Leftrightarrow |\pm 1/2\rangle$ and $|-1/2\rangle \Leftrightarrow |+1/2\rangle$ spin states. The non-degeneracy of the $\Delta M_S = 1$ transition contributes to inhomogeneous line broadening. In absences of inhomogeneous line broadening, the line widths (ΔH) are the sum of two contributions $\Delta H = \Delta H_I + \Delta H_D$ where ΔH_I is the intrinsic line width due to ion collisional relaxation processes (35, 20) and ΔH_D is the width due to dipolar interactions between neighboring Mn^{2+} neighbors (17). The ΔH_D term is concentration dependent because the dipolar interactions are proportional to r^{-3}, where r is the average Mn^{2+}-Mn^{2+} distance. In dilute solution (<0.01 M, r>55 Å), the lines are narrow and are determined exclusively by ΔH_I.

As can be seen from Fig. 8-10, the line widths for Mn^{2+} in fully hydrated hectorite are appreciably greater than those for Mn^{2+} in dilute solution. Since the average interlayer distance is in the range 10-14 Å, the line widths of Mn^{2+} in clays should be determined mainly by ΔH_D. This is verified by the plot in Fig. 8-11 of line width versus distance for Mn^{2+} in four clays and Mn^{2+} in solution. The increase in line width of Mn^{2+} with decreasing degree of hydration is due to restricted mobility and increasing dipolar interactions.

As can be seen by comparing spectra in Fig. 8-11 and Fig. 8-12, the dipolar interactions in Mn^{2+}-saturated hectorite can be eliminated by doping the ion into Mg^{2+}-hectorite at the 5% level. Theory indicates (9, 10, 11, 16) that τ, the correlation time for collision of Mn^{2+} ions with bulk water molecules, is directly proportional to the width of the $|-1/2\rangle \Leftrightarrow |+1/2\rangle$ transition, 9i.e., the fourth highest field line) provided that dipolar interactions are absent. The line width of Mn^{2+}-doped Mg^{2+}-hectorite is 28.7 G versus 22 G for Mn^{2+} in dilute solution. Thus τ is only ~ 30% longer in the interlayer region of fully wetted Mg^{2+}-hectorite ($d_{001} \simeq 21$ Å) than in bulk solution where it has been estimated to be 3.2×10^{-12}

sec (35). We may conclude, therefore, that the interlayers are very much solution like and the Mn^{2+} ions tumble rapidly even when the interlayers are only ~12 Å thick (see Fig. 8-13).

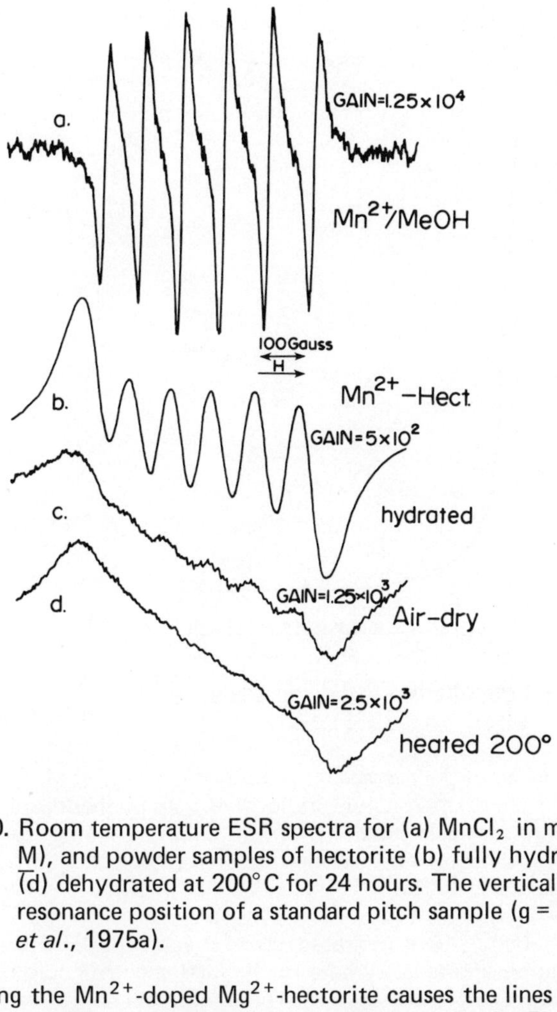

Figure 8-10. Room temperature ESR spectra for (a) $MnCl_2$ in methanol (5.0 x 10^{-5} M), and powder samples of hectorite (b) fully hydrated, (c) air-dried, and (d) dehydrated at 200°C for 24 hours. The vertical lines represent the resonance position of a standard pitch sample (g = 2.0028) from McBride *et al.*, 1975a).

Drying the Mn^{2+}-doped Mg^{2+}-hectorite causes the lines to broaden due to a decrease in mobility as illustrated by spectra b and c in Fig. 8-12. Spectrum c consists of six main lines (ΔM_I = 0) and five weaker pairs of doublets due to forbidden transitions with ΔM_I = 1. This type of spectrum is characteristic of Mn^{2+} in certain crystalline matrices and in frozen glasses in absence of dipolar coupling (3). As shown in Fig. 8-14 the six main resonances for thermally dehydrated Mn^{2+} in the doped Mn^{2+}/Mg^{2+}-hectorite system are split, possibly by hyperfine coupling to either OH protons or framework fluoride ions (24), or possibly by fine structure created by the crystal field.

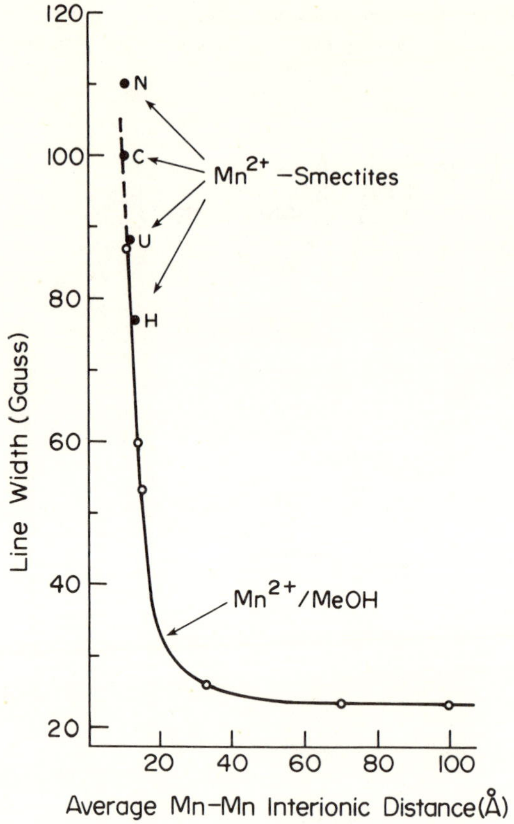

Figure 8-11. Dependence of the average $m_I = \pm 5/2$ line widths of Mn^{2+} on interionic distance. Open points are for $MnCl_2$ in methanol solution, solid points are for nontronite (N), Upton (U) and Chambers (C) montmorillonites, and hectorite (H) under fully hydrated conditions (from McBride *et al.,* 1975b).

Vanadyl Ion, VO^{2+}. The hydrated vanadyl ion, $VO(H_2O)_5{}^{2+}$, has a d^1 electronic configuration and axial symmetry. Its ESR spectroscopic properties resemble those of $Cu(H_2O)_6{}^{2+}$, with $g_\parallel \neq g_\perp$. Under normal conditions the $g_\parallel$ and $g_\perp$ resonances exhibit hyperfine splitting due to coupling of the S=1/2 electron spin with the I=7/2 nuclear spin.

$VO(H_2O)_6{}^{2+}$–hectorite in the fully wetted state exhibits the blue color characteristic of the ion in aqueous solution. The blue color is lost and a tan color develops when the sample is air-dried or freeze-dried, indicating that a surface reaction takes place which depends on moisture content. Subsequent exposure of the tan-colored material to water vapor results in fairly rapid conversion to a bright yellow clay, perhaps containing V(V). The color changes cannot be reversed by vacuum drying over P_2O_5 or by heating. These surface reactions have been investigated, in part, by ESR spectroscopy (22, 33).

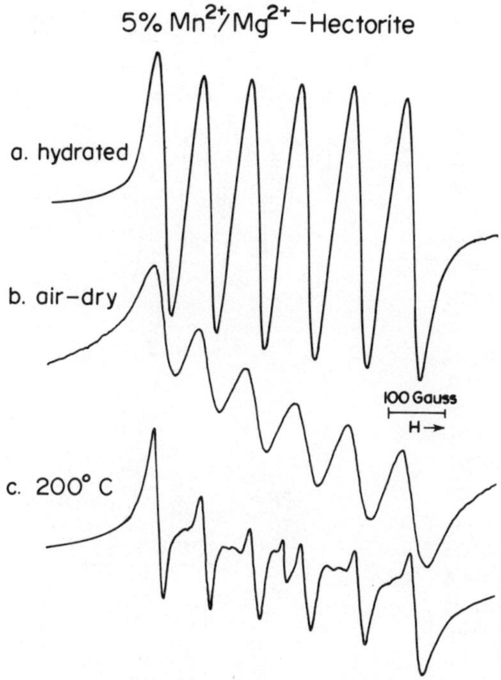

Figure 8-12. Room temperature ESR spectra of powder samples of 5% Mn^{2+}-doped Mg^{2+}-hectorite (a) fully hydrated, (b) air-dried, and (c) thermally dehydrated at 200°C (from McBride *et al.*, 1975b).

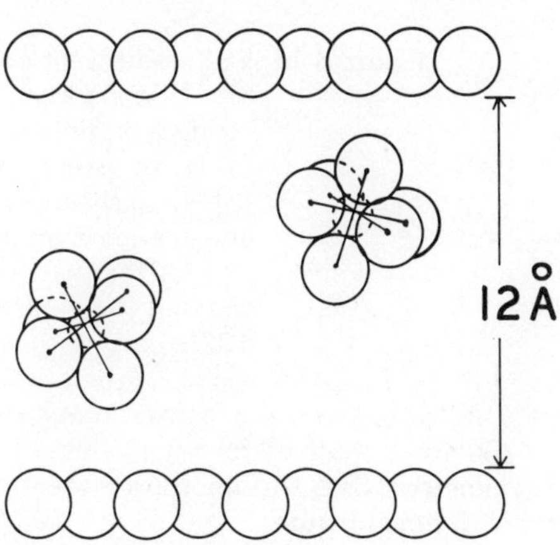

Figure 8-13. Schematic representation of the tumbling of $Mn(H_2O)_6{}^{2+}$ ions in the interlayers of fully wetted hectorite.

Fully wetted VO^{2+}-hectorite gives an isotropic ESR spectrum with A_O = 114.5 G (versus 115.1 G for VO^{2+} in aqueous solution). The spectrum remains isotropic when the ion is doped into Mg^{2+}-hectorite at the 50% level. As shown in Fig. 8-15, the spectrum of VO^{2+} in fully wetted hectorite is analogous to the solution spectrum. However, when VO^{2+} is doped into Mg^{2+}-hectorite at the 5% level, a quite different spectrum is observed, as illustrated in spectrum a, Fig. 8-16. This spectrum is anisotropic and indicative of an immobile ion ($A_{\parallel}$ = 194 G, $A_{\perp}$ = 71 G). At the 20% exchange level, a mixture of immobile and mobile VO^{2+} species is indicated (see spectrum b, Fig. 8-16). The spectrum of the immobile VO^{2+} is orientation independent in the magnetic field, suggesting that its symmetry axis is inclined near 45° to the silicate sheets or is randomly oriented. The results have been rationalized in terms of dissociation of blue $VO(H_2O)_5{}^{2+}$ to grey-brown $VO(OH)_2(H_2O)_3$ at low surface coverages where the hydrogen ion concentration is low. At higher loadings the surface acidity increases and the equilibrium is shifted toward $VO(H_2O)_6{}^{2+}$ (22).

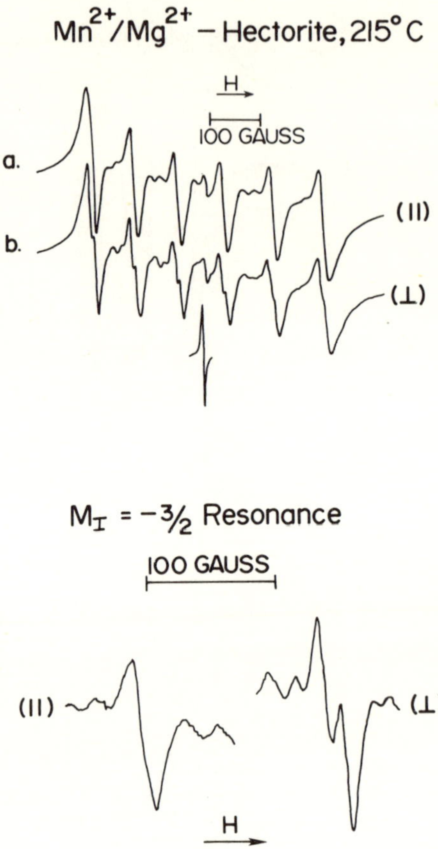

Figure 8-14. ESR spectra for oriented film of Mn^{2+}-doped Mg^{2+}-hectorite after dehydration at 215° for 6 hr. (a) $H_{\parallel}$ to sheets (b) $H_{\perp}$ to sheets (b) $H_{\perp}$ to sheets lower spectrum are an expansion of the $M_I = -3/2$ line for the two orientations of the film (from McBride *et al.*, 1975a).

Air-dried VO^{2+}-saturated hectorite gives the orientation-dependent spectra shown in Fig. 8-17. Since $A_{\parallel}$ (204 G) is observed for the ⊥ orientation and $A_{\perp}$ (81 G) for the ∥ orientation, the symmetry axis is ⊥ to the silicate sheets. Based on the value of $|A_{\parallel}\text{-}A_{\perp}|$ (12) the correlation time for tumbling (τ_R) is $> 10^{-9}$ sec.

The M_I = +7/2 linewidth of the VO^{2+} resonance has been shown to be proportional to rotational correlation time τ_R of the ion in solution (13). Since the line width of fully hydrated VO^{2+}/Mg^{2+}-hectorite free of excess iron impurities is

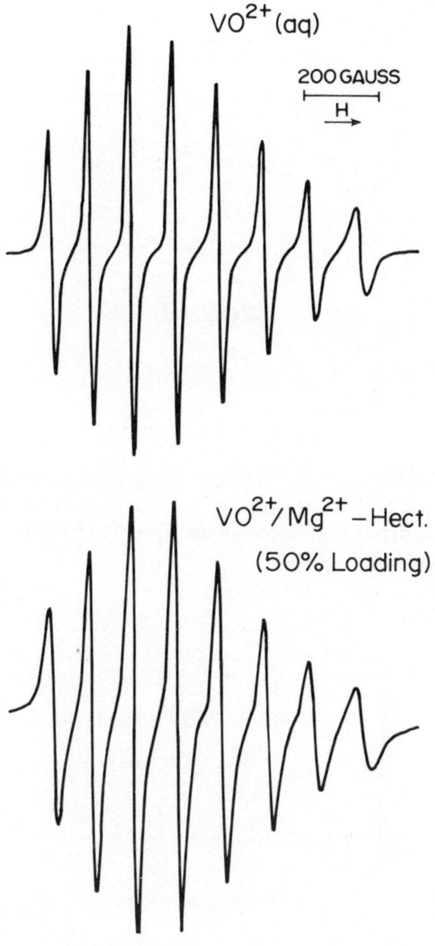

Figure 8-15. ESR spectra of VO^{2+} at ambient temperature in aqueous solution (2 x 10^{-2} M $VOSO_4$, pH = 1.5 and adsorbed on a fully wetted Mg^{2+}-hectorite film (50% mole ratio of VO^{2+}) (from McBride, 1979).

35 G, and the width for VO^{2+} in solution at room temperature is 23 G, the correlation time is 1.5 times larger in the clay than in dilute solution. Since the correlation time in solution is 5 x 10^{-11} sec, τ_R for the clay environment, where $d_{001} \simeq 20$ Å, is 7.5 x 10^{-11} sec. Thus, the VO^{2+} ion, like $Mn(H_2O)_6{}^{2+}$ in fully wetted hectorite, tumbles rapidly in a solution-like environment.

Replacing the water with methanol ($d_{001} \simeq 17$ Å) decreases the mobility of the VO^{2+} ion and leads to an intermediate rate of tumbling. As can be seen from Fig. 8-18, the motion is not sufficiently fast to completely average $A_{\parallel}$ and $A_{\perp}$. However, the cation is not totally oriented in the interlayers, as evidenced by the relative lack of orientation dependence of the spectra.

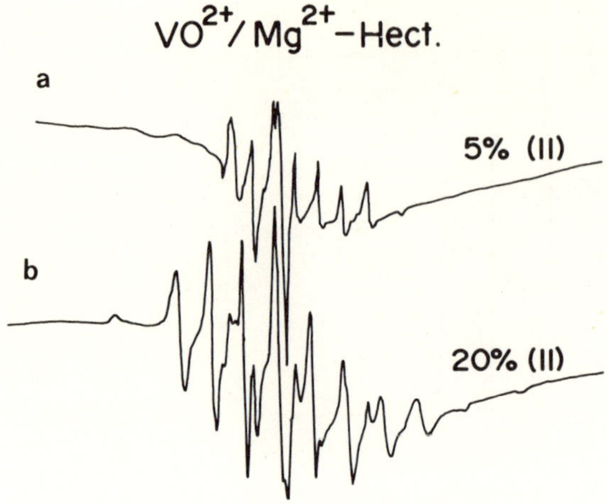

Figure 8-16. ESR spectra of VO^{2+} at ambient temperature adsorbed on fully wetted Mg^{2+}-hectorite films (‖ orientation) at the ~ 5% level of exchange and adsorbed on fully wetted Mg^{2+}-hectorite films (‖ orientation at the ~ 20% level (from McBride, 1979).

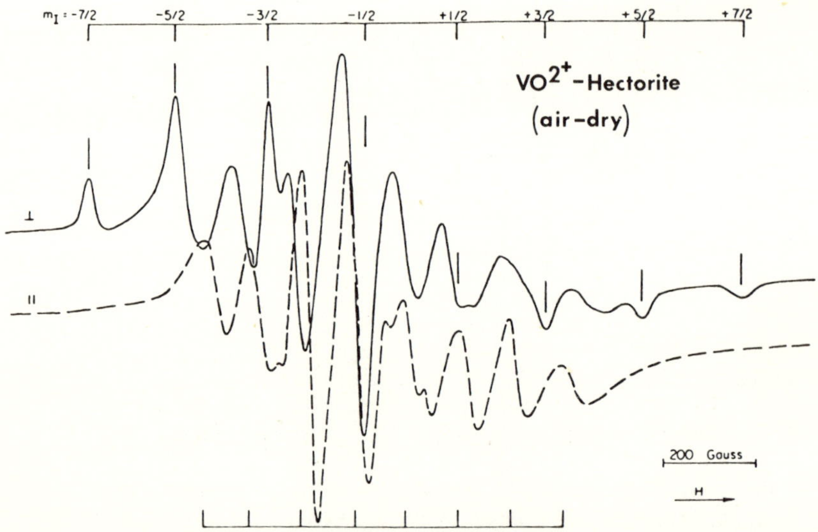

Figure 8-17. ESR spectra of an air-dry VO^{2+}-saturated hectorite film oriented perpendicular (⊥) and parallel (‖) to H. The hyperfine resonance positions for the ⊥ and ‖ orientations are indicated by the eight markers at the top and bottom of the figure respectively (from McBride, 1979).

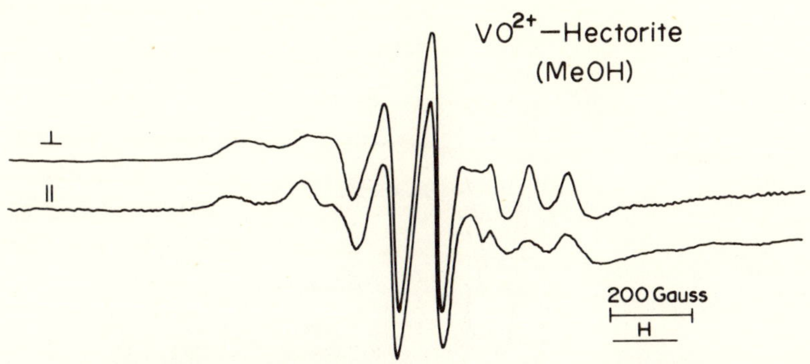

Figure 8-18. ESR spectra of a Mg^{2+}-hectorite film ≈ 50% saturated with VO^{2+} solvated in excess methanol and oriented ⊥ and ∥ to H (from McBride, 1979).

8-2.2. Metal Complexes

ESR can be a powerful means of observing the formation of metal complexes formed between exchange cations and adsorbate molecules on clay surfaces. Any complex formed on the surface should have ESR parameters ($g_{\parallel}$, $g_{\perp}$, $A_{\parallel}$, $A_{\perp}$) different from those of the simple hydrated ion, except in those cases where tumbling is sufficiently fast to average the parameters.

Excellent examples of the formation of distinct copper complexes in the interlayers of montmorillonite upon solvation with various swelling solvents have been provided by Berkheiser and Mortland (5). Figs. 8-19 and 8-20 illustrate the spectra obtained when $Cu(H_2O)_x{}^+$-montmorillonite dried at 100°C is solvated by nitrobenzene ($d_{001} \simeq 14.9$ Å), pyridine ($d_{001} \simeq 19.5$ Å), and dimethylsulfoxide (DMSO) ($d_{001} \simeq 18.6$ Å). An oriented $Cu(py)_4{}^{2+}$ complex, with $g_{\parallel} = 2.24$, $g_{\perp} = 2.06$ and $A_{\parallel} = 0.0139$, is formed with Cu^{2+}-saturated or Cu^{2+}-doped montmorillonite. With DMSO, the solvated complex is sufficiently small to tumble rapidly in the interlayers giving $\langle g \rangle = 2.15$. With nitrobenzene, the signal is too broad to observe at room temperature, suggesting that very efficient relaxation mechanisms are operating in this case.

The chemistry of surface metal complexes prepared directly by ion-exchange can also be examined by ESR (6, 37). Fig. 8-21 illustrates the dependence of the ESR spectrum of $Cu(phen)_3{}^{2+}$-hectorite (phen = 1, 10-phenanthroline) on the degree of hydration. The fully wetted $Cu(phen)_3{}^{2+}$-exchange form gives a nearly isotropic spectrum, indicating that the ion is quite mobile in the interlayer. As the degree of hydration of the clay is decreased, an anisotropic spectrum is observed for an ordered Cu/phen complex. The ESR parameters obtained after heating to 200° ($g_{\parallel} = 2.240$, $g_{\perp} = 2.058$; $A_{\parallel} = 0.0172$ cm^{-1}) are characteristic of $Cu(phen)_2{}^{2+}$. That is, the ESR data show that the ligand dissociation $Cu(phen)_3{}^{2+} \rightarrow Cu(phen)_2{}^{2+}$ + phen occurs as the interlayers collapse.

Spectrum A in Fig. 8-22 was obtained for a hectorite sample prepared by ion exchange with a solution that contained mainly the *mono*-ethylenediamine(en)

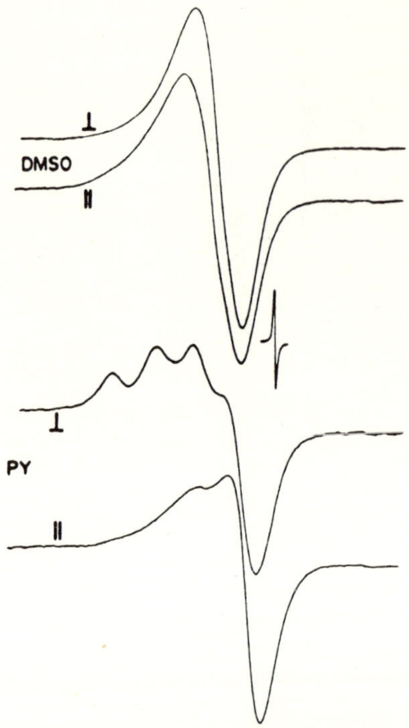

Figure 8-19. ESR spectra of exchangeable Cu(II) in oriented films of Cu(II) smectite solvated in DMSO and PY (from Berkheiser and Mortland, 1975).

complex in solution. However, it can be seen from the number of hyperfine lines for the $g_\parallel$ resonance that more than one copper complex must be present on the surface. The observed spectrum can be accounted for by the presence of both $Cu(en)_2{}^{2+}$ ($g_\parallel$ = 2.181, $A_\parallel$ = 0.0204 cm^{-1}, $g_\perp$ = 2.030, $A_\perp$ = 0.0019 cm^{-1}) and $Cu(en)^{2+}$ ($g_\parallel$ = 2.261, $A_\parallel$ = 0.0182 cm^{-1}, $g_\perp$ = 2.053 and $A_\perp$ = 0.0013). A spectrum much like that discussed above for $Cu(phen)_2{}^{2+}$-hectorite is observed for $Cu(en)_2{}^{2+}$-saturated hectorite. The adsorption of excess en vapor onto $Cu(en)_2$-hectorite gave spectrum B in Fig. 8-22. ($g_\parallel$ = 2.20, $A_\parallel$ = .0183 cm^{-1}, $g_\perp$ = 2.048, $A_\perp$ = .0007 cm^{-1}) indicative of the presence of $Cu(en)_3{}^{2+}$ (d_{001} = 14.6 Å). The spectrum of a film sample of $Cu(en)_3{}^{2+}$-hectorite is independent of orientation in the magnetic field, indicating that the symmetry axis is inclined near 45° to the silicate sheets.

The ESR parameters of $Cu(en)^{2+}$, $Cu(en)_2{}^{2+}$ and $Cu(en)_3{}^{2+}$ on hectorite surfaces are very similar to those for the ion in dilute aqueous solutions. Schoonheydt (36), using different theoretical models, has shown that the extent of out-of-plane π-bonding is slightly increased on clay surfaces relative to solution. The effect is small, however, as might be expected when the π-bonding for the free complex is already relatively weak.

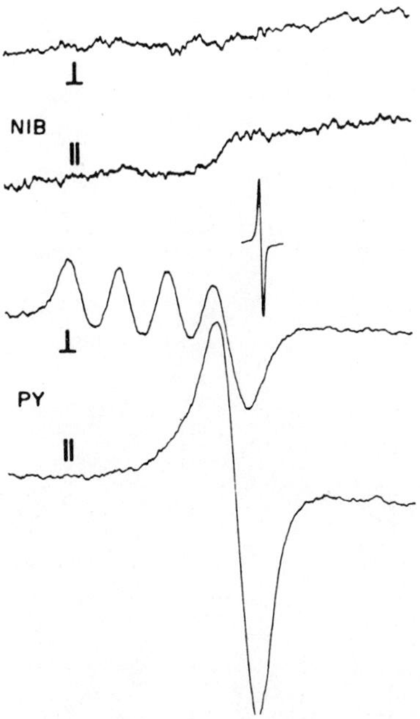

Figure 8-20. ESR spectra of exchangeable Cu(II) in oriented films of Cu(II)-doped Ca(II) smectite solvated in nitrobenzene (NIB) and pyridine (py) (from Berkheiser and Mortland, 1975).

8-3. FRAMEWORK PARAMAGNETIC CENTERS.

Natural clays may contain a variety of paramagnetic ions, some of which may be present on the exchange sites or in octahedral or tetrahedral positions of the oxygen framework. Others may be present as a separate impurity phase, such as hydrous iron oxides. The impurity phases can sometimes be removed by sedimentation or by a suitable chemical extraction, an example being the citrate/dithionite method used for removing iron from clay samples (31).

The most abundant, ESR-observable paramagnetic ion in natural clays is Fe^{3+}. Other ions such as Mn^{2+} or VO^{2+} may be observed in smaller quantities. Some paramagnetic ions (e.g., Fe^{2+}, Ti^{2+}) are ESR silent because of short relaxation times or other factors.

8-3.1. Smectites

Among all of the readily available natural clays, hectorite contains one of the lowest concentrations of framework Fe^{3+}. Consequently, dipolar interactions which broaden the signals of surface exchange ions are minimal in this clay. This is the reason it has been extensively used to investigate the surface chemistry of exchange ions. It must be pointed out, however, that some hectorite samples, depending on the exact location, can contain especially high concentrations of iron

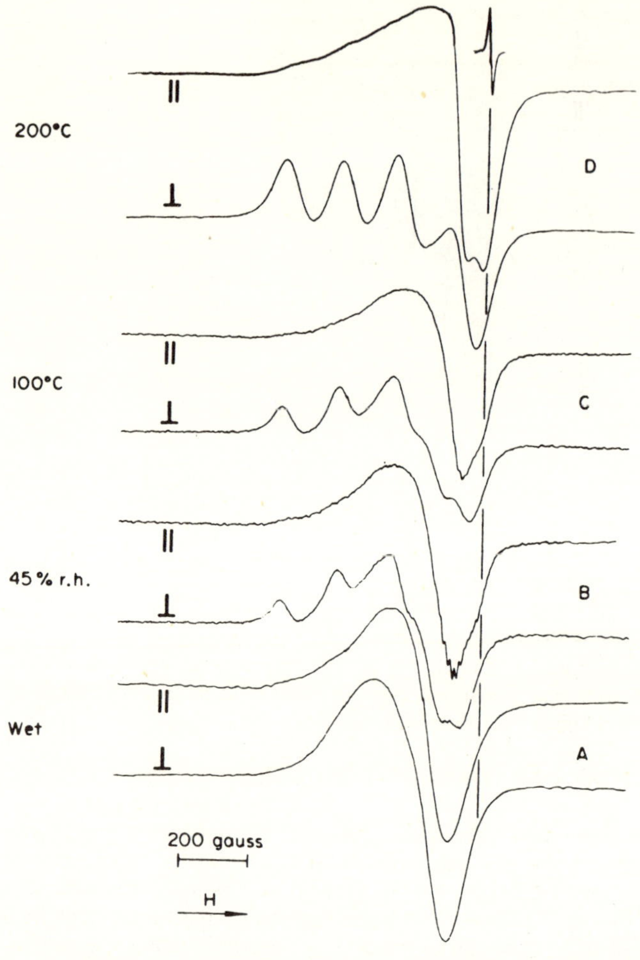

Figure 8-21. ESR spectra of oriented thin films of Cu(II) in $Cu(phen)_3{}^{2+}$hectorite at different levels of hydration. The free electron signal indicates g = 2.0028; films were oriented parallel (∥) and perpendicular (⊥) to the magnetic field H (from Berkheiser and Mortland, 1977).

oxides. Fig. 8-23 illustrates the spectra for a hectorite sample before and after citrate/dithionite treatment. The broad ($\Delta H > 1000$ G), intense line centered near g= 2.0 is an iron oxide impurity. The weaker set of lines near g = 4.2-4.3 arises from framework Fe^{3+} (see discussion below). Most clays containing framework Fe^{3+} exhibit a resonance near g = 4.2.

Fig. 8-24 shows ESR spectra for two micas and a vermiculite, kindly provided by M.B. McBride. The spectra for unweathered and weathered phlogopite indicate that weathering causes oxidation of Fe^{2+} to Fe^{3+} by O_2 which greatly increases the concentration of Fe^{3+} in the octahedral layer (7, 32). As a result,

Fe^{3+}-Fe^{3+} spin interactions become significant and domains of ferromagnetism are created (possibly by the expulsion of Fe^{3+} from the structure to form hydrous oxides) which produce the very broad resonance. A similar phenomenon may cause the broad Fe^{3+} signal in vermiculite. It is noteworthy that the Mn^{2+} in both phlogopite and muscovite can be observed as a six line spectrum near g = 2.

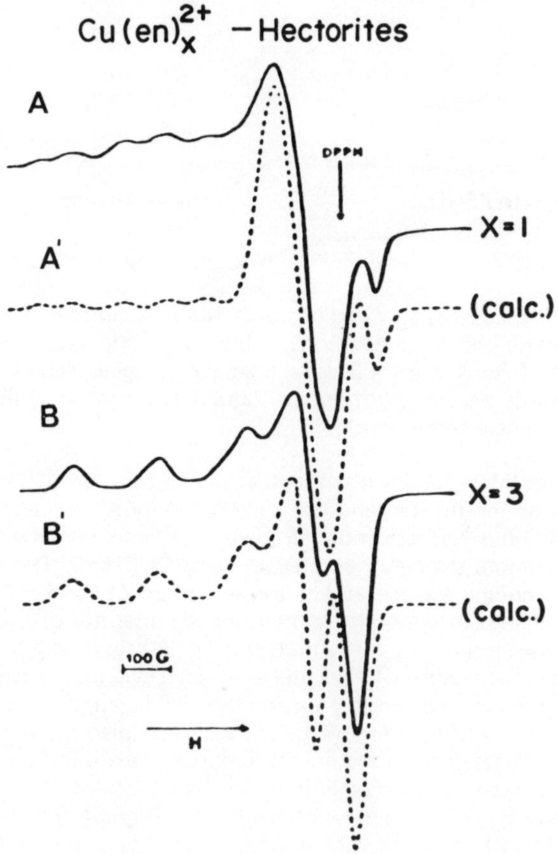

Figure 8-22. EPR spectra (77° K) of Cu^{2+} complexes in hictorite: A, after exchange with a solution containing the mono complex; A′, simulated spectrum; B, after adsorption of excess en vapor, forming the tris complex; B′, simulated spectrum of the tris complex (from Velghe *et al.*, 1977).

The framework Fe^{3+} resonance in smectites which occurs near g = 4.3 is anisotropic (5). The orientation dependence for this resonance in Ca^{2+}-montmorillonite solvated by pyridine is shown in Fig. 8-25. The anisotropic component of the Fe^{3+}, i.e., the higher field component, is sensitive to the position of the exchange cation in smectites (5, 26, 27). As illustrated in Fig. 8-26, the high field component is clearly present in hydrated K^+-, Na^+-, Li^+-, and Ca^{2+}-montmoril-

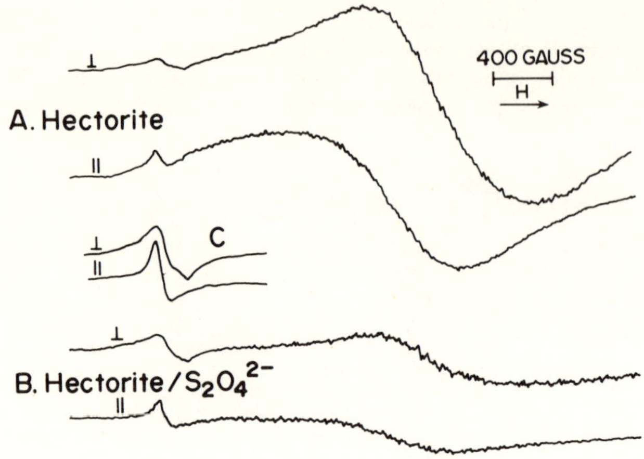

Figure 8-23. The ESR spectra of air-dried hectorite films: A, untreated; B, citrate-bicarbonate-dithionite treated. The films were oriented ⊥ and ∥ to the magnetic field. The g = 4.3 signal is shown at higher gain in spectrum C. (The free electron position, g = 2.0023, is indicated by the narrow resonance near center field.)

lonite. At 0% relative humidity its intensity decays to zero for K^+- and Na^+-exchange forms, but not for the Li^+ and Ca^{2+} exchange forms. Apparently, at 0% relative humidity, K^+ and Na^+ are more strongly associated with the framework oxygens than are the more strongly hydrated Li^+ and Ca^{2+} ions. The position of the exchange cation on the surface affects the symmetry of the Fe^{3+} ion in the framework and the change in symmetry diminishes the intensity of the high field component. It is known from x-ray diffraction studies (34) that the b-dimension of smectites depends on the nature of the interlayer exchange ion. It has been suggested (25, 26) that the Fe^{3+} responsible for the high-field component is located in octahedral positions adjacent to charge deficient sites occupied by Mg^{2+}, whereas Fe^{3+} adjacent to Al^{3+} gives rise to the orientation independent line at lower field. A simpler and more likely possibility is that the two types of Fe^{3+} are distinguished by the *cis* and *trans* orientations of the hydroxyl groups of the octahedra which they occupy (7, 32).

8-3.2. Kaolinite

The ESR spectra of kaolinites vary markedly, depending on their locality. Typical examples are shown in Fig. 8-27. However, all natural kaolinites have two principal ESR features in common. They all exhibit a group of broad lines near g = 4 which is attributed to framework Fe^{3+} substituting for Al^{3+} in octahedral position. They also exhibit a second group of lines near g = 2 which arises from lattice defects (19, 30).

Fig. 8-28 illustrates the triplet of lines near g = 4 observed at x-band frequency. The ratio of the outer lines to the inner line, as defined by the line shape parameter (LSP = AB/CD), is variable among different kaolinites, indicating that at

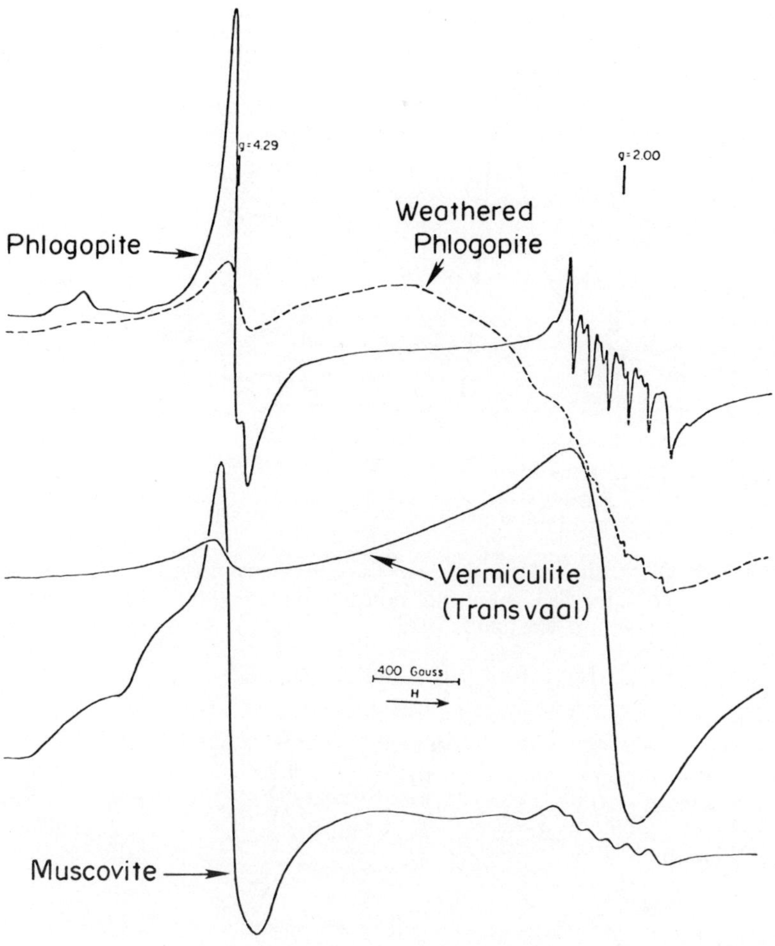

Figure 8-24. ESR spectra of micas and their weathering products: a) phlogopite (high Fe content) before (solid line) and after (broken line) several months weathering by Na^+-tetraphenylboron-NaCl solution (15-20 μ particle size). b) vermiculite (Transvaal), c) muscovite (provided by M.B. McBride).

least two types of Fe^{3+} centers (designated I and II by Jones *et al.* (19) give rise to the triplet. The magnitude of LSP increases with increasing degree of crystallinity. For a given kaolin, LSP decreases with increasing pressure (*cf*, Fig. 8-29). The value of LSP also decreases when DMSO is intercalated into the kaolin structure (*cf*, Fig. 8-30). Based on these and other observations (19, 30) the central line (Center I) is attributed to rhombic Fe^{3+} in a strong crystal field in layers with stacking disorder. The remaining lines (Center II) are attributed to rhombic Fe^{3+} in regions of high crystallinity.

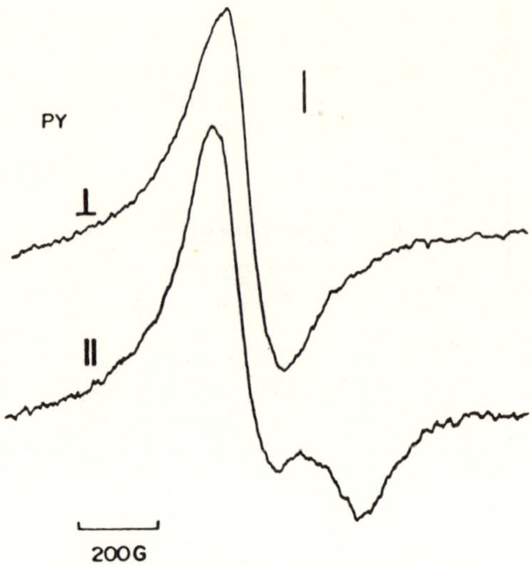

Figure 8-25. Orientation dependence of the ESR signal near g = 4.3 of framework Fe^{3+} for Ca^{2+}-montmorillonite solvated with pyridine (from Berkheiser and Mortland, 1975).

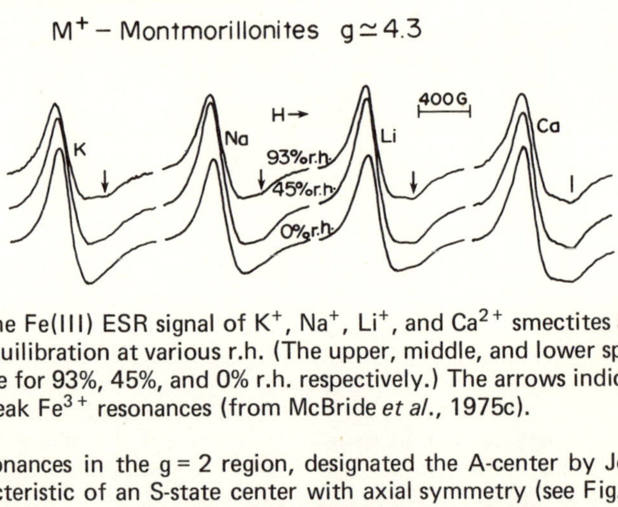

Figure 8-26. The Fe(III) ESR signal of K^+, Na^+, Li^+, and Ca^{2+} smectites after equilibration at various r.h. (The upper, middle, and lower spectra are for 93%, 45%, and 0% r.h. respectively.) The arrows indicate the weak Fe^{3+} resonances (from McBride *et al.*, 1975c).

The resonances in the g = 2 region, designated the A-center by Jones *et al.* (19) are characteristic of an S-state center with axial symmetry (see Fig. 8-31). As expected for axial symmetry, the $g_\parallel$ = 2.049 and $g_\perp$ = 2.003 components are orientation dependent. Two models (Fig. 8-32) have been proposed (19) for the defect center. One involves an O^+ center bound to Mg^{2+} substituting for Al^{3+}. The other, less likely, possibility involves incorporation of superoxide ions into the structure. Meads and Malden (30), however, have found evidence for hyperfine splitting of $g_\parallel$ and $g_\perp$ by Al^{3+}, suggesting that the defect can also be associated with Al^{3+} substituting for Si^{4+} in tetrahedral layers.

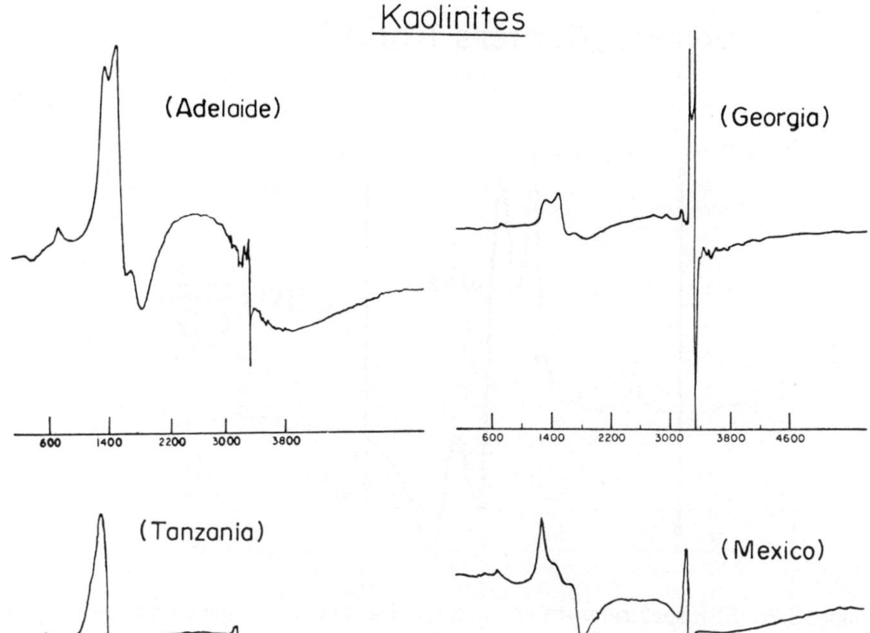

Figure 8-27. X-band ESR spectra of natural kaolins from different localities. In each case the magnetic field strength is in gauss (from Meads and Malden, 1975).

Angel *et al.* (4) have prepared a series of synthetic kaolinites which reproduce the principal ESR features of the natural kaolins (see Fig. 8-33). Clearly, the triplet of lines near g = 4 is due to incorporation of Fe^{3+} in the framework, while the g = 2 lines are induced by the presence of Mg^{2+}.

Using the orientation dependence of the anisotropic g = 2 signals, Swartz *et al.* (37) have calculated the distribution of platelets in a kaolinite pellet prepared under axial stress. The platelet distribution function is illustrated in Fig. 8-34. As expected, the platelets tend to align with the silicate sheets ⊥ to the stress direction. This application could be useful in deducing the direction of geological forces acting on a natural bed of kaolin-containing clay.

Acknowledgements. I wish to thank Professor Brian Hoffman of Northwestern University and Professor M.B. McBride of Cornell University for making available to me preprints of papers prior to publication.

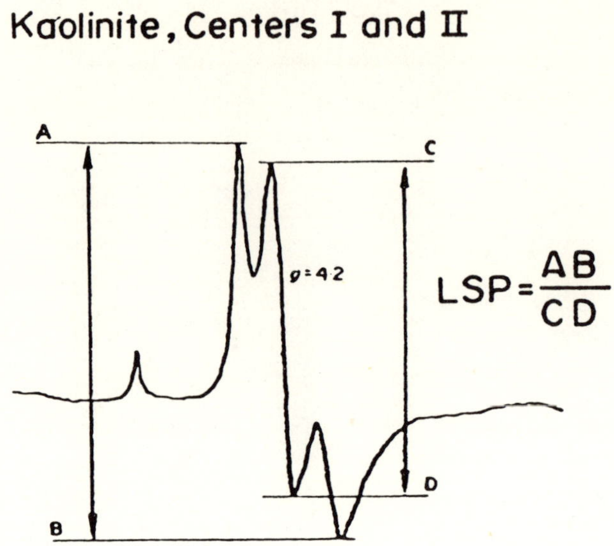

Figure 8-28. ESR spectrum of Fe^{3+} in kaolinite. The line shape parameter (LSP) = AB/CD (from Jones *et al.*, 1974).

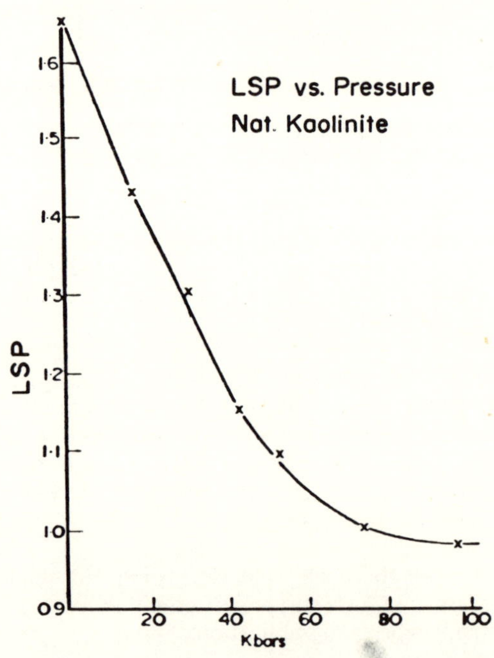

Figure 8-29. LSP vs. applied pressure for a natural kaolinite (from Jones *et al.*, 1974).

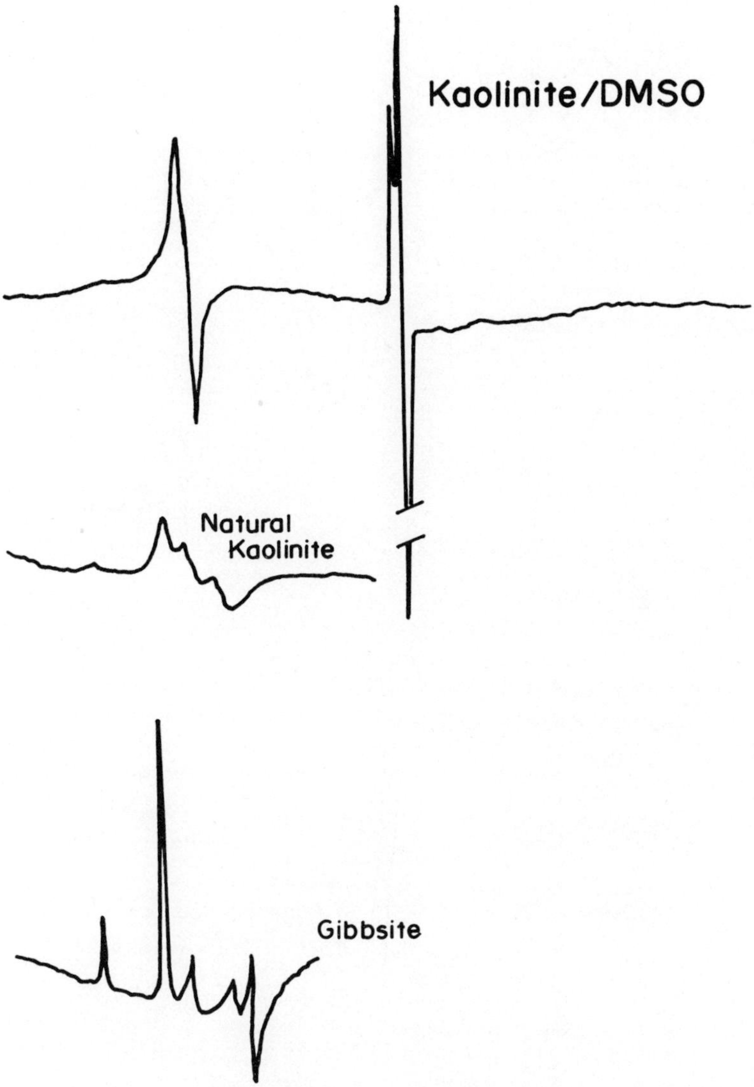

Figure 8-30. ESR spectra: (a) kaolinite intercalated with DMSO, (b) natural kaolinite, (c) gibbsite (from Jones *et al.*, 1974).

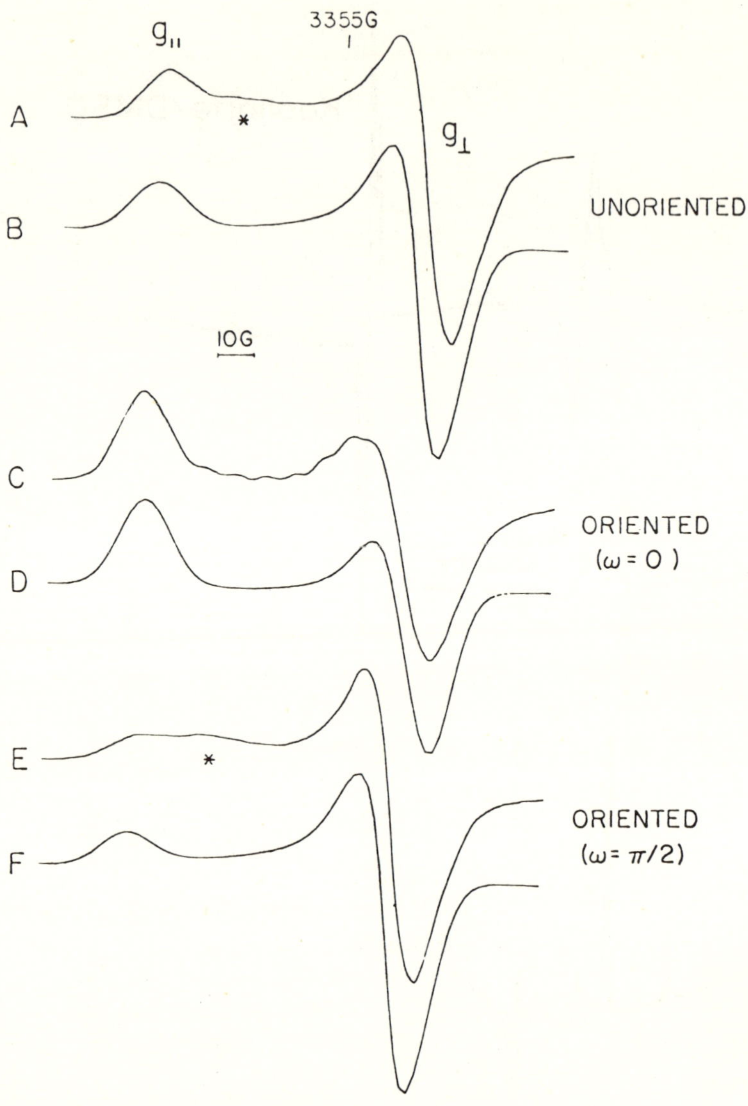

Figure 8-31. EPR of kaolinite at 300° K, υ = 9451 MHz. (A) Kaolinite powder, unoriented and (B) computer simulation; (C) consolidated kaolinite pellet with the stress direction (S) ‖ to the magnetic field (ω = 0) and (D) computer simulation; (E) consolidated kaolinite with S perpendicular to the magnetic field ($\omega = \pi/2$) and (F) computer simulation. Asterisk (*) indicates small contributions from other paramagnetic centers which have been ignored in the computer simulation (from Swartz *et al.*, 1979).

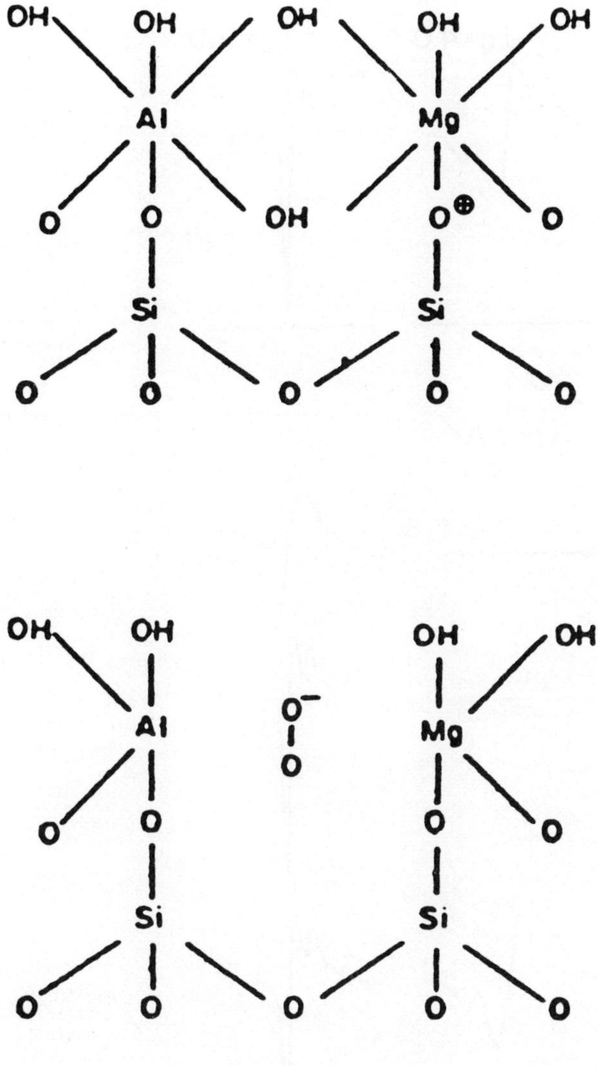

Figure 8-32. Proposed models for the A-center in kaolinite.

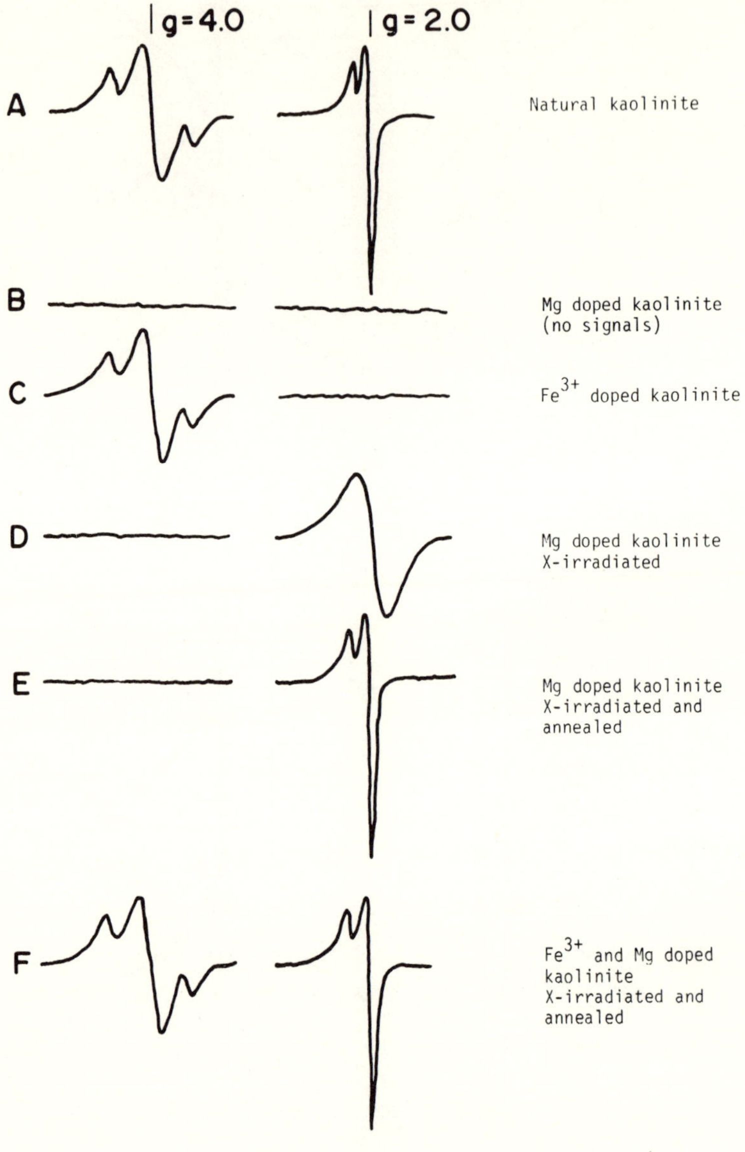

Figure 8-33. ESR spectra of synthetic kaolinites.

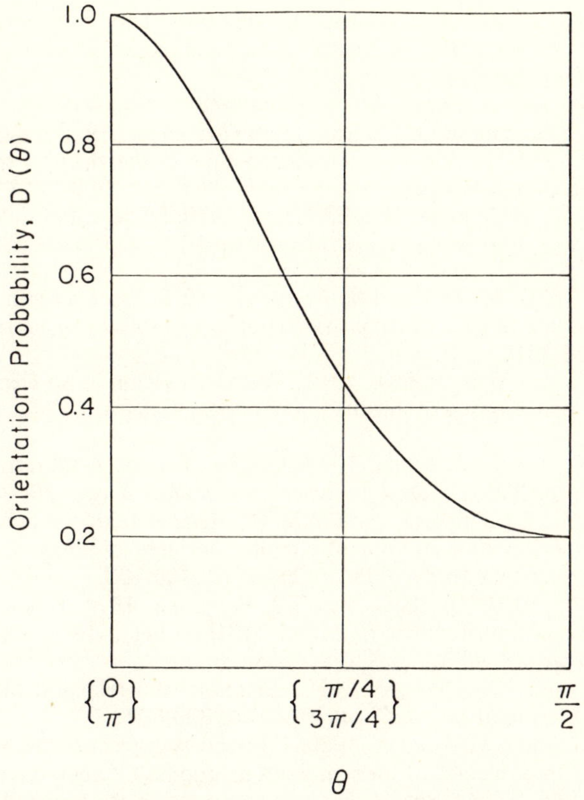

Figure 8-34. Relative probability for P, the normal to a clay particle, to lie at an angle θ with respect to $\underline{S}$, the stress direction (from Swartz *et al.,* 1979).

REFERENCES

1. Abragam, A. and B. Bleany. 1970. Electron paramagnetic resonance of transition metal ions. Oxford University Press, London. 911 pp.
2. Adrian, F.J. 1968. Guidelines for interpreting electron spin resonance spectra of paramagnetic species absorbed on surfaces. *J. Colloid Interface Sci.* **26**: 317-354.
3. Allen, B.T. and D.W. Nebert. 1964. Hyperfine structure in the EPR spectrum of the manganous ion in frozen solutions. *J. Chem. Phys.* **41**: 1983-1985.
4. Angel, B.R., K. Richards, and J.P.E. Jones. 1976. The synthesis, morphology, and general properties of kaolinites specifically doped with metallic ions, and defects generated by irradiation. *In* S.W. Bailey, ed., *Proc. Inter. Clay Conf.,* Mexico City . Applied Publishing Ltd., Wilmette, IL. pp. 297-304.
5. Berkheiser, V. and M.M. Mortland. 1975. Variability in exchange ion position in smectite: dependence on interlayer solvent. *Clays Clay Miner.* **23**: 404-410.
6. Berkheiser, V. and M.M. Mortland. 1977. Hectorite complexes with Cu(II) and Fe(II)-1,10-Phenanthroline chelates. *Clays Clay Miner.* **25**: 105-112.

7. Besson, G., H. Estrade, L. Gatineau, C. Tchoubar, and J. Mering. 1975. A kinetic survey of the cation exchange and of the oxidation of a vermiculite. *Clays Clay Miner.* **23**: 318-322.
8. Brindley, G.W. and G. Ertem. 1971. Preparation and solvation properties of some variable charge montmorillonites. *Clays Clay Miner.* **19**: 399-404.
9. Burlamacchi, L. 1971. Motional correlation time in the electron spin relaxation of ^6S spin state ions in solution. *J. Chem. Phys.* **55**: 1205-1212.
10. Burlamacchi, L., G. Martini, and E. Tiezzi. 1970. Solvent and ligand dependence of electron spin relaxation of Manganese(II) in solution. *J. Phys. Chem.* **74**: 3980-3987.
11. Burlamacchi, L., G. Martini, and M. Romanelli. 1973. Electron spin relaxation and hyperfine line shape of manganese(II) in mixed-solvent systems. *J. Chem. Phys.* **59**: 3008-3014.
12. Cambell, R.F. and M.W. Hanna. 1976. The vanadyl ion as an electron paramagnetic resonance probe of micelle-liquid crystal systems. *J. Phys. Chem.* **80**: 1892-1898.
13. Chasteen, N.D. and M.W. Hanna. 1972. Electron paramagnetic resonance line widths of vanadyl(IV) α-hydroxycarboxylates. *J. Phys. Chem.* **76**: 3951-3958.
14. Clementz, D.M., T.J. Pinnavaia, and M.M. Mortland. 1973. Stereochemistry of hydrated copper(II) ions on the intermellar surfaces of layer silicates. An electron spin resonance study. *J. Phys. Chem.* **77**: 196-200.
15. Clementz, D.M., M.M. Mortland, and T.J. Pinnavaia. 1974. Properties of reduced charge montmorillonites: hydrated Cu(II) ions as a spectroscopic probe. *Clays Clay Miner.* **22**: 49-57.
16. Garrett, B.B. and L.O. Morgan. 1966. Electron spin relaxation in solvated manganese(II) ion solutions. *J. Chem. Phys.* **44**: 890-897.
17. Hinckley, C.C. and L.O. Morgan. 1966. Electron spin resonance linewidths of manganese(II) ions in concentrated aqueous solutions. *J. Chem. Phys.* **44**: 898.
18. Hudson, A. 1966. The effects of dynamic exchange on the electron resonance line shapes of octahedral copper complexes. *Mol. Phys.* **10**: 575-581.
19. Jones, J.P.E., B.R. Angel, and P.L. Hall. 1974. Electron spin resonance studies of doped synthetic kaolinite II. *Clay Miner.* **10**: 257-270.
20. Luckhurst, G.R. and G.F. Pedulli. 1971. Research notes electron spin relaxation in solutions of manganese(II) ions. *Mol. Phys.* **22**: 931.
21. McBride, M.B. 1976. Origin and position of exchange sites in kaolinite: an ESR study. *Clays Clay Miner.* **24**: 88-92.
22. McBride, M.B. 1979. Mobility and reactions of VO^{2+} on hydrated smectite surfaces. *Clays Clay Miner.* **27**: 91-96.
23. McBride, M.B. and M.M. Mortland. 1974. Copper(II) interactions with montmorillonite: evidence from physical methods. *Soil Sci. Soc. Am. Proc.* **38**: 408-415.
24. McBride, M.B., T.J. Pinnavaia, and M.M. Mortland. 1975a. Electron spin resonance studies of cation orientation in restricted water layers on phyllosilicate (smectite) surfaces. *J. Phys. Chem.* **79**: 2430-2435.
25. McBride, M.B. .T.J. Pinnavaia, and M.M. Mortland. 1975b. Electron spin relaxation and the mobility of manganese(II) exchange ions in smectites. *Am. Mineral.* **60**: 66-72.
26. McBride, M.B., T.J. Pinnavaia, and M.M. Mortland. 1975c. Perturbation of structural Fe^{3+} in smectites by exchange ions. *Clays Clay Miner.* **23**: 103-107.
27. McBride, M.B., T.J. Pinnavaia and M.M. Mortland. 1975d. Exchange ion posi-

tions in smectite: effects on electron spin resonance of structural iron. *Clays Clay Miner.* **23**: 162-163.

28. McGarvey, B.R. 1966. Electron spin resonance of transition-metal complexes. *In* R.L. Carlin, ed. *Transition metal chemistry.* Vol. 3. Marcel Dekker, Inc., New York. pp. 89-201.
29. McGarvey, B.R. 1969. Charge transfer in the metal-ligand bond as determined by electron spin resonance. *In* T.F. Yen, ed. *Electron spin resonance of metal complexes.* Plenum Publishing Corp., New York.
30. Meads, R.E. and P.J. Malden. 1975. Electron spin resonance in natural kaolinites containing Fe^{3+} and other transition metal ions. *Clay Miner.* **10**: 313-345.
31. Mehra, O.P. and M.L. Jackson. 1960. Iron oxide removal from soils and clays by a dithionite-citrate system buffered with sodium bicarbonate. *Clays Clay Miner.* **7**: 317-327.
32. Olivier, D., J.C. Vedrine, and H. Pezerat. 1975. Resonance paramagnetique electronique du Fe^{3+} dans les argiles alteres artificiellement et dans le milieu naturel. *In* S.W. Bailey, ed., *Proc. Inter. Clay Conf.,* Mexico City. Allied Publishing Ltd., Wilmette, IL. pp. 231-238.
33. Pinnavaia, T.J., P.L. Hall, S.S. Cady, and M.M. Mortland. 1974. Aromatic radical cation formation on the intracrystal surfaces of transition metal layer lattice silicates. *J. Phys. Chem.* **78**: 994-999.
34. Ravina, I. and P.F. Low. 1977. Change of b-dimension with swelling of montmorillonite. *Clays Clay Miner.* **25**: 201-204.
35. Rubinstein, M., A. Baram, and Z. Lug. 1971. Electronic and nuclear relaxation in solutions of transition metal ions with spin = 3/2 and 5/2. *Mol. Phys.* **20**: 67.
36. Schoonheydt, R.A. 1978. Analysis of the electron paramagnetic resonance spectra of Bis (ethylenediamine) copper(II) on the surfaces of zeolites X and Y and of a Camp Berteau montmorillonite. *J. Phys. Chem.* **82**: 497-498.
37. Swartz, J.C., B.M. Hoffman, R.J. Krizek, and D.K. Atmatzidis. 1979. A general procedure for simulating EPR spectra of partially oriented paramagnetic centers. *J. Mag. Res.* **36**: 259-268.
38. Velghe, F., R.A. Schoonheydt, J.B. Uytterhoeven, P. Peigneus, and J.H. Lunsford. 1977. Spectroscopic characterization and thermal stability of copper(II) ethylenediamine complexes on solid surfaces. 2. Montmorillonite. *J. Phys. Chem.* **81**: 1187-1194.
39. Wertz, J.E. and J.R. Bolton. 1972. Electron spin resonance: elementary theory and practical applications. McGraw-Hill, New York. Chaps. 11, 12.

Chapter 9

APPLICATION OF SPIN PROBES TO ESR STUDIES OF ORGANIC-CLAY SYSTEMS

Murray B. McBride
Department of Agronomy
Cornell University

9-1. NITROXIDE SPIN PROBES – ORIGIN OF THE ESR SPECTRUM.

The nitroxide free radicals, used in ESR spectroscopy as spin "probes" or "labels," all contain the paramagnetic group

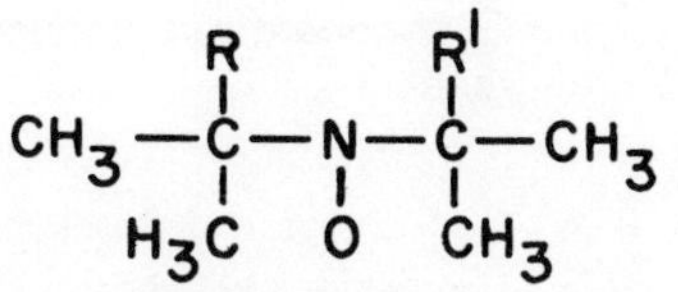

which is unusually stable and inert because of the protective effect provided by the four methyl groups. The usual molecular axis system chosen for nitroxides has the z-axis along the nitrogen $2p\pi$-orbital and the x-axis along the N-O bond, as depicted below

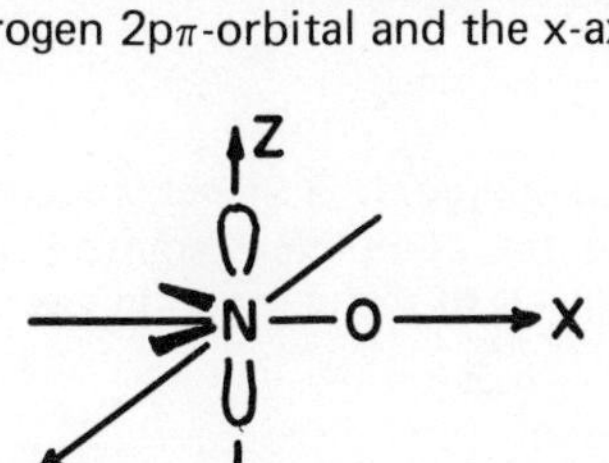

The unpaired electron is largely localized on the $2p\pi$ orbital, so that the magnetic interaction between the electron and nuclear spin of ^{14}N (I=1) produces the hyperfine Hamiltonian given by

$$\mathcal{H}_{hf} = \vec{I} \cdot A \cdot \vec{S} \qquad [9\text{-}1]$$

In the principal axis system of the hyperfine tensor, this can be rewritten:

J. W. Stucki and W. L. Banwart (eds.), Advanced Chemical Methods for Soil and Clay Minerals Research, 423–450.

$$\mathcal{H}_{hf} = A_{xx}\hat{I}_x\hat{S}_x + A_{yy}\hat{I}_y\hat{S}_y + A_{zz}\hat{I}_z\hat{S}_z \quad [9\text{-}2]$$

where A_{xx}, A_{yy}, A_{zz} are the principal values of the hyperfine tensor A; $\hat{I}_x$, $\hat{I}_y$, $\hat{I}_z$ and $\hat{S}_x$, $\hat{S}_y$, $\hat{S}_z$ are the principal axis operators of the nuclear spin vector $\vec{I}$ and the electron spin vector $\vec{S}$. The principal values of the hyperfine tensor, A, include a dipolar and a contact term:

$$A_{jj} = A'_{jj} + \delta_{jj}a. \quad [9\text{-}3]$$

The A'_{jj} or dipolar interaction term results from the classical magnetic interaction between electron and nuclear spins, and is symmetrical. The magnitude of the dipolar interaction depends on the orientation of the molecule relative to the applied magnetic field of the spectrometer. If the unpaired electron were in a spherically symmetrical (i.e., s-type) orbital, the dipolar term would vanish as the electron averaged the local magnetic field of the nucleus to zero. However, in p, d, and f-orbitals, the dipolar interaction is observable. The second term in equation [9-3] is the contact interaction, a quantum mechanical phenomenon resulting from a finite electron density at the nucleus. The isotropic coupling constant, a, should have non-zero values only for s-type atomic orbitals, since all other orbitals have zero electron density at the nucleus. Most organic free radicals in solution exhibit hyperfine structure due to interactions with nuclei. Since rapid rotational motion in solution necessarily averages the dipolar term of equation [9-3] to zero, the contact interaction term must be non-zero. It can be shown that the hyperfine structure results from a slight induced unpaired electron density in an s-orbital with spin opposite to that of the unpaired electron in the p-orbital.

In a strong magnetic field, $\vec{I}$ and $\vec{S}$ are quantized along the magnetic field direction, z, so that equation [9-2] reduces to

$$\mathcal{H}_{hf} = A_{zz}\hat{I}_z\hat{S}_z. \quad [9\text{-}4]$$

For the nitroxide molecules, the component of the ^{14}N nuclear spin is allowed the values M_I=1,0,−1 along the magnetic field, H, while the electron spin is allowed values of M_s=+1/2, −1/2. Equation [9-4] can then be re-expressed as

$$E_{hf} = hA_oM_sM_I \quad [9\text{-}5]$$

where E_{hf} is the hyperfine splitting energy, h is Planck's constant, and A_o is the hyperfine coupling constant. Since the complete Hamiltonian must include the electron Zeeman term, the energy levels of the electron in the nitroxide radical are described by:

$$\begin{aligned} E &= E_{Zeeman} + E_{hf} \\ E &= gBHM_s + hA_oM_sM_I \end{aligned} \quad [9\text{-}6]$$

where g, B and H are the g-factor for the electron, the Bohr magneton, and the applied magnetic field, respectively. Substituting all of the possible combinations of values of M_s (+1/2, −1/2) and M_I (1, 0, −1) into equation [9-6] produces six energy levels, *viz.*

$$
\begin{aligned}
E_1 &= 1/2\ gBH + 1/2\ hA_o \\
E_2 &= 1/2\ gBH \\
E_3 &= 1/2\ gBH - 1/2\ hA_o \\
E_4 &= -1/2\ gBH - 1/2\ hA_o \\
E_5 &= -1/2\ gBH \\
E_6 &= -1/2\ gBH + 1/2\ hA_o .
\end{aligned} \qquad [9\text{-}7]
$$

Thus, nitroxide radicals are capable of possessing six different alignments of electron and nuclear spins in a strong magnetic field. The energy levels produced from these alignments are shown in Fig. (9-1), demonstrating the three allowed electron transitions ($\Delta M_s = \pm 1$, $\Delta M_I = 0$) which produce the observed, three-line spectrum when the magnetic field is scanned. The allowed transition energies are

$$
\begin{aligned}
\Delta E_1 &= E_1 - E_4 = gBH + hA_o \\
\Delta E_2 &= E_2 - E_5 = gBH \\
\Delta E_3 &= E_3 - E_6 = gBH - hA_o .
\end{aligned} \qquad [9\text{-}8]
$$

Since ΔE must equal $h\upsilon$ for resonance (absorption of microwave radiation) to occur in the ESR experiment, and the frequency (υ) is generally held constant, the resonance condition is met by varying the magnetic field, H, until absorption of energy occurs (i.e., $\Delta E = h\upsilon_o$). It is, therefore, more meaningful to convert the above energy levels into magnetic field positions of the resonances:

$$
\begin{aligned}
H_1 &= H' + \frac{hA_o}{gB} = H' + a \\
H_2 &= H' \\
H_3 &= H' - \frac{hA_o}{gB} = H' - a
\end{aligned} \qquad [9\text{-}9]
$$

The ESR spectrum consists of three absorption peaks separated by the hyperfine splitting constant, a, measured in gauss (Fig. 9-1). The resonances are of equal intensity because the three M_I values of +1, 0, and −1 are equally probable, and therefore the three allowed transitions are equally probable. In practice, the nuclear spins of structural hydrogen ($I=1/2$) in the nitroxide molecule also contribute magnetic interactions, which have the effect of broadening the resonance lines (3). For this reason, nitroxides are often synthesized in the deuterated form, since the nuclear spin of ^2H ($I=1$) has a much weaker magnetic interaction with the unpaired electron. Weak satellite lines due to the natural occurrence of ^{13}C ($I = 1/2$) and ^{15}N ($I = 1/2$) isotopes may also be observed in ESR spectra of nitroxides (9).

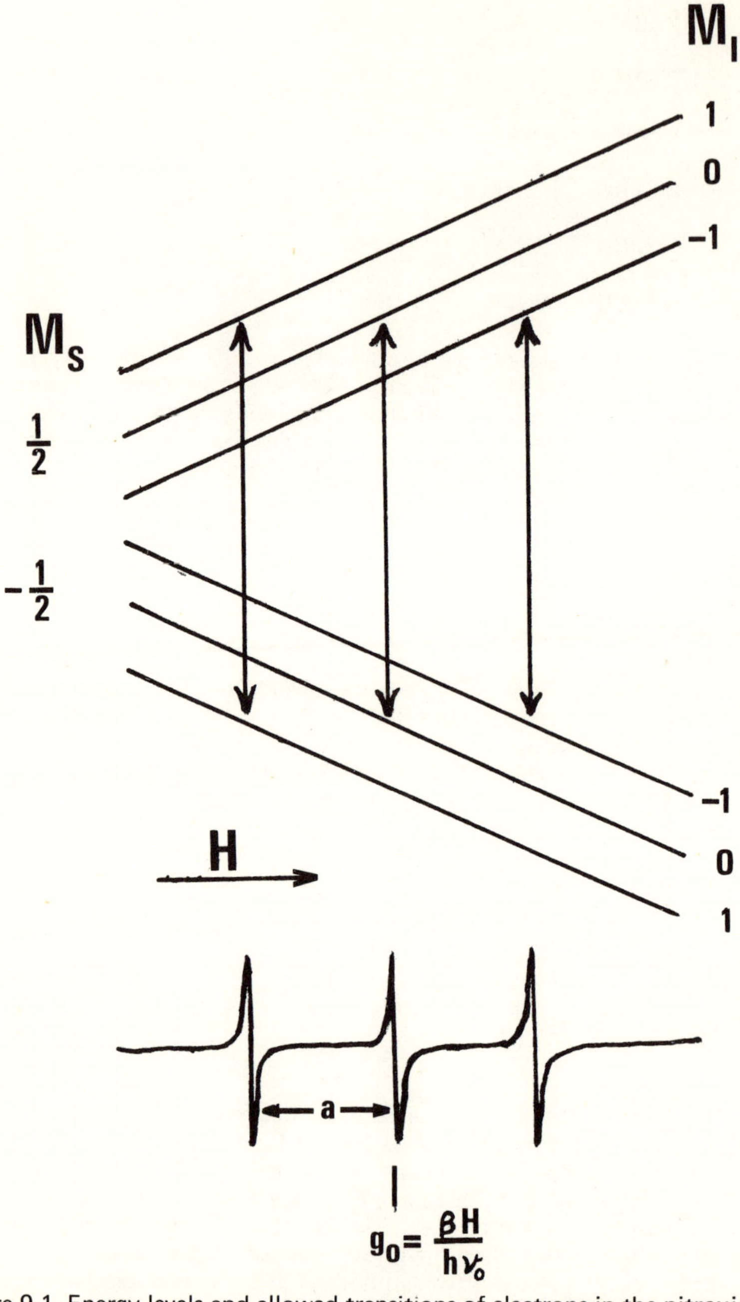

Figure 9-1. Energy levels and allowed transitions of electrons in the nitroxide radical as a function of applied magnetic field, H. The observed first-derivative ESR spectrum is shown at the bottom of the figure.

9-2. NITROXIDES IN LOW-VISCOSITY MEDIA – RAPID ISOTROPIC MOTION

In solvents of low viscosity, the molecular rotation of the nitroxide molecule is rapid enough to average the dipolar interaction term in equation [9-3] to zero, leaving only the contact interaction to produce hyperfine splitting. With an X-band spectrometer ($H \approx 3.3 \times 10^3$ gauss), the dipolar interaction will only be averaged if the rotational frequency is much greater than the frequency corresponding to the largest differences between the principal components of the hyperfine coupling tensor, A ($>> |A_{zz} - A_{xx}| \approx 73$ MHz). Similarly, differences in the principal components of the g-tensor will only be averaged if the rotational frequency is much greater than $|g_{xx} - g_{zz}| \frac{BH}{h} \approx 29$ MHz (13). Thus, with rotation rates much greater than about 10^8 sec^{-1}, anisotropies in the A and g tensors are averaged out, and isotropic hyperfine splitting (A_o) and g-value (g_o) are observed. The three resonance lines are narrow and of similar width (Fig. 9-1), with isotropic parameters given by $A_o = \frac{1}{3}(A_{xx}+A_{yy}+A_{zz})$ and $g_o = \frac{1}{3}(g_{xx}+g_{yy}+g_{zz})$. However, in moderately viscous solvents at room temperature, the three lines are not of the same height, which is not a result of a change in peak intensities, but rather a change in linewidth. Thus, the spectra of nitroxide radicals in water have slightly different peak heights at room temperature (Fig. 9-2). The linewidth variations arise from the incompletely averaged anisotropic terms of the magnetic Hamiltonian, which produce measurable line broadening even when the measured magnetic parameters of the spectrum (A_o, g_o) are essentially unaffected (10). If a correlation time, τ_c, is defined as the time required for the molecule to reorient in solution by random isotropic tumbling, then for relatively rapid rotation ($\tau_c < 5 \times 10^{-9}$ sec.), an approximate theoretical solution for the relative widths of the three nitroxide ESR lines is given by (13):

$$\frac{\Delta H(M_\ell)}{\Delta H(0)} = 1 - \frac{\tau_c}{\sqrt{3}\,\pi\,\Delta\upsilon(0)}\left[c_1 M_I + c_2 M_I^2\right] \qquad [9\text{-}10]$$

In this equation, $\Delta H(M_I)$ is the peak-to-peak linewidth (in gauss units) of the low (M_I=+1), center (M_I=zero) and high (M_I=–1) field resonance lines (see Fig. 9-2), and $\Delta\upsilon(0)$ is the peak-to-peak linewidth of the center resonance expressed in Hz units $[\Delta\upsilon(0) = (gB/h)\Delta H(0)]$. The constants c_1 and c_2 are determined by the principal components of the **A** and g tensors by the following relations:

$$c_1 = \frac{-16\pi|B|H}{45h}\left[A_{zz}-A_{xx}\right]\left[g_{zz}-1/2\,(g_{xx}+g_{yy})\right]$$
$$c_2 = \frac{2\pi^2}{9}\left[A_{zz}-A_{xx}\right]^2 \qquad [9\text{-}11]$$

The A_{jj} components are expressed in MHz (megahertz). Since the intensities of ESR lines are generally proportional to the product of their peak-to-peak heights and their widths squared, equation [9-10] can be expressed in terms of heights which are more easily measured than widths, hence

$$\sqrt{\frac{h(0)}{h(M_I)}} = 1 - \frac{\tau_c}{\sqrt{3}\,\pi\Delta\upsilon(0)}\left[c_1 M_I + c_2 M_I^2\right]. \qquad [9\text{-}12]$$

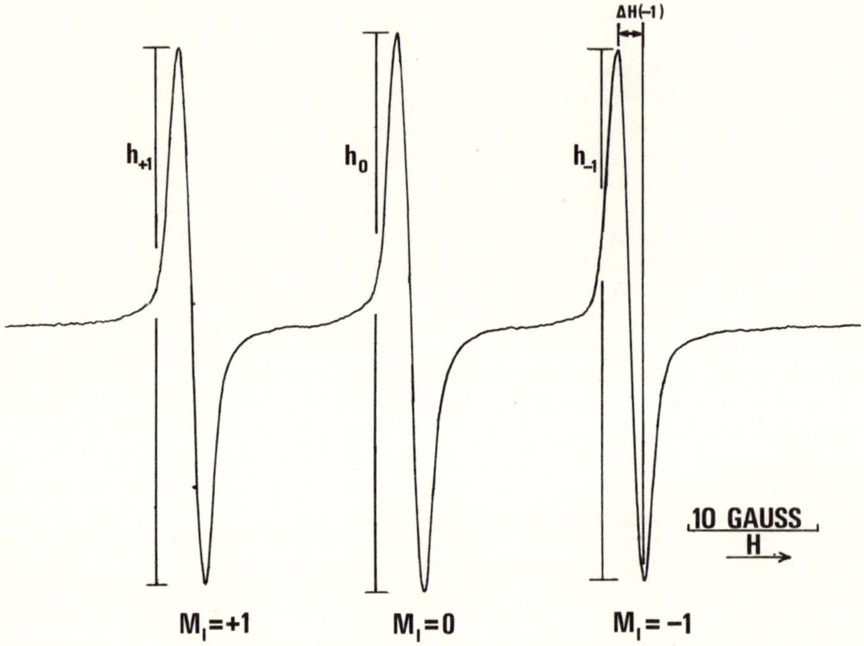

Figure 9-2. ESR spectrum of 10^{-4} M TEMPO-phosphate in aqueous solution, indicating measured peak heights (h) and linewidths (ΔH).

If the difference and sum of $\sqrt{\frac{h(0)}{h(-1)}}$ and $\sqrt{\frac{h(0)}{h(+1)}}$ are taken, two independent equations for the determination of τ_c can be obtained – one containing the c_1 term and the other containing the c_2 term:

$$\tau_c = \frac{\left(\sqrt{\frac{h(0)}{h(-1)}} - \sqrt{\frac{h(0)}{h(+1)}}\right)\sqrt{3}\,\pi\,\Delta\upsilon(0)}{2\,c_1} \qquad [9\text{-}13]$$

$$\tau_c = \frac{\left(\sqrt{\frac{h(0)}{h(-1)}} + \sqrt{\frac{h(0)}{h(+1)}} - 2\right)\sqrt{3}\,\pi\,\Delta\upsilon(0)}{2\,c_2} \qquad [9\text{-}14]$$

These two estimates of τ_c are not generally identical, τ_c in equation [9-13] being sensitive to the applied microwave power (13). Thus, it is suggested that τ_c be estimated from equation [9-14]. From equations [9-13] and [9-14], simplified equations have been developed for rapid estimation of correlation times from nitroxide probe spectra (10):

$$\tau_c = \frac{-2211 \times \Delta H(0) \times R_-}{H_o} \qquad [9\text{-}15]$$

$$\tau_c = .65 \times \Delta H(0) \times (R_+ - 2) \qquad [9\text{-}16]$$

where $$R_\pm = \sqrt{\frac{h(0)}{h(+1)}} \pm \sqrt{\frac{h(0)}{h(-1)}} \qquad [9\text{-}17]$$

and H_o is the magnetic field corresponding to the central resonance line. Equations [9-15] and [9-16] are derived from [9-13] and [9-14], respectively, with the values of c_1 and c_2 determined from the known principal components (A_{jj}, g_{jj}) of the probe in use. Although different nitroxide probes may have somewhat different A and g values, these do not vary enough to markedly change equations [9-15] and [9-16] (see Table 9.1). However, when dissolved in solvents of greatly different polarity, the A and g values of a given probe are generally shifted slightly.

The changing ESR lineshapes as a function of solution viscosity are shown in Fig. 9-3, where the rotational correlation time, as calculated from the Stokes law is

$$\tau_c = \frac{4\pi\eta r^3}{3kT}. \qquad [9\text{-}18]$$

The correlation time is varied by changing temperature or viscosity (η) of the solvent, and r is the effective, spherical molecular radius. Equation [9-18] can be expected to apply to molecules having spherical symmetry. Typical values of τ_c for small molecules in low-viscosity solvents are in the range of 10^{-10} to 10^{-11} sec. For aqueous solutions at 20° C, $\eta = 1.005 \times 10^{-2}$ g cm^{-1} sec^{-1}, and for a small nitroxide probe such as protonated 4-amino-2,2,6,6-tetramethylpiperidine N-oxide ($TEMPAMINE^+$), r is about 0.3 nm. From equation [9-18], τ_c should then be about 3×10^{-11} sec. This agrees well with the value of 5×10^{-11} sec. obtained from the ESR spectrum of $TEMPAMINE^+$ in water using equations [9-15] and [9-16] (6, 7). Clearly, in non-viscous liquids, the nitroxides can be very useful probes of microscopic viscosity, allowing the calculation of viscosity of very small liquid volumes. For this reason, the method has been utilized extensively to study biological systems at the cellular level.

9-3. NITROXIDES IN HIGH-VISCOSITY MEDIA

When the mobility of nitroxides is restricted by viscous fluids or attachment to large macromolecules, a spectrum is obtained which is no longer a simple superposition of Lorentzian lines. The range of correlation times, 10^{-9} sec $\leqslant \tau \leqslant 10^{-6}$ sec, represents the slow motional region of Fig. 9-3 where motion is too slow to permit the use of equations [9-15] and [9-16] in estimating τ_c, but too rapid to yield a proper rigid-limit spectrum (3). The range of spectral shapes (Fig. 9-3) demonstrates that the general shape of the spectrum tends toward two well-separated, outer hyperfine peaks and an overlapped central region as τ_c becomes longer.

Table 9-1. Principal Values of the g and hyperfine (**A**) tensors for commonly used Nitroxides*

STRUCTURE	g_{xx}	g_{yy}	g_{zz}	A_{xx}(G)	A_{yy}(G)	A_{zz}(G)
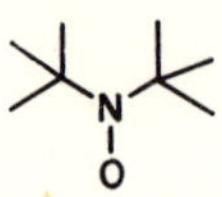	2.0088	2.0062	2.0027	7.6	6.0	31.8
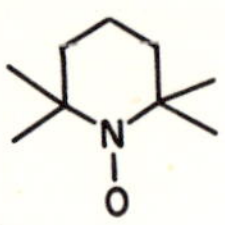	2.0103	2.0069	2.0030	––	––	––
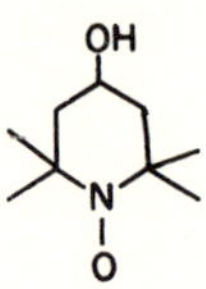	2.0095	2.0064	2.0027	––	––	––
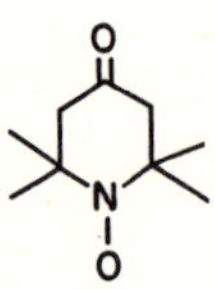	2.0104	2.0074	2.0026	5.2	5.2	31
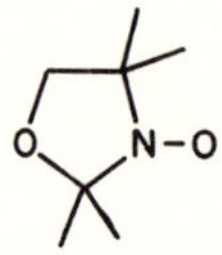	2.0088	2.0058	2.0022	5.9	5.4	32.9
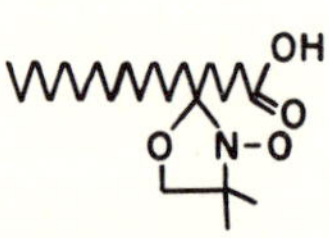	2.0088	2.0061	2.0027	6.3	5.8	33.6

*Parameters were determined for probes doped into single crystal hosts. Taken from Berliner, 1976, Appendix II.

An analysis of the spectral lineshapes in the slow motional region is rather difficult. However, from the Stokes-Einstein relationship, the rotational diffusion coefficient, R, of the nitroxide is defined as

$$R = \frac{kT}{8\pi\eta r^3}. \qquad [9\text{-}19]$$

Not all probe molecules show isotropic rotational diffusion in isotropic liquids, thereby requiring the definition of $R_{x'x'}$, $R_{y'y'}$ and $R_{z'z'}$, the rotational diffusion coefficients about the axes x′, y′ and z′, respectively, which are fixed with respect to the molecule. For completely asymmetric Brownian rotation, $R_{x'x'} \neq R_{y'y'} \neq$

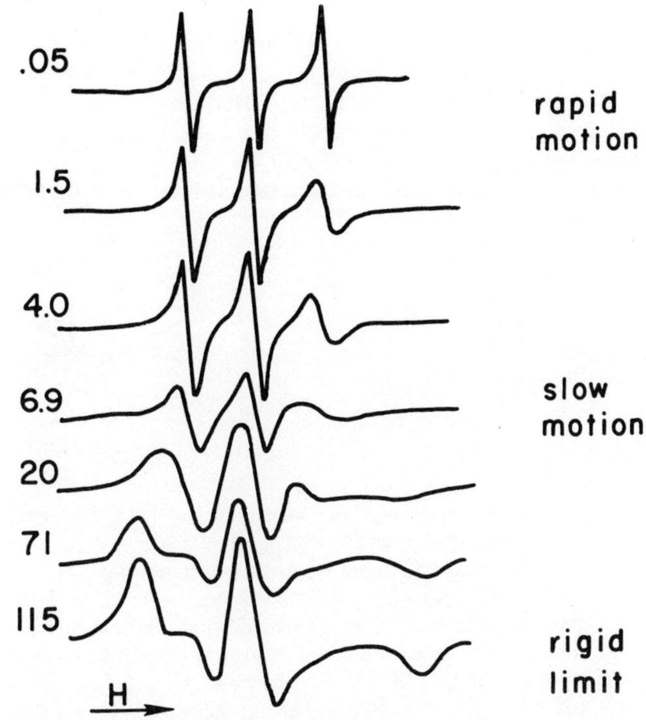

Figure 9-3. Spectra of a small molecular weight nitroxide at different τ_c values (calculated from $\tau_c = 4\pi\eta r^3/3kT$) in solutions of varying viscosity η (Adapted from Smith, 1972).

$R_{z'z'}$. For axially symmetric rotation about the z'-axis, $R_{x'x'} = R_{y'y'} = R_\perp$ and $R_{z'z'} = R_\parallel$. The general method for calculating τ_c values in the slow motion region involves the calculation of the experimental values of B and C (3) in the equation

$$\Delta H(M_I) = A + BM_I + CM_I^2, \qquad [9\text{-}20]$$

which is derived from equation [9-10] by the introduction of parameters A, B, and C. B and C must then be determined from motionally narrowed spectra where equations [9-10] and [9-20] apply. $R_\parallel$ and $R_\perp$ are calculated over a range of solution temperatures where motional narrowing theory applies, and the following parameters of the system are obtained at each temperature:

$$\tau_{\overline{R}} = (6\,\overline{R})^{-1} \qquad [9\text{-}21a]$$

$$\overline{R} = (R_\parallel R_\perp)^{1/2} \qquad [9\text{-}21b]$$

$$N = {}^{R_\parallel}/_{R_\perp}. \qquad [9\text{-}21c]$$

Since log ($\tau_{\bar{R}}$) has an almost linear relationship with $1/T$, as expected for a rotation with an activation energy, and N is found to be independent of temperature, the information obtained on $\tau_{\bar{R}}$ and N can be extrapolated into the slow motional region at lower temperatures (3). For example, it has been found that peroxylamine disulfonate has anisotropic rotational diffusion in aqueous glycerol solvents with N=4.7; that is, rotational diffusion about the z′-axis is faster than that about the x′- and y′-axes. However,

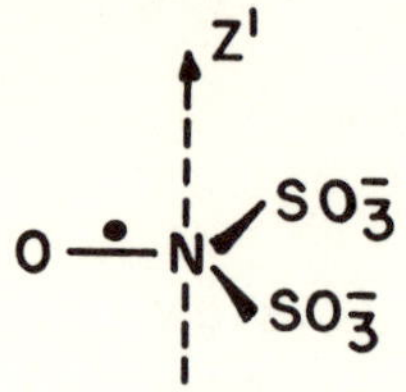

peroxylamine disulfonate

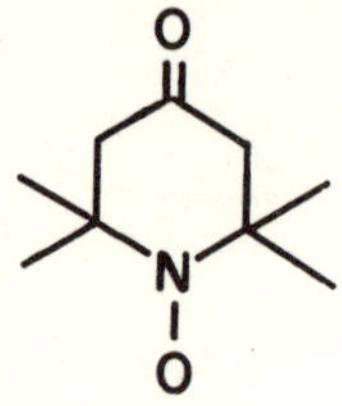

TEMPONE

2,2,6,6-tetramethyl-4-piperidone-1-oxyl (TEMPONE) has isotropic rotation (N=1) in glycerol solvents (3). It is likely, then, that approximate isotropic rotation can be assumed for the small nitroxide molecules in isotropic media.

Because of the complexity of the above method for measurement of rotational correlation times in the slow motional region, a simplified method has been suggested as a desirable alternative (3). The parameter, A'_{zz}, defined as one-half the separation of the outer hyperfine extrema, can be measured for the slow-tumbling spectrum and for the rigid limit of the same nitroxide molecule in the same solvent (Fig. 9-4). The latter value is usually obtained by lowering the temperature of the system until there is no motional averaging of the A and g tensors ($\tau_c > 10^{-6}$ sec).

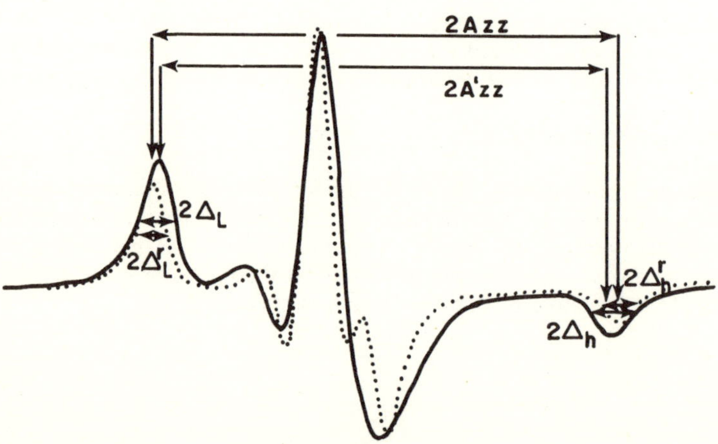

Figure 9-4. Rigid-limit nitroxide spectrum (broken line) and slow-tumbling spectrum with $\tau_c = 5 \times 10^{-8}$ sec (solid line), showing half-height linewidths (2Δ) (Adapted from Freed, 1976).

If the nitroxide is oriented in a crystal, the principal **A** splitting values (A_{xx}, A_{yy}, A_{zz}) and g-values (g_{xx}, g_{yy}, g_{zz}) are obtained by orienting the external magnetic field along the x, y and z axes of the nitroxide group, respectively (Figs. 9-5a, 9-5b, 9-5c). However, in rigid glasses (frozen solvents) or powders, the x, y and z spectral components are summed to produce the rigid-limit spectrum because of the random orientation of molecular axes relative to the magnetic field direction (Fig. 9-5d). The value of A_{zz} from this spectrum is the true hyperfine splitting for the magnetic field aligned parallel to the nitrogen π-orbital (note that for nitroxides, $A_{zz}>A_{xx}\approx A_{yy}$ and $g_{xx}\approx g_{yy}>g_{zz}$). Under ideal conditions with minimal line-broadening, A_{xx} and A_{yy} can also be measured on rigid-limit spectra (Fig. 9-6a). Although this degree of resolution is generally obtained only with deuterated probe molecules (4), certain types of probe motion allow the estimation of parameters other than A_{zz} from the near-rigid-limit spectra. For example, a long-chain fatty acid labeled with the nitroxide moeity (shown below),

$$CH_3\,(CH_2)_{17}-\overset{\displaystyle O\quad N\!-\!O}{C}-(CH_2)_3COOH$$

when doped into membranes, gives a sharper spectrum than expected for true rigid-limit conditions (Fig. 9-6b) as a result of rapid motion about the long axis of the molecule. This axial motion averages g_{xx} and g_{yy} as well as A_{xx} and A_{yy}, giving an effective hyperfine splitting, $A_{\perp}$ (the splitting value with the z axis of the N-O group perpendicular to the magnetic field). The A_{zz} value from the spectrum can then be defined as $A_{\parallel}$.

For near-rigid-limit spectra, a parameter S can be defined as

$$S = \frac{A'_{zz}}{A_{zz}} \qquad [9\text{-}22]$$

which is a measure of the extent to which the nitroxide has approached the rigid limit. The value of S is obtained from equation [9-22], with the values of A'_{zz} and A_{zz} obtained from the slow tumbling and rigid-limit spectra, respectively, as shown in Fig. 9-4. Since S is a sensitive increasing function of τ_c, it is possible to calculate τ_c for isotropic diffusion using the following expression (3):

$$\tau_c = m(1-S)^n \qquad [9\text{-}23]$$

where the m and n parameters depend on whether diffusion is best described by Brownian, strong-jump, or free-rotational motion. The relationship between τ_c and S is shown in Fig. 9-7 with the assumption of a peak-to-peak spectral linewidth (ΔH) of 3 gauss. It is obvious that S becomes a very insensitive function of τ_c when $\tau_c > 10^{-7}$ sec (3). Thus, measurement of S is a useful indicator of τ_c only in the range of about $10^{-8} \leqslant \tau_c \leqslant 10^{-7}$ sec.

A related, more sensitive method uses the width of the outer hyperfine extrema, indicated by Δ in Fig. 9-4. In the rigid limit, these outer peaks are produced by nitroxide radicals having the $2p\pi$ orbital of the N atom nearly parallel to the applied magnetic field and an M_I quantum number of +1 or −1. The rigid-limit linewidths of the outer extrema, as measured by the half-width at half-height (Fig. 9-4), Δ, have been found to be (3):

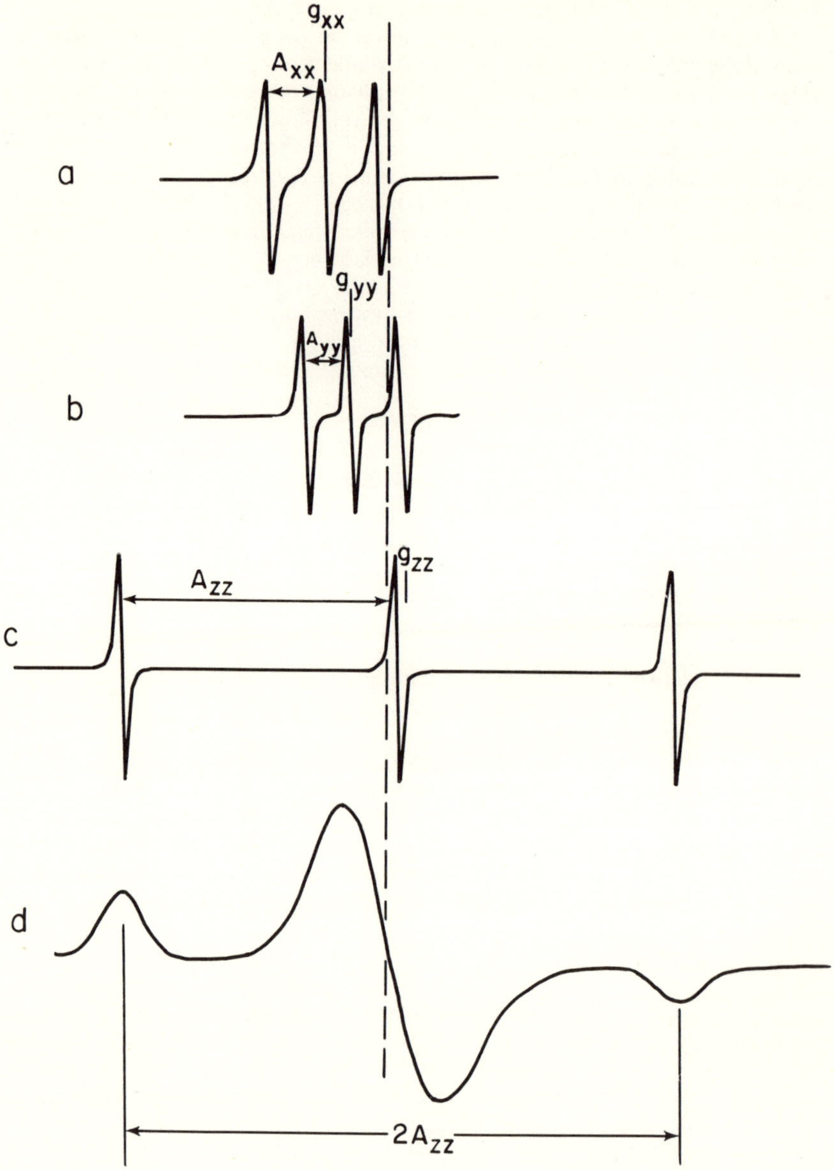

Figure 9-5. Spectrum of the nitroxide obtained with the magnetic field oriented (a) along the x axis, (b) along the y axis, and (c) along the z axis. The spectrum of randomly oriented nitroxides in the rigid-limit is shown in d.

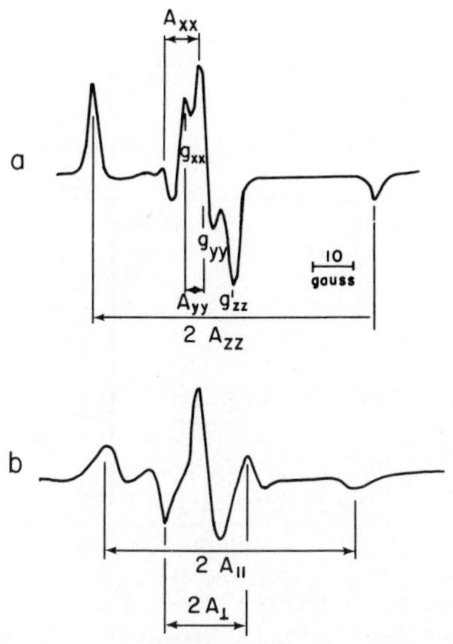

Figure 9-6. Rigid limit spectra of nitroxide for the cases of (a) unusually high resolution resulting from deuteration of the probe (adapted from Hwang *et al.*, 1975), and (b) axial motion which averages A_{xx} and A_{yy} (adapted from Smith, 1972).

$$2\Delta_{\ell}^{r} = 1.59\ \Delta H$$
$$2\Delta_{h}^{r} = 1.81\ \Delta H \qquad [9\text{-}24]$$

where ΔH is the peak-to-peak spectral linewidth, the subscripts ℓ and h refer to the low- and high-field extrema, respectively, and r refers to the rigid-limit values. In the slow motional region, the value of Δ is composed of two contributions, the inhomogeneous linewidth component given by equations [9-24] and the excess motional width. This second broadening arises from slow rotational motion that can be considered to alter the orientation of the nitroxide and change the ESR resonance frequency. It is this motional effect that causes S to be less than unity. The spectra of the slow motional region can then be described by the parameters:

$$W_{\ell} = \frac{\Delta_{\ell}}{\Delta_{\ell}^{r}} \qquad [9\text{-}25]$$

$$W_{h} = \frac{\Delta_{h}}{\Delta_{h}^{r}} \qquad [9\text{-}26]$$

A value of W_{ℓ} or W_{h} near unity indicates that the nitroxide is near the rigid limit, and the relationship between W_i and τ_c (Fig. 9-8) can be used to obtain estimates of τ_c. However, a good estimate of ΔH is necessary to obtain reliable values of τ_c, and this can be done using equations [9-24] for rigid-limit spectra. Near the rigid-limit, the method of using W_i as a measure of τ_c is much more sensitive than the

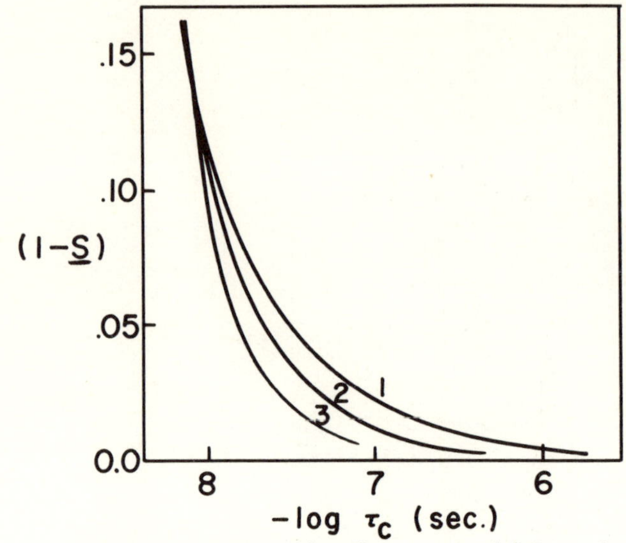

Figure 9-7. Relationship between $(1-S)$ and τ_c for (1) Brownian diffusion, (2) free diffusion and (3) strong jump diffusion. Peak-to-peak derivative line-width is assumed to be 3.0 gauss (adapted from Freed, 1976).

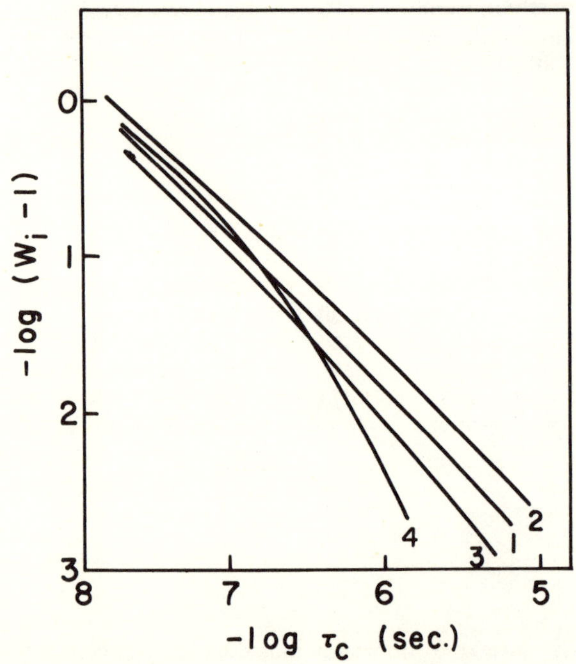

Figure 9-8. Relationship between (W_i-1) and τ_c for isotropic rotation of nitroxide assuming peak-to-peak linewidth of 3.0 gauss. Curves 1 and 2 represent free diffusion for W_ℓ and W_h, respectively. Curves 3 and 4 represent Brownian diffusion for W_ℓ amd W_h, respectively (adapted from Freed, 1976).

method using *S*, since motional broadening can double the widths of the outer extrema without having a very great effect on the separation of the extrema as measured by *S* (3). However, with motion considerably more rapid than the "near rigid-limit" case, the approximate ΔH needed to establish the relationship between W_i and τ_c will become smaller as heterogeneity of the environment is averaged.

9-4. NITROXIDES ADSORBED ON CLAY SURFACES.

Adsorption of organics on clays has been a subject of great interest to soil chemists, but investigations of the surface interactions have generally been restricted by the methods used. ESR offers inherent advantages over a number of other more commonly used physical methods of investigation (e.g., infrared spectroscopy), since clay-organic systems can be analyzed without excessively loading the systems with organics or removing the solvent. This allows the study of adsorption of small quantities of organics from aqueous (or non-aqueous) solution without having to perturb the solid-liquid equilibrium. Since very few organic molecules in nature are stable free radicals possessing ESR spectra, spin probes must generally be synthesized with the desired properties. For example, the amine form of 2,2,6,6-tetramethylpiperidine N-oxide (TEMPAMINE) is readily protonated to form a cation (see equation [9-26]). This cation is readily adsorbed

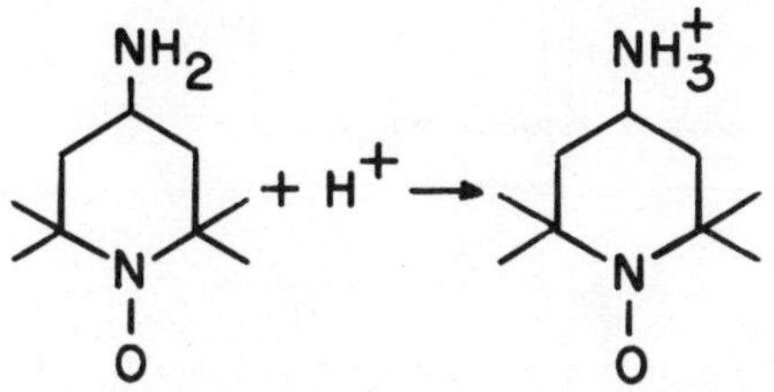

by montmorillonites, and based on the previous discussion of the dependence of the ESR spectrum on τ_c, might be expected to report the "microscopic viscosity" of the interlayer regions of clays. However, it is found that strong interaction with the clay surfaces in fully hydrated systems not only reduces the rotational mobility but also partially orients the molecule (5, 6). The ESR spectra show different hyperfine splitting values when wetted clay films are aligned parallel (∥) and perpendicular (⊥) to the magnetic field of the spectrometer (Fig. 9-9). The orientation dependence arises from anisotropic tumbling of the ion at the surfaces, where there is a tendency for the z axis of the nitroxide to spend a greater fraction of time normal or nearly normal to the plane of the clay surfaces than expected for random tumbling. The greater hyperfine splitting then occurs with the magnetic field ⊥ to the plane of the clay film, since the largest hyperfine splitting value will occur with the magnetic field parallel to the z axis (Fig. 9-5). The spectra are consistent with an average molecular alignment shown in Fig. 9-10. This orientation may result from the attraction of the protonated amine and methyl groups to the surface because of electrostatic and hydrophobic attractive forces, respectively.

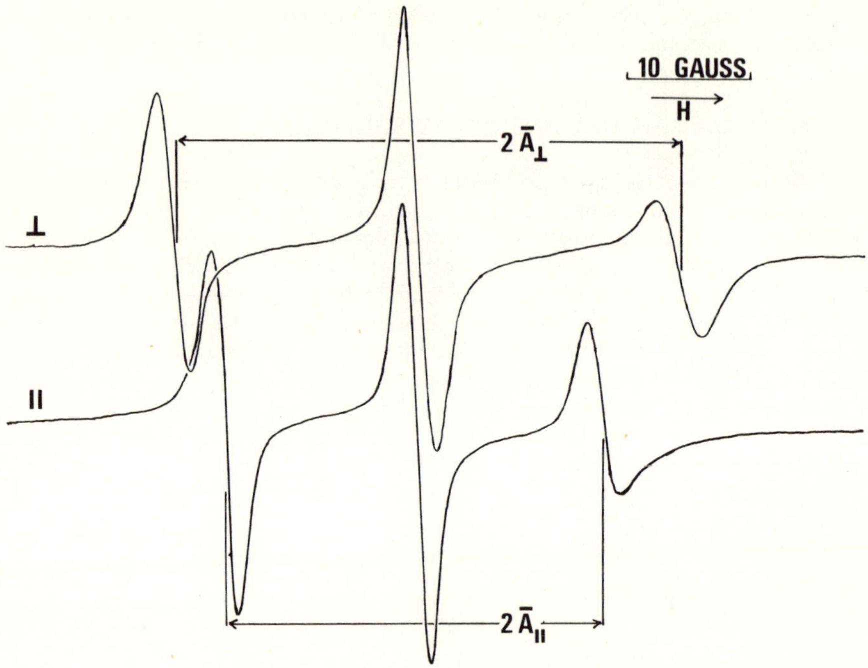

Figure 9-9. ESR spectrum of TEMPAMINE$^+$ adsorbed on fully H_2O-wetted K^+-hectorite films at about the 1% exchange level, showing the effect of film orientation on the value of $\bar{A}$. For the ⊥ orientation, $\bar{A}_\perp = 20.5$ gauss, and for the ∥ orientation, $\bar{A}_\parallel$ 15.2 gauss.

From the above discussion, it is clear that near-perfect surface alignment of the probe would produce the greatest orientation dependence of the spectra. One can quantify measurements of degree of orientation by defining an order parameter,

$$s = \frac{\bar{A}_\perp - \bar{A}_\parallel}{A_{zz} - A_{xx}} = \frac{\Delta \bar{A}}{A_{zz} - A_{xx}} \qquad [9\text{-}27]$$

where $\bar{A}_\perp$ and $\bar{A}_\parallel$ are half the field distance in gauss between the low field and high field resonances for the $\perp$ and $\parallel$ orientations of the clay films, respectively (Fig. 9-9), and A_{zz} and A_{xx} are the rigid-limit hyperfine splitting parameters of the probe molecules for alignment of the magnetic field along the z and x axes of the nitroxide, respectively. For dried TEMPAMINE$^+$-doped hectorite films, there is little thermal motion of the probe because the interlayer spacing is too small to permit free rotation, and a near rigid-limit type of spectrum is obtained with $\bar{A}_\perp$ and $\bar{A}_\parallel$ values of 31.4 and 9.7 gauss, respectively (Fig. 9-11). Assuming that these are essentially the rigid-limit values for air-dried clays, A_{zz} and A_{xx} can be taken to be equal to 31.4 and 9.7 gauss, respectively. Alignment is very greatly enhanced by removing water and collapsing the interlayers (Fig. 9-12).

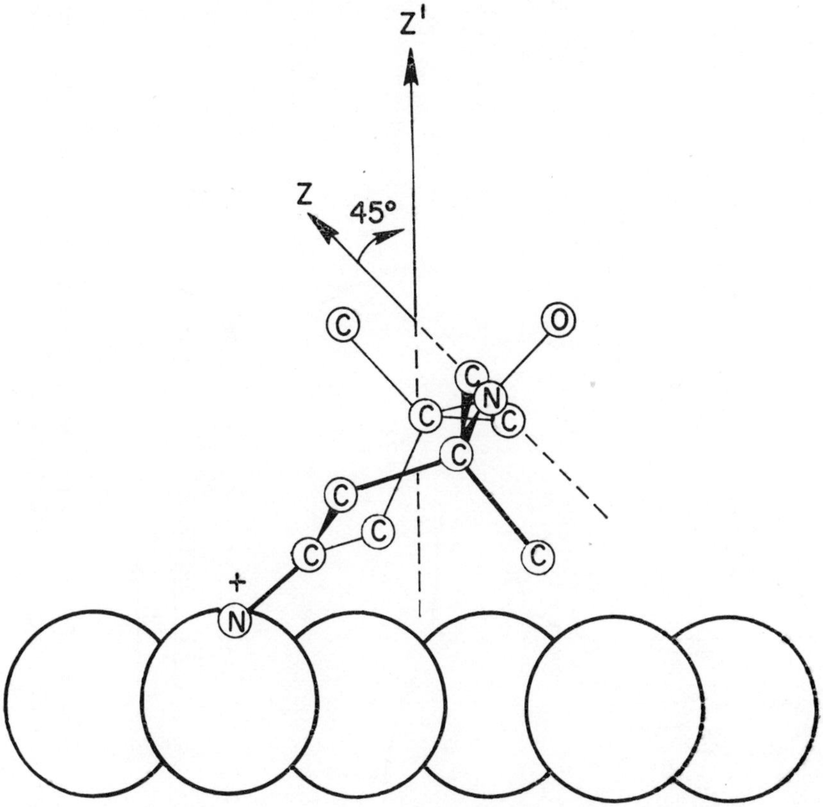

Figure 9-10. Average orientation of TEMPAMINE$^+$ on fully H_2O-wetted hectorite. The framework structure of the molecule is shown with hydrogen atoms omitted, and the silicate surface oxygens are represented by circles. (From McBride, 1979a).

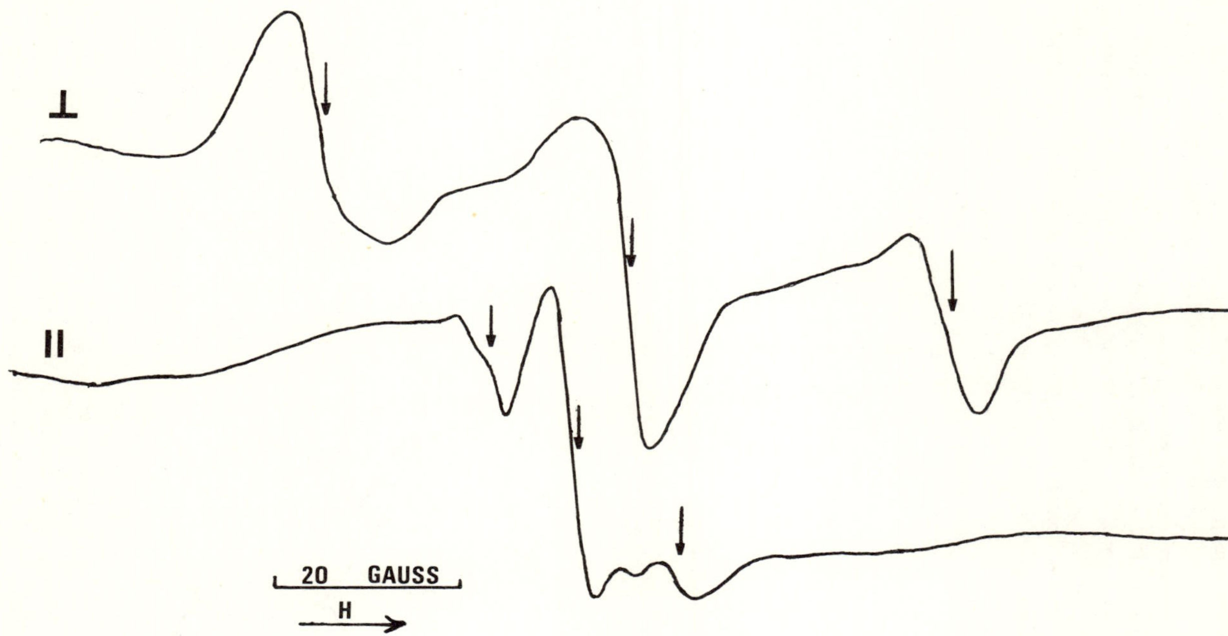

Figure 9-11. ESR spectrum of TEMPAMINE$^+$ adsorbed at the 1% level on K^+-hectorite dried at 110°C. The arrows indicate the approximate positions of the three resonances for the ⊥ and ∥ orientations of the hectorite film relative to the magnetic field.

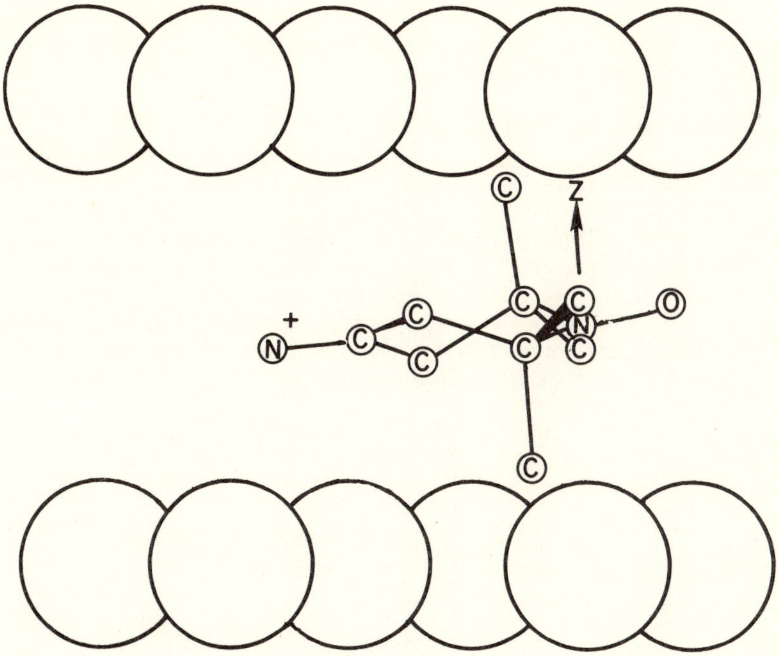

Figure 9-12. Probable rigid orientation of TEMPAMINE$^+$ in dried hectorite, with the z axis of the nitroxide approximately perpendicular to the plane of the silicate oxygens.

For fully hydrated K^+-hectorite doped at low TEMPAMINE$^+$ levels (1-2% of the CEC), $\overline{A}_\perp - \overline{A}_\parallel$ is about 5 gauss, so that the value of s from equation [9-27] is 0.23. The value of s appears to be somewhat reduced for M^{2+}-montmorillonites compared to M^+-montmorillonites (7), possibly because of an inhibiting effect of the more structured hydration water of M^{2+}-exchange ions upon surface contact by the organic cation. In any event, the organic probe appears to align in such a way as to maximize methyl group interactions with the hydrophobic siloxane surfaces of hydrated montmorillonites (Fig. 9-10).

In describing relatively rapid anisotropic motion of spin probes at surfaces ($\tau_c < 3 \times 10^{-9}$ sec) such as is observed for fully hydrated montmorillonites, an external, fixed axis system (x″, y″, z″) can be defined in addition to the molecule-fixed axes (x, y, z). The angle, θ, between the z and z″ axes is given by

$$\cos \theta = \vec{z} \cdot \vec{z}''. \qquad [9\text{-}28]$$

In experiments on layer silicate minerals, the z″ axis can be chosen ⊥ to the plane of the silicate sheets of the clay. The angle θ fluctuates rapidly with time (Fig. 9-13), but if rotational symmetry of molecular motion about the z″ axis can be assumed, the time-averaged system is axially symmetric, (Note that the random orientation of the a and b axes of the clay platelets in the x″y″ plane will spatially average out any anisotropic motion about z″ even if it existed within individual interlamellar regions). The following equations can then be derived for orientations of the magnetic field ⊥ and ∥ to the z″-axis (11):

$$A_\perp = 1/2\,(1 - \langle\cos^2\theta\rangle)\,(A_{zz} - A_{xx}) + A_{xx} \qquad [9\text{-}29]$$

$$A_\parallel = \langle\cos^2\theta\rangle\,(A_{zz} - A_{xx}) + A_{xx} \qquad [9\text{-}30]$$

where $\langle\cos^2\theta\rangle$ is the time-averaged value of $\cos^2\theta$. It should be noted that when the magnetic field is oriented ∥ to the plane of the clay film, it is directed ⊥ to the z″-axis. As a result, $A_\parallel = \bar{A}_\perp$ and $A_\perp = \bar{A}_\parallel$. Using values of $A_\perp$ = 15.2 gauss and $A_\parallel$ = 20.2 gauss for TEMPAMINE$^+$ adsorbed on fully wetted K^+-hectorite in equations [9-29] and [9-30], *apparent* average θ values of about 45° are obtained, accounting for the orientation angle shown in Fig. 9-10.

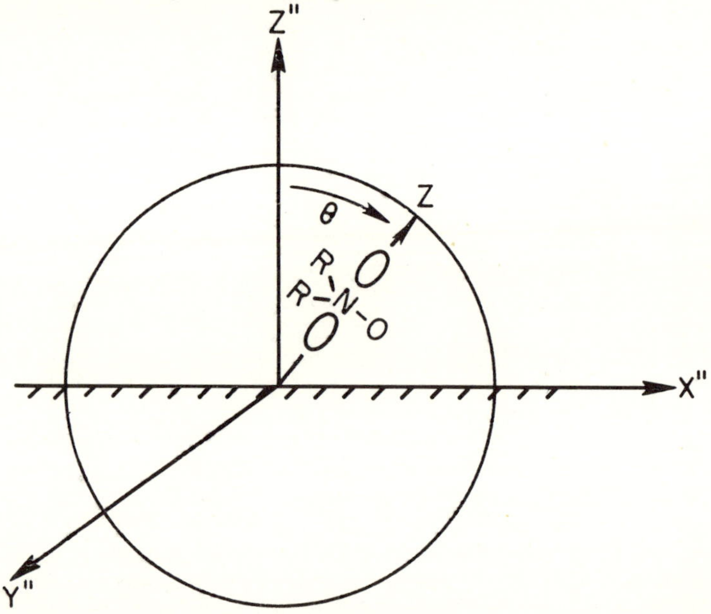

Figure 9-13. Fluctuation of θ for a surface-adsorbed nitroxide as $\vec{z}$ reorients relative to the fixed axes (x″, y″, z″).

The value of the order parameter, *s* is not only determined by the degree of order of the probe in the system, but also by the geometry of the spin probe. Therefore, one can only directly compare values of *s* for the same spin probe, since the angle between the N-O bond axis and the symmetry axes of the molecule to which it is attached may be different for different probes. It is the general shape of the molecule which is likely to determine its preferred orientation at surfaces. As a result, different *s* values can arise for probes with different N-O group orientations, even though the degree of molecular order is the same. However, for a given spin probe with a rigid molecular framework, *s* is a measure of the anisotropy of the motion. If an adsorbed probe were tumbling randomly on a surface, then $A_\parallel = A_\perp = A_o$ and $g_\parallel = g_\perp = g_o$ where A_o and g_o are the isotropic hyperfine splitting constant and isotropic g-factor, respectively; hence,

$$A_o = 1/3\ (A_{xx} + A_{yy} + A_{zz}) \quad [9\text{-}31]$$

$$g_o = 1/3\ (g_{xx} + g_{yy} + g_{zz}) \quad [9\text{-}32]$$

From equations [9-27], [9-29], and [9-30], it follows that rapid isotropic motion produces a value of $s = o$ and $\langle \cos^2 \theta \rangle = 1/3$. This can be considered to correspond to an *apparent* average angular fluctuation of 54.7°, but the average of $\cos^2 \theta$ (not θ) is the quantity that is actually measured. Positive or negative deviations from $s =$ 0 indicate anisotropic motion of the molecule, but a value of $s = 0$ may result from a particular combination of anisotropic motion and orientation of the N-O group as well as from isotropic motion.

The value of s is a useful indicator of clay-organic interaction, with strong surface bonding of the probe expected to produce non-zero values of s. For example, at moderate loading levels of TEMPAMINE$^+$ on methanol solvated K^+-hectorite films, $\overline{A}_\perp$ and $\overline{A}_\parallel$ are about 16.4 and 16.1 gauss, respectively. This corresponds to an s value of about 0.01, representing virtually random motion despite the fact that the interlayer spacings are smaller than in the case of hydrated clays. The less polar solvent has a greater ability than water to solvate the probe, thereby minimizing probe-surface interaction. At very low loading levels of TEMPAMINE$^+$ on methanol-solvated hectorites, a much more strongly oriented probe is detectable with $\overline{A}_\perp$ and $\overline{A}_\parallel$ values of 11.7 and 19.8 gauss, respectively (8). The value of s for this species is −0.37, indicating a greater degree of alignment relative to the surfaces compared with the hydrated hectorite, but in the opposite direction. It is likely that such an orientation, corresponding to an apparent average θ value of 73°, represents probes motionally restricted by the relatively narrow interlayer space (7-8 Å) and simultaneously contacting adjacent surfaces (Fig. 9-14).

The ESR spectra of adsorbed probes clearly provide useful information on mechanisms of clay-organic interactions. They do not necessarily report interlayer solvent viscosities, since their rotational correlation times are modified by specific surface interactions in addition to the properties of the solvent. For example, using equations [9-15] and [9-16] to estimate τ_c for adsorbed TEMPAMINE$^+$, values of about $1\text{-}3 \times 10^{-9}$ sec. are obtained, an estimated 20-60 times reduction in rate of molecular rotation relative to the solution state (7). However, longer values for τ_c are invariably found for the $\perp$ orientation of the clay films in the magnetic field compared to the $\parallel$ orientation, a result of anisotropic rotation. In addition, M^{2+}-hectorites show τ_c values about twice as long as M^+-hectorites, possibly a reflection of the limited interlamellar volume in the former case. It is interesting that the probe doped at low levels in ethanol-solvated hectorite has an apparent τ_c of $2\text{-}4 \times 10^{-9}$ sec, or an approximate two orderof magnitude reduction in rotational motion resulting from adsorption, despite relatively isotropic motion in the adsorbed state. The equations used to obtain τ_c are strictly valid only for isotropic rotation; however, they are useful for comparative purposes in studying anisotropic rotation at surfaces.

Besides motional information, spin probes report the polarity of their immediate chemical environment through small changes in A_o and g_o. The magnetic

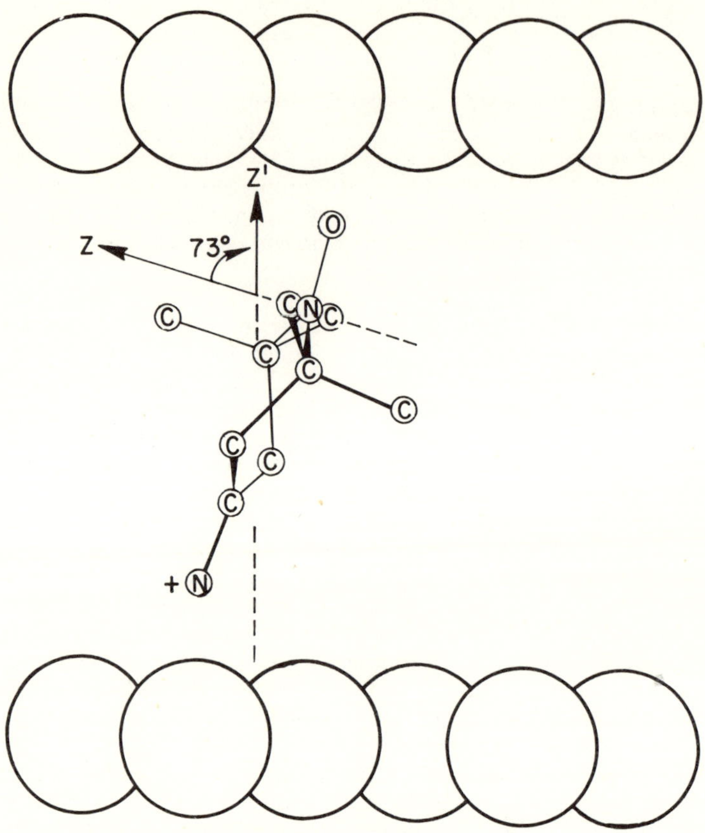

Figure 9-14. Probable average alignment of strongly oriented TEMPAMINE^{+} in the ~7 Å interlayers of methanol-solvated hectorite (From McBride, 1979a).

parameters of radical species are sensitive functions of the electronic distribution in the molecule. Hydrophilic solvents produce isotropic coupling constants more than 10% greater than hydrophobic solvents. Polar solvents interact with the lone pairs of electrons on the oxygen atom of the nitroxide, lowering the energy of the nonbonded electrons and increasing the electron affinity of the oxygen atom. The three π-electron distribution of the N-O group is shifted toward maximum unpaired electron density on the nitrogen atom:

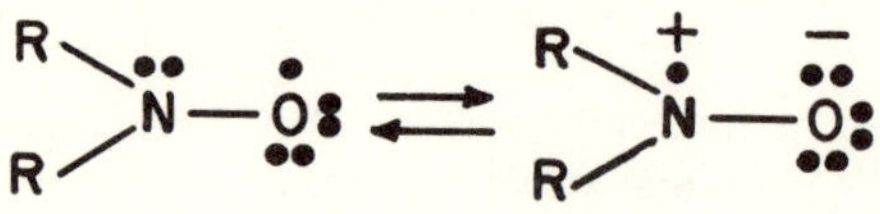

[9-33]

and the ^{14}N splitting constant is thereby increased. Using this feature of A_o, it should be possible to measure approximate solvent compositions at surfaces that are different from bulk composition. For example, adsorbed TEMPAMINE$^+$ on hydrated hectorite has $\bar{A}_\perp = 20.4$, $\bar{A}_\parallel = 15.2$, and $\bar{A} = \frac{1}{3}(\bar{A}_\perp + 2\bar{A}_\parallel) = 16.9$, the same value as the isotropic hyperfine splitting for TEMPAMINE$^+$ in aqueous solution. However, the value of $\bar{A}$ is about 15.7 and 15.4 gauss for TEMPAMINE$^+$ adsorbed on methanol- and isopropanol-solvated hectorites, respectively. In cases where the probe is strongly immobilized by adsorption, it is possible that $\bar{A}$ values would reflect the polarizing nature of the adsorbing surface if the N-O group directly interacts with the surface rather than solvent molecules.

It has generally been found that the doping level affects the spectrum of probe adsorbed on clay. As loading levels of TEMPAMINE$^+$ on solvated montmorillonite are increased, the spectrum of the probe suggests more mobility and less anisotropic rotation. This may result from a fast exchange of the probe between surface sites and solution or solution-like environments. If the probe moves sufficiently rapidly between these sites, the result will be a single spectrum that reports an environment which is a weighted average of the solution and surface environments. From equation [9-13], τ_c is shown to be proportional to the square root of peak heights:

$$\tau_c \propto \left[\sqrt{\frac{h(0)}{h(-1)}} - \sqrt{\frac{h(0)}{h(+1)}}\right]\Delta\upsilon(0) \qquad [9\text{-}34]$$

A relaxation time, T_2, may be defined as

$$1/T_2 = k\gamma_e\Delta\upsilon$$

where $\Delta\upsilon$ is the resonance linewidth, γ_e is the electronic magnetogyric ratio, and k is a constant. Since $1/T_2$ is a linear function of linewidth, it includes broadening introduced by fast exchange between two or more states (lifetime broadening), and other broadening processes of homogenous nature described by a relaxation time, T_2', called the spin-spin relaxation time. The former type of broadening (lifetime) results from the exchange of energy between spin states, and is described by a spin-lattice relaxation time, T_1. T_1 is characteristic of the mean lifetime of a given spin state, since the Heisenberg uncertainty principle requires that a very short lifetime of a given state results in uncertainty in the energy level of that state, i.e.

$$\Delta\upsilon\,\Delta t \geqslant \frac{h}{2\pi}\,. \qquad [9\text{-}36]$$

If T_1 is used as an estimate of Δt, then a small T_1 will lead to a large $\Delta\upsilon$ and a measurable uncertainty in energy levels that will be observed as a broadening of lines in the ESR spectrum. The spectral linewidth, then, includes two main components of broadening, *viz.*

$$\frac{1}{T_2} = \frac{1}{T_2'} + \frac{1}{2T_1} \qquad [9\text{-}37]$$

Rewriting equation [9-34] by substituting relaxation times for the square root of the peak heights, h, (since for a given resonance intensity, $\sqrt{h} \propto 1/\Delta\upsilon \propto T_2$) one obtains

$$\tau_c \; \alpha \left[\frac{1}{T_{-1}} - \frac{1}{T_{+1}}\right] T_o \Delta\upsilon(0) \qquad [9\text{-}38]$$

where the T values represent relaxation times for the corresponding nuclear spin states (−1, 0, +1) and $\Delta\upsilon(0)$ is the linewidth of the center line. The value of $T_o\Delta\upsilon(0)$ is constant (since $T_o \propto \frac{1}{\Delta\upsilon(0)}$). For the case where a probe is undergoing fast exchange between two sites, A and B, the relation is

$$\frac{1}{\overline{T}_2} = \frac{f_A}{T_{2A}} + \frac{f_B}{T_{2B}} \qquad [9\text{-}39]$$

where $\overline{T}_2$ is the mean relaxation time, T_{2A} and T_{2B} are the relaxation times at the A and B sites, and f_A, f_B are the fractions of time spent at the two sites, respectively. Equations [9-38] and [9-39] can then be manipulated to show that

$$\overline{\tau_c} = f_A\tau_A + f_B\tau_B \qquad [9\text{-}40]$$

where $\overline{\tau_c}$ is the average correlation time. The result is that the mean correlation time is simply a weighted average of the correlation times, τ_A and τ_B, in each environment. The nature of equation [9-40] is such that, even if the probe spends a small fraction of its time in a more rigid environment, the average correlation time as observed by the spectral linewidth may be considerably lengthened. For example, TEMPAMINE$^+$ in solution has a τ_c value near 5×10^{-11} sec, but adsorbed at the 1% exchange level on hydrated montmorillonite the probe is motionally restricted with a τ_c of about 10^{-9} sec. Even if the probes on the average spent 95% of their time in the solution phase (f_A = .95), line broadening would occur to produce an apparent doubling of the rotational correlation time relative to solution. For this reason, great care must be taken in interpreting correlation times obtained from systems that contain more than one phase, since rapid exchange between phases can produce a single spectrum with an apparent τ_c that is not a true measure of the rotational correlation time in any one phase. Very high viscosities have been suggested by spectra of spin probes in cellular protoplasm (1), but these are likely a result of binding of a fraction of probes to membranes. It should also be stressed that the above analysis of linewidths is only applicable in the range of τ_c values for which linewidths are linearly proportional to the measured τ_c values (i.e. equations [9-13] and [9-14] are applicable). Linewidths will be proportional to τ_c only in the fast motional region, where the linewidths are largely determined by the degree to which spectral anisotropies are averaged out (14).

Based upon the above analysis, the lower apparent values of τ_c at high doping levels of TEMPAMINE$^+$ in montmorillonite can be considered to reflect the shift of equilibrium toward solution as more probe ions are added to the system. The partial loss of the anisotropy of the signal that is observed appears to arise from convergence of the solution and adsorbed-state signals to a position determined by the weighted mean. For example, Na$^+$-hectorite with a small amount of

TEMPAMINE$^+$ added (1-2%) and solvated in ethanol, has $\overline{A}_\perp$ = 15.2, $\overline{A}_\parallel$ = 16.8 gauss and an apparent $\tau_c \approx 4 \times 10^{-9}$ sec. However, with about 10% of the exchange sites occupied by TEMPAMINE$^+$, $A_\perp$ = 15.5, $A_\parallel$ = 15.6 and τ_c is about 4×10^{-10} sec. The fact that two superimposed spectra are not seen, one for the adsorbed state and one for the solution-like state, indicates that the condition of fast exchange must apply; that is, the mean lifetime of the probes in states A and B must be much less than $\frac{1}{\gamma_e \delta H}$, where δH is the line separation (in gauss) between the corresponding resonances due to probes in the A and B states in the absence of fast exchange. Since the high field line of TEMPAMINE$^+$ adsorbed at low levels on hydrated Na^+-hectorite is shifted by about 2 gauss (or 5.6 MHz) relative to the high field line of solution TEMPAMINE$^+$, the lifetimes in the adsorbed and solution states must be much less than 1.8×10^{-7} sec. However, this is a sufficiently long time to allow diffusion of probes between the surface and a much less restrictive environment near the surface or in bulk solution, thereby producing a single averaged spectrum.

9-5. EXPERIMENTAL CONSIDERATIONS IN USING NITROXIDE SPIN PROBES

An important phenomenon to consider in choosing the doping level of probe on surfaces is dipole-dipole interaction. At high concentrations, spectra of spin probes are broadened by interaction of the dipole moments of two unpaired electrons. If both dipoles reorient rapidly in the applied magnetic field, this interaction is averaged to zero regardless of concentration. If they reorient at an intermediate rate relative to the frequency corresponding to the electron resonance energy, line broadening will occur. With increasing temperature, more rapid molecular reorientation in the magnetic field will decrease dipole-dipole line broadening. At higher concentrations, spectra of spin probes may be further broadened by the interaction of two radical molecules with an exchange of spin states. This phenomenon, called electron spin exchange, arises from the overlap of the orbitals on separate radicals containing the unpaired electrons, and is shown in Fig. 9-15b. Even higher concentrations cause the three resonance lines to coalesce into a single line, which will become narrowed as the concentration is further increased (Figs. 9-15c, d). The latter effect is termed exchange narrowing, since the electron spins are exchanged between molecules so rapidly that the time average of the hyperfine field is almost zero. On clay surfaces, adsorbed TEMPAMINE$^+$ is often strongly concentrated so that probe-probe distances are small enough to produce spin-exchanged spectra, even when only about 10% of the exchange sites are occupied by probes. It is likely that spin exchange is enhanced by concentration of probe ions in certain interlamellar regions, a phenomenon of ion segregation (demixing) that is not uncommon. Very low loading levels on the exchange sites (1% of CEC or less) are therefore preferable to avoid unnecessary line broadening and spin exchange.

Dipole-dipole interaction can also result from paramagnetic impurities in clays, broadening ESR spectra noticeably. For example, small amounts of free iron oxides in hectorites have been observed to broaden spectra of adsorbed vanadyl ion (9). These impurities can be removed by a standard citrate-dithionite treatment. However, care must be taken to choose clay minerals with little or no structural paramagnetic ion content (e.g., Fe^{3+}, Mn^{2+}, Ni^{2+}) when conducting ESR studies of surface-adsorbed species. Although ESR spectra can be obtained for species adsorbed on clays such as Upton montmorillonite, they are considerably broadened compared with the spectra of the same species on hectorite (a clay with only trace quantities of structural Fe^{3+}).

a

b

c

d

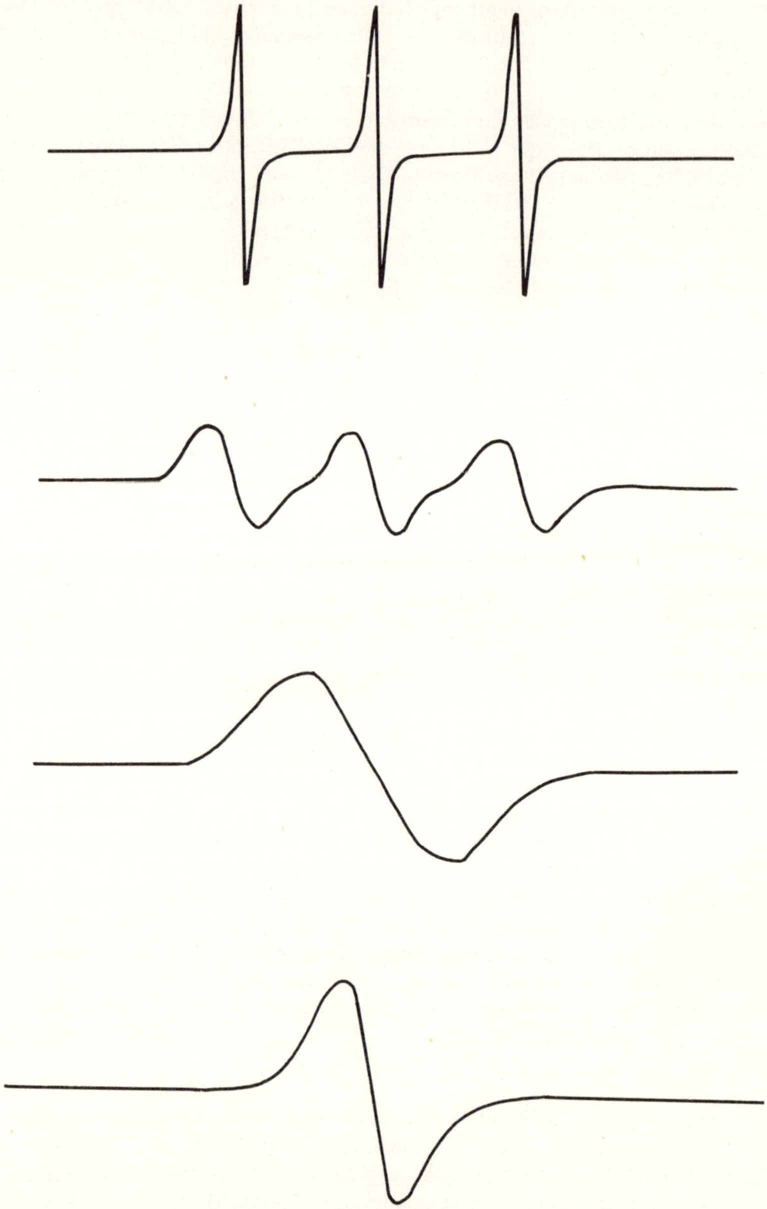

Figure 9-15. ESR spectra of a nitroxide spin probe in aqueous solution at room temperature at concentrations of (a) 10^{-4} M, (b) 10^{-2} M, and (c) 10^{-1} M. The spectrum of the pure nitroxide (undiluted in solvent) is shown in d (Adapted from Wertz and Bolton, 1972).

Molecular oxygen is a triplet; and is therefore paramagnetic. Dissolved oxygen may interact with spin probes through spin exchange and dipole mechanisms, thereby producing spectral broadening. This is especially critical in the case of hydrocarbon solvents such as methanol, not being as evident in water. Oxygen may be removed from water by bubbling pure nitrogen or argon gas through the samples.

Experimentally, it is sometimes useful to know whether the probe is in a protected environment of the system of interest. Since sodium ascorbate chemically reduces nitroxides almost instantaneously at room temperature, the rapid disappearance of the probe spectrum after addition of ascorbate to the system is evidence that the probe is accessible. Whether a cationic probe such as TEMPAMINE$^+$ is on external surfaces or in interlamellar regions of layer silicates would be expected to influence the rate of reduction.

Stable radicals of very different molecular structure are available as spin probes. For example, the long-chain fatty acid types of probe with the nitroxide moiety attached at different positions of the structure may be useful in studying fatty acid-oxide interactions. Preliminary experiments show that the carboxylic acid group is attached to hydrous alumina surfaces, thereby greatly reducing the rotational mobility of the N-O group of probe II (shown below), but having less effect on the N-O group of probe I. Thus, the site of

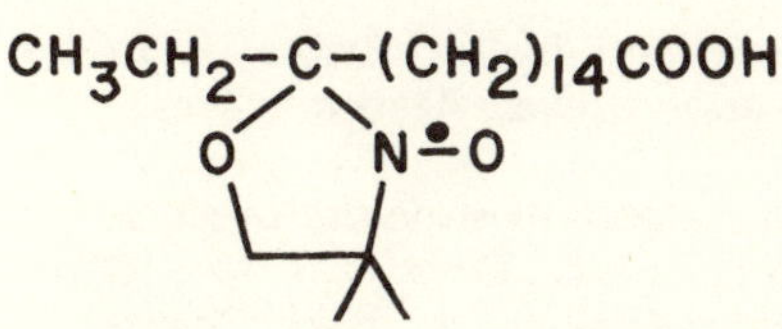

I

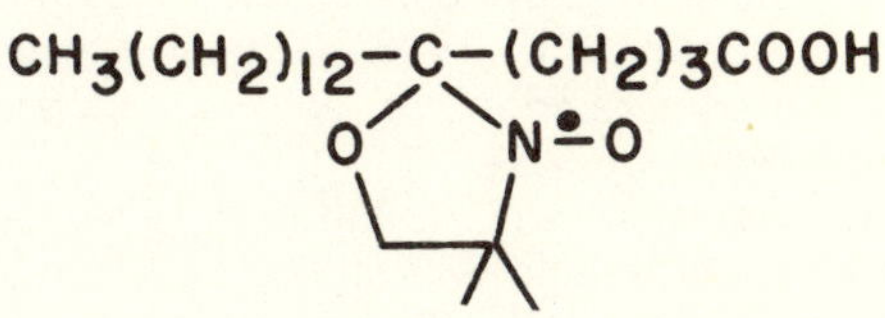

II

attachment of organic molecules on surfaces can be deduced from comparison of the behavior of probe molecules with different molecular geometry or functional groups. However, attempts to immobilize TEMPO-phosphate (shown below) on alumina

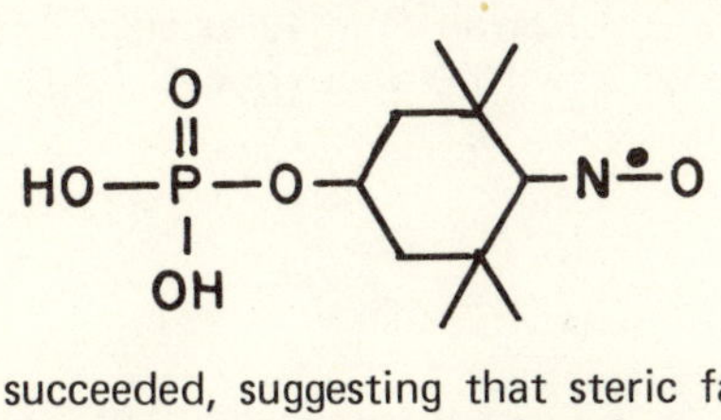

have not succeeded, suggesting that steric factors are involved in preventing the ligand displacement of OH^- by phosphate on alumina. Although probes can be synthesized to model certain properties of ions and molecules, they may not react with surfaces as expected, especially if active functional groups are not well separated from the methylated hydrophobic (nitroxide) portion of the molecule.

REFERENCES

1. Berliner, L.J., ed. 1976. Spin labeling – theory and applications. Academic Press, New York. 592 pp.
2. Finch, E.D. and J.F. Harmon. 1974. Viscosity of cellular protoplasm: What do spin probes tell us? *Science* **186**:157-158.
3. Freed, J.H. 1976. Theory of slow tumbling ESR spectra for nitroxides. Chapter 3. *In* L.J. Berliner, ed., *Spin labeling – theory and applications.* Academic Press, New York. pp. 53-132.
4. Hwang, J.S., R.P. Mason, L.P. Hwang and J.H. Freed. 1975. ESR studies of anisotropic rotational reorientation and slow tumbling in liquid and frozen media III: Perdeuterotempone and an analysis of fluctuating torques. *J. Phys. Chem.* **79**:489-511.
5. McBride, M.B. 1976. Nitroxide spin probes on smectite surfaces. Temperature and solvation effects on the mobility of exchange cations. *J. Phys. Chem.* **80**:196-203.
6. McBride, M.B. 1977a. Adsorbed molecules on solvated layer silicates: surface mobility and orientation from ESR studies. *Clays Clay Miner.* **25**:6-13.
7. McBride, M.B. 1977b. Exchangeable cation and solvent effects upon the interlamellar environment of smectites: ESR spin probe studies. *Clays Clay Miner.* **25**:205-210.
8. McBride, M.B. 1979a. Cationic spin probes on hectorite surfaces: demixing and mobility as a function of adsorption level. *Clays Clay Miner.* **27**:97-104.
9. McBride, M.B. 1979b. Mobility and reactions of VO^{2+} on hydrated smectite surfaces. *Clays Clay Miner.* **27**:91-96.
10. Nordio, P.L. 1976. General magnetic resonance theory. Chapter 2. *In* L.J. Berliner, ed., *Spin labeling – theory and applications.* Academic Press, New York. pp.5-52.
11. Sachs, F. and R. Latorre. 1974. Cytoplasmic solvent structure of single barnacle muscle cells studies by electron spin resonance. *Biophys. J.* **14**:316-326.
12. Seelig, J. 1976. Anisotropic motion in liquid crystalline structures. Chapter 10. *In* L.J. Berliner, ed., Spin labeling – theory and applications. Academic Press, New York. pp. 373-409.
13. Smith, I.C.P. 1972. The spin label method. Chapter 11. *In* H.M. Swartz, J.R. Bolton and D.C. Borg, (eds.). Biological applications of electron spin resonance. Wiley-Interscience, New York. pp. 483-539.
14. Snipes, W. and A.D. Keith. 1974. Response to "Viscosity of cellular protoplasm: What do spin probes tell us?". *Science* **186**:158.
15. Wertz, J.E. and J.R. Bolton. 1972. Electron Spin Resonance – Elementary theory and practical applications. McGraw-Hill, New York. 497 pp.

Chapter 10

APPLICATIONS OF PHOTOACOUSTIC SPECTROSCOPY TO THE STUDY OF SOILS AND CLAY MINERALS

Raymond L. Schmidt

Senior Research Chemist
Chevron Oil Field Research Company
La Habra, CA 90631

10-1. INTRODUCTION

Photoacoustic spectroscopy (PAS, also called optoacoustic spectroscopy) although over 100 years old, has experienced renewed interest with the detection of trace levels of atmospheric pollutants. More recently PAS of condensed systems has resulted from a better understanding of the underlying physics of the PA effect. In this communication, I wish to illustrate the use of PAS to obtain the absorption spectra of soil and clay mineral solids, which are especially difficult to study by more conventional transmission or reflectance spectroscopy due to sample heterogeniety and light scattering artifacts.

Several good references are available for more detailed study of the theory and application of PAS (5, 6). This paper will only conceptually outline the underlying physics and experimental procedures, and will show studies on soil and clay samples.

Fig. 10-1 illustrates how the PAS experiment is carried out. A tunable narrow band light source is chopped (mechanically or electrically) at a frequency ω_o and illuminates the sample under study. The sample is contained in a fixed volume cell with a transparent window and the microphone detector. In the photoacoustic experiment one listens with the microphone to the intensity of sound being generated as the sample absorbs light of wavelength λ_o. How is this sound generated by the light absorption process?

As the sample chromophores absorb λ_o energy, they go from a ground to an excited quantum state of energy $\Delta E_{absorption} = E_{excited} - E_{ground} = hc/\lambda_o$ above the ground state. This excited-state excess energy can then be dissipated by any of three major relaxation mechanisms: (1) spontaneous emission of light of the

J. W. Stucki and W. L. Banwart (eds.), Advanced Chemical Methods for Soil and Clay Minerals Research, 451–465.

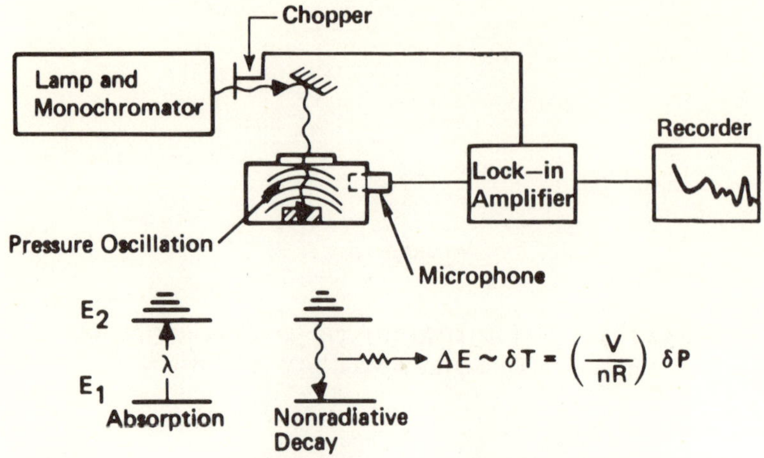

Region	Wavelength	Adsorbed Species
Ultraviolet	250 – 400 nm	Organics
Visible	400 – 700 nm	Metal Complexes
Near Infrared	700 – 2600 nm	Clay OH, H_2O Organic CH

Figure 10-1. Photoacoustic Spectroscopy.

same or slightly lower energy after various excited state internal conversion mechanisms; (2) stimulated emission under restrictive conditions which lead to laser action; or (3) conversion to heat via vibrational, rotational and collisional interaction of the chromophore with its near neighbor environment. It is the last, nonradiative deactivation channel of the excited state which permits the absorption step to be detected by the acoustic microphone. If the heat generated (ΔT) by the nonradiative decay is related to the amount of light absorbed (β), the modulated light source generates a periodic temperature fluctuation (δT) in the sample. This sample temperature fluctuation causes a proportional temperature fluctuation in the gas (air) in contact with the sample in the fixed volume sample cell. Finally, the gas phase thermal fluctuation results in a pressure fluctuation [$\delta P = (nR/V)\delta T$] which is then detected by the microphone and lock-in amplifier set at the modulation frequency, ω_O.

From this brief description three different length scales are apparent which must be considered in the underlying physics of the PAS effects of solids: (1) the physical length of the optical absorber, ℓ; (2) the absorption length scale, $\mu_\beta = 1/\beta$, which defines the amount of incident light absorbed and the depth of penetration of the light into the solid; and (3) the thermal length scales of the absorber and gas $\mu_S = \sqrt{2\alpha/\omega_O}$ where α is the thermal diffusivity and μ_S characterizes the rate and depth of heat production which is sampled by the acoustic detection.

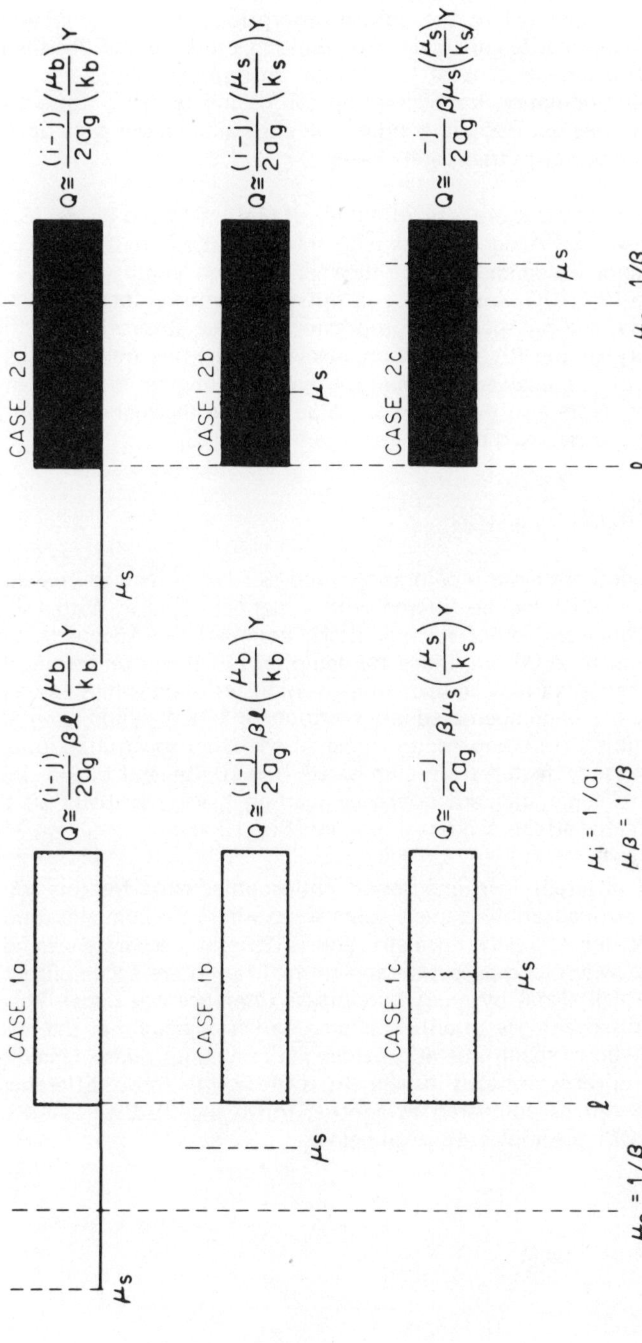

Figure 10-2. Schematic Representation of Special Cases for the Photoacoustic Effect in Solids (From Pao, 1977).

By considering the relative magnitude of these length scales, various theories for PAS of solids have emerged. Several limiting cases are illustrated in Fig. 10-2. Note that in all cases except 2a and 2b (Fig. 10-2), the observed photoacoustic output is proportional to the optical absorption coefficient, β, times either the physical or thermal length scale. For limiting cases 2a and 2b, the PA output is saturated and independent of the optical absorption coefficient; by use of sufficiently high modulation frequencies μ_S can be shortened to cause cases 2a and 2b to approach case 2c. Particle grinding can also cause an optically opaque sample to approach the optically transparent cases.

Results of the more refined theory of McDonald and Wetsel (3) show quantitatively how the PA signal varies with the absorption coefficient β and the chopping frequency which controls the thermal diffusion length, $\sqrt{2\alpha/\omega_O}$. The behavior is shown in Fig. 10-3. An increase in optical absorption increases the PA signal at fixed ω, thus the PA spectra is proportional to the absorption spectra. At fixed β the intensity of the PA signal decreases with chopping frequency as less and less thermal depth is allowed to heat the gas resulting in lower intensity pressure fluctuations. Note also that since ω_O modifies the thermal diffusion depth, depth profiling spectroscopy is possible for layered samples.

10-2. INSTRUMENTATION

A xenon arc lamp is commonly used because of its high intensity and broadband output which can be filtered with a monochromator. With a single lamp and various gratings and order sorting filters, commercial PAS spectrometers operate from 200 nm to 2800 nm. Since the lamp output is not uniformly intense across this entire range, various source compensation techniques must be employed. Fig. 10-4 shows the uncompensated lamp output for a 1KW xenon lamp. By employing an electronic source compensator, most of the lamp spectral and monochromator grating efficiency features are eliminated (Fig.10-4b) and finally, by normalizing the signal to some standard reference (carbon black), virtually all the lamp and monochromator effects can be eliminated (Fig. 10-4c).

Two different instrumentation philosophies exist for this source compensation and normalization: a single beam setup where the compensation can be done directly and the reference normalization is done in a separate step; or the double beam setup where compensation and normalization are accomplished by dividing the sample cell signal by a nearly identical reference cell signal. There are advantages and disadvantages to either scheme, but it appears that the microprocessor-based single-beam spectrometer provides the best compromise. One distinct feature of the microprocessor-based instrument is the ability to do difference PAS, where one sample can be compared to another (difference PAS = [SAMPLE-BLANK] / REFERENCE). Examples are given below.

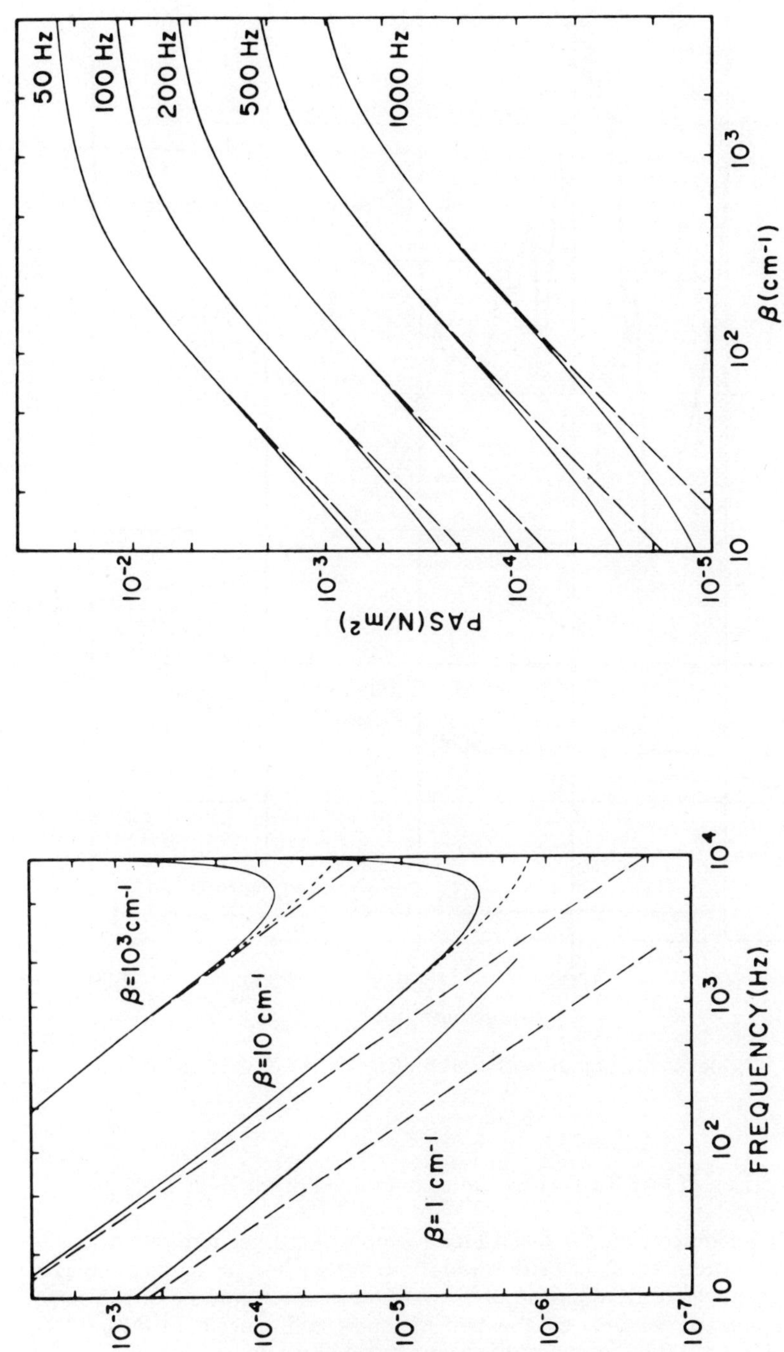

Figure 10-3. Theoretical Values of Photoacoustic Signal for Various Values of Absorption Coefficient, β, and Modulation Frequency (From McDonald and Wetsel, 1978).

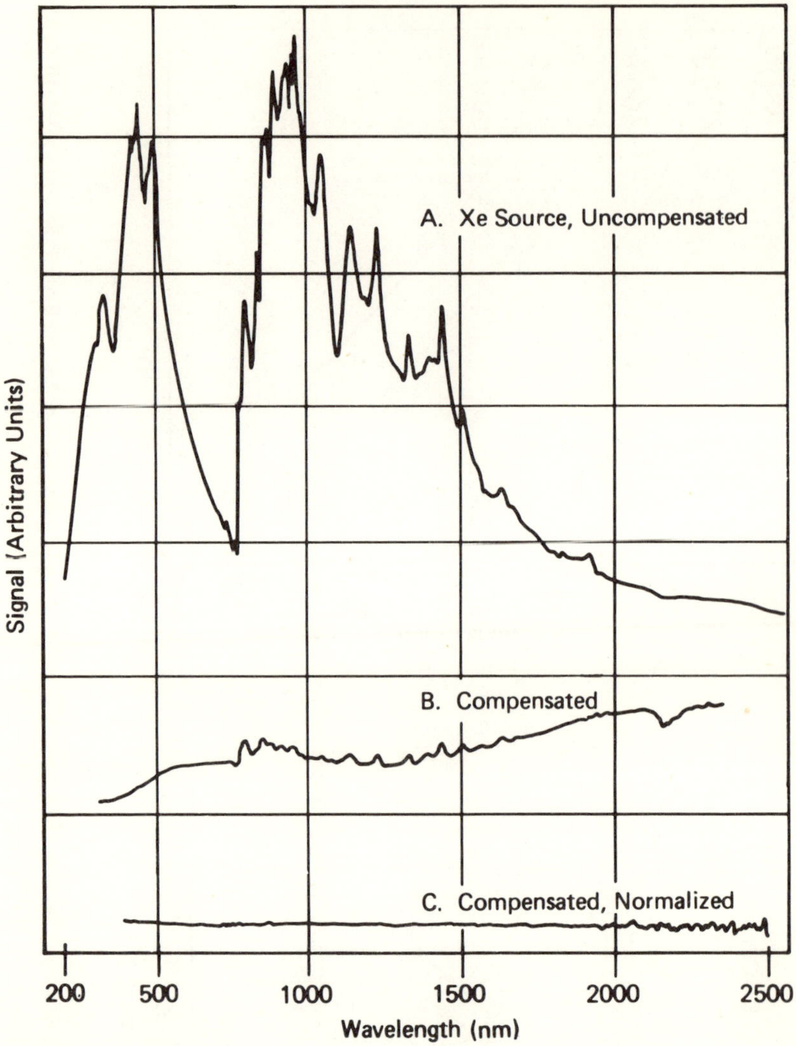

Figure 10-4. Photoacoustic Signal from Carbon Black with Xenon Lamp Source.

10-3. RESULTS

10-3.1. Comparison of PAS To Diffuse Reflectance Spectra

Fig. 10-5 compares the PA spectra for a sample of crushed natural muscovite with published diffuse reflectance spectra for both natural and synthetic samples in the near infrared region. Comparison of the natural samples shows a direct correlation of the spectral features by the two methods; bands occur at 870, 1100, 1380, 1880, 2180 and 2410 nm in both spectra and with analogous intensities. The PA spectra in fact shows more structural detail in the 2300-2500 nm region. The

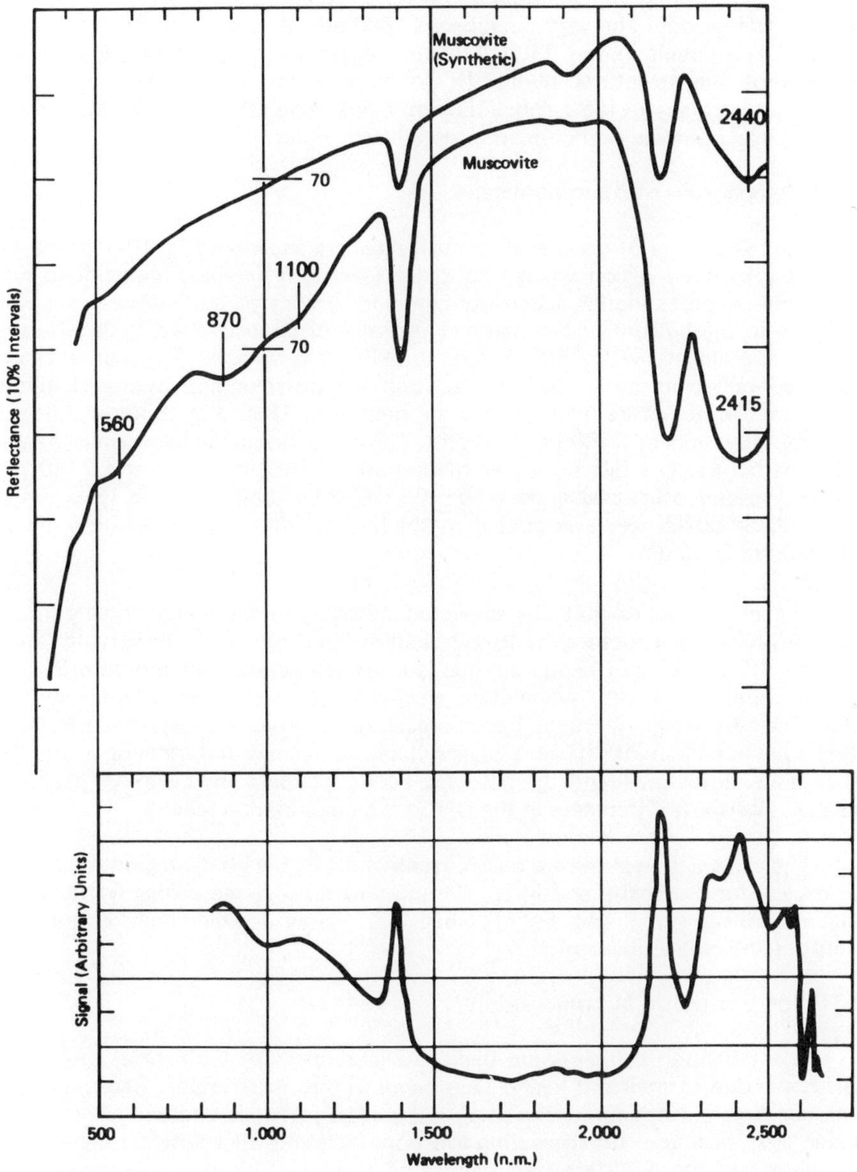

Figure 10-5. Comparison of Diffuse Reflectance and Photoacoustic Spectra of Muscovite.

higher energy bands, 870-1100 nm, are probably impurities of transition metals (possibly Fe^{+2}, Fe^{+3}), while the bands at 1380 nm and 2180 nm result from the structural OH groups. The very weak broad 1800 nm band is due to small amounts of water. The details in the 2300-2500 nm region are most likely combination overtones of the structural hydroxyl stretching mode with underlying lattice modes. The very sharp 1380 and 2180 nm bands suggest that the hydroxyls are located in homogenous, well defined, ordered lattice sites.

10-3.2. Comparisons of Aluminosilicates

The PA spectra of various aluminosilicates are shown in Fig. 10-6, primarily in the near infrared region where the major spectral differences are evident. For comparison purposes the PA spectra of commercial silica gel and magnesium oxide are shown in the bottom of the figure. The major differences have to do with the presence of molecular water and the structural hydroxyl groups. The zeolites (mordenite) contain clathrate water of hydration but no structural hydroxyl in the lattice network; therefore, water bands are present at 1380 and 1880 nm, but the structural OH band at 2180 nm is absent. Micas, on the other hand (muscovite), contain structural OH but no water of hydration; thus the 1380 and 2180 nm hydroxyl stretch overtones appear but the water bending mode at 1880 nm is absent. Major differences also appear in the hydroxyl/lattice combination region from 2300 to 2500 nm.

The clays also exhibit the expected behavior in the water and hydroxyl bands. All have the structural hydroxyl band at 2180 nm but with varying intensity. The 1880 nm water bending mode is most pronounced for montmorillonite with the interlayer water, while illite and kaolinite contain only outer surface water. The water and hydroxyl bands are all quite broad and assymmetric, indicating a heterogeneity of different lattice sites as compared to muscovite where the OH sites are all identical and the observed bands are sharp and quite symmetrical. The clays also show differences in the OH/lattice combination region.

The clays and zeolites are nearly transparent in the visible region, 500-1000 nm, except for muscovite and illite. The muscovite bands are probably transition metal impurities (Fe^{+2} and Fe^{+3}), while illites show a rather high, continuous adsorption into the visible region.

10-3.3. Ion-Exchanged Montmorillonite

The transition elements are highly variable in color with large absorption coefficients due to their d-d ligand field bands in the visible region. Because of the favorably large extinction coefficients, visible absorption spectroscopy has been a routine analytical tool for measuring low concentrations of transition metal ions. We determined the photoacoustic spectra of a series of montmorillonite samples which were ion-exchanged with different transition metal cations. The samples were run as fine powders; the chromophore was the transition metal cation at the concentration of the clay's cation exchange capacity (of the order of 35 meq/100g).

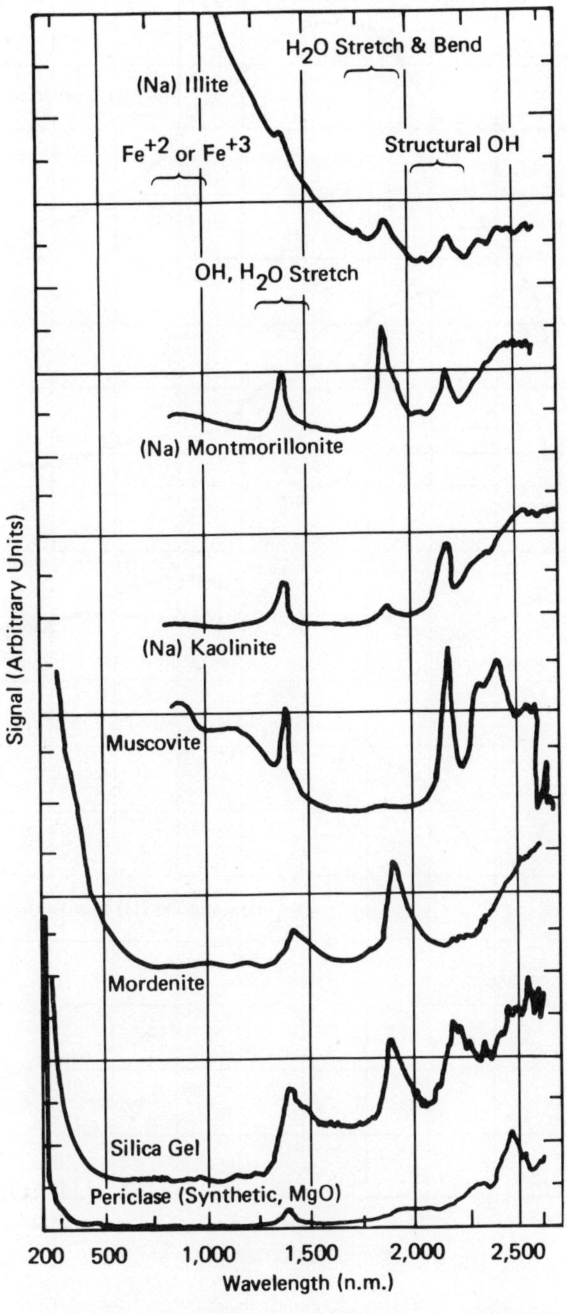

Figure 10-6. Near Infrared Photoacoustic Spectra of Aluminosilicates.

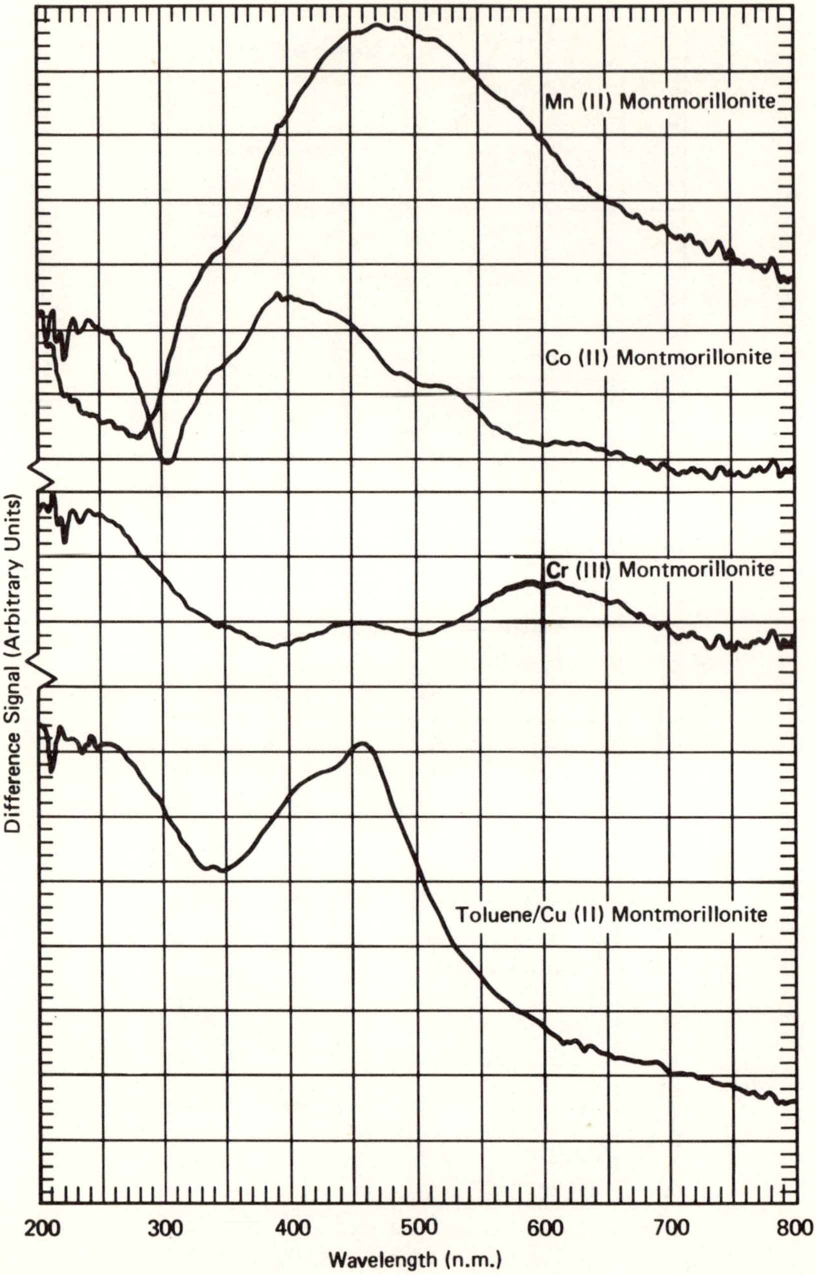

Figure 10-7. Difference Photo Acoustic Spectra of Transition Metal Ion Exchanged Montmorillonite (Na-Montmorillonite Use as Blank).

Typical results are shown in the top three spectra in Fig. 10-7 for the Mn(II), Co(II) and Cr(III) ion-exchanged clays. The spectra shown are difference spectra (calculated by the instrument's microprocessor) in which a sample of the Na(I)-montmorillonite was used as a blank. The photoacoustic signal for Na-montmorillonite alone has a rather steep rise toward lower wavelengths, thus difference PAS proved to be the only way to detect the low concentrations of the transition metal ion. Typical spectral assignments for PAS peaks can be found by comparison with polarized absorption spectra of various minerals, as reported by Burns and Vaughan (1).

True organometallic complexes can also be prepared within the interlayer space of expandable clays. Using the Cu(II) cation, Doner and Mortland (2) prepared the benzene/Cu(II) π complex with montmorillonite, which is characterized by a dramatic color shift from the normal blue Cu(II) to an orange-yellow for the Cu-benzene complex. The lower spectrum in Fig. 10-7 displays the difference PAS of the Cu-toluene complex upon intercalation with montmorillonite. The clay/organometallic complex appears yellow due to the spectral shift of the Cu(II) bands. Cu(II) in solution absorbs in the near ultraviolet and the yellow-red region. Upon complexation with toluene the longer wavelength band (600 nm) is shifted toward the blue region and the complex appears yellow.

10-3.4. Surface Adsorbed Metalloporphyrin

The porphyrins represent another set of molecules which contain unique features in their absorption spectra useful for identification. Their UV/visible spectra are characterized by a rather strong Soret band (400-500 nm) plus a set of two weaker bands, called the α, β bands, at longer wavelength (450-700 nm). The exact position of these bands depends on the porphyrin ring substitution and more uniquely on the nature of the transition metal held in the porphyrin chelation pocket. Fig. 10-8 shows some representative spectra for the mesoporphyrin IX dimethyl ester in solution and adsorbed on various substrate materials. Table I presents the wavelength and bandwidths for both the nickel and vanadyl porphyrin complexes. In these qualitative runs no attempt has been made to determine the surface loading, however, these results are near maximum loading conditions. It is evident that spectra of the clay/porphyrin complex are possible.

Van Damme *et al.* (7) have recently reported a study of clay-induced demetallation of porphyrins based on the acid-base properties of the clay. They report a strong shift of the Soret bands to longer wavelength when the porphyrin base is protonated by the acidic interlayer water of montmorillonite. The data in Table I indicate that the PAS technique shows similar results. As we proceed from the solution to the relatively neutral, filter paper substrate little change is observed except for a broadening of the vanadyl Soret band. However, for the silica gel which contains some acidic SiOH groups, and for the Ca^{+2} montmorillonite complex, there is a measurable red shift and broadening of the bands for the vanadyl complex. The band position of the nickel complex does not change but there is a measurable broadening of the Soret band. Note should also be made of the splitting of the vanadyl Soret band in the Ca^{+2} montmorillonite case. This has also been observed by Van Damme *et al.* (7) in strong acid (1N HCl) solution.

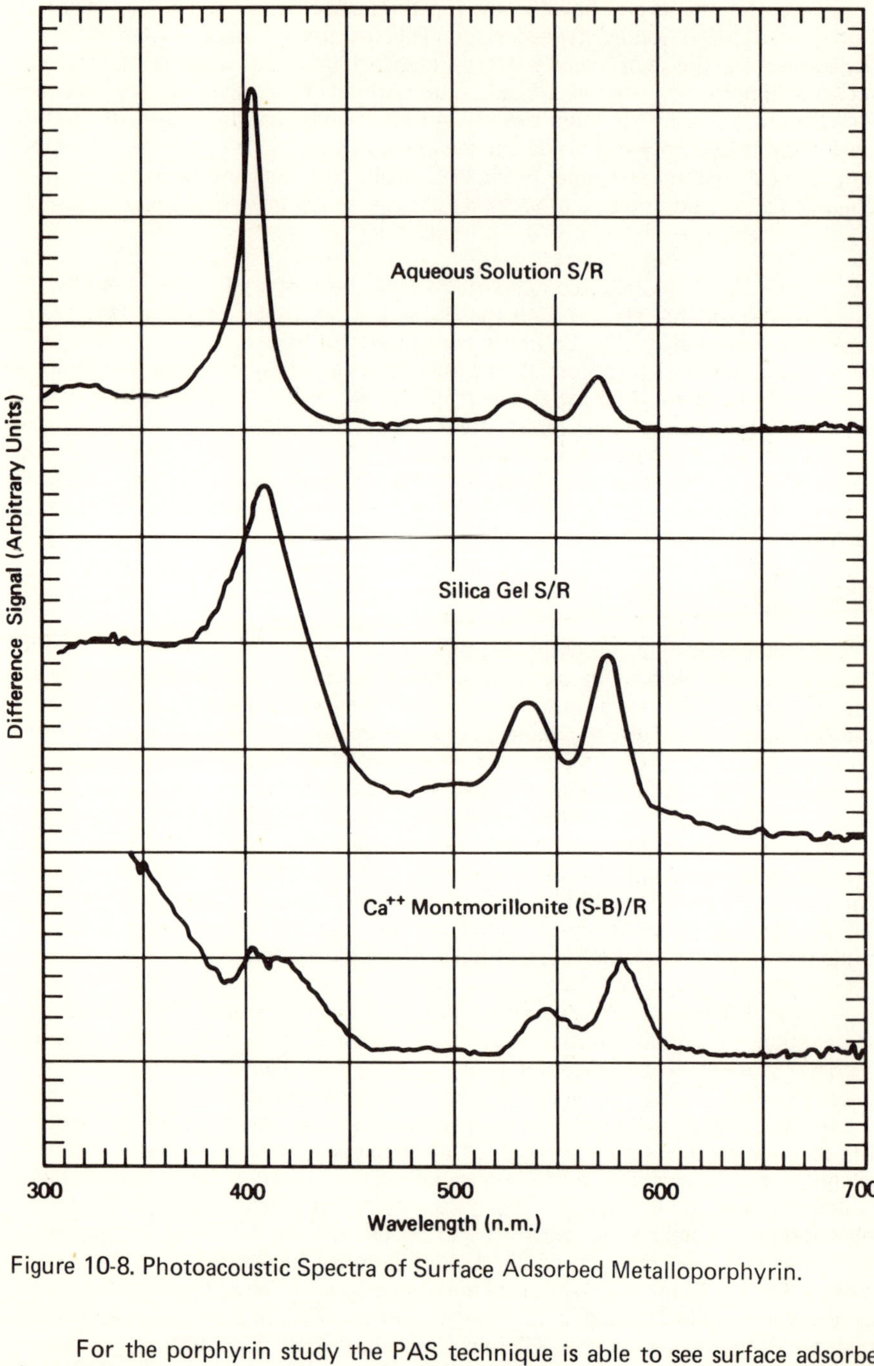

Figure 10-8. Photoacoustic Spectra of Surface Adsorbed Metalloporphyrin.

For the porphyrin study the PAS technique is able to see surface adsorbed chromophores and to monitor their perturbations by the acidic nature of the surface.

TABLE 10-1. SPECTRAL PARAMETERS FOR SURFACE ADSORBED METALLOPORPHYRINS.

	Nickel			Vanadyl		
Substrate	λ^*_{Soret}	λ^*_{α}	λ^*_{β}	λ^*_{Soret}	λ^*_{α}	λ^*_{β}
Aqueous Solution	387(19)	512	548(17)	402(13)	528	567(16)
Filter Paper	388(16)	512	548(17)	403(23)	530	568(17)
Silica Gel	388(26)	511(24)	548(18)	405(40)	532(27)	570(21)
Ca^{++} Montmorillonite		511	549(17)	407	535	575(20)
Ca^{++} Montmorillonite				398 409(40)	539	575(23)

*All wavelengths in nanometers, nm.
( ) indicates the bandwidth at half-maximum.

10-3.5. CORE DIFFERENCES

A real power of the PAS technique lies in the ability to detect small differences between samples. The key to this technique (as in any difference technique), however, lies in the ability to select the proper blank. The results of comparing different samples of Boise and Berea sandstone samples are shown in Fig. 10-9. The lower curve is the instrumental noise between two carbon black spectra.

The upper curve shows the differences between an old and a new sample of Boise from the same outcrop, the older sample having been fired to remove hydration water and stabilize the clay minerals. The effect of the firing process is clearly visible with the unfired sample showing more water. The two Boise samples also show measurable differences in Fe^{+2} and Fe^{+3} and in structural OH groups. The firing process should not effect the iron but could reduce the structural hydroxyl concentrations.

Old and new Berea samples, again from the same outcrop, are compared in the second curve. The newer sample contains more Fe^{+2} and apparently more structural hydroxyl, while the water content is apparently the same. This would suggest that there is a somewhat higher amount of nonexpandable clay, perhaps kaolinite, in the newer samples.

The third curve shows no detectable spectroscopic differences between the Big Block and Yesterday's Berea samples except in the UV region.

10-4. CONCLUSIONS

The photoacoustic technique permits a convenient method for obtaining a good approximation to the true optical absorption spectra for powdered soil and clay mineral samples throughout the ultraviolet, visible and near infrared region. There are some restrictions on various length scales to insure the PA signal is

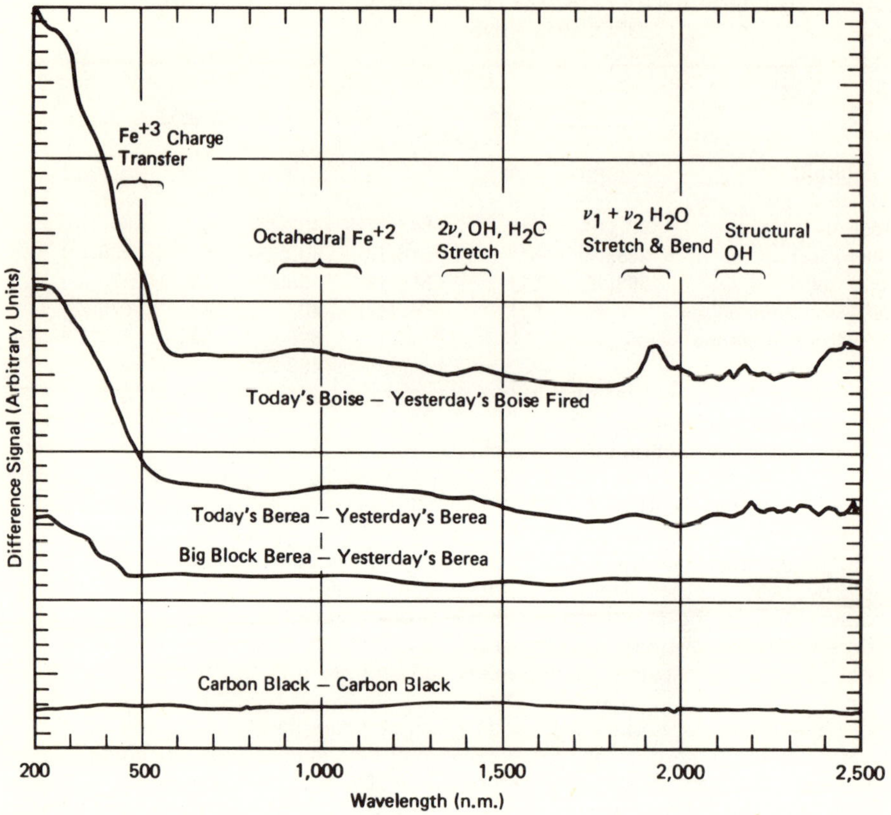

Figure 10-9. Difference PAS Technique for Samples of Boise and Berea Sandstone Material.

proportional to the optical absorption coefficients. Difference PAS techniques proves to be a valuable method for detecting minor amounts of chromophores in soil-type samples and to see perturbations of surface adsorbed organometallics. There is every reason to anticipate a bright future of photoacoustic spectroscopy in soils and clay minerals research.

ACKNOWLEDGMENTS

The author would like to thank Princeton Applied Research and Gilford Instruments for use of their equipment, and the management of Chevron Oil Field Research Company for permission to publish this work.

REFERENCES

1. Burns, R.G. and D.J. Vaughan. 1975. Polarized electronic spectra. *In* C.Karr, Jr., ed., *Infrared and Raman spectroscopy of lunar and terrestrial minerals.* Academic Press, New York. pp. 39-72.
2. Doner, H.E. and M.M. Mortland. 1969. Benzene complexes with copper(II) montmorillonite. *Science* **166**: 1406-1407.
3. McDonald, F.A. and G.C. Wetsel, Jr. 1978. Generalized theory of the photoacoustic effect. *J. Applied Phys.* **49**: 2313-2322.
4. Pao, Y-H. 1977. *Optoacoustic spectroscopy and detection.* Academic Press, New York. 239 pp.
5. Rosencwaig, A. 1977. Solid state photoacoustic spectroscopy. *In* Y-H Pao, ed., *Optoacoustic spectroscopy and detection.* Academic Press, New York. pp. 193-239.
6. Somoano, R.B. 1978. Photoacoustic spectroscopy of condensed matter. *Angew. Chem. Int. Ed.* **17**: 238-245.
7. Van Damme, H., M. Crespin, F. Obrecht, M.I. Cruz and J.J. Fripiat. 1978. Acid-base and complexation behavior of porphyrins on the intracrystal surface of swelling clays: meso-tetraphenylporphyrin and meso-tetra(4-pyridyl)porphyrin on montmorillonites. *J. Colloid Interface Sci.* **66**: 43-54.

INDEX